Multimedia Content and the Semantic Web

Multimedia Content and the Semantic Web

METHODS, STANDARDS AND TOOLS

Edited by

Giorgos Stamou and Stefanos Kollias

Both of
National Technical University of Athens, Greece

John Wiley & Sons, Ltd

Telephone (+44) 1243 779777

Email (for orders and customer service enquiries): cs-books@wiley.co.uk
Visit our Home Page on www.wileyeurope.com or www.wiley.com

Other Wiley Editorial Offices

John Wiley & Sons Inc., 111 River Street, Hoboken, NJ 07030, USA

Jossey-Bass, 989 Market Street, San Francisco, CA 94103-1741, USA

Wiley-VCH Verlag GmbH, Boschstr. 12, D-69469 Weinheim, Germany

John Wiley & Sons Australia Ltd, 33 Park Road, Milton, Queensland 4064, Australia

John Wiley & Sons (Asia) Pte Ltd, 2 Clementi Loop #02-01, Jin Xing Distripark, Singapore 129809

John Wiley & Sons Canada Ltd, 22 Worcester Road, Etobicoke, Ontario, Canada M9W 1L1

Wiley also publishes its books in a variety of electronic formats. Some content that appears
in print may not be available in electronic books.

British Library Cataloguing in Publication Data

A catalogue record for this book is available from the British Library

ISBN-13 978-0-470-85753-3 (HB)
ISBN-10 0-470-85753-6 (HB)

Typeset in 10/12pt Times by TechBooks, New Delhi, India.
Printed and bound in Great Britain by Antony Rowe Ltd, Chippenham, Wiltshire.
This book is printed on acid-free paper responsibly manufactured from sustainable forestry
in which at least two trees are planted for each one used for paper production.

Contents

List of Contributors

Kiyoharu Aizawa
Departments of Electrical Engineering and Frontier Informatics, Faculty of Engineering, University of Tokyo, 7-3-1 Hongo, Bunkyo, Tokyo 113 8656, Japan

Yannis Avrithis
Image, Video and Multimedia Laboratory, Institute for Computer and Communication Systems, Department of Electrical and Computer Engineering, National Technical University of Athens, Iroon Polytexneiou 9, 15780 Zografou, Greece

Jenny Benois-Pineau
LaBRI, Domaine Universitaire, l'Université Bordeaux 1, 351 cours de la Libération, 33405 Talence Cédex, France

George Carayannis
Institute for Language and Speech Processing, Artemidos 6 & Epidavrou, GR-151 25 Maroussi, Greece

Jean Charlet
Mission de Recherche STIM, 91 Boulevard de l'Hôpital, 75634 Paris Cédex 13, France

Ira Cohen
HP Labs, Palo Alto, 1501 Page Mill Road, Palo Alto, CA 94304, USA

Shishir Garg
France Telecom R&D, LLC, 801 Gateway Boulevard, South San Francisco, CA 94080, USA

Xavier Giró
Universitat Politècnica de Catalunya, Jordi Girona 1–3, 08034 Barcelona, Spain

Tetsuro Hori
Departments of Electrical Engineering and Frontier Informatics, Faculty of Engineering, University of Tokyo, 7-3-1 Hongo, Bunkyo, Tokyo 113 8656, Japan

Thomas S. Huang
Beckman Institute for Advanced Science and Technology, Department of Electrical and Computer Engineering, University of Illinois, 405 N. Mathews Avenue, Urbana, IL 61801, USA

Jane Hunter
DSTC Pty Ltd, Level 7, General Purpose South, University of Queensland, Brisbane QLD 4072, Australia

Philippe Joly
Institut de Recherche en Informatique de Toulouse, Université Toulouse 3, 118 route de Narbonne, 31062 Toulouse Cédex 4, France

Stefanos Kollias
Image, Video and Multimedia Laboratory, Institute for Computer and Communication Systems, Department of Electrical and Computer Engineering, National Technical University of Athens, Iroon Polytexneiou 9, 15780 Zografou, Greece

Ioannis Kompatsiaris
Information Processing Laboratory, Electrical and Computer Engineering Department, Aristotle University of Thessaloniki, Thessaloniki 54124, Greece, and Centre for Research and Technology Hellas (CERTH), Informatics and Telematics Institute (ITI), Thessaloniki, Greece

Alain Léger
France Telecom R&D, 4 rue du Clos Courtel, 35512 Cesson, France

Alexander Maedche
FZI Research Center for Information Technologies at the University of Karlsruhe, Haid-und-Neu-Straße 10–14, 76131 Karlsruhe, Germany

Ferran Marqués
Universitat Politècnica de Catalunya, Jordi Girona 1–3, 08034 Barcelona, Spain

Vasileios Mezaris
Information Processing Laboratory, Electrical and Computer Engineering Department, Aristotle University of Thessaloniki, Thessaloniki 54124, Greece, and Centre for Research and Technology Hellas (CERTH), Informatics and Telematics Institute (ITI), Thessaloniki, Greece

Boris Motik
FZI Research Center for Information Technologies at the University of Karlsruhe, Haid-und-Neu-Straße 10–14, 76131 Karlsruhe, Germany

Pramila Mullan
France Telecom R&D, LLC, 801 Gateway Boulevard, South San Francisco, CA 94080, USA

Milind R. Naphade
Pervasive Media Management Group, IBM Thomas J. Watson Research Center, 19 Skyline Drive, Hawthorne, NY 10532, USA

Harris Papageorgiou
Institute for Language and Speech Processing, Artemidos 6 & Epidavrou, GR-151 25 Maroussi, Greece

Fernando Pereira
Instituto Superior Técnico, Instituto de Telecomunicações, Universidade Técnica de Lisboa, Av. Rovisco Pais, 1049–001 Lisboa, Portugal

Nemanja Petrovic
Beckman Institute for Advanced Science and Technology, Department of Electrical and Computer Engineering, University of Illinois, 405 N. Mathews Avenue, Urbana, IL 61801, USA

Prokopis Prokopidis
Institute for Language and Speech Processing, Artemidos 6 & Epidavrou, GR-151 25 Maroussi, Greece

Athanassios Protopapas
Institute for Language and Speech Processing, Artemidos 6 & Epidavrou, GR-151 25 Maroussi, Greece

Philippe Salembier
Universitat Politècnica de Catalunya, Jordi Girona 1–3, 08034 Barcelona, Spain

John R. Smith
Pervasive Media Management Group, IBM Thomas J. Watson Research Center, 19 Skyline Drive, Hawthorne, NY 10532, USA

Giorgos Stamou
Image, Video and Multimedia Laboratory, Institute for Computer and Communication Systems, Department of Electrical and Computer Engineering, National Technical University of Athens, Iroon Polytexneiou 9, 15780 Zografou, Greece

Michael G. Strintzis
Information Processing Laboratory, Electrical and Computer Engineering Department, Aristotle University of Thessaloniki, Thessaloniki 54124, Greece, and Centre for Research and Technology Hellas (CERTH), Informatics and Telematics Institute (ITI), Thessaloniki, Greece

Vassilis Tzouvaras
Image, Video and Multimedia Laboratory, Institute for Computer and Communication Systems, Department of Electrical and Computer Engineering, National Technical University of Athens, Iroon Polytexneiou 9, 15780 Zografou, Greece

Rik Van de Walle
Multimedia Laboratory, Department of Electronics and Information Systems, University of Ghent, Sint-Pietersnieuwstraat 41, B-9000 Ghent, Belgium

Verónica Vilaplana
Universitat Politècnica de Catalunya, Jordi Girona 1–3, 08034 Barcelona, Spain

Raphael Volz
FZI Research Center for Information Technologies at the University of Karlsruhe, Haid-und-Neu-Straße 10–14, 76131 Karlsruhe, Germany

Manolis Wallace
Image, Video and Multimedia Laboratory, Institute for Computer and Communication Systems, Department of Electrical and Computer Engineering, National Technical University of Athens, Iroon Polytexneiou 9, 15780 Zografou, Greece

Foreword

The global vision of the development of the Semantic Web is to make the contents of the Web machine interpretable. To achieve this overall goal, ontologies play an important role since they provide the means for associating precisely defined semantics with the content that is provided by the Web. A major international effort resulted in the specification of the Web ontology language OWL that has recently become a W3C recommendation.

A lot of activities have been devoted to develop methods and tools that exploit ontologies to associate relational metadata with textual Web documents. In essence, such metadata are specified as instances of ontology classes and properties and thus gain the formal semantics from the underlying ontology. By combining information extraction techniques with machine learning approaches, such a metadata specification process can be partially automated.

Up-to-now, these methods and tools did not find their way into the area of multimedia content that is more and more found on the Web. Obviously, the extraction of semantic metadata from multimedia content is much more difficult when compared to textual sources. On the other hand, the integration of techniques for the generation of audio or visual descriptors with methods for generating ontology-based metadata seems to be a very promising approach to address these challenges.

Only few research and development projects have yet addressed such an integrated approach and a lot of open issues have to be investigated in the future. In that setting, this book provides an excellent overview about the current state of the art in this exciting area by covering these two research and development areas. Given the rather broad coverage of topics, this book is a highly interesting source of information for both researchers and practicioners.

Rudi Studer
Institute AIFB, University of Karlsruhe

http://www.aifb.uni-karlsruhe.de/Personen/viewPersonenglish?id_db=57

Foreword

Production and consumption of multimedia content and documents have become ubiquitous, which makes efficient tools for metadata creation and multimedia indexing essential for effective multimedia management. Metadata can be textual and/or content-based in the form of visual or audio descriptors. Such metadata is usually presented in a standard compliant machine readable format, e.g., International Standards Organization MPEG-7 descriptions in an XML file or a binary file, to facilitate consumption by search engines and intelligent agents.

Computer vision and video processing communities usually employ shot-based or object-based structural video models, and associate low-level (signal domain) descriptors such as color, texture, shape and motion, and semantic descriptions in the form of textual annotations, with these structural elements. Database and information retrieval communities, on the other hand, employ entity-relationship (ER) or object-oriented models to model the semantics of textual and multimedia documents. An important difference between annotations versus semantic models is that the latter can support entities, such as objects and events, and relations between them, which make processing complex queries possible.

There are very few works that aim to bridge the gap between the signal-level descriptions and semantic-level descriptions to arrive at a well integrated structural-semantic video model together with algorithms for automatic initialization and matching of such models for efficient indexing and management of video data and databases.

This book is a significant contribution to bring the fields of multimedia signal processing and semantic web closer, by including multimedia in the semantic web and by using semantics in multimedia content analysis. It is prepared by some of the most authoritative individuals playing key roles in the subjects of semantic web and multimedia content analysis. It not only provides a general introduction to the critical technologies in both fields, but also important insight into the open research problems in bridging the two fields. It is a must read for scientists and practitioners.

A. Murat Tekalp
Distinguished Professor
Department of Electrical and
Computer Engineering
University of Rochester

http://www.ece.rochester.edu/users/tekalp/

Introduction

The emerging idea of the Semantic Web is based on the maximum automation of the complete knowledge lifecycle processes, i.e. knowledge representation, acquisition, adaptation, reasoning, sharing and use. The success of this attempt depends on the ability of developing systems for acquiring, analysing and processing the knowledge embedded in multimedia content. In the multimedia research field, there has been a growing interest for analysis and automatic annotation of semantic multimedia content. Based on metadata information, the Moving Pictures Expert Group (MPEG) developed the Multimedia Content Description Interface (MPEG-7) and is now developing MPEG-21. One of the main goals of the above standards is to develop a rich set of standardised tools to enable machines to generate and understand audiovisual descriptions for retrieval, categorisation, browsing and filtering purposes. However, it has become clear that in order to make the multimedia semantic content description useful and interoperable with other Semantic Web domains, a common language, able to incorporate all the appropriate characteristics (enriched description, knowledge representation, sharing and reuse) should be used.

Part One of the book forms the basis for interweaving the two fields of multimedia and the semantic web. It introduces and analyzes both multimedia representations and knowledge representations. It includes four chapters. The first introduces the MPEG-7 and MPEG-21 multimedia standards, while the second refers to semantic web technologies, focusing on ontology representations. Interweaving of the two fields is described, first, in the third chapter, which presents a multimedia MPEG-7 ontology; then, the fourth chapter defines fuzzy knowledge representations that can be used for reasoning on multimedia content.

In particular:

F. Pereira and R. Van de Walle, in chapter 1, describe and analyze multimedia standards, focusing on MPEG-7 and MPEG-21. They present an overview of context, objectives, technical approach, work plan and achievements of both standards, with emphasis on content description technologies. Following a presentation of objectives and tools of MPEG-7, emphasis is given to descriptors and descriptor tools that have an important role in creating knowledge representations of multimedia information. Presentation is then turned to MPEG-21 objectives and tools, with the emphasis given to declaration, identification and adaptation of digital items.

In chapter 2, B. Motik, A. Maedche and R. Volz explore representations of ontological knowledge for semantic-driven applications. They first present an ontology representation that is capable of handling constraints and meta-concept modeling. Then, they analyse ontology querying, focusing on conceptual querying and describe the implementation of the Karlsruhe Ontology and Semantic Web tool suite.

J. Hunter presents, in chapter 3, a novel framework for interweaving multimedia analysis and related standardization activities with recent semantic web standards and concepts. She first points at the need for defining the semantics of MPEG-7 metadata terms in an ontology, so as to generate multimedia content that can be effectively processed, retrieved and re-used by services, agents and devices on the Web. She then describes how to build an MPEG-7 ontology based on the Web Ontology Language (OWL) and uses this ontology for inferencing domain-specific semantic descriptions of multimedia content from automatically extracted low level features.

In the next chapter, V. Tzouvaras, G. Stamou and S. Kollias extend this blending of technologies, by proposing fuzzy representation and reasoning for handling the inherent imprecision in multimedia object analysis and retrieval. A hybrid knowledge base is presented, consisting of general knowledge pertaining to the domain of interest and of propositional rules that are used to infer implicit knowledge. A hybrid neurofuzzy network is proposed for implementation of the inference engine, supported by an adaptation algorithm that operates on the rules of the system.

Part Two of the book focuses on multimedia content analysis, presenting the technologies that form the basis for semantic multimedia analysis based on the above-described representations. It contains five chapters. The first tackles identification of structure in multimedia documents, mainly dealing with video shot segmentation. Video object extraction is the topic of the next chapter, which examines spatio-temporal segmentation and extraction of MPEG-7 descriptors from both uncompressed and MPEG compressed audiovisual data. Semantic object detection using region-based structural image representation is the topic of the third chapter. The last two chapters deal with probabilistic graphical models for mining the semantics of video concepts with context and video understanding.

In the first chapter of Part Two, P. Joly investigates structure identification in multimedia content. He starts by stating that, most of the time, video shot segmentation remains a prerequisite for more complex video 'macrosegmentation' tasks which exploit large sets of low-level features. This chapter, thus, first considers shot segmentation evaluating the corresponding state-of-the-art, and then extends its results to macrosegmentation in well-defined contexts.

J. Benois-Pineau starts chapter 6 by considering the 'macro' view of video content, in terms of chapters characterized by low-level (signal-based) homogeneity or high level (semantic) uniformity. The 'micro' view of content focuses on sets of objects evolving inside the chapters and is of much interest in particular for retrieval tasks. The usual query by object implies matching of descriptors of an example object or of a generic prototype imagined by the user with the description of objects in an indexed content. Although, MPEG-7 supplies a normalized framework for object description in multimedia content, it does not stipulate methods for automatic extraction of objects for the purpose of generating their normalized description. Spatio-temporal segmentation is described for this purpose, combining both grey-level/color and motion information from the video stream. Object extraction from video is presented in the framework of a new 'rough indexing' paradigm.

In the following chapter, X. Giró, V. Vilaplana, F. Marqués and P. Salembier further investigate automatic analysis tools that work on low-level image representations and are able to detect the presence of semantic objects. Their approach focuses on still images and relies on combining two types of models: a perceptual and a structural model. The algorithms that are proposed for both types of models make use of a region-based description of the image relying on a Binary Partition Tree. Perceptual models link the low-level signal description with semantic classes of limited variability. Structural models represent the common structure of all instances by decomposing the semantic object into simpler objects and by defining the relations between them.

In chapter 8, M. Naphade and J. Smith examine the learning of models for semantics of concepts, context and structure and then use these models for mining purposes. They propose a hybrid framework that can combine discriminant or generative models for concepts with probabilistic graphical network models for context. They show that robust models can be built for several diverse visual semantic concepts, using the TREC Video 2002 benchmark corpus and a novel factor graphical to model inter-conceptual context for semantic concepts of the corpus. Moreover, the sum-product algorithm is used for approximate or exact inference in these factor graph multinets. As a consequence, errors made during isolated concept detection are corrected by forcing high-level constraints and a significant improvement in the overall detection performance is achieved.

Graphical models constitute the main research topic of the next chapter, by N. Petrovic, I. Cohen and T. Huang, which analyses their use in multimedia understanding and computer vision. They start by mentioning that extracting low level features and trying to infer their relation to high-level semantic concepts can not, in general, provide semantics that describe high level video concepts, which is the well known ‘semantic gap’ between features and semantics. Modelling difficult statistical problems in multimedia and computer vision often requires building complex probabilistic models; probabilistic inference as well as learning model parameters are the basic tasks associated with such models. It is certain properties of models (factorization) that make them suitable for graphical representation, that proves to be useful for manipulation of underlying distributions. Graphical probabilistic models are, thus, described as a powerful and prosperous technique for tackling these problems.

Part Three of the book deals with multimedia content management systems and semantic web applications. The first chapter introduces a variety of semantic web applications that are relevant or deal with multimedia. The next four chapters concentrate on one of the most important applications, multimedia indexing and retrieval, combining the afore-mentioned multimedia and knowledge representations and technologies.

The semantic web technology is more and more applied to a large spectrum of applications, within which domain knowledge is conceptualized and formalized as a support for reasoning purposes. Moreover, these representations can be rendered understandable by human beings so that a subtle coupling between human reasoning and computational power is possible. At the crossroad of a maturing technology and a pressing industry anticipating real benefits for cost reduction and market expansion, objective evaluation of the expectations via benchmarking is a real necessity. A first concrete step towards this evaluation is to present the most prominent prototypical applications either deployed or simply fielded. To this end, A. Léger, P. Mullan, S. Garg, and J. Charlet tentatively trace some of the most significant semantic applications in chapter 10.

Chapter 11, by H. Papageorgiou, P. Prokopidis, A. Protopapas and G. Carayannis, introduces the reader into multimedia semantic indexing and retrieval. An extensive set of technologies on image, speech and text is presented, particularly suited for multimedia content analysis, so as to automatically generate metadata annotations associated with digital audiovisual segments. A multifaceted approach for the location of important segments within multimedia material is presented, followed by generation of high-level semantic descriptors in the metadata space, that serve for indexing and retrieval purposes.

In the following chapter, M. Wallace, Y. Avrithis, G. Stamou and S. Kollias exploit the advances in handling multimedia content related metadata, introduced by MPEG-7/MPEG-21 and knowledge technologies, to offer advanced access services characterized by semantic phrasing of the query, unified handling of multimedia documents and personalized response. The above are included in a system that targets intelligent extraction of semantic information from multimedia document descriptions, taking into account knowledge about queries that users may issue and the context determined by user profiles.

I. Kompatsiaris, V. Mezaris and M. Strintzis, in chapter 13, propose a multilevel descriptor approach for retrieval in generic multimedia collections, where there does not exist a domain-specific knowledge base. The low-level includes features extracted from spatial or spatio-temporal objects using unsupervised segmentation algorithms. These are automatically mapped to qualitative intermediate-level descriptors, that form a semantic object ontology. The latter is used to allow the qualitative definition of high-level concepts queried by the user. Relevance feedback, using support vector machines and the low-level features, is also used to produce the final query results.

In everyday life, digitization of personal experiences is being made possible by continuous recordings that people make, using portable or wearable video cameras. It is evident that the resulting amount of video content is enormous. Consequently, to retrieve and browse desired scenes, a vast quantity of video data must be organized using structural information. In the last chapter of the book, K. Aizawa and T. Hori describe the architecture and functionality of a context-based video retrieval system for such life-log applications, that can incorporate the multimedia and knowledge representations and technologies described in the earlier book chapters.

Giorgos Stamou and Stefanos Kollias

Part One

Knowledge and Multimedia

1

Multimedia Content Description in MPEG-7 and MPEG-21[1]

Fernando Pereira and Rik Van de Walle

1.1 Multimedia Content Description

1.1.1 Context and Motivation

The amount of digital multimedia information accessible to the masses is growing every day, not only in terms of consumption but also in terms of production. Digital still cameras directly storing in JPEG format have hit the mass market and digital video cameras directly recording in MPEG-1 format are also available. This transforms every one of us into a potential content producer, capable of creating content that can be easily distributed and published using the Internet. But if it is today easier and easier to acquire, process and distribute multimedia content, it should be equally easy to access the available information, because huge amounts of digital multimedia information are being generated, all over the world, every day. In fact, there is no point in making available multimedia information that can only be found by chance. Unfortunately, the more information becomes available, the harder it is to identify and find what you want, and the more difficult it becomes to manage the information.

People looking for content are typically using text-based browsers with rather moderate retrieval performance; often, these search engines yield much noise around the hits. However, the fact that they are in widespread use indicates that a strong need exists. These text-based engines rely on human operators to manually describe the multimedia content with keywords and free annotations. This solution is increasingly unacceptable for two major reasons. First, it is a costly process, and the cost increases quickly with the growing amount of content. Second, these descriptions are inherently subjective and their usage is often confined to the specific application domain for which the descriptions were created. Thus, it is necessary to

[1] Some of the MPEG-7 related sections of this chapter are adapted from Fernando Pereira and Rob Koenen (2001) MPEG-7: a standard for multimedia content description. *International Journal of Image and Graphics*, **1** (3), 527–546, with permission from World Scientific Publishing Company.

Multimedia Content and the Semantic Web Edited by Giorgos Stamou and Stefanos Kollias

automatically and objectively describe, index and annotate multimedia information (notably audiovisual data), using tools that automatically extract (possibly complex) features from the content. This would substitute or complement the use of manually determined, text-based, descriptions. Automatically extracted audiovisual features will have three principal advantages over human annotations: (i) they will be automatically generated, (ii) in general, they will be more objective and domain-independent, and (iii) they can be native to the audiovisual content. Native descriptions would use non-textual data to describe content, notably features such as colour, shape, texture, melody and timbre, in a way that allows the user to search by comparing non-textual descriptions. Even though automatically extracted descriptions will be very useful, it is evident that descriptions, the 'bits about the bits', will always include textual components. There are in fact many features about the content that can only be expressed through text, e.g. author names and titles.

It should also be noted that, in advanced multimedia applications, the 'quality of service' is not only determined by the characteristics of the multimedia content itself (and its descriptions). Indeed, the quality of service is also determined by the characteristics of the terminals that end users are using to render and experience their multimedia content, by the characteristics of the network that links multimedia content consumers with content providers, by end-user preferences, etc. This leads, for example, to a need for tools that allow for the adaptation of multimedia content, taking into account terminal characteristics, network conditions, natural environment features and user preferences.

Many elements already exist to build an infrastructure for the delivery and consumption of multimedia content, including, besides numerous media resource codecs, intellectual property management and protection (IPMP) and digital rights management (DRM) tools, terminal and network technologies, and tools for the expression of user preferences. But until recently there was no clear view on how these elements relate to each other and how they can efficiently be used in an interoperable way. Making the latter possible is the main goal of the MPEG-21 project. In the MPEG-21-related sections of this chapter, it will be shown how the concept of multimedia content description plays a crucial role in this project.

1.1.2 Why do we Need Standards?

There are many ways to describe multimedia content, and, indeed, today many proprietary ways are already in use in various digital asset management systems. Such systems, however, do not allow a search across different repositories for a certain piece of content, and they do not facilitate content exchange between different databases using different systems. These are interoperability issues, and creating a standard is an appropriate way to address them.

The MPEG [1] standards address this kind of interoperability, and offer the prospect of lowering product costs through the creation of mass markets, and the possibility to make new, standards-based services explode in terms of number of users. To end users, a standard will enable tools allowing them to easily surf on the seas and filter the floods of multimedia information, in short managing the information. To consumer and professional users alike, MPEG content description tools will facilitate management of multimedia content. Of course, in order to be adopted, standards need to be technically sound. Matching the needs and the technologies in multimedia content description was thus the task of MPEG in the MPEG-7 and also partly in the MPEG-21 standardization processes.

1.1.3 MPEG Standardization Process

Two foundations of the success of the MPEG standards so far are the toolkit approach and the 'one functionality, one tool' principle [2]. The toolkit approach means setting a horizontal standard that can be integrated with, for example, different transmission solutions. MPEG does not set vertical standards across many layers in the Open Systems Interconnection (OSI) model [3], developed by the International Organization for Standardization (ISO). The 'one functionality, one tool' principle implies that no two tools will be included in the standard if they provide essentially the same functionality. To apply this approach, the standards' development process is organized in three major phases:

- Requirements phase
  1. *Applications*: identify relevant applications using input from the MPEG members; inform potential new participants about the new upcoming standard.
  2. *Functionalities*: identify the functionalities needed by the applications above.
  3. *Requirements*: describe the requirements following from the envisaged functionalities in such a way that common requirements can be identified for different applications.
- Development phase
  4. *Call for proposals*: a public call for proposals is issued, asking all interested parties to submit technology that could fulfil the identified requirements.
  5. *Evaluation of proposals*: proposals are evaluated in a well-defined, adequate and fair evaluation process, which is published with the call itself; the process may entail, for example, subjective testing, objective comparison or evaluation by experts.
  6. *Technology selection*: following the evaluation, the technologies best addressing the requirements are selected. MPEG usually does not choose one single proposal, but typically starts by assembling a framework that uses the best ranked proposals, combining those (so-called 'cherry picking'). This is the start of a collaborative process to draft and improve the standard.
  7. *Collaborative development*: the collaboration includes the definition and improvement of a *Working Model*, which embodies early versions of the standard and can include also non-normative parts (this means parts which do not need to be normatively specified to provide interoperability). The Working Model typically evolves by having alternative tools challenging those already in the Working Model, by performing Core Experiments (CEs). *Core Experiments* are technical experiments carried out by multiple independent parties according to predefined conditions. Their results form the basis for technological choices. In MPEG-7, the Working Model is called eXperimentation Model (XM) and in MPEG-21 it is called sYstems Model (YM).
  8. *Balloting*: when a certain level of maturity has been achieved, national standardization bodies review the Draft Standard in a number of ballot rounds, voting to promote the standard, and asking for changes.
- Verification phase
  9. *Verification*: verify that the tools developed can be used to assemble the target systems and provide the desired functionalities. This is done by means of Verification Tests. For MPEG-1 to MPEG-4, these tests were mostly subjective evaluations of the decoded quality. For MPEG-7, the verification tests had to assess efficiency in identifying the right content described using MPEG-7 tools. For MPEG-21, no verification tests had been performed by March 2004.

Because MPEG always operates in new fields, the requirements landscape keeps moving and the above process is not applied rigidly. Some steps may be taken more than once, and iterations are sometimes needed. The time schedule, however, is always closely observed by MPEG. Although all decisions are taken by consensus, the process maintains a fast pace, allowing MPEG to provide timely technical solutions.

To address the needs expressed by the industry, MPEG develops documents with the technical specification of the standard, called *International Standards* (IS); this corresponds to the collaborative development step mentioned above. The progress towards an International Standard is:

- *New Work Item Proposal (NP)*: 3 months ballot (with comments)
- *Working Draft (WD)*: no ballot by National Bodies
- *Committee Draft (CD)*: 3 months ballot (with comments)
- *Final Committee Draft (FCD)*: 4 months ballot (with comments)
- *Final Draft International Standard (FDIS)*: 2 months binary (only yes/no) ballot. Failing this ballot (no vote) implies going back to WD stage
- *International Standard (IS)*.

The addition of new tools to an International Standard may be performed by issuing *Amendments* to that standard. To correct a technical defect identified in a standard, a *Corrigendum* has to be issued. Besides standards, amendments and corrigenda, MPEG may also issue *Technical Reports*, which are documents containing information of a different kind from that normally published as an International Standard, such as a model/framework, technical requirements and planning information, a testing methodology, factual information obtained from a survey carried out among the national bodies, information on work in other international bodies or information on the state-of-the-art regarding national body standards on a particular subject [4].

1.2 MPEG-7: Multimedia Content Description Interface

1.2.1 Objectives

The anticipated need to efficiently manage and retrieve multimedia content, and the foreseeable increase in the difficulty of doing so was recognized by the Moving Picture Experts Group (MPEG) in July 1996. At the Tampere meeting, MPEG [1] stated its intention to provide a solution in the form of a 'generally agreed-upon framework for the description of audiovisual content'. To this end, MPEG initiated a new work item, formally called Multimedia Content Description Interface, generally known as MPEG-7 [5]. MPEG-7 specifies a standard way of describing various types of multimedia information, irrespective of their representation format (e.g. analogue or digital) or storage support (e.g. paper, film or tape). Participants in the development of MPEG-7 represent broadcasters, equipment and software manufacturers, digital content creators, owners and managers, telecommunication service providers, publishers and intellectual property rights managers, as well as university researchers. MPEG-7 is quite a different standard compared to its predecessors. MPEG-1, -2 and -4 all represent the content itself—'the bits'—while MPEG-7 represents information about the content—'the bits about the bits'.

Like the other members of the MPEG family, MPEG-7 will be a standard representation of multimedia information satisfying a set of well-defined requirements [6], which, in this case, relate to the description of multimedia content. 'Multimedia information' includes still pictures, video, speech, audio, graphics, 3D models and synthetic audio. The emphasis is on audiovisual content, and the standard will not specify new description tools for describing and annotating text itself, but will rather consider existing solutions for describing text documents, such as HyperText Markup Language (HTML), Standardized General Markup Language (SGML) and Resource Description Framework (RDF) [6], supporting them as appropriate. While MPEG-7 includes statistical and signal processing tools, using textual descriptors to describe multimedia content is essential for information that cannot be derived from the content either by automatic analysis or human viewing. Examples include the name of a movie and the date of acquisition, as well as more subjective annotations. Moreover, MPEG-7 will allow linking multimedia descriptions to any relevant data, notably the described content itself.

MPEG-7 has been designed as a generic standard in the sense that it is not tuned to any specific application. MPEG-7 addresses content usage in storage, online and offline, or streamed, e.g. broadcast and (Internet) streaming. MPEG-7 supports applications operating in both real-time and non-real-time environments. It should be noted that, in this context, a 'real-time environment' corresponds to the case where the description information is created and associated with the content while that content is being captured.

MPEG-7 descriptions will often be useful as stand-alone descriptions, e.g. if only a quick summary of the multimedia information is needed. More often, however, they will be used to locate and retrieve the same multimedia content represented in a format suitable for reproducing the content: digital (and coded) or even analogue. In fact, as mentioned above, MPEG-7 data is intended for content *identification and managing* purposes, while other representation formats, such as MPEG-1, MPEG-2 and MPEG-4, are mainly intended for content *reproduction (visualization and hearing)* purposes. The boundaries may be less sharp sometimes, but the different standards fulfil substantially different sets of requirements. MPEG-7 descriptions may be physically co-located with the corresponding 'reproduction data', in the same data stream or in the same storage system. The descriptions may also live somewhere else. When the various multimedia representation formats are not co-located, mechanisms linking them are needed. These links should be able to work in both directions: from the description data to the reproduction data, and vice versa.

Because MPEG-7 intends to describe multimedia content regardless of the way the content is made available, it will depend neither on the reproduction format nor the form of storage. Video information could, for instance, be available as MPEG-4, -2, or -1, JPEG, or any other coded form—or not even be coded at all: it is even possible to generate an MPEG-7 description for an analogue movie or for a picture that is printed on paper. However, there is a special relationship between MPEG-7 and MPEG-4, as MPEG-7 is grounded on an object-based data model, which is also used by MPEG-4 [7]. Like MPEG-4, MPEG-7 can describe the world as a composition of multimedia objects with spatial and temporal behaviour, allowing object-based multimedia descriptions. As a matter of fact, each object in an MPEG-4 scene can have an MPEG-7 description (stream) associated with it; this description can be accessed independently.

Normative versus non-normative tools

A standard should seek to provide interoperability while trying to keep the constraints on the freedom of the user to a minimum. To MPEG, this means that a standard must offer the maximum of advantages by specifying the minimum necessary, thus allowing for competing implementations and for evolution of the technology in the so-called non-normative areas. MPEG-7 only prescribes the multimedia description format (syntax and semantics) and usually not the extraction and encoding processes. Certainly, any part of the search process is outside the realm of the standard. Although good analysis and retrieval tools are essential for a successful MPEG-7 application, their standardization is not required for interoperability. In the same way, the specification of motion estimation and rate control is not essential for MPEG-1 and MPEG-2 applications, and the specification of segmentation is not essential for MPEG-4 applications. Following the principle of specifying the minimum for maximum usability, MPEG concentrates on standardizing the tools to express the multimedia description. The development of multimedia analysis tools—automatic or semi-automatic—as well as tools that will use the MPEG-7 descriptions—search engines and filters—are tasks for the industries that build and sell MPEG-7-enabled products. This strategy ensures that good use can be made of the continuous improvements in the relevant technical areas. New automatic analysis tools can always be used, also after the standard is finalized, and it is possible to rely on competition for obtaining ever better results. In fact, it will be these very non-normative tools that products will use to distinguish themselves, which only reinforces their importance.

Low-level versus high-level descriptions

The description of content may typically be done using two broadly defined types of features: those expressing information about the content such as creation date, title and author, and those expressing information present in the content. The features expressing information present in the content may be rather low level, signal-processing-based, or rather high level, associated with the content semantics. The so-called low-level features are those like colour and shape for images, or pitch and timbre for speech. High-level features typically have a semantic value associated with what the content means to humans, e.g. events, or genre classification. Low-level features have three important characteristics:

- They can be extracted automatically, and thus not specialists but machinery will worry about the great amount of information to describe.
- They are objective, thus eliminating problems such as subjectivity and specialization.
- They are native to the audiovisual content, allowing queries to be formulated in a way more suited to the content in question, e.g. using colours, shapes and motion.

Although low-level features are easier to extract (they can typically be extracted fully automatically), most (especially non-professional) consumers would like to express their queries at the semantic level, where automatic extraction is rather difficult. If high-level features are not directly available in the content description, it may be the browser's task to perform the semantic mapping between the high-level query expressed by the user and the available low-level description features, all this in a way transparent to the user. This becomes easier when a specific application/content domain is targeted (e.g. express what a goal is in the context of a soccer match). One of MPEG-7's main strengths is that it provides a description framework that supports the combination of low-level and high-level features in a single description, leaving to the content creators and to the querying engines' developers the task of choosing which features to include in the descriptions and in the query matching process; both processes are fully non-normative. In combination with the highly structured nature of MPEG-7 descriptions, this capability constitutes one of the major differences between MPEG-7 and other available or emerging multimedia description solutions.

Extensibility

There is no single 'right' description for a piece of multimedia content. What is right depends strongly on the application domain. MPEG-7 defines a rich set of core description tools. However, it is impossible to have MPEG-7 specifically addressing every single application. Therefore, it is essential that MPEG-7 be an open standard, extensible in a normative way to address description needs, and thus application domains, that are not fully addressed by the core description tools. The power to build new description tools (possibly based on the standard ones) is achieved through a standard description language, the Description Definition Language (DDL).

1.2.2 Applications and Requirements

MPEG-7 targets a wide range of application environments and it will offer different levels of granularity in its descriptions, along axes such as time, space and accuracy. Descriptive features must be meaningful in the context of an application, so the descriptions for the same content can differ according to the user domain and application. This implies that the same material can be described in various ways, using different features, and with different levels of abstraction for those features. It is thus the task of the content description generator to choose the right features and corresponding granularity. From this, it becomes clear that no single 'right' description exists for any piece of content; all descriptions may be equally valid from their own usage point of view. The strength of MPEG-7 is that these descriptions will all be based on the same description tools, and can be exchanged in a meaningful, interoperable way. MPEG-7 requirements are application driven. The relevant applications are all those that should be enabled by the MPEG-7 toolbox. Addressing new applications, i.e. those that do not exist yet but will be enabled by the standard, has the same priority as improving the functionality of existing ones. There are many application domains that could benefit from the MPEG-7 standard, and no application list drawn up today can be exhaustive.

The MPEG-7 Applications document [8] includes examples of both improved existing applications as well as new ones that may benefit from the MPEG-7 standard, and organizes the example applications into three sets:

- *Pull applications*: applications such as storage and retrieval in audiovisual databases, delivery of pictures and video for professional media production, commercial musical applications, sound effects libraries, historical speech databases, movie scene retrieval by memorable auditory events, and registration and retrieval of trademarks.
- *Push applications*: applications such as user-agent-driven media selection and filtering, personalized television services, intelligent multimedia presentations and information access facilities for people with special needs.
- *Specialized professional applications*: applications that are particularly related to a specific professional environment, notably tele-shopping, biomedical, remote sensing, educational and surveillance applications.

For each application listed, the MPEG-7 Applications document gives a description of the application, the corresponding requirements, and a list of relevant work and references. The set of applications in the MPEG-7 Applications document [8] is a living set, which will be augmented in the future, intended to give the industry—clients of the MPEG work—some hints about the application domains addressed. If MPEG-7 will enable new and unforeseen applications to emerge, this will show the strength of the toolkit approach.

Although MPEG-7 intends to address as many application domains as possible, it is clear that some applications are more important than others due to their relevance in terms of foreseen business, research investment etc.

In order to develop useful tools for the MPEG-7 toolkit, functionality requirements have been extracted from the identified applications. The MPEG-7 requirements [6] are currently divided into five sections associated with descriptors, description schemes, Description Definition Language, descriptions and systems requirements. Whenever applicable, visual and audio requirements are considered separately. The requirements apply, in principle, to both real-time and non-real-time systems as well as to offline and streaming applications, and they should be meaningful to as many applications as possible.

1.2.3 Basic Elements

MPEG-7 specifies the following types of tools [6]:

- *Descriptors*: a descriptor (D) is a representation of a feature; a feature is a distinctive characteristic of the data that signifies something to somebody. A descriptor defines the syntax and the semantics of the feature representation. A descriptor allows an evaluation of the corresponding feature via the descriptor value. It is possible to have several descriptors representing a single feature, i.e. to address different relevant requirements/functionalities, e.g. see colour descriptors. Examples of descriptors are a time-code for representing duration, colour moments and histograms for representing colour, and a character string for representing a title.
- *Description schemes*: a description scheme (DS) specifies the structure and semantics of the relationships between its components, which may be both descriptors and description schemes. A DS provides a solution to model and describe multimedia content in terms of structure and semantics. A simple example is a movie, temporally structured as scenes and shots, including some textual descriptors at the scene level, and colour, motion and audio amplitude descriptors at the shot level.
- *Description Definition Language*: the Description Definition Language (DDL) is the language used for the definition of descriptors and description schemes; it also allows the creation of new description schemes or just the extension and modification of existing description schemes.
- *Systems tools*: tools related to the binarization, synchronization, transport and storage of descriptions, as well as to the management and protection of intellectual property associated with descriptions.

These are the normative elements of the standard. In this context, 'normative' means that if these elements are used, they must be used according to the standardized specification since this is essential to guarantee interoperability. Feature extraction, similarity measures and search engines are also relevant, but will not be standardized since this is not essential for interoperability.

1.2.4 MPEG-7 Standard Organization

For the sake of legibility, organization and easier usage, the MPEG-7 standard is structured in ten parts [5]:

- *ISO/IEC 15938-1 or MPEG-7 Part 1—Systems*: specifies the tools that are needed to prepare MPEG-7 descriptions for efficient transport and storage, to allow synchronization between content and descriptions, and the tools related to managing and protecting intellectual property of descriptions [9].
- *ISO/IEC 15938-2 or MPEG-7 Part 2—Description Definition Language*: specifies the language for defining the descriptors and description schemes; it also allows the definition of new or extended description schemes [10].
- *ISO/IEC 15938-3 or MPEG-7 Part 3—Visual*: specifies the descriptors and description schemes dealing only with visual information [11].
- *ISO/IEC 15938-4 or MPEG-7 Part 4—Audio*: specifies the descriptors and description schemes dealing only with audio information [12].
- *ISO/IEC 15938-5 or MPEG-7 Part 5—Generic Entities and Multimedia Description Schemes*: specifies the descriptors and description schemes dealing with generic (non-audio- or video-specific) and multimedia features [13].
- *ISO/IEC 15938-6 or MPEG-7 Part 6—Reference Software*: includes software corresponding to the specified MPEG-7 tools [14].
- *ISO/IEC 15938-7 or MPEG-7 Part 7—Conformance Testing*: defines guidelines and procedures for testing conformance of MPEG-7 descriptions and terminals [15].
- *ISO/IEC 15938-8 or MPEG-7 Part 8—Extraction and Use of MPEG-7 Descriptions*: technical report (not normative) providing informative examples that illustrate the instantiation of description tools in creating descriptions conforming to MPEG-7, and detailed technical information on extracting descriptions automatically from multimedia content and using them in multimedia applications [16].
- *ISO/IEC 15938-9 or MPEG-7 Part 9—Profiles and levels*: defines profiles and levels for MPEG-7 descriptions [17].
- *ISO/IEC 15938-10 or MPEG-7 Part 10—Schema Definition*: includes the mechanism which specifies the MPEG-7 schema definition across all parts of the MPEG-7 standard. This schema definition shall evolve through versions as the various parts of the MPEG-7 standard are amended. The MPEG-7 schema definition shall specify the schema using the MPEG-7 Description Definition Language [18].

Parts 1–6 and 9–10 specify the core MPEG-7 technology, while Parts 7 and 8 are 'supporting parts'. Although the various MPEG-7 parts are rather independent and thus can be used by themselves, or in combination with proprietary technologies, they were developed in order that the maximum benefit results when they are used together. Contrary to previous MPEG standards, where profiles and levels were defined together with the tools, MPEG-7 part 9 is specifically dedicated to this type of specification.

1.2.5 MPEG-7 Schedule

After an initial period dedicated to the specification of objectives and the identification of applications and requirements, MPEG-7, issued in October 1998 a Call for Proposals [19] to gather the best available technology fitting the MPEG-7 requirements. By the deadline (1 December 1998) 665 proposal pre-registrations had been received [20]. Of these, 390 (about 60%) were actually submitted as proposals by the 1 February 1999 deadline. Out of these 390 proposals there were 231 descriptors and 116 description schemes. The proposals for normative elements were evaluated by MPEG experts, in February 1999, in Lancaster (UK), following the procedures defined in the MPEG-7 Evaluation documents [21,22]. A special set of audiovisual content was provided to the proponents for usage in the evaluation process; this content set has also being used in the collaborative phase. The content set consists of 32 compact discs with sound tracks, pictures and video [23]. It has been made available to MPEG under the licensing conditions defined in [24]. Broadly, these licensing terms permit usage of the content exclusively for MPEG-7 standard development purposes. While fairly straightforward methodologies were used for the evaluation of the audiovisual description tools in the MPEG-7 competitive phase, more powerful methodologies were developed during the collaborative phase in the context of tens of core experiments. After the evaluation of the technology received, choices and recommendations were made and the collaborative phase started with the most promising tools [25]. The 'collaboration after competition' approach concentrates the efforts of many research teams throughout the world on further improving the technology that has already been demonstrated to be top-ranking.

For MPEG-7, the standardization process described in Section 1.1.3 translates to the work plan presented in Table 1.1.

In the course of developing the standard, additional calls may be issued when not enough technology is available within MPEG to meet the requirements, but there must be indications that the technology does indeed exist; this has happened already for some of the systems tools included in the first amendment to the MPEG-7 Systems part. After issuing the first version of the various parts of the standard, additional tools addressing new functionalities or significant improvements to available functionalities may be included in the standard by developing amendments to the relevant parts of the standard. This is the way to further complete the standard without delaying the first version too much, thus including in a timely way a substantial number of tools. Amendment 1 to the various parts of the standard is often referred to as Version 2.

1.2.6 MPEG-7 Description Tools

Since March 1999, MPEG has developed a set of multimedia description tools addressing the identified MPEG-7 requirements [6]. These tools are specified in the core parts of the MPEG-7 standard.

Systems tools

The MPEG Systems subgroup is in charge of developing a set of systems tools and the Description Definition Language (parts 1 and 2 of the standard) [9,10]. The systems tools allow preparing the MPEG-7 descriptions for efficient transport and storage (binarization), to synchronize

Table 1.1 MPEG-7 work plan

Part	Title	Committee Draft or Proposed Draft Technical Report	Final Committee Draft	Final Draft International Standard or Draft Technical Report
1	Systems	00/10	01/03	01/07
	Amendment 1	02/12	03/07	03/12
2	Description Definition Language	00/10	01/03	01/07
3	Visual	00/10	01/03	01/07
	Amendment 1	03/03	03/07	03/12
4	Audio	00/10	01/03	01/07
	Amendment 1	02/05	02/10	03/07
5	Multimedia Description Schemes	00/10	01/03	01/07
	Amendment 1	02/05	02/10	03/07
6	Reference Software	00/10	01/03	01/07
	Amendment 1	02/05	03/03	03/10
7	Conformance Testing	01/12	02/03	02/07
	Amendment 1	03/07	03/10	04/03
8	Extraction and Use of MPEG-7 Descriptions	02/03	—	02/05
9	Profiles and Levels	03/10	04/07	05/01
10	Schema Definition	03/10	04/03	04/07

content and the corresponding descriptions, and to manage and protect intellectual property. The DDL is the language for defining new description schemes and extending existing ones.

MPEG-7 data can be represented either in textual format, in binary format or a mixture of the two formats, depending on the application [9]. MPEG-7 defines a unique bidirectional mapping between the binary format and the textual format. This mapping can be lossless either way. The syntax of the binary format—BiM (Binary format for MPEG-7 data)—is defined in part 1 of the standard [9]. The syntax of the textual format—TeM (Textual format for MPEG-7 data)—is defined in part 2 of the standard [10]. Description schemes are defined in parts 3–5 of the standard [11–13].

There are two major reasons for having a binary format (besides a textual format) for MPEG-7 data. First, in general the transmission or storage of the textual format requires a much higher bandwidth than necessary from a theoretical point of view (after binarization, compression gains of 98% may be obtained for certain cases); an efficient compression of the textual format is applied when converting it to the binary format. Second, the textual format is not very appropriate for streaming applications since it only allows the transmission of a description tree in the so-called *depth-first tree order*. However, for streaming applications more flexibility is required with respect to the transmission order of the elements: the BiM provides this flexibility. The BiM allows randomly searching or accessing elements of a binary MPEG-7 description directly on the bitstream, without parsing the complete bitstream before these elements. At the description-consuming terminal, the binary description can be either converted to the textual format or directly parsed.

The BiM is designed in a way that allows fast parsing and filtering at binary level, without decompressing the complete description stream beforehand. This capability is particularly

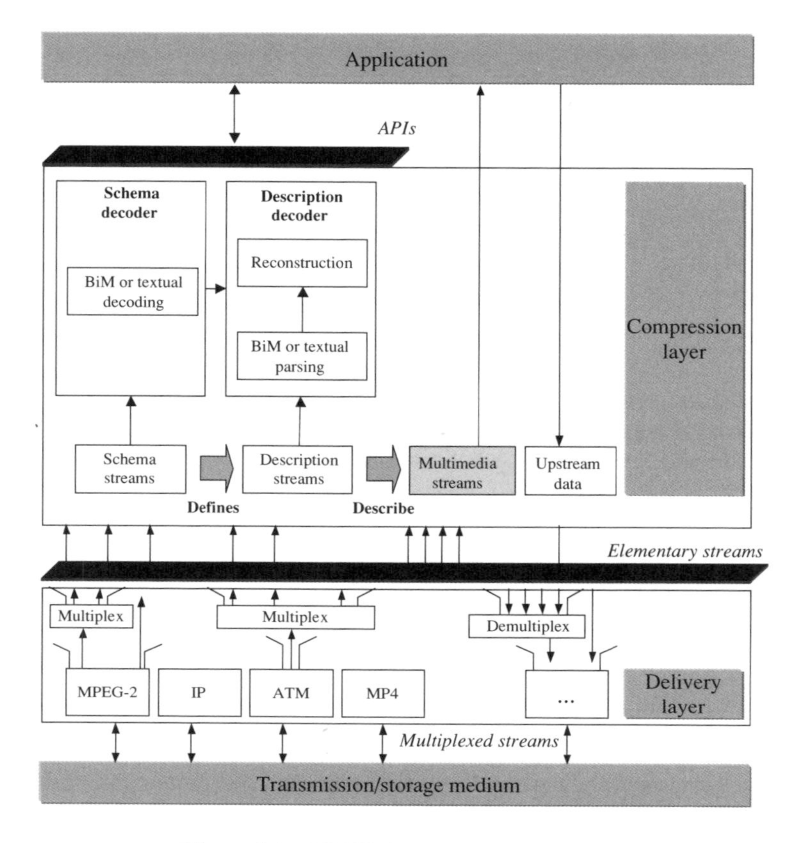

Figure 1.1 MPEG-7 systems architecture [9]

important for small, mobile, low-power devices with restricted CPU and memory capacity. The binary format is composed of one global header, the *Decoding Modes,* which specify some general parameters of the encoding, and a set of consecutive and nested coding patterns. These patterns are nested in the same way the elements are nested in the original DDL file, i.e. as a tree.

MPEG-7 descriptions may be delivered independently or together with the content they describe. The MPEG-7 architecture [9] (see Figure 1.1) allows conveying data back from the terminal to the transmitter or server, such as queries. The Systems layer encompasses mechanisms allowing synchronization, framing and multiplexing of MPEG-7 descriptions, and may also be capable of providing the multimedia content data if requested. The delivery of MPEG-7 content on particular systems is outside the scope of the Systems specification. Existing delivery tools, such as TCP/IP, MPEG-2 Transport Stream (TS) or even a CD-ROM may be used for this purpose.

MPEG-7 elementary streams consist of consecutive individually accessible portions of data named *access units*; an access unit is the smallest data entity to which timing information can be attributed. MPEG-7 elementary streams contain information of a different nature: (i) description schema information defining the structure of the MPEG-7 description, and (ii) description information that is either the complete description of the multimedia content or fragments of the description.

Following the specification of MPEG-7 Systems Version 1 [9], further work is in progress targeting the decoding of fragment references and the use of optimized binary decoders. This means decoders dedicated to certain encoding methods better suited than the generic ones; an optimized decoder is associated with a set of simple or complex types [5]. Future work related to MPEG-7 Systems may include the development of more efficient binary coding methods for MPEG-7 data, and the support of the transmission of MPEG-7 descriptions using a variety of transmission protocols [5].

DDL tools

The Description Definition Language (the textual format) is based on W3C's XML (eXtensible Markup Language) Schema Language [26]; however, some extensions to XML Schema were developed in order that all the DDL requirements [6] are fulfilled by the MPEG-7 DDL [10]. In this context, the DDL can be broken down into the following logical normative components:

- *XML Schema structural language components*: these components correspond to part 1 of the XML Schema specification [26], and provide facilities for describing and constraining the content of XML 1.0 documents.
- *XML Schema data type language components*: these components correspond to part 2 of the XML Schema specification [26], and provide facilities for defining data types to be used to constrain the data types of elements and attributes within XML Schemas.
- *MPEG-7-specific extensions*: these extensions correspond to the features added to the XML Schema Language to fulfil MPEG-7-specific requirements, notably new data types [10].

MPEG-7 DDL specific parsers add the validation of the MPEG-7 additional constructs to standard XML Schema parsers. In fact, while a DDL parser is able to parse a regular XML Schema file, a regular XML Schema parser may parse an MPEG-7 textual description, although with a reduced level of validation due to the MPEG-7-specific data types that cannot be recognized.

The BiM format [9] is a DDL compression tool, which can also be seen as a general XML compression tool since it also allows the efficient binary encoding of XML files in general, as long as they are based on DDL or XML Schema and the respective DDL or XML schema definition is available [27].

No DDL-related work targeting Version 2 is foreseen as of March 2004.

Visual tools

The MPEG Video subgroup is responsible for the development of the MPEG-7 Visual description tools (part 3 of the standard) [11]. MPEG-7 Visual description tools include basic structures and descriptors or description schemes enabling the description of some visual features of the visual material, such as colour, texture, shape and motion, as well as the localization of the described objects in the image or video sequence. These tools are defined by their syntax in DDL and binary representations and semantics associated with the syntactic elements. For each tool, there are normative and non-normative parts: the normative parts specify the textual and binary syntax and semantics of the structures, while the non-normative ones propose relevant associated methods such as extraction and matching.

Basic structures: elements and containers
The MPEG-7 Visual basic elements are:

- ***Spatial 2D coordinates***: this descriptor defines a 2D spatial coordinate system to be used by reference in other Ds/DSs, when relevant. It supports two kinds of coordinate systems: local and integrated. In a local coordinate system, the coordinates used for the creation of the description are mapped to the current coordinate system applicable; in an integrated coordinate system, each image (frame) of e.g. a video sequence may be mapped to different areas with respect to the first frame of a shot or video.
- ***Temporal interpolation***: the `Temporal_Interpolation` descriptor characterizes temporal interpolation using connected polynomials to approximate multidimensional variable values that change with time, such as object position in a video sequence. The descriptor size is usually much smaller than describing all position values.

The MPEG-7 Visual containers—structures allowing the combination of visual descriptors according to some spatial/temporal organization—are:

- ***Grid layout***: the grid layout is a splitting of the image into a set of rectangular regions, so that each region can be described separately. Each region of the grid can be described in terms of descriptors such as colour or texture.
- ***Time series***: the `TimeSeries` structure describes a temporal series of descriptors in a video segment and provides image to video frame and video frame to video frame matching functionalities. Two types of `TimeSeries` are defined: `RegularTimeSeries and IrregularTimeSeries`. In the `RegularTimeSeries`, descriptors are located regularly (with constant intervals) within a given time span; alternatively, in the `IrregularTimeSeries`, descriptors are located irregularly along time.
- ***Multiple view***: the `MultipleView` descriptor specifies a structure combining 2D descriptors representing a visual feature of a 3D object seen from different view angles. The descriptor forms a complete 3D-view-based representation of the object, using any 2D visual descriptor, such as shape, colour or texture.

Descriptors and description schemes
The MPEG-7 Visual descriptors cover five basic visual features—colour, texture, shape, motion and localization—and there is also a face recognition descriptor.

Color
There are seven MPEG-7 **colour** descriptors:

- ***Colour space***: defines the colour space used in MPEG-7 colour-based descriptions. The following colour spaces are supported: RGB, YCbCr, HSV, HMMD, linear transformation matrix with reference to RGB, and monochrome.
- ***Colour quantization***: defines the uniform quantization of a colour space.
- ***Dominant colour***: specifies a set of dominant colours in an arbitrarily shaped region (maximum 8 dominant colours).
- ***Scalable colour***: defines a colour histogram in the HSV colour space, encoded by a Haar transform; its binary representation is scalable in terms of bin numbers and bit representation accuracy over a broad range of data rates.

- ***Color layout***: specifies the spatial distribution of colours for high-speed retrieval and browsing; it can be applied either to a whole image or to any part of an image, including arbitrarily shaped regions.
- ***Color structure***: captures both colour content (similar to that of a colour histogram) and the colour structure of this content via the use of a structuring element composed of several image samples.
- ***GoF/GoP colour***: defines a structure required for representing the colour features of a collection of (similar) images or video frames by means of the scalable colour descriptor defined above. The collection of video frames can be a contiguous video segment or a non-contiguous collection of similar video frames.

Texture
There are three MPEG-7 texture descriptors; texture represents the amount of structure in an image such as directionality, coarseness, regularity of patterns etc:

- ***Homogeneous texture***: represents the energy and energy deviation values extracted from a frequency layout where the 2D frequency plane is partitioned into 30 channels; the frequency plane partitioning is uniform along the angular direction (equal step size of 30 degrees) but not uniform along the radial direction.
- ***Texture browsing***: relates to the perceptual characterization of the texture, similar to a human characterization, in terms of regularity (irregular, slightly regular, regular, highly regular; see Figure 1.2), directionality (0°, 30°, 60°, 90°, 120°, 150°) and coarseness (fine, medium, coarse, very coarse).
- ***Edge histogram***: represents the spatial distribution of five types of edges in local image regions as shown in Figure 1.3 (four directional edges and one non-directional edge in each local region, called sub-image).

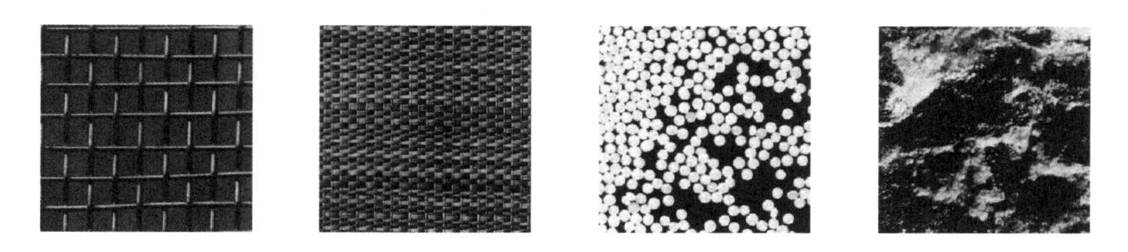

Figure 1.2 Examples of texture regularity from highly regular to irregular [11]

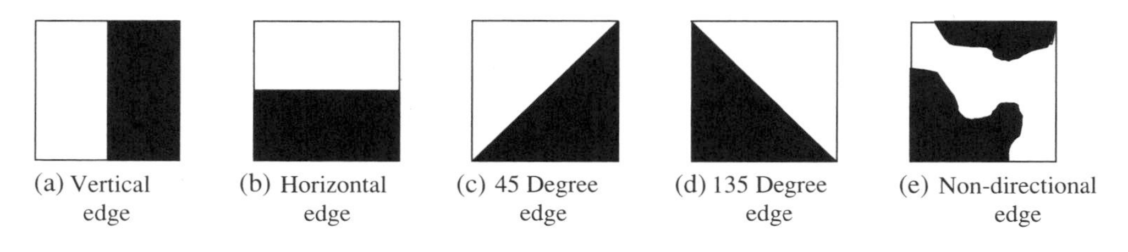

Figure 1.3 The five types of edges used in the MPEG-7 edge histogram descriptor [11]

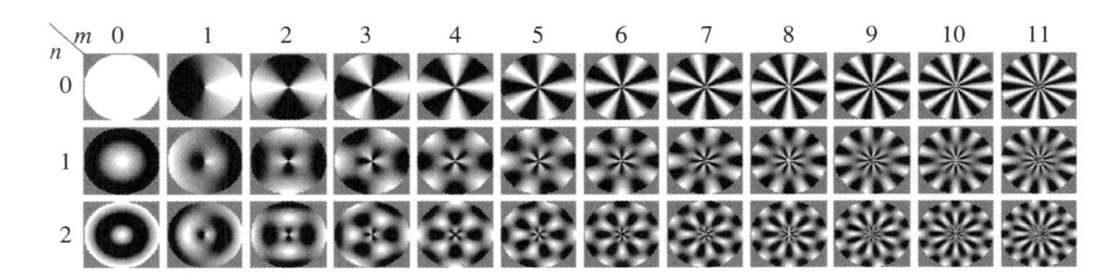

Figure 1.4 Real parts of the ART basis functions [11]

Shape

There are three MPEG-7 shape descriptors:

- ***Region-based shape—angular radial transform***: makes use of all pixels constituting the shape and can describe any shape, i.e. not only a simple shape with a single connected region but also a complex shape consisting of several disjoint regions. The region-based shape descriptor is a set of angular radial transform (ART) coefficients. The ART is a 2D complex transform defined on a unit disk in polar coordinates:

$$F_{nm} = \langle V_{nm}(\rho,\theta), f(\rho,\theta) \rangle = \int_0^{2\pi} \int_0^1 V_{nm}^*(\rho,\theta)\, f(\rho,\theta)\, \rho \mathrm{d}\rho \mathrm{d}\theta$$

  where F_{mn} is an ART coefficient of order n and m, $f(\rho, \theta)$ is an image function in polar coordinates and $V_{nm}(\rho, \theta)$ is the ART basis function (Figure 1.4).
- ***Contour-based shape—curvature scale space representation***: describes a closed contour of a 2D object or region in an image or video sequence (Figure 1.5). This descriptor is based on the curvature scale space (CSS) representation of the contour, typically making very compact descriptions (below 14 bytes in size, on average). The idea behind the CSS representation is that the contour can be represented by the set of points where the contour curvature changes and the curvature values between them; the curvature value is defined as the tangent angle variation per unit of contour length. For each point in the contour, it is possible to compute the curvature of the contour at that point, based on the neighbouring points; a point whose two closest neighbours have different curvature values is considered a curvature change. The CSS representation of the contour (CSS image) basically corresponds to a plot where the arc length parameter value (relatively to an arbitrary starting point) is the

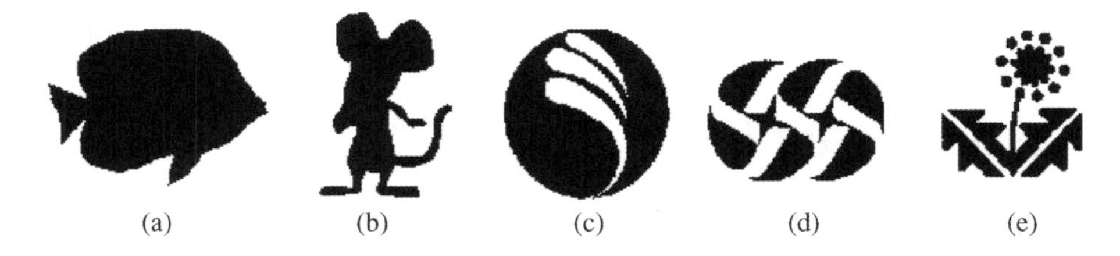

Figure 1.5 Example shapes: the ART descriptor can describe all these shapes but the CSS descriptor can only describe the shapes with one single connected region [11]

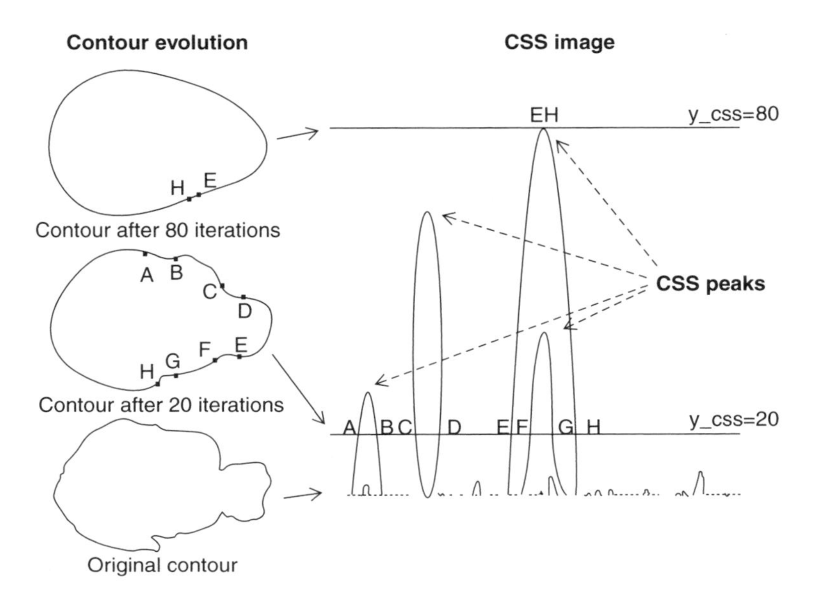

Figure 1.6 CSS image formation [11]

x-coordinate and the number of low-pass filter passes (or iterations) is the y-coordinate (see example in Figure 1.6).

- ***Shape 3D***: provides an intrinsic shape description of 3D mesh models, based on the histogram of 3D shape indexes representing local curvature properties of the 3D surface.

Motion

There are four MPEG-7 motion descriptors:

- ***Camera motion***: characterizes 3D camera motion based on 3D camera motion parameter information, which can be automatically extracted or generated by capture devices. This descriptor supports the following basic camera operations (Figure 1.7): fixed, panning (horizontal rotation), tracking (horizontal transverse movement, also called travelling in the film industry), tilting (vertical rotation), booming (vertical transverse movement), zooming

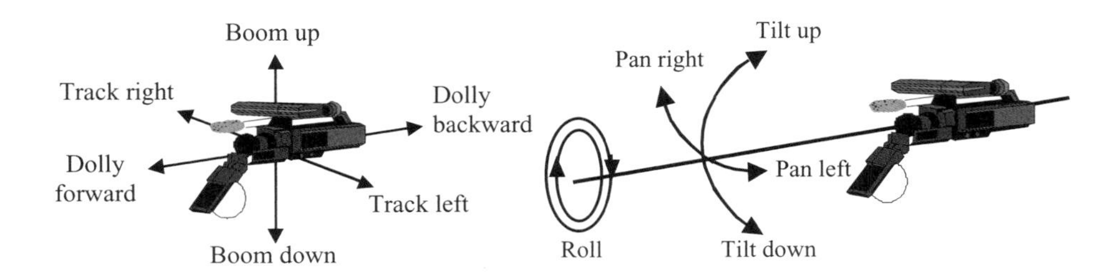

Figure 1.7 Types of MPEG-7 camera motion [11]

(change of the focal length), dollying (translation along the optical axis) and rolling (rotation around the optical axis).
- ***Motion trajectory***: defines a spatio-temporal localization of a representative point of a moving region, for example the centroid.
- ***Parametric motion***: characterizes the evolution of arbitrarily shaped regions over time in terms of a 2D geometric transform (translational, rotation/scaling, affine, perspective, quadratic). This descriptor addresses the motion of objects in video sequences, as well as global motion; if it is associated with a region, it can be used to specify the relationship between two or more feature point motion trajectories, according to the underlying motion model.
- ***Motion activity***: captures the intuitive notion of 'intensity of action' or 'pace of action' in a video segment. The activity descriptor includes the following five attributes: intensity of activity, direction of activity, spatial distribution of activity, spatial localization of activity and temporal distribution of activity.

Localization
There are two MPEG-7 localization descriptors:

- ***Region locator***: enables localization of regions within images or frames by specifying them with a brief and scalable representation of a `Box` or a `Polygon`.
- ***Spatio-temporal locator***: describes spatio-temporal regions in a video sequence by one or several sets of reference regions and motion using two description schemes: `FigureTrajectory` and `ParameterTrajectory`. These two description schemes are selected according to moving object conditions: if a moving object region is rigid and the motion model is known, `ParameterTrajectory` is appropriate; if a moving region is non-rigid, `FigureTrajectory` is appropriate.

Face recognition
Finally, there is also a face recognition descriptor based on principal components analysis (PCA). The face recognition descriptor represents the projection of a (normalized) face vector onto a set of 49 basis vectors (derived from the eigenfaces of a set of training faces) that span the space of all possible face vectors.

Although the set of MPEG-7 Visual description tools presented above provides a very powerful solution in terms of describing visual content for a large range of applications, new tools are needed to support additional features. Examples of additional to Version 1 visual description tools under consideration or already approved by March 2004 were [5]:

- ***GoF/GoP feature***: generic and extensible container to include several description tools defined in MPEG-7 Visual Version 1 to describe representative features regarding a group of frames (GoF) or group of pictures (GoP).
- ***Colour temperature***: specifies the perceptual temperature feeling of illumination colour in an image or video for browsing and display preferences control purposes to improve the user's experience. Four perceptual temperature categories are provided: hot, warm, moderate and cool.
- ***Illumination invariant colour***: wraps the colour descriptors in MPEG-7 Visual Version 1, which are dominant colour, scalable colour, colour layout, and colour structure.

- ***Shape variation***: describes shape variations in time by means of a shape variation map and the statistics of the region shape description of each binary shape image in the collection.
- ***Advanced face recognition***: describes a face in a way robust to variations in pose and illumination conditions.
- ***Media-centric description schemes***: include three visual description schemes designed to describe several types of visual contents: `StillRegionFeatureType` contains several elementary descriptors to describe the characteristics of arbitrary shaped still regions while `VideoSegmentFeatureType` and `MovingRegionFeatureType` describe moving pictures; the former supports ordinary video without shape and the latter arbitrary shaped video sequences.

Audio tools

The MPEG Audio subgroup had the task of specifying the MPEG-7 Audio description tools (part 4 of the standard). MPEG-7 Audio description tools are organized in the following areas [12]:

- ***Scalable series***: efficient representation for series of feature values; this is a core part of MPEG-7 Audio.
- ***(Low-level) Audio framework***: collection of low-level audio descriptors, many built upon the scalable series, e.g. waveform, spectral envelope, loudness, spectral centroid, spectral spread, fundamental frequency, harmonicity, attack time, spectral basis.
- ***Silence***: descriptor identifying silence.
- ***Spoken content***: set of description schemes representing the output of automatic speech recognition (ASR).
- ***Timbre description***: collection of descriptors and description schemes describing the perceptual features of instrument sounds.
- ***Sound effects***: collection of descriptors and description schemes defining a general mechanism suitable for handling sound effects.
- ***Melody contour***: description scheme allowing retrieval of musical data.
- ***Melody***: more general description framework for melody.

As for MPEG-7 Visual, additional audio description tools were under consideration or already approved by March 2004 [5]:

- ***Spoken content***: includes a modification to the Version 1 spoken content descriptor.
- ***Audio signal quality***: collection of descriptors expressing quality attributes of the audio signal.
- ***Audio tempo***: describes the tempo of a musical item according to standard musical notation; its scope is limited to describing musical material with a dominant musical tempo and only one tempo at a time.

Multimedia tools

The MPEG Multimedia Description Schemes (MDS) subgroup had the task of specifying a set of description tools, dealing with generic as well as multimedia entities (part 5 of the standard)

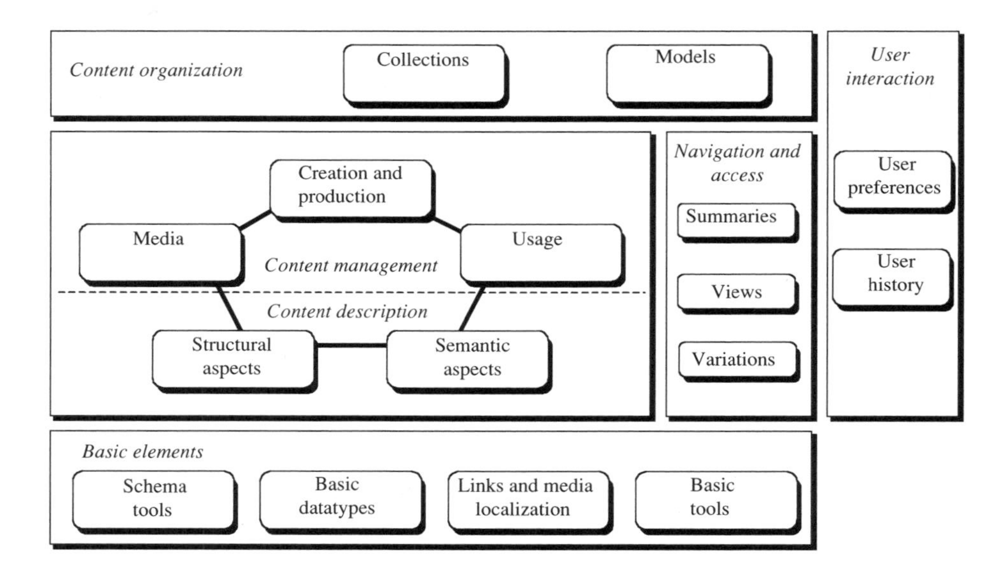

Figure 1.8 Overview of the MDS description tools [13]

[13]. Generic entities are those that can be used in audio, visual and textual descriptions, and therefore are 'generic' to all media, e.g. vector, histogram, time etc. Multimedia entities are those that deal with more than one medium, e.g. audio and video. These description tools can be grouped into six different classes according to their functionality (Figure 1.8):

- ***Content description—structural and conceptual aspects***: representation of perceivable information in terms of structural and conceptual aspects; the visual and audio descriptors and descriptions schemes presented above allow the structural description of the content. The structural and semantic trees may be linked together, leading to a graph, which provides a powerful description of the content since it combines a more low-level description with a more abstract description.
- ***Content management—media, usage, creation and production***: information related to the management of the content, notably information that cannot be automatically extracted from the content, e.g. rights holders (content usage), author's name (creation and production).
- ***Content organization—collection, classification and modelling***: organization and classification of content as well as analytic modelling.
- ***Navigation and access—summary and variation***: summaries enabling fast visualization and sonification as well as generating variations of the content.
- ***User—user preferences***: limited set of descriptors for user preferences pertaining to multimedia content.
- ***Basic elements—data type and structures, schema tools, link and media localization, and basic DSs***: tools such as basic data types, mathematical structures, schema tools (root element, top-level element and packages) etc. which are found as elementary components of more complex DSs. The root element is a wrapper of the description, which may be a complete description (in the sense of an application) or a partial/incremental description.

Regarding additional tools to the MDS Version 1 specification, the following tools were under consideration or already approved by March 2004 [5]:

- ***New base types***: `audioDType` and `AudioDSType` are extended by providing each with an optional attribute that denotes which audio channels are used in computing the values of the Audio Descriptors and Audio Description Schemes, respectively.
- ***Stream locator data type***: describes the location of data within a stream.
- ***Subject classification scheme***: allows encoding of standard thesauri and classification schemes such as the Library of Congress Thesaurus of Graphical Material (TGM)-I and Library of Congress Subject Headings (LCSH). This Description Scheme accommodates more complex thesauri and classification schemes than those supported by the classification scheme DS defined in MDS Version 1.
- ***General relation DS***: extends the Relation DS in MDS Version 1 (by providing a typelist attribute to accommodate multiple relation terms, and `generalSource` and `general-Target` attributes to accommodate `termReferenceType` arguments).
- ***Linguistic description tools***: tools to create linguistic descriptions: a new multimedia content entity tool, a new linguistic Description Scheme and new structure description tools for linguistic descriptions.

Reference software

Since the overall value of the MPEG-7 standard will not only depend on the quality of each set of tools but also on the quality of their interaction and complementary capabilities, the MPEG-7 standard has been developed using an intense collaborative framework, where synergies could be fully exploited. The best example of this practice is the MPEG-7 eXperimentation Model (XM), which integrates the Systems, DDL, Audio, Visual and MDS tools into one comprehensive experimental framework. The XM became part 6 of MPEG-7: Reference Software [14].

Version 2 of the reference software will include the software corresponding to the tools defined in Version 2 of the Systems, Audio, Visual and MDS parts of the MPEG-7 standard.

1.2.7 Profiling

Since MPEG-7 standardizes a large number of description tools, which can be used requiring more or less computational resources at the receiver, it is essential to provide a mechanism by which interoperability is assured with an acceptable degree of complexity. Moreover the same mechanism will be useful to check whether and how MPEG-7 devices comply with the standard (conformance testing) [15].

This mechanism—so-called *profiling*—is based on two major concepts: profiles and levels [17]. For MPEG-7, MPEG defines profiles and levels for the descriptions but not for the description consumption terminal since there will be many different ways to consume descriptions (with the exception of binary decoders, if needed). MPEG does not fix any relation between applications and profiles and levels. This is a choice left for the implementors, taking into account their specific needs and constraints. Description profiles and levels provide a means of defining subsets of the syntax and semantics of the MPEG-7 schema.

The keyword to define profiles is 'functionality'; a profile is a set of tools providing a set of functionalities for a certain class of applications. A new profile should be defined if it provides a significantly different set of functionalities. Description profiles provide a means of defining restrictions on the MPEG-7 schema, thereby constraining conforming descriptions in their content. A description profile generally limits or mandates the use of description tools to subsets of the description tools defined in MPEG-7. The description tools in a description profile support a set of functionalities for a certain class of applications. MPEG-7 description profiles will be defined across MPEG-7 standard parts, notably Visual (part 3), Audio (part 4) and MDS (part 5). The keyword to define levels is 'complexity'; a level of a description profile defines further constraints on conforming descriptions, changing their complexity. A new level should be defined if it is associated with a significantly different implementation complexity.

Profiling of MPEG-7 descriptions is made using a schema-based approach through the so-called *profile schema*. A profile schema is a restriction of the MPEG-7 schema in the following sense: any description that is valid against the profile schema shall also be valid against the MPEG-7 schema, while any description that is valid against the MPEG-7 schema may or may not be valid against the profile schema. The profile schema provides the basis for determining conformance to a description profile (similar to the case of conformance to the MPEG-7 schema; see MPEG-7 part 7 [15]), namely by testing the validity of a description against the profile schema.

A level within a description profile, referred to as profile@level, implies a restriction of the profile schema in the same sense that a description profile implies a restriction of the MPEG-7 schema. A level of a description profile defines further constraints on conforming descriptions, changing their complexity.

Following requests from the industry, MPEG will define some description profiles and levels following a predefined process. By March 2004, three description profiles were being considered for adoption: Simple Metadata, User Description and Core Description profiles [17]. Unlike previous MPEG standards, MPEG-7 will include a part especially dedicated to the specification of profiles and levels (part 9).

1.3 MPEG-21: Multimedia Framework

1.3.1 Objectives

Nowadays, multimedia content creators and consumers can already rely on a huge number of coding technologies, content delivery models and architectures, multimedia application models, different types of network infrastructures etc. As a result, end users expect to have access to multimedia data and services anywhere and anytime. In addition, ease of use is extremely important, since non-professional individuals are producing more and more digital media for their personal use and for sharing with family and friends (which is evidenced by the large number of amateur music, photo and video sharing websites). At the same time, these amateur content providers have many of the same concerns as commercial content providers. These include management of content, re-purposing content based on consumer/device capabilities, rights protection, protection from unauthorized access/modification, protection of privacy of providers and consumers etc.

Besides, recent developments in the field of multimedia systems and applications have been pushing the boundaries of existing business models for trading physical goods. As a result, new models for electronically distributing and trading digital content are required. For example, it is becoming increasingly difficult for legitimate users of content to identify and interpret the different intellectual property rights that are associated with multimedia content. Additionally, there are some users who freely exchange content with disregard for the rights associated with this content. At the same time, rights holders are often powerless to prevent them from doing this. In this context, delivery of a wide variety of media types should be considered: audio (music and spoken words), accompanying artwork (graphics), text (lyrics), video (visual data), synthetic data etc. New solutions are required for the access, delivery, management and protection of these different content types in an integrated and harmonized way. They should be implemented in a manner that is entirely transparent to all users of multimedia services.

The need for technological solutions to these challenges motivated MPEG to start developing the MPEG-21 standard, formally called 'Multimedia Framework' [28]. This initiative aims to enable the transparent and augmented use of multimedia data across a wide range of networks and devices. In order to realize this goal, MPEG-21 provides a normative open framework for multimedia delivery and consumption. This framework will be of use for all players in the multimedia delivery and consumption chain. It will provide content creators, producers, distributors and service providers with equal opportunities in the MPEG-21-enabled open market. This will also be to the benefit of content consumers, providing them access to a large variety of content in an interoperable manner.

Today, many elements exist to build an infrastructure for the delivery and consumption of multimedia content. There is, however, no 'big picture' to describe how these elements, either in existence or under development, relate to each other. The first aim for MPEG-21 was therefore to describe how the various elements fit together. Where relevant standards are already in place, MPEG-21 provides the means to incorporate them within the Multimedia Framework. For example, existing MPEG-7 tools could be used for describing multimedia content in the context of MPEG-21. Where gaps exist, MPEG-21 recommends which new standards are required. MPEG-21 then develops new standards as appropriate while other relevant specifications may be developed by other bodies. These specifications will be integrated into the Multimedia Framework through collaboration between MPEG and these bodies. The envisaged result is an open framework for multimedia delivery and consumption, with both the content creator and content consumer as focal points.

MPEG-21 is based on two essential concepts: the fundamental unit of distribution and transaction, the Digital Item, and the entity that interacts with Digital Items, the User. Digital Items can be considered the 'what' of the Multimedia Framework (e.g. a video collection, a music album etc.) and the Users can be considered the 'who' of the Multimedia Framework. Both concepts will be discussed in detail in Section 1.3.3. The goal of MPEG-21 can thus be rephrased as defining the technology needed for Users to exchange, access, consume, trade and otherwise manipulate Digital Items in an efficient, transparent and interoperable way.

In Section 1.3.4, an overview of all existing MPEG-21 standard parts will be given. From this overview, it will become clear that MPEG-21 is strongly relying on description tools. More specifically, the description of the 'structure' of Digital Items is specified via the MPEG-21 Digital Item Declaration (DID) standard, while these Digital Items can be identified by applying the MPEG-21 Digital Item Identification (DII) specification. Besides, the usage environment in which multimedia content is to be consumed can be described using some of the tools defined in MPEG-21 Digital Item Adaptation (DIA). MPEG-21 DIA tools not only provide means for describing the usage environment; for example, they also provide means for describing the syntax of multimedia bitstreams and for adapting these bitstreams according to end-user preferences, network conditions, the availability (or lack) of terminal resources etc.

1.3.2 Applications and Requirements

As stated before, MPEG-21 technologies can be used within a very wide range of applications. They are *not* focused on a particular media type (like still images, video, audio etc.) or multimedia data representation (like JPEG2000, GIF, MPEG-4 etc.); MPEG-21 supports the use of any media type or media format. It is not focused on a particular application model either; it supports both peer-to-peer and client–server applications. In addition, and this is probably the most important difference between the scope of the MPEG-21 specifications and the scope of existing standards, MPEG-21 is explicitly addressing *all* players (such as content providers and content consumers) and key elements in the Multimedia Framework: the declaration and identification of digital items, intellectual property management and protection, digital rights management, terminals and networks, content handling and usage (providing interfaces and protocols that enable creation, manipulation, search, access, storage, delivery and use of content across the content distribution and consumption value chain), content representation/coding, event reporting (e.g. financial event reporting, usage tracking) etc.

While developing the MPEG-21 specifications, many applications have been studied in detail. This study is important, since it forces MPEG-21 proponents to take into account the requirements of relevant applications while developing the new MPEG-21 standard. In the MPEG-21 Use Case Scenarios document [29], several applications are addressed. They can be categorized into the following application classes:

- *Multimedia data distribution and archiving*: this class of applications includes scenarios such as the release of digital albums (e.g. digital music albums), the setup and management of picture archives, the distribution of news; real-time financial data distribution and several e-commerce scenarios.
- *Universal Media Access (UMA)*: considers scenarios such as dynamic content adaptation to terminal characteristics, network resources and/or User preferences, streaming services/applications for 3G mobile devices, terminal quality of service (QoS) management, description and enforcement of aggregation conditions, content adaptation based on limiting rights; description and enforcement of spatio-temporal links between different media data, device use tracking via event reporting, and enabling or disabling access to fragments of multimedia data.

As for MPEG-7, functionalities and requirements have been extracted from these application examples. These requirements are consolidated in the MPEG-21 Requirements document [30], which includes the requirements for all key elements that have been mentioned before. During the MPEG-21 standardization process, Calls for Proposals based upon these requirements have been issued by MPEG, which resulted in different parts of the MPEG-21 standard (i.e. ISO/IEC 21000-x, where x is the part number).

1.3.3 Basic Elements

Within the MPEG-21 framework, two basic elements can be identified: MPEG-21 Digital Items and MPEG-21 Users. Both elements are introduced below.

MPEG-21 digital items

A lot of advanced multimedia applications are not solely relying on the consumption of isolated multimedia data anymore (like a single MP3 file or a single MPEG-4 movie). Instead, they are based on the delivery and presentation of complex media packages.

For example, consider the consumption of a digital music album (Figure 1.9). This digital music album may consist of several sub-items, such as the different music tracks, the corresponding lyrics, a biography of the performers, some video clips, commercial information, copyright information, digital signatures etc. Each of these sub-items may have its own representation (e.g. MP3 for music tracks, MPEG-4 for video clip data, an MPEG-21 Rights Expression for copyright information, a 128-bit hash for digital signatures etc.), and its own descriptive metadata (e.g. using MPEG-7). Moreover, the author of such a digital music album may want to express its structure. For example, he wants to be able to express the relationship between the lyrics and the sound tracks ('lyric A corresponds to music track X'). He may also be willing to attach descriptive information to the digital music album as a whole ('the title of the album is *The Best of Puccini*; the composer is Giacomo Puccini'), or to sub-items in particular ('the title of the fourth track is *Nessun Dorma*'). In addition, he may want to express conditional access to certain sub-items, like 'music track Y can only be played provided that the network bandwidth is at least 192 kbps' or 'music track Z can only be stored provided that payment has been completed'.

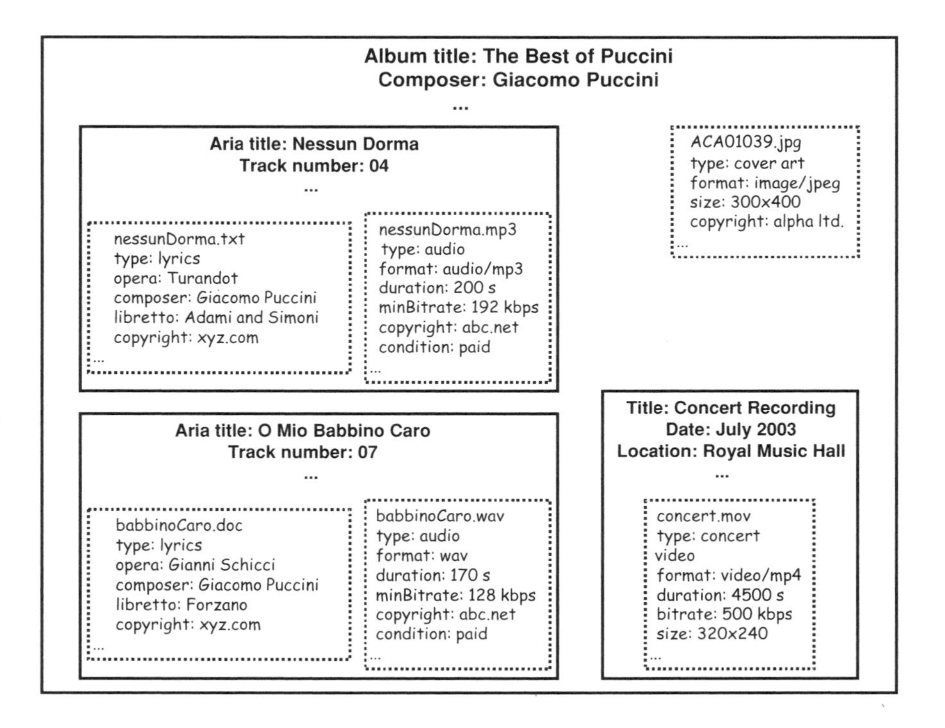

Figure 1.9 Example MPEG-21 music album Digital Item

A more detailed example, illustrating some more complex issues about the Digital Item concept, is a dynamic web page. A web page typically consists of an HTML document with embedded links to (or dependencies on) various image files (e.g. JPEG or GIF files), possibly some layout information (e.g. stylesheets) or even rendering software. In this simple case, it is a straightforward exercise to inspect the HTML document and deduce that this Digital Item consists of the HTML document itself, plus all of the other multimedia data upon which it depends.

Now let us modify the example to assume that the web page contains some custom scripted logic (e.g. JavaScript) to determine the preferred language of the viewer among a predefined set of choices, and to either build/display the page in that language, or to revert to a default choice if the preferred language is not available. The addition of the scripting code changes the declarative links of the simple web page into links that (in the general case) can be determined only by running the embedded script on a specific platform. As a result, in this case, it should be possible for authors of Digital Items to express choices, like the language choice.

The Digital Item Declaration specification (part 2 of MPEG-21) provides such flexibility for representing Digital Items. This specification was developed based on a set of clear definitions and requirements, including the following: a Digital Item is a structured digital object with a standard representation, identification and description within the MPEG-21 framework. This entity is also the fundamental unit of distribution and transaction within the MPEG-21 framework.

MPEG-21 user

Within MPEG-21, a User (with a capital U) is any entity that interacts in the MPEG-21 environment or makes use of a Digital Item. Such Users include individuals, consumers, communities, organizations, corporations, consortia, governments, agents acting on behalf of end users etc. Users are identified specifically by their relationship to another User for a certain interaction. From a purely technical perspective, MPEG-21 makes no distinction between a content provider and a consumer; both are Users. A single entity may use content in many ways (publish, deliver, consume etc.); all parties interacting within MPEG-21 are categorized as Users equally. However, MPEG-21 Users may assume specific or even unique rights and responsibilities according to their interaction with other Users within MPEG-21.

At its most basic level, MPEG-21 provides a framework in which one User interacts with another User, and the object of that interaction is a Digital Item commonly called content. Examples of such interactions are creating content, providing content, archiving content, rating content, enhancing and delivering content, aggregating content, delivering content, syndicating content, retail selling of content, consuming content, subscribing to content, regulating content, and facilitating and regulating transactions that occur from any of the above. Any of these are addressed by MPEG-21, and the parties involved are always called MPEG-21 Users.

1.3.4 MPEG-21 Standard Organization

Like the MPEG-7 standard, and again for the sake of legibility, organization and easier usage, the MPEG-21 standard is structured in different parts:

- ***ISO/IEC 21000-1 or MPEG-21 Part 1—Vision, Technologies and Strategy***: describes the Multimedia Framework and its architectural elements together with the functional requirements for their specification [31,32].
- ***ISO/IEC 21000-2 or MPEG-21 Part 2—Digital Item Declaration (DID)***: provides a uniform and flexible abstraction and interoperable schema for declaring Digital Items [33].
- ***ISO/IEC 21000-3 or MPEG-21 Part 3—Digital Item Identification (DII)***: defines the framework for the identification of any entity regardless of its nature, type or granularity [34].
- ***ISO/IEC 21000-4 or MPEG-21 Part 4—Intellectual Property Management and Protection (IPMP)***: provides the means to enable content to be persistently and reliably managed and protected across networks and devices [35].
- ***ISO/IEC 21000-5 or MPEG-21 Part 5—Rights Expression Language (REL)***: specifies a machine-readable language that can declare rights and permissions using the terms as defined in the Rights Data Dictionary [36].
- ***ISO/IEC 21000-6 or MPEG-21 Part 6—Rights Data Dictionary (RDD)***: specifies a dictionary of key terms which are required to describe rights of all Users [37].
- ***ISO/IEC 21000-7 or MPEG-21 Part 7—Digital Item Adaptation (DIA)***: defines description tools for usage environment and content format features that may influence the transparent access to the multimedia content, notably terminals, networks, users and the natural environment where users and terminals are located [38].
- ***ISO/IEC 21000-8 or MPEG-21 Part 8—Reference Software***: includes software implementing the tools specified in the other MPEG-21 parts; this software can be used for free in MPEG-21-compliant products as ISO waives its copyright [39].
- ***ISO/IEC 21000-9 or MPEG-21 Part 9—File Format***: defines a file format for the storage and distribution of Digital Items [40].
- ***ISO/IEC 21000-10 or MPEG-21 Part 10—Digital Item Processing (DIP)***: defines mechanisms that provide for standardized and interoperable processing of the information in Digital Items [41].
- ***ISO/IEC 21000-11 or MPEG-21 Part 11—Evaluation Tools for Persistent Association Technologies***: documents best practice in the evaluation of persistent association technologies, i.e. technologies that link information to identify and describe content using the content itself [42].
- ***ISO/IEC 21000-12 or MPEG-21 Part 12—Test Bed for MPEG-21 Resource Delivery***: provides a (software-based) test bed for (i) the delivery of scalable media delivery and (ii) testing/evaluating this scalable media delivery in streaming environments (e.g. by taking into account varying network environments) [43].
- ***ISO/IEC 21000-13 or MPEG-21 Part 13—Scalable Video Coding***: defines scalable video coding (SVC) technology with high compression performance. Examples of potential SVC applications include Internet video, mobile wireless video for conversational, video on demand and live broadcasting purposes, multi-channel content production and distribution, surveillance-and-storage applications, and layered protection of contents [44].
- ***ISO/IEC 21000-14 or MPEG-21 Part 14—Conformance***: defines conformance testing for other parts of MPEG-21 [45].
- ***ISO/IEC 21000-15 or MPEG-21 Part 15—Event Reporting***: standardizes the syntax and semantics of a language to express Event Report Requests and Event Reports. It will also address the relation of Event Reports Requests and Event Reports with other parts of MPEG-21. This enables Users within the multimedia framework to, for example, monitor the use of Digital Items, Resources and their descriptions, and to monitor the load of networks [46].

• ***ISO/IEC 21000-16 or MPEG-21 Part 16—Binary Format***: supplies a binary format, based on tools in MPEG-7 Systems [9], which allows the binarization, compression and streaming of some or all parts of an MPEG-21 Digital Item [47].

Part 1 gives an overview of the overall MPEG-21 Multimedia Framework project. It is formally a Technical Report and not a Standard. Parts 11 and 12 are also Technical Reports. Technical reports are 'supporting parts' as well as conformance testing (part 14), while all other parts (2–10, 13, 15, 16) specify the core MPEG-21 technology.

1.3.5 MPEG-21 Schedule

The different parts of MPEG-21 have their own time schedule. These parts have been started following the needs identified, while considering also the limited amount of human resources available. An overview of this schedule is given in Table 1.2 [48].

Table 1.2 MPEG-21 work plan

Part	Title	Committee Draft or Proposed Draft Technical Report	Final Committee Draft	Final Draft International Standard or Draft Technical Report
1	Vision, Technologies and Strategy	01/01	—	01/07
1	Vision, Technologies and Strategy (2nd edition)	—	—	04/03
2	Digital Item Declaration	01/07	01/12	02/05
2	Digital Item Declaration (2nd edition)	04/03	04/07	05/01
3	Digital Item Identification	01/12	02/03	02/07
4	Intellectual Property Management and Protection	04/07	05/01	05/07
5	Rights Expression Language	02/07	02/12	03/07
6	Rights Data Dictionary	02/07	02/12	03/07
7	Digital Item Adaptation	02/12	03/07	03/12
7	Digital Item Adaptation (Amendment 1: DIA conversions and permissions)	04/07	04/10	05/04
8	Reference Software	04/03	04/07	05/01
9	File Format	04/03	04/07	05/01
10	Digital Item Processing	03/12	04/07	04/10
11	Evaluation Methods for Persistent Association Technologies	03/10	—	04/03
12	Test Bed for MPEG-21 Resource Delivery	03/12	—	04/07
13	Scalable Video Coding	05/10	06/03	06/07
14	Conformance	04/03	04/07	05/01
15	Event Reporting	04/10	05/04	05/10
16	Binary Format	04/07	04/10	05/04

1.3.6 MPEG-21 Description Tools

In the following sections, those parts of MPEG-21 that are directly related to multimedia content description are discussed: DID, DII and DIA. An overview of all MPEG-21 parts can be found in [28].

Digital item declaration

Before MPEG-21, there was no standard model or representation for complex Digital Items. Specific models and some representations did exist for certain application areas and/or media types, but there was no generic, flexible and interoperable solution for all kinds of content in any context.

As part of the above situation, there was also no uniform way of linking all types of descriptive information to any kind of multimedia data (or other descriptive information). Since the very concept of a Digital Item is built upon the notion of explicitly capturing the relationship between media data and descriptive data (so that complex relationships can be made unambiguous), it was a serious limitation that no standard model or representation for this capability did exist.

These limitations made it very difficult to implement key applications across all content types. An example would be content delivery that incorporates important capabilities such as the ability to define and process highly configurable content packages in a standard way (especially where the configuration may be desirable at multiple points along a delivery chain). Another more consumer motivated example is the ability to intelligently manage collections of content of diverse types and from multiple sources.

This presented a strong challenge to lay out a powerful and flexible model for Digital Items which could accommodate the myriad of forms that content and its descriptions can take (and the new forms it will assume in the future). Such a model is only truly useful if it yields a format that can be used to represent any Digital Items defined within the model unambiguously and communicate them successfully, together with information about them.

The MPEG-21 Digital Item Declaration (DID) technology is described in three normative sections [33]:

- ***Model***: the Digital Item Declaration model describes a set of abstract terms and concepts to form a useful model for defining Digital Items. Within this model, a Digital Item is the digital representation of 'a work', and as such, it is the thing that is acted upon (managed, described, exchanged, collected etc.) within the MPEG-21 Multimedia Framework. The goal of this model is to be as flexible and general as possible, while providing for the 'hooks' that enable the use of Digital Items within all parts of MPEG-21. It specifically does *not* imply the use of a particular language. Instead, the model provides a set of abstract concepts and terms that can be used to define such a language, or to perform mappings between existing languages capable of Digital Item Declaration, for comparison purposes.
- ***Representation***: the Digital Item Declaration Language (DIDL) includes the normative description of the syntax and semantics of each of the Digital Item Declaration elements, as represented in XML. This section also contains some non-normative examples for illustrative purposes.
- ***Schema***: the Schema includes the normative XML schema comprising the entire grammar of the Digital Item Declaration representation in XML.

Below, the semantic meaning of the major Digital Item Declaration model elements is described. Note that in the definitions below, the elements in italics are intended to be unambiguous terms within this model [33].

- ***Container***: a *container* is a structure that allows *items* and/or *containers* to be grouped. These groupings of *items* and/or *containers* can be used to form logical *packages* (for transport or exchange) or logical *shelves* (for organization). *Descriptors* allow for the labelling of *containers* with information that is appropriate for the purpose of the grouping (e.g. delivery instructions for a *package* or category information for a *shelf*). It should be noted that a *container* itself is not an *item; containers* are groupings of *items* and/or *containers*.
- ***Item***: an *item* is a grouping of sub-*items* and/or *components* that are bound to relevant *descriptors*. *Descriptors* contain information about the *item*, as a representation of a work. *Items* may contain *choices*, which allow them to be customized or configured. *Items* may be conditional (on *predicates* asserted by *selections* defined in the *choices*). An *item* that contains no sub-items can be considered an entity—a logically indivisible work. An *item* that does contain sub-*items* can be considered a compilation—a work composed of potentially independent sub-parts. *Items* may also contain *annotations* to their sub-parts. The relationship between *items* and MPEG-21 Digital Items could be stated as follows: *items* are declarative representations of MPEG-21 Digital Items.
- ***Component***: a *component* is the binding of a *resource* to all of its relevant *descriptors*. These *descriptors* are information related to all or part of the specific *resource* instance. Such *descriptors* will typically contain control or structural information about the *resource* (such as bit rate, character set, start points or encryption information) but not information describing the content within. It should be noted that a *component* itself is not an *item*; *components* are building blocks of *items*.
- ***Anchor***: an *anchor* binds *descriptors* to a *fragment*, which corresponds to a specific location or range within a *resource*.
- ***Descriptor***: a *descriptor* associates information with the enclosing element. This information may be a *component* (such as a thumbnail of an image or a text *component*) or a textual *statement*.
- ***Condition***: a *condition* describes the enclosing element as being optional, and links it to the *selection(s)* that affect its inclusion. Multiple *predicates* within a *condition* are combined as a conjunction (an AND relationship). Any *predicate* can be negated within a *condition*. Multiple *conditions* associated with a given element are combined as a disjunction (an OR relationship) when determining whether to include the element.
- ***Choice***: a *choice* describes a set of related *selections* that can affect the configuration of an *item*. The *selections* within a *choice* are either exclusive (choose exactly one) or inclusive (choose any number, including all or none).
- ***Selection***: a *selection* describes a specific decision that will affect one or more *conditions* somewhere within an *item*. If the *selection* is chosen, its predicate becomes true; if it is not chosen, its *predicate* becomes false; if it is left unresolved, its *predicate* is undecided.
- ***Annotation***: an *annotation* describes a set of information about another identified element of the model without altering or adding to that element. The information can take the form of *assertions*, *descriptors* and *anchors*.

- ***Assertion***: an *assertion* defines a full or partially configured state of a *choice* by asserting true, false or undecided values for some number of *predicates* associated with the *selections* for that *choice*.
- ***Resource***: a *resource* is an individually identifiable asset such as a video or audio clip, an image or a textual asset. A *resource* may also potentially be a physical object. All *resources* must be locatable via an unambiguous address.
- ***Fragment***: a *fragment* unambiguously designates a specific point or range within a *resource*. A *fragment* may be *resource* type specific.
- ***Statement***: a *statement* is a literal textual value that contains information, but not an asset. Examples of likely *statements* include descriptive, control, revision tracking or identifying information.
- ***Predicate***: a *predicate* is an unambiguously identifiable Declaration that can be true, false or undecided.

From the above, it is clear that the MPEG-21 Digital Item Declaration standard provides a very generic tool for multimedia content description. It should be noted that, in this context, 'description' must be interpreted broadly. Indeed, MPEG-21 DID technology not only allows for static description of individual resources or resource collections; it also provides, for example, a very flexible (but still standardized and hence interoperable) mechanism for defining decision trees on which (re)configuration of Digital Items can be based. Given the broadness of its scope and its support for advanced features like (re)configuration of Digital Items, the MPEG-21 DID tool can be seen as an important milestone in the field of multimedia content description.

As an illustration, consider the most basic form of a Digital Item: a content Digital Item. This type of Digital Item is suited for consumption by an end user and usually consists of a set of resources such as audio or video streams, a set of choices such as the audio and video bit rates or resolutions available, and a set of descriptors containing metadata regarding the various resources and the DI as a whole. In Figure 1.10, a content DI that consists of several video streams is declared using the DID language. The video streams contained in the Digital Item have different spatial resolutions and different bit rates. This Digital Item also contains two `Choices` that allow a user to choose between the different bit rates and between the different resolutions.

Digital item identification

The MPEG-21 Digital Item Identification specification itself does not specify new identification systems for Digital Items. For example, MPEG-21 part 3, formally ISO/IEC 21000-3, does not attempt to replace the International Standard Recording Code (ISRC) for sound recordings, as defined in ISO 3901 [49], but allows ISRCs to be used within MPEG-21. In fact, DII provides tools for identifying MPEG-21 Digital Items via existing identification schemes that are defined by other bodies than MPEG. More precisely, the DII standard provides tools to uniquely identify Digital Items and parts thereof (including resources), to uniquely identify intellectual property related to the Digital Items (and parts thereof), to uniquely identify Description Schemes (e.g. MPEG-7 or RDF [50] based Description Schemes) and the relationship between Digital Items

```
<?xml version="1.0" encoding="UTF-8"?>
<DIDL xmlns="urn:mpeg:mpeg21:2002:01-DIDL-NS"
      xmlns:xsi="http://www.w3.org/2001/XMLSchema-instance"
      xsi:schemaLocation="urn:mpeg:mpeg21:2002:01-DIDL-NS .\DIDL.xsd">
  <Item id="example_cdi">
    <Choice choice_id="resolution">
      <Selection select_id="qcif"/>
      <Selection select_id="cif"/>
    </Choice>
    <Choice choice_id="bitrate">
      <Selection select_id="kbps_128"/>
      <Selection select_id="kbps_256"/>
    </Choice>
    <Component>
      <Condition require="qcif kbps_128"/>
      <Resource mimeType="video/mp4v-es" ref="foreman_qcif_128.mp4"/>
    </Component>
    <Component>
      <Condition require="qcif kbps_256"/>
      <Resource mimeType="video/mp4v-es" ref="foreman_qcif_256.mp4"/>
    </Component>
    <Component>
      <Condition require="cif kbps_128"/>
      <Resource mimeType="video/mp4v-es" ref="foreman_cif_128.mp4"/>
    </Component>
    <Component>
      <Condition require="cif kbps_256"/>
      <Resource mimeType="video/mp4v-es" ref="foreman_cif_256.mp4"/>
    </Component>
  </Item>
</DIDL>
```

Figure 1.10 An example content Digital Item

(and parts thereof) and relevant description schemes, and to express the relationship between Digital Items (and parts thereof) and existing identification systems [34].

Identifiers covered by this specification can be associated with Digital Items by including them in a specific place in the Digital Item Declaration. This place is the `Statement` element. Examples of likely `Statements` include descriptive, control, revision tracking and/or identifying information. Figure 1.11 shows this relationship. The shaded boxes are subject to the DII specification while the bold boxes are defined in the DID specification.

Several elements within a Digital Item Declaration can have zero, one or more `Descriptors` (as specified in MPEG-21 part 2, DID). Each `Descriptor` may include one `Statement` which can contain one identifier relating to the parent element of the `Statement`. In Figure 1.11, the two statements shown are used to identify a `Component` (left-hand side of the diagram) and an `Item` (right-hand side of the diagram). Given that, within MPEG-21, Digital Item Identifiers are defined via the concepts of `Descriptors/Statements` (as defined in the DID specification), the DII tool can be considered to be yet another illustration of the importance of multimedia content description.

Digital Items and their parts within the MPEG-21 Multimedia Framework are identified via Uniform Resource Identifiers (URIs). A URI is a compact string of characters for identifying an abstract or physical resource, where a resource is very broadly defined as 'anything that has identity'.

The requirement that an MPEG-21 Digital Item Identifier be a URI is consistent with the concept that the MPEG-21 identifier may be a Uniform Resource Locator (URL). The term URL refers to a specific subset of the URIs that are in use today as pointers to information on the Internet; it allows for long-term to short-term persistence depending on the business case.

MPEG-21 DII allows any identifier in the form of a URI to be used as identifier for Digital Items (and parts thereof). The specification also provides the ability to register identification schemes, whereby unambiguous names are formally associated with specific objects. This process is coordinated by a so-called Registration Authority (RA), which may be an automated facility [51].

Figure 1.12 shows how a music album—and its parts—can be identified through DII. The music album contains several music tracks and the corresponding lyrics. The tracks are encoded using the MPEG-4 Advanced Audio Coding (AAC) format, while the lyrics are encoded as

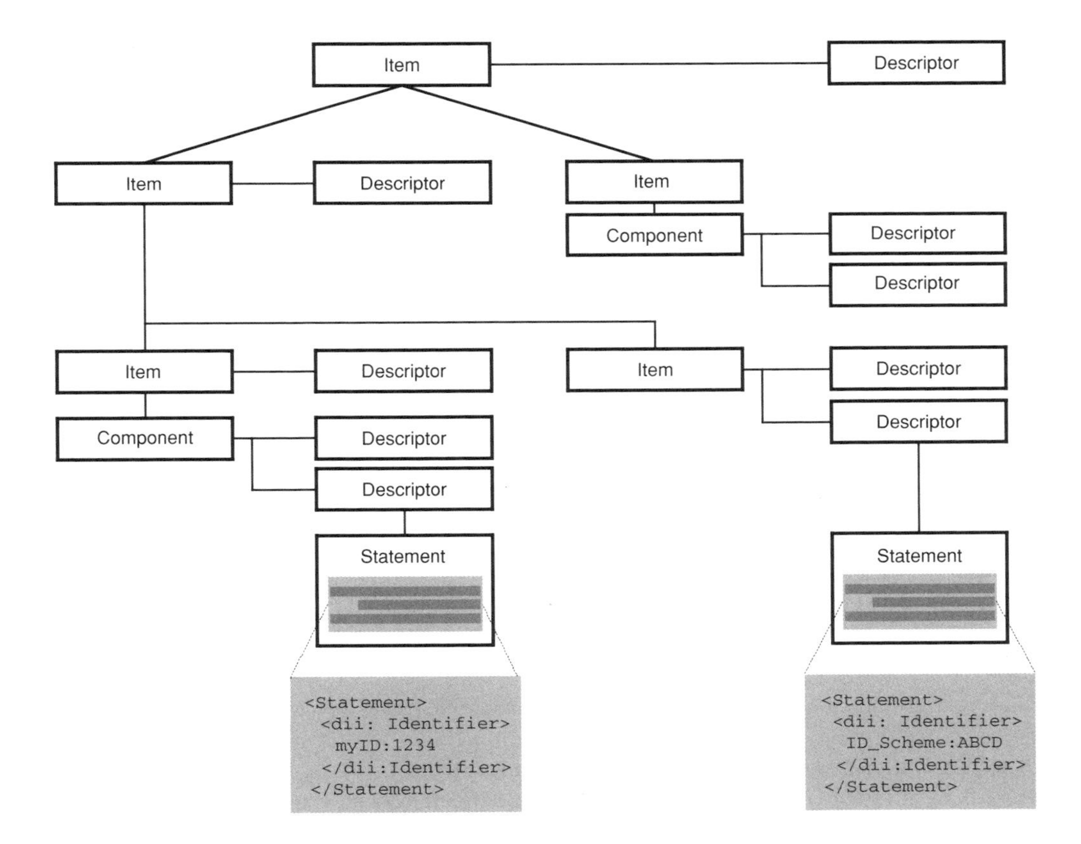

Figure 1.11 Relationship between Digital Item Declaration and Digital Item Identification [34]

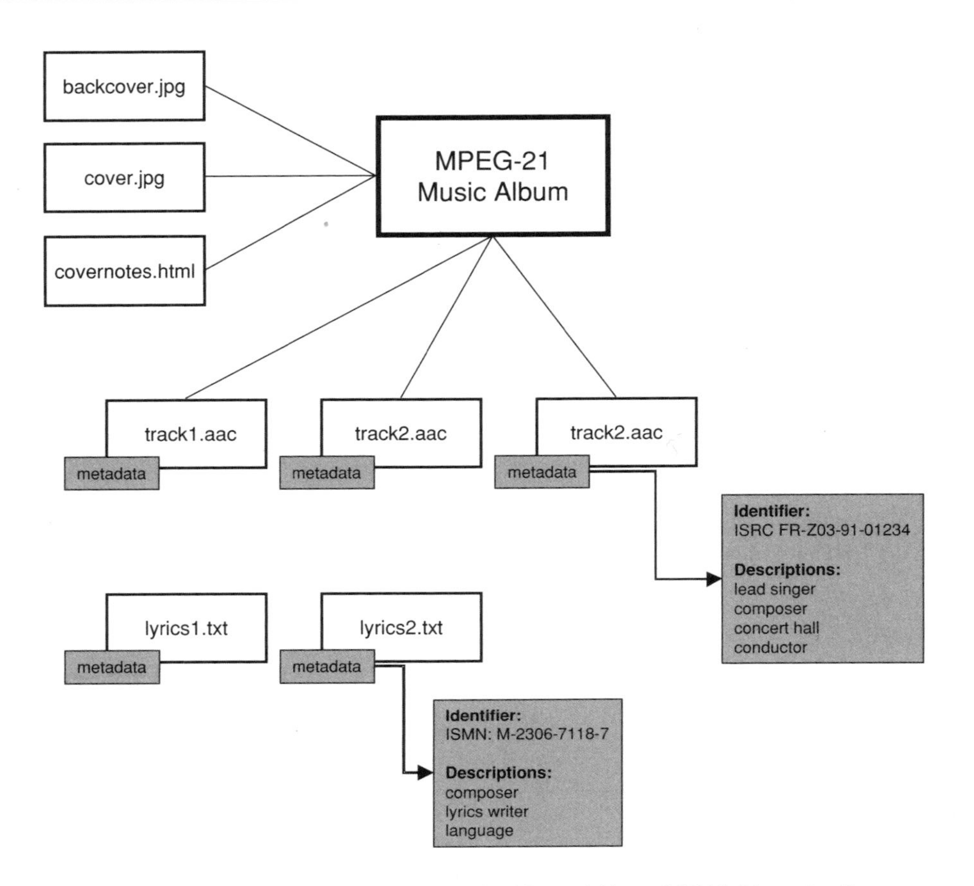

Figure 1.12 Example of metadata and identifiers within an MPEG-21 music album

plain text (txt format). Besides, the digital music album contains two images corresponding to a back cover and a cover (both encoded in the JPEG format), and some cover notes that are delivered as a HTML page. Each of these sub-items of the music album can be identified by using the DII tool. Of course, different sub-items can be identified via different (existing) identification schemes, as illustrated in Figure 1.12: the audio tracks are identified via the ISRC specification, while the lyrics are identified via the International Standard Music Number (ISMN) specification [52]. In the example, no metadata is associated to the image and to the HTML cover notes. Note that, in some cases, it may be necessary to use an automated resolution system to retrieve the Digital Item (or parts thereof) or information related to a Digital Item from a server (e.g. in the case of an interactive online content delivery system). Such a resolution system allows for submitting an identifier to a network service and receiving in return one or more pieces of information (which includes resources, descriptions, another identifier, Digital Item etc.) that is related to that identifier.

Also, as different Users of MPEG-21 may have different schemes to describe their content, it is necessary for MPEG-21 DII to allow differentiation of such different schemes. MPEG-21 DII utilizes the XML mechanism of namespaces to do this.

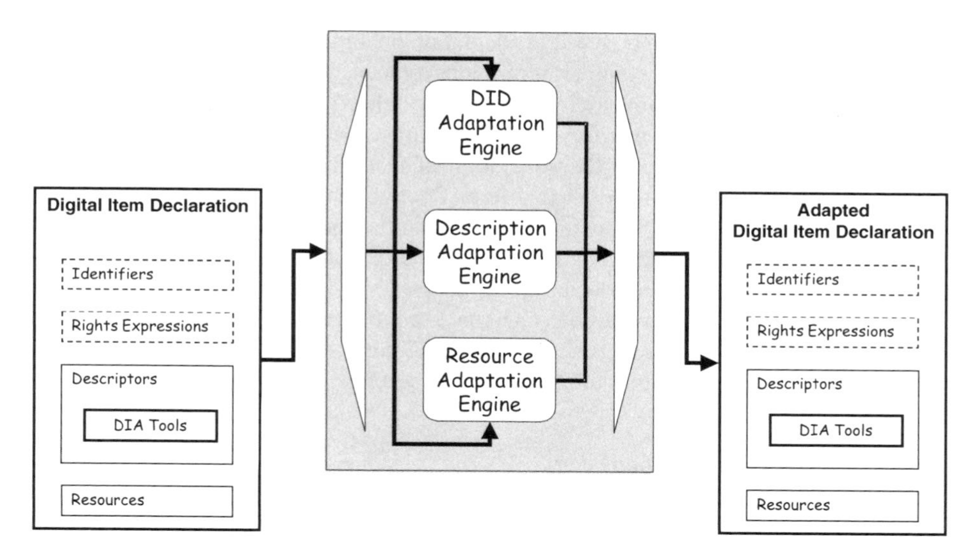

Figure 1.13 Digital Item Adaptation architecture, including the Digital Item Adaptation engine (grey box)

Digital item adaptation

One of the main goals of the MPEG-21 Multimedia Framework is to achieve interoperable transparent access to (distributed) multimedia content by shielding end users from network and terminal installation, management and implementation issues. To achieve this goal, the adaptation of Digital Items may be required. Figure 1.13 illustrates the overall architecture of Digital Item Adaptation. MPEG-21 DIA standardizes several tools, allowing for the adaptation of Digital Items, or, more precisely, the adaptation of resources within DIs, the adaptation of descriptions within DIs and/or the adaptation of the structure of DIs (the latter possibly without changing individual resources or descriptions). This is illustrated in Figure 1.13. From this figure, it is also clear that DIA is strongly relying on the use of descriptive information. Indeed, DIA tools are hosted by the `Descriptor` element as defined within the DID language.

The Digital Item Adaptation tools are clustered into three major categories according to their functionality and use for Digital Item Adaptation [38]:

- ***Usage Environment Description Tools***: this category includes tools to describe MPEG-21 User characteristics, terminal capabilities, network characteristics and natural environment characteristics. These tools provide descriptive information about these various dimensions of the usage environment to accommodate, for example, the adaptation of Digital Items for transmission, storage and consumption.
- ***Digital Item Resource Adaptation Tools***: this category includes tools targeting the adaptation of individual resources within a Digital Item. In the case of bitstream syntax description, tools are specified to facilitate the scaling of bitstreams in a format-independent manner. Such tools are useful for the adaptation of resources contained within a Digital Item at network nodes that are not knowledgeable about the specific resource representation format. The Digital Item Adaptation specification is also providing a method to describe unambiguously the

relationship between constraints, feasible adaptation operations satisfying these constraints, and associated qualities. For example, it provides a means to make trade-offs among these parameters so that an adaptation strategy can be formulated.

- ***Digital Item Declaration Adaptation Tools***: in contrast to the set of Digital Item Resource Adaptation Tools, which targets the adaptation of resources in a Digital Item, the tools in this category aim to adapt the Digital Item Declaration as a whole. For example, for session mobility, the configuration state information that pertains to the consumption of a Digital Item on one device is transferred to a second device. This enables the Digital Item to be consumed on the second device in an adapted way. Another tool in this set, DID configuration preferences, provides mechanisms to configure the `Choices` within a Digital Item Declaration according to a User's intention and preference. This enables operations to be performed on elements within the DID, such as sorting and deleting.

Bitstream Syntax Description Language

The Bitstream Syntax Description Language (BSDL) is a normative language based on XML Schema, which makes it possible to design specific Bitstream Syntax Schemas (BSs) describing the syntax of particular scalable media resource formats. These schemas can then be used by a generic processor/software to automatically parse a bitstream and generate its syntax description, or, vice versa, to generate an adapted bitstream using an adapted syntax description. Using such a description, a Digital Item resource adaptation engine can transform a (somehow scalable) bitstream and the corresponding syntax description using editing-style operations such as data truncation and simple modifications. This transformation can, for instance, be performed under control of an eXtensible Stylesheet Language Transform (XSLT) [53]. It is important to note that bitstream syntax descriptions can be defined for any scalable multimedia resource format. This enables adaptation of a binary resource to take place at nodes—for instance, network gateways—that are not knowledgeable about the specific resource representation format.

Figure 1.14 depicts the architecture of BSDL-driven resource adaptation. This architecture includes the original scalable bitstream and its description, description transformations, the

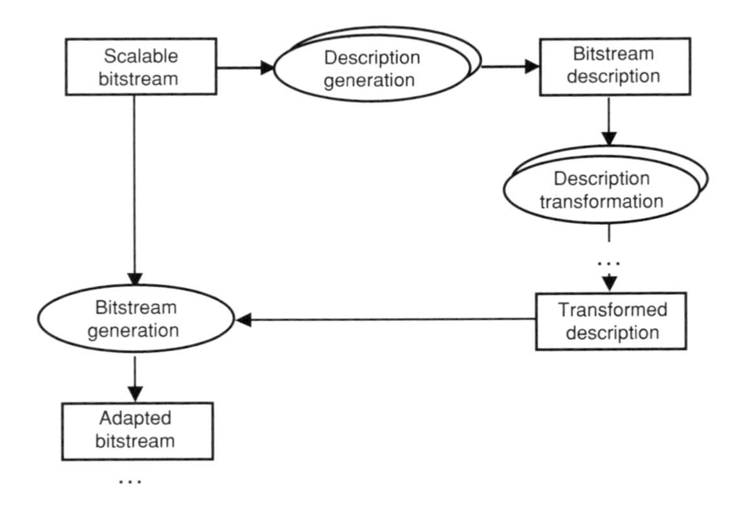

Figure 1.14 Architecture of BSDL-driven resource adaptation

adapted bitstream and its description, and two processors: a description generator (which generates bitstream descriptions corresponding to given bitstreams) and a bitstream generator (which does the opposite: generating bitstreams that correspond to given bitstream descriptions). The input for BSDL-driven resource adaptation is a scalable bitstream. From this bitstream, the corresponding bitstream description is derived using the description generator. Next, depending on the usage environment characteristics and thus on the adaptation that needs to be performed, one or more transformations are applied to the bitstream description, which leads to a new bitstream syntax description (the transformed description), which fits the needs of the usage environment. Finally, by applying the bitstream generator on the latter description, the output bitstream is obtained from the original bitstream (and the latter description). From this architecture, it is clear that the actual transformations are performed at the level of the Bitstream Syntax Descriptions, and not at the level of the bitstreams themselves. This is advantageous for certain applications, especially for those applications that are running on devices with limited capabilities. Indeed, one can assume that modifying bitstream descriptions at the XML level is less complex than modifying bitstreams themselves. For example, mobile phones should be able to perform basic XML processing, although they are often not able to perform complex adaptations of video bitstreams directly.

Consider, as an example use case, a mobile phone receiving a video bitstream that needs to be adapted before it can be shown to the end user (e.g. to decrease the spatial and temporal resolution). The phone, having only very limited capabilities, is not able to perform the bitstream adaptation itself. However, the phone can perform the corresponding transformation at the bitstream description level. The resulting adapted bitstream description is then sent back to the video bitstream provider, asking the latter to generate the bitstream that corresponds to the adapted bitstream description. Finally, the adapted bitstream is sent to the mobile phone, which will be able to show it to the end user. Another example use case is performing transformations of bitstream descriptions in a network node along the content delivery chain. Although the network node may not know the complete usage environment of the end user, it may know the status of the network itself. The network node could, for example, 'see' congestion in the network while an end user is requesting a high-bitrate video stream from his content provider. Due to the network congestion, the high-bitrate video stream cannot be transferred to the end user; an adaptation of the video bitstream is therefore required. The network node could then perform the corresponding transformation at the bitstream description level. Based on the resulting adapted bitstream description, the node could more easily produce an adapted (lower bitrate) video bitstream and send it to the end user (instead of the originally requested high-bitrate version which the congested network could not deliver).

While BSDL makes it possible to design specific Bitstream Syntax Schemas describing the syntax of *particular* media resource formats, generic Bitstream Syntax Schema (gBS Schema) enables resource-format-*independent* bitstream syntax descriptions (gBSDs) to be constructed. As a consequence, it becomes possible for resource-format-agnostic adaptation engines to transform bitstreams and associated syntax descriptions. In addition, the gBS Schema provides means to describe bitstreams at different levels of detail using a hierarchical approach; for instance, to describe both individual frames and scenes in a video sequence. Finally, resource-format-specific information as well as semantic handles can be included in a gBSD; for example, frames can be marked by their frame type and a scene can be associated with a marker. The marker can be referred to in descriptions to identify, for instance, the violence level. The latter could be used by an adaptation engine to cut out violent scenes in a lightweight manner.

1.4 Final Remarks

MPEG-1 and MPEG-2 have been successful standards, and it is expected that MPEG-4 will also set new frontiers in terms of multimedia representation. Following these projects, MPEG has moved on to address the problem of identifying, selecting and managing various types of multimedia material. To achieve that goal, MPEG-7 offers a standardized way of describing multimedia content. In comparison with other available or emerging audiovisual description frameworks, MPEG-7 can be characterized by its:

1. Genericity: this means the capability to describe multimedia content for many application environments.
2. Object-based data model: this means the capability to independently describe individual objects, e.g. scenes, shots, objects within a scene, be they MPEG-4 or not.
3. Integration of low-level and high-level features/descriptors into a single architecture, allowing the power of both types of descriptors to be combined.
4. Extensibility provided by the Description Definition Language, which allows MPEG-7 to keep growing, to be extended to new application areas, to answer to newly emerging needs and to integrate novel description tools.

The next work item that was started within MPEG, the MPEG-21 Multimedia Framework, is still under development. However, the first parts of MPEG-21 had already reached International Standard status by March 2004, notably Digital Item Declaration, Digital Item Identification and Rights Expression Language. Other parts will reach this status in the near future. Compared with other standards, MPEG-21:

1. provides, for the first time, a standardized way to describe unambiguously the structure of generic Digital Items, including descriptive metadata and (re)configuration information, via the Digital Item Declaration Language;
2. addresses *all* components of the multimedia data distribution and consumption chain, instead of just a single component (such as multimedia data encoding, metadata etc.);
3. provides tools for integrating both existing and future technologies that are/will be developed for these components;
4. offers a consistent and complete set of tools related to Digital Rights Management (DRM);
5. offers a rich set of tools for the adaptation of Digital Items, including adaptation based on User preferences, or, more generally, adaptation based on the environment in which these DIs are to be consumed.

After the definition of the technologies (which may continue with future developments and amendments), MPEG-7 and MPEG-21 are now in a phase where promotion within the industry is very important in order that adoption in products may increase. For this purpose, the MPEG-4 Industry Forum (M4IF), which was created in 2000 to further the adoption of the MPEG-4 standard, was transformed, in July 2003, into the MPEG Industry Forum (MPEGIF) [54]. The change was made so that the Forum could address the interests of MPEG-7 and MPEG-21 users as well. The MPEG Industry Forum is a not-for-profit organization with the following goal: 'To further the adoption of MPEG Standards, by establishing them as well accepted and widely used standards among creators of content, developers, manufacturers, providers of services, and end-users'.

Finally, it should be noted that, since there were companies claiming to have patents on several MPEG-7 and MPEG-21 tools, discussions on licensing between interested companies (but always outside MPEG) should reach a conclusion as soon as possible in order to allow companies to start deployment of their MPEG-7- and MPEG-21-based products and services. In fact, MPEG itself is not allowed to deal with any issues beside the development of the technical specifications, notably patents identification and patents licensing; these issues have to be addressed by the industry players interested in the deployment of products and applications, and, of course, by the patent owners.

Acknowledgments

The authors would like to thank all the MPEG members for the numerous interesting and fruitful discussions in meetings and by email, which substantially enriched their technical knowledge.

References

[1] MPEG Home Page, http://www.chiariglione.org/mpeg.
[2] L. Chiariglione, The challenge of multimedia standardization, *IEEE Multimedia*, **4**(2), 79–83, 1997.
[3] ISO/IEC 7498-1:1994, *Information Technology—Open Systems Interconnection—Basic Reference Model: The Basic Model*. ISO, Geneva, 1994.
[4] JTC1 Home Page, http://www.jtc1.org.
[5] MPEG Requirements Group, MPEG-7 Overview, Doc. ISO/MPEG N5525, Pattaya MPEG Meeting, March 2003.
[6] MPEG Requirements Group, MPEG-7 Requirements, Doc. ISO/MPEG N4981, Klagenfurt MPEG Meeting, July 2002.
[7] F. Pereira, T. Ebrahimi (eds), *The MPEG-4 Book*. Prentice Hall, Upper Saddle River, NJ, 2002.
[8] MPEG Requirements Group, MPEG-7 Applications, Doc. ISO/MPEG N4676, Jeju MPEG Meeting, March 2002.
[9] ISO/IEC 15938-1:2002, '*Information Technology—Multimedia Content Description Interface—Part 1: Systems*', Version 1. ISO, Geneva, 2002.
[10] ISO/IEC 15938-2:2002, *Information Technology—Multimedia Content Description Interface—Part 2: Description Definition Language*, Version 1. ISO, Geneva, 2002.
[11] ISO/IEC 15938-3:2002, *Information Technology—Multimedia Content Description Interface—Part 3: Visual*, Version 1. ISO, Geneva, 2002.
[12] ISO/IEC 15938-4:2002, *Information Technology—Multimedia Content Description Interface—Part 4: Audio*, Version 1. ISO, Geneva, 2002.
[13] ISO/IEC 15938-5:2003, *Information Technology—Multimedia Content Description Interface—Part 5: Multimedia Description Schemes*, Version 1. ISO, Geneva, 2003.
[14] ISO/IEC 15938-6:2003, *Information Technology—Multimedia Content Description Interface—Part 6: Reference Software*, Version 1. ISO, Geneva, 2003.
[15] ISO/IEC 15938-7:2003, *Information Technology—Multimedia Content Description Interface—Part 7: Conformance Testing*, Version 1. ISO, Geneva, 2003.
[16] ISO/IEC 15938-8:2003, *Information Technology—Multimedia Content Description Interface—Part 8: Extraction and Use of MPEG-7 Descriptions*, Version 1. ISO, Geneva, 2002.
[17] MPEG Requirements Group, MPEG-7 Profiles and Levels, Committee Draft, Doc. ISO/MPEG N6040, Brisbane MPEG Meeting, October 2003.

[18] MPEG MDS Group, MPEG-7 Schema Definition, Committee Draft, Doc. ISO/MPEG N5929, Brisbane MPEG Meeting, October 2003.
[19] MPEG Requirements Group, MPEG-7 Call for Proposals, Doc. ISO/MPEG N2469, Atlantic City MPEG Meeting, October 1998.
[20] MPEG Requirements Group, MPEG-7 List of Proposal Pre-registrations, Doc. ISO/MPEG N2567, Rome MPEG Meeting, December 1998.
[21] MPEG Requirements Group, MPEG-7 Evaluation Process, Doc. ISO/MPEG N2463, Atlantic City MPEG Meeting, October 1998.
[22] MPEG Requirements Group, MPEG-7 Proposal Package Description (PPD), Doc. ISO/MPEG N2464, Atlantic City MPEG Meeting, October 1998.
[23] MPEG Requirements Group, Description of MPEG-7 Content Set, Doc. ISO/MPEG N2467, Atlantic City MPEG Meeting, October 1998.
[24] MPEG Requirements Group, Licensing Agreement for MPEG-7 Content Set, Doc. ISO/MPEG N2466, Atlantic City MPEG Meeting, October 1998.
[25] MPEG Requirements Group, Results of MPEG-7 Technology Proposal Evaluations and Recommendations, Doc. ISO/MPEG N2730, Seoul MPEG Meeting, March 1999.
[26] W3C, XML Schema (Primer, Structures, Datatypes), W3C Working Draft, April 2000.
[27] B.S. Manjunath, P. Salembier, T. Sikora (eds), *Introduction to MPEG-7: Multimedia Content Description Interface*. John Wiley & Sons, Chichester, 2002.
[28] MPEG Requirements Group, MPEG-21 Overview, Doc. ISO/MPEG N5231, Shanghai MPEG Meeting, October 2002.
[29] MPEG Requirements Group, MPEG-21 Use Case Scenarios, Doc. ISO/MPEG N4991, Klagenfurt MPEG Meeting, July 2002.
[30] MPEG Requirements Group, MPEG-21 Requirements, Doc. ISO/MPEG N6264, Waikaloa MPEG Meeting, December 2003.
[31] ISO/IEC 21000-1:2001, *Information Technology—Multimedia Framework (MPEG-21)—Part 1: Vision, Technologies and Strategy*. ISO, Geneva, 2001.
[32] MPEG Requirements Group, *Information Technology—Multimedia Framework (MPEG-21)—Part 1: Vision, Technologies and Strategy*, second edition, Doc. ISO/MPEG N6388, Munich MPEG meeting, March 2004.
[33] ISO/IEC 21000-2:2003, *Information Technology—Multimedia Framework (MPEG-21)—Part 2: Digital Item Declaration*. ISO, Geneva, 2003.
[34] ISO/IEC 21000-3:2003, *Information Technology—Multimedia Framework (MPEG-21)—Part 3: Digital Item Identification*. ISO, Geneva, 2003.
[35] MPEG Requirements Group, Requirements for MPEG-21 IPMP, Doc. ISO/MPEG N6389, Munich MPEG Meeting, March 2004.
[36] ISO/IEC 21000-5:2004, *Information Technology—Multimedia Framework (MPEG-21)—Part 5: Rights Expression Language*. ISO, Geneva, 2004.
[37] MPEG Multimedia Description Schemes Group, MPEG-21 Rights Data Dictionary: Final Draft International Standard, Doc. ISO/MPEG N5842, Trondheim MPEG Meeting, July 2003.
[38] MPEG Multimedia Description Schemes Group, MPEG-21 Digital Item Adaptation: Final Draft International Standard, Doc. ISO/MPEG N6168, Waikaloa MPEG Meeting, December 2003.
[39] MPEG Integration Group, MPEG-21 Reference Software: Committee Draft, Doc. ISO/MPEG N6470, Munich MPEG Meeting, March 2004.
[40] MPEG Systems Group, MPEG-21 File Format: Committee Draft, Doc. ISO/MPEG N6331, Munich MPEG Meeting, March 2004.
[41] MPEG Multimedia Description Schemes Group, MPEG-21 Digital Item Processing: Committee Draft, Doc. ISO/MPEG N6173, Waikaloa MPEG Meeting, December 2003.
[42] MPEG Requirements Group, Evaluation Tools for Persistent Association Technologies: Draft Technical Report, Doc. ISO/MPEG N6392, Munich MPEG Meeting, March 2004.
[43] MPEG Integration Group, Test Bed for MPEG-21 Resource Delivery: Proposed Draft Technical Report, Doc. ISO/MPEG N6255, Waikaloa MPEG Meeting, December 2003.
[44] MPEG Video Group, Scalable Video Model, Doc. ISO/MPEG N6372, Munich MPEG Meeting, March 2004.
[45] MPEG Integration Group, MPEG-21 Conformance: Working Draft, Doc. ISO/MPEG N6472, Munich MPEG Meeting, March 2004.

[46] MPEG Multimedia Description Schemes Group, MPEG-21 Event Reporting: Working Draft, Doc. ISO/MPEG N6419, Munich MPEG Meeting, March 2004.
[47] MPEG Systems Group, MPEG-21 Binary Format: Working Draft, Doc. ISO/MPEG N6333, Munich MPEG Meeting, March 2004.
[48] ISO/IEC JTC 1/SC 29, Programme of Work, http://www.itscj.ipsj.or.jp/sc29/29w42911.htm.
[49] ISO/IEC 3901:2001, Information and Documentation—International Standard Recording Code, September 2001.
[50] W3C, Resource Description Framework (RDF) Model and Syntax Specification, W3C Recommendation, February 1999.
[51] ISO/IEC JTC 1/SC 29, JTC 1 Directives: Procedures for the Technical Work of ISO/IEC JTC 1 on Information Technology (4th edition), http://www.itscj.ipsj.or.jp/sc29/directives.pdf.
[52] ISO/IEC 10957:1993, *Information and Documentation—International Standard Music Number*. ISO, Geneva, 1999.
[53] W3C, XSL Transformations (XSLT) Version 2.0, W3C Working Draft, May 2003.
[54] MPEG Industry Forum Home Page, http://www.m4if.org/.

2

Ontology Representation and Querying for Realizing Semantics-Driven Applications

Boris Motik, Alexander Maedche and Raphael Volz

2.1 Introduction

The application of ontologies is increasingly seen as key to enabling semantics-driven information access. There are many applications of such an approach, e.g. automated information processing, information integration and knowledge management, to name just a few. Especially after Tim Berners-Lee coined the vision of the Semantic Web [1], where web pages are annotated by ontology-based metadata, the interest in ontology research increased, in the hope of finding ways to offload processing of large volumes of information from human users to autonomous agents.

Many ontology languages have been developed, each aiming to solve particular aspects of ontology modelling. Some of them, such as RDF(S) [2,3], are simple languages offering elementary support for ontology modelling for the Semantic Web. There are other, more complex languages with roots in formal logic, with particular focus on advanced inference capabilities—mechanisms to automatically infer facts not explicitly present in the model. For example, the F-logic language [4] offers ontology modelling through an object-oriented extension of Horn logic. On the other hand, various classes of description logic languages (e.g. OIL [5] or OWL [6]) are mainly concerned with finding an appropriate subset of first-order logic with decidable and complete subsumption inference procedures.

Despite a large body of research in improving ontology management and reasoning, features standardized and widely adopted in the database community (such as scalability or transactions) must be adapted and re-implemented for logic-based systems. Support for ontology modularization is typically lacking in most systems. Logic-based approaches often focus primarily on the expressivity of the model, often neglecting the impact that this expressivity has

Multimedia Content and the Semantic Web Edited by Giorgos Stamou and Stefanos Kollias

on performance and ease of integration with other systems. Because of these problems, up until today there has not been a large number of successful enterprise applications of ontology technologies.

On the other hand, relational databases have been developed over the past 20 years to a maturity level unparalleled with that of ontology management systems, incorporating features critical for business applications, such as scalability, reliability and concurrency support. However, database technologies alone are not appropriate for handling semantic information. This is mainly due to the fact that the conceptual model of a domain (typically created using entity-relationship modelling [7]) for the actual implementation must be transformed into a logical model. After the transformation, the structure and the intent of the original model are not obvious. Therefore, the conceptual and the logical model tend to diverge. Further, operations that are natural within the conceptual model, such as navigation between objects or ontology querying, are not straightforward within the logical model.

In this chapter we present an approach for ontology representation and querying that tries to adapt the expressivity of the model with other requirements necessary for successful realization of business-wide applications. We adjust the expressiveness of traditional logic-based languages to sustain tractability, thus making realization of enterprise-wide ontology-based systems possible using existing well-established technologies, such as relational databases. Other critical features are modularization and modelling meta-concepts with well-defined semantics. Based on our ontology structure, we give an approach for conceptual querying of ontologies. We present the current status of the implementation of our approach within KAON [8]—Ontology and Semantic Web tool suite used as basis for our research and development.

The rest of this chapter is structured as follows. In Section 2.2 we present the requirements based on which we derive our ontology representation and querying approach. Section 2.3 presents our ontology language in more detail, by giving a mathematical definition, denotational semantics and some examples. In Section 2.4 we present our approach for answering queries over ontologies. In Section 2.5 we present the current implementation status. In Section 2.6 we contrast our work with related approaches. Finally, in Section 2.7 we conclude the chapter and give our directions for future research.

2.2 Requirements

In this section we discuss the requirements based on our observations we gathered while working on several research and industry projects. For example, Ontologging is focused on applying ontologies to knowledge management in the hope of improving searching and navigation within a large document base. Harmonise tries to apply ontologies to provide interoperability among enterprise systems in business-to-business (B2B) scenarios, with tourism as the application field. Finally, Semantic Web enabled Web Services (SWWS) explores applying ontologies to improve and automate web service-based interaction between businesses on the Internet.

Object-oriented Modelling Paradigm

In the past decade the object-oriented paradigm has become prevalent for conceptual modelling. A wide adoption of UML [9,10] as syntax for object-oriented models has further increased its acceptance.

The object-oriented modelling paradigm owes its success largely to the fact that it is highly intuitive. Its constructs match well with the way people think about the domain they are modelling. Object-oriented models can easily be visualized, thus making understanding conceptual models much simpler. Hence, any successful ontology modelling approach should follow the object-oriented modelling paradigm.

Meta-concepts

When trying to create complex models of the real world, often it may be unclear whether some element should be represented as a concept or as an instance. An excellent example in [11] demonstrates problems with meta-concept modelling. While developing a semantics-driven image retrieval system, it was necessary to model the relationship between notions of species, ape types and particular ape individuals. The most natural conceptualization is to introduce the SPECIES concept (representing the set of all species), with instances such as APE. However, APE may be viewed as a set of all apes. It may be argued that APE may be modelled as a subconcept of SPECIES. However, if this is done, other irregularities arise. Since APE is a set of all apes, SPECIES, being a superconcept of APE, must contain all apes as its members, which is clearly wrong. Further, when talking about the APE species, there are many properties that may be attached to it, such as habitat, type of food etc. This is impossible to do if APE is a subconcept of SPECIES, since concepts cannot have instantiated property values.

There are other examples where meta-concept modelling is necessary:

- Ontology mapping is a process of mapping ontology entities in different ontologies in order to achieve interoperability between information in both ontologies. As described in [12], it is useful to represent ontology mapping as a meta-ontology that relates concepts of the ontologies being mapped.
- It is beneficial to represent ontology changes by instantiating a special evolution log ontology [13]. This (meta-)ontology consists of concepts reflecting various types of changes in an ontology.
- As described in [14], to guide ontology modelling decisions, it is beneficial to annotate ontology entities with meta-information reflecting fundamental properties of ontology constructs. To attach this meta-information to classes in an ontology, classes must be viewed as instances of meta-concepts that define these properties.

As already noted by some researchers [15,16], support for modelling meta-concepts is important for adequate expression of complex knowledge models. There are logic languages (e.g. HiLog [17]) with second-order syntax but with first-order semantics for which it has been shown that a sound and complete proof theory exists. Many running systems, such as Protege-2000 [18], incorporate this functionality into their knowledge model.

Modularization

It is a common engineering practice to extract a well-encapsulated body of information in a separate module, that can later be reused in different contexts. However, modularization of ontologies has some special requirements: both instances and schematic definitions may be subjected to modularization.

For example, a concept CONTINENT will have exactly seven instances. In order to include information about continents in some ontology, it is not sufficient to reuse only the CONTINENT concept—to be able to talk about particular continents, such as EUROPE, one must reuse the instances of CONTINENT as well. We consider modularization—on both ontologies and instances—to be an important aspect of reuse. Supporting modularization introduces some requirements on the ontology modelling language, but also on the systems implementing the language as well.

Lexical Information

Many applications extensively depend on lexical information about entities in an ontology, such as labels in different languages. Hence, consistent way of associating lexical information with ontology entities is mandatory. Lexical information can be thought of as meta-information about an ontology, since it talks about the ontology elements, so it makes sense to represent this information as just another type of metadata. This has the benefit that the ontology and lexical information can be manipulated in the same way. For example, querying the ontology by the lexical information is done in the same way as the usual ontology queries.

Adaptable Inference Capabilities

Inference mechanisms for deduction of information not explicitly asserted is an important characteristic of ontology-based systems. However, systems with very general inference capabilities often do not take into account other needs, such as scalability and concurrency.

For example, in the RDFS and OWL ontology languages it is possible to make some classes the domain or the range of some property. This statement can be interpreted as an axiom saying that for any property instance in the ontology, the source and target instances can be inferred to be members of the domain and target concepts, respectively. From our experience, this is not how users, especially those with a strong background in database and object-oriented systems, typically think of domains and ranges. In OO and database systems, domains and ranges simply specify schema constraints that must be satisfied for the property to be instantiated in the first place. They do not infer new things, but guide the user in constructing the ontology by determining what can be explicitly said at all and provide guidance in what to do when the ontology changes.

Using domains and ranges as inference rules impacts the performance significantly—every property typically has at least one domain and range concept, which means that for every property there are two new inference rules that must be taken into account. If treating domains and ranges as schema constraints is enough, but only the inference rule semantics is provided, then the performance and the scalability of the system will suffer. Hence, for the successful application of ontologies it is necessary to be able to precisely control the inference capabilities according to the requirements at hand.

Conceptual Querying

Adequate query facilities are of critical importance for any ontology system. It is important that results of such queries reflect the original semantics of the model. For example, a typical query to a system managing knowledge about a set of documents is to retrieve information about some documents and associated authors. If the query returns a relational-style answer (a table with all document–author pairs in each line), then the semantics of information is destroyed by the query: the model contains an n:m relationship, which is in the query result represented as a table with repeated rows. Hence, we stress the requirement for a query facility which does not destroy the semantics of the model.

Technical Issues

In order to be applicable for real-world enterprise applications, our ontology representation approach must make it easy to fulfil the following technical requirements:

- scalability: systems must be able to cope with large quantities of information;
- concurrency support: it must be possible for several users to use and change information at the same time;
- reliability: the system must under no circumstances lose or corrupt information;
- easy integration with existing data sources.

These requirements are not trivial to fulfil, largely due to the fact that ontology management infrastructure has not reached the maturity of the relational databases. For example, many existing tools are still file-oriented. This limits the size of ontologies that can be processed, as the whole ontology must be read into main memory. Further, the multi-user support and transactions are typically not present, so the whole infrastructure realizing these requirements must be created from scratch.

2.3 Ontology Representation

In this section we present the mathematical definition of our modelling language, followed by the presentation of denotational semantics in standard Tarski style [19]. Finally, we present an ontology example.

2.3.1 Mathematical Definition

We present our approach on an abstract, mathematical level that defines the structure of our models, which can be supported using several different syntaxes.

Definition 1 (OI model structure). *An OI model (ontology–instance model) structure is a tuple* $OIM := (E, INC)$ *where:*

- *E is the set of entities of the OI model,*
- *INC is the set of included OI models.*

An OI model represents a self-contained unit of structured information that may be reused. An OI model consists of entities and may include a set of other OI models (represented through the set INC). Definition 5 lists the conditions that must be fulfilled when an OI model includes another model.

Definition 2 (Ontology structure). *An ontology structure of an OI model is a structure* $O(OIM) := (C, P, R, S, T, INV, H_C, H_P, domain, range, mincard, maxcard)$ *where:*

- $C \subseteq E$ *is a set of concepts,*
- $P \subseteq E$ *is a set of properties,*
- $R \subseteq P$ *is a set of relational properties (properties from the set* $A = P \setminus R$ *are called attribute properties),*
- $S \subseteq R$ *is a subset of symmetric properties,*
- $T \subseteq R$ *is a subset of transitive properties,*
- $INV \subseteq R \times R$ *is a symmetric relation that relates inverse relational properties; if* $(p_1, p_2) \in INV$*, then* p_1 *is an inverse relational property of* p_2*,*
- $H_C \subseteq C \times C$ *is an acyclic relation called concept hierarchy; if* $(c_1, c_2) \in H_C$ *then* c_1 *is a subconcept of* c_2 *and* c_2 *is a superconcept of* c_1*,*
- $H_P \subseteq P \times P$ *is an acyclic relation called property hierarchy; if* $(p_1, p_2) \in H_P$ *then* p_1 *is a subproperty of* p_2 *and* p_2 *is a superproperty of* p_1*,*
- *function domain*: $P \rightarrow 2^C$ *gives the set of domain concepts for some property* $p \in P$*,*
- *function range*: $R \rightarrow 2^C$ *gives the set of range concepts for some relational property* $p \in R$*,*
- *function mincard*: $C \times P \rightarrow N_0$ *gives the minimum cardinality for each concept–property pair,*
- *function maxcard*: $C \times P \rightarrow (N_0 \cup \{\infty\})$ *gives the maximum cardinality for each concept–property pair.*

Each OI model has an ontology structure associated with it. The ontology structure is a particular view of the OI model, containing definitions specifying how instances should be constructed. It consists of concepts (to be interpreted as sets of elements) and properties (to be interpreted as relations between elements). Properties can have domain concepts, and relation properties can have range concepts, which constrain the types of instances to which the properties may be applied. If these constraints are not satisfied, then an ontology is inconsistent. Domain and range concepts define schema restrictions and are treated conjunctively—all of them must be fulfilled for each property instantiation. This has been done in order to maintain compatibility with various description logics dialects (e.g. OWL), which treat domains and ranges conjunctively. Relational properties may be marked as transitive and/or symmetric, and it is possible to say that two relational properties are inverse of each other. For each class–property pair it is possible to specify the minimum and maximum cardinalities, defining how many times a property can be instantiated for instances of that class. Concepts (properties) can be arranged in a hierarchy, as specified by the H_C (H_P) relation. This relation relates directly connected concepts (properties), whereas its reflexive transitive closure follows from the semantics, as defined in the next subsection.

Definition 3 (instance pool structure). *An instance pool associated with an OI model is a 4-tuple* $IP(OIM) := (I, L, instconc, instprop)$ *where:*

- $I \subseteq E$ *is a set of instances,*
- L *is a set of literal values,* $L \cap E = \emptyset$*,*
- *function instconc*: $C \rightarrow 2^I$ *relates a concept with a set of its instances,*
- *function instprop*: $P \times I \rightarrow 2^{I \cup L}$ *assigns to each property–instance pair a set of instances related through a given property.*

Each OI model has an instance pool associated with it. An instance pool is constructed by specifying instances of different concepts and by establishing property instantiation between instances. Property instantiations must follow the domain and range constraints, and must obey the cardinality constraints, as specified by the semantics.

Definition 4 (root OI model structure). *Root OI model is defined as a particular, well-known OI model with structure ROIM* := ({KAON:ROOT}, ∅). KAON:ROOT *is the root concept; every other concept must subclass* KAON:ROOT *(it may do so indirectly).*

Every other OI model must include ROIM and thus gain visibility to the root concept. Many knowledge representation languages contain the TOP concept, which is a superconcept of all other concepts. This is also similar to many object-oriented languages—for example, in Java, every class extends the java.lang.Object class.

Definition 5 (modularization constraints). *If OI model OIM imports some other OI model OIM_1 (with elements marked with subscript 1), that is, if $OIM_1 \in INC(OIM)$, then the following modularization constraints must be satisfied:*

- $E_1 \subseteq E, C_1 \subseteq C, P_1 \subseteq P, R_1 \subseteq R, T_1 \subseteq T, INV_1 \subseteq INV, H_{C1} \subseteq H_C, H_{P1} \subseteq H_P$,
- $\forall p \in P_1 \;\; domain_1(p) \subseteq domain(p)$,
- $\forall p \in P_1 \;\; range_1(p) \subseteq range(p)$,
- $\forall p \in P_1, \forall c \in C_1 \;\; mincard_1(c, p) = mincard(c, p)$,
- $\forall p \in P_1, \forall c \in C_1 \;\; maxcard_1(c, p) = maxcard(c, p)$,
- $I_1 \subseteq I, L_1 \subseteq L$,
- $\forall c \in C_1 \;\; instconc_1(c) \subseteq instconc(c)$,
- $\forall p \in P_1, i \in I_1 \;\; instprop_1(p, i) \subseteq instprop(p, i)$.

In another words, if an OI model imports some other OI model, it contains all information from the included OI model—no information may be lost. It is possible to add new definitions, but not possible to remove existing definitions (for example, it is possible to add a domain concept to a property in the included model, but it is not possible to remove one). Modularization constraints just specify structural consequences of importing an OI model, independently from the implementation. For example, in some cases imported OI models may be physically duplicated, whereas in other cases they may be linked to the importing model without making a copy.

Definition 6 (meta-concepts and meta-properties). *In order to introduce meta-concepts, the following constraint is stated: $C \cap I$ may, but does not need to be ∅. Also, $P \cap I$ may, but does not need to be ∅.*

The same element may be used as a concept and as an instance, or as a property and as an instance in the same OI model. This has significant impact on the semantics of the model, which is described in Section 2.3.2.

Definition 7 (lexical OI model structure). *Lexical OI model structure LOIM is a well-known OI model with the structure presented in Figure 2.1. We present a graphical view of LOIM rather than a formal definition because we consider it to be more informative.*

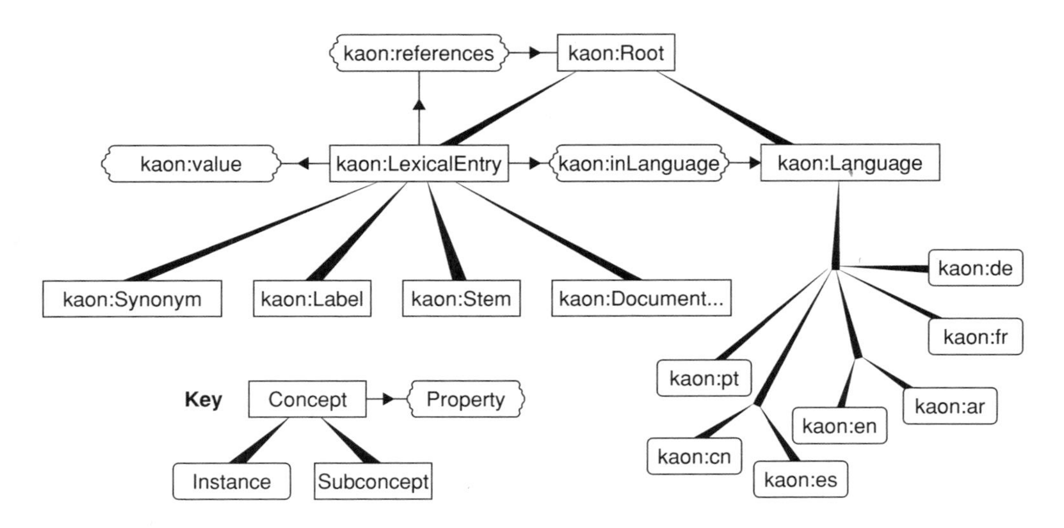

Figure 2.1 Lexical OI model structure

Lexical entries (instances of the KAON:LEXICALENTRY concept) reflect various lexical properties of ontology entities, such as a label, stem or textual documentation. There is an $n{:}m$ relationship between lexical entries and instances, established by the KAON:REFERENCES property. Thus, the same lexical entry may be associated with several elements (e.g. jaguar label may be associated with an instance representing a Jaguar car or a jaguar cat). The value of the lexical entry is given by property KAON:VALUE, whereas the language of the value is specified through the KAON:INLANGUAGE property. Concept KAON:LANGUAGE represents the set of all languages, and its instances are defined by the ISO standard 639. The lexical structure is not closed, as is the case in most languages. On the contrary, it is possible to define other types of lexical information by providing additional subclasses of the KAON:LEXICALENTRY concept.

A careful reader may have noted that LOIM defines the KAON:REFERENCES property to have the KAON:ROOT concept as the domain. In another words, this means that each instance of KAON:ROOT may have a lexical entry. This excludes concepts from having lexical entries—concepts are subclasses, and not instances of root. However, it is possible to view each concept as an instance of some other concept (e.g. the KAON:ROOT concept), and thus to associate a lexical value with it.

2.3.2 Denotational Semantics

In this subsection we give meaning to OI models through denotational semantics in the spirit of description logics. The main distinction of our definition lies in the support for meta-concept modelling, the semantics of domains and ranges, and treatment of cardinalities. These distinctions are discussed after the formal presentation of the semantics.

Definition 8 (OI model interpretation). *An interpretation of an OI model OIM is a structure* $I = (\Delta^I, \Delta_D, E^I, L^I, C^I, P^I)$ *where:*

- $\triangle^I$ *is the set of interpretations of objects,*
- $\triangle_D$ *is the concrete domain for data types,* $\triangle^I \cap \triangle_D = \emptyset$,
- $E^I : E \to \triangle^I$ *is an entity interpretation function that maps each entity to a single element in a domain,*
- $L^I : L \to \triangle_D$ *is a literal interpretation function that maps each literal to an element of the concrete domain,*
- $C^I : \triangle^I \to 2^{\triangle^I}$ *is a concept interpretation function by treating concepts as subsets of the domain,*
- $P^I : \triangle^I \to 2^{\triangle^I \times (\triangle^I \cup \triangle_D)}$ *is a property interpretation function by treating properties as relations on the domain.*

An interpretation is a model of OIM if it satisfies the following properties:

- $C^I(E^I(kaon{:}Root)) = \triangle^I$,
- $\forall c, i \;\; i \in instconc(c) \Rightarrow E^I(i) \in C^I(E^I(c))$,
- $\forall c_1, c_2 \;\; (c_1, c_2) \in H_C \Rightarrow C^I(E^I(c_1)) \subseteq C^I(E^I(c_2))$,
- $\forall p, i, i_1 \;\; i_1 \in instprop(p, i) \wedge i_1 \in E \Rightarrow (E^I(i), E^I(i_1)) \in P^I(E^I(p))$,
- $\forall p, i, x \;\; x \in instprop(p, i) \wedge x \in L \Rightarrow (E^I(i), L^I(x)) \in P^I(E^I(p))$,
- $\forall p, x, y \;\; p \in R \wedge (x, y) \in P^I(E^I(p)) \Rightarrow y \in \triangle^I$,
- $\forall p, x, y \;\; p \in P \setminus R \wedge (x, y) \in P^I(E^I(p)) \Rightarrow y \in \triangle_D$,
- $\forall p_1, p_2 \;\; (p_1, p_2) \in H_P \Rightarrow P^I(E^I(p_1)) \subseteq P^I(E^I(p_2))$,
- $\forall s \;\; s \in S \Rightarrow P^I(E^I(s))$ *is a symmetric relation,*
- $\forall p, ip \;\; (p, ip) \in INV \Rightarrow P^I(E^I(ip))$ *is an inverse relation of* $P^I(E^I(p))$,
- $\forall t \;\; t \in T \Rightarrow P^I(E^I(t))$ *is a transitive relation,*
- $\forall p, c, i \;\; c \in domain(p) \wedge (\exists x \;\; (E^I(i), x) \in P^I(E^I(p))) \wedge E^I(i) \notin C^I(E^I(c)) \Rightarrow$ *ontology is inconsistent,*
- $\forall p, c, i \;\; c \in range(p) \wedge (\exists x \;\; (x, E^I(i)) \in P^I(E^I(p))) \wedge E^I(i) \notin C^I(E^I(c)) \Rightarrow$ *ontology is inconsistent,*
- $\forall p, c, i \;\; E^I(i) \in C^I(E^I(c)) \wedge mincard(c, p) > |\{ y \mid (E^I(i), y) \in P^I(E^I(p))\}| \Rightarrow$ *ontology is inconsistent,*
- $\forall p, c, i \;\; E^I(i) \in C^I(E^I(c)) \wedge maxcard(c, p) < |\{ y \mid (E^I(i), y) \in P^I(E^I(p))\}| \Rightarrow$ *ontology is inconsistent.*

OIM is unsatisfiable if it does not have a model. The following definitions say what can be inferred from an OI model:

- $H_C^* \subseteq C \times C$ *is the reflexive transitive closure of the concept hierarchy if: in all models* $C^I(E^I(c_1)) \subseteq C^I(E^I(c_2)) \Leftrightarrow (c_1, c_2) \in H_C^*$,
- $H_P^* \subseteq P \times P$ *is the reflexive transitive closure of the property hierarchy if: in all models* $P^I(E^I(p_1)) \subseteq P^I(E^I(p_2)) \Leftrightarrow (p_1, p_2) \in H_P^*$,
- $instconc^* : C \to 2^I$ *represents inferred information about instances of a concept if: in all models* $E^I(i) \in C^I(E^I(c)) \Leftrightarrow i \in instconc^*(c)$,
- $instprop^* : P \times I \to 2^{I \cup L}$ *represents the inferred information about instances if: in all models* $i_2 \in instprop^*(p, i_1) \Leftrightarrow (E^I(i_1), E^I(i_2)) \in P^I(E^I(p)) \wedge$ *in all models* $l \in instprop^*(p, i) \Leftrightarrow (E^I(i), L^I(l)) \in P^I(E^I(p))$,
- $domain^*(p) = \bigcup_{(p, p_1) \in H_P^*} domain(p_1)$ *denotes all domain concepts of a property,*
- $range^*(p) = \bigcup_{(p, p_1) \in H_P^*} range(p_1)$ *denotes all range concepts of a property.*

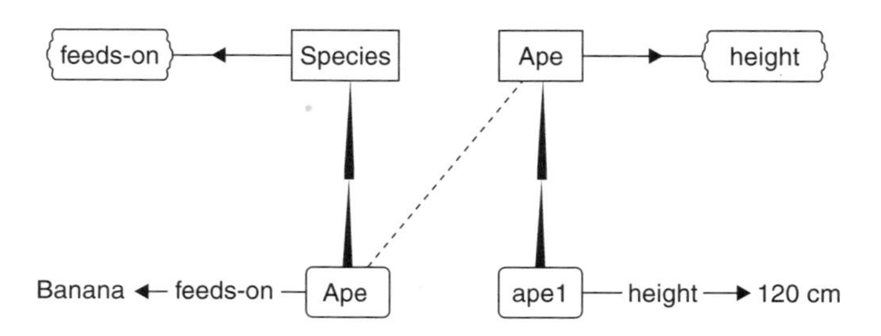

Figure 2.2 Meta-modelling example

Meta-modelling

In the usual presentation of first-order semantics, the sets of concept, property and instance symbols are disjoint. The interpretation assigns an element of the domain set to each instance, a subset of the domain set to each concept and a relation on the domain set to each property symbol.

We, on the other hand, allow for each symbol to be interpreted as a concept, a property or an instance, and thus obtain a first-order logic with second-order flavor. In [17] it has been proven that such logic has a sound and complete proof theory, which justifies its usage for ontology modelling. Hence, function E^I associates a domain object with each entity symbol, and functions C^I and P^I provide concept and property interpretations to these objects.

Returning to the example presented in Section 2.2, information about species, ape species and apes may be modelled as in Figure 2.2. In this model element APE plays a dual role. Once it is treated as a concept, in which it has the semantics of a set, and one can talk about the members of the set, such as APE1. However, the same object may be treated as an instance of the (meta-)concept SPECIES, thus allowing information such as the type of food to be attached to it. Both interpretations of the element SPECIES are connected by the dotted line, which represents the so-called spanning object relationship between the concept and the instance interpretation of the APE symbol.

In [16] the problems of considering concepts as instances are well explained. The proposed solution is to isolate different domains of discourse. What is a concept in one domain may become an instance in a higher-level domain. Elements from two domains are related through so-called spanning objects. Our approach builds on that, however, without explicit isolation of domains of discourse. This has subtle consequences on how an OI model should be interpreted. It is not allowed to ask 'what does entity e represent?'. Instead, one must ask a more specific question: 'what does e represent if it is considered as either a concept, a property or an instance?'. Thus, the interpreter must filter out a particular view of the model—it is not possible to consider multiple interpretations simultaneously. However, it is possible to move from one interpretation to another—if something is viewed as a concept, it is possible to switch to a different view and to look at the same thing as an instance.

Our approach is clearly different from that presented in [20], where the initial RDFS semantics has been criticized for its infinite meta-modelling architecture that may under some circumstances cause Russell's paradox. That paper proposes a fixed four-layer meta-modelling architecture called RDFS(FA) that introduces a strict separation between concepts and instances. Concepts are part of the ontology layer, whereas instances are part of the instance layer.

Our approach to defining semantics is more similar to the RDFS-MT [21], which avoids Russell's paradox in a similar way as we do—the same name can be interpreted in many ways, but these interpretations are still isolated from one another. However, contrary to RDFS-MT, we still separate the modelling primitives, such as subconcept or subproperty relations, into an ontology language layer. In another words, the RDFS:SUBCONCEPTOF property is not available within the model, but exists in the ontology language layer. This approach has been chosen to avoid ambiguities when modelling primitives themselves are redefined (e.g. what semantics does a subproperty of RDFS:SUBCONCEPTOF have?). In effect, our approach is similar to RDFS-FA, but with only three layers—the RDFS-FA ontology and instance layer have been merged into one OI model layer and multiple interpretations of some symbols are allowed.

Domains and ranges

Our definition of domains and ranges differs from that of RDFS and OWL. In these languages, domain and range specifications are axioms specifying sufficient conditions for an instance to be a member of some class. For example, for the ontology from Figure 2.3, although A is not explicitly stated to be an instance of DOMAIN, because it has PROPERTY instantiated and because PROPERTY has DOMAIN as domain, it can be inferred that A is an instance of DOMAIN. The semantics of this approach would be captured by replacing the domain and range restrictions with the following conditions:

$$\forall p, c, i \quad c \in \text{domain}(p) \wedge (\exists x \quad (E^I(i), x) \in P^I(E^I(p))) \Rightarrow E^I(i) \in C^I(E^I(c)),$$

$$\forall p, c, i \quad c \in \text{range}(p) \wedge (\exists x \quad (x, E^I(i)) \in P^I(E^I(p))) \Rightarrow E^I(i) \in C^I(E^I(c)).$$

From our experience, while sometimes such inferencing may indeed be useful, often it is not needed, or even desired, in closed environments, such as e.g. presented by most knowledge management applications. Most users without a formal background in logic, but with a strong background in databases and object-oriented systems, intuitively expect domains and ranges to specify the constraints that must be fulfilled while populating ontology instances. In another words, unless A is known to be an instance of DOMAIN, then PROPERTY cannot be instantiated for A in the first place, or the ontology becomes inconsistent. This approach has the following benefits:

- Treating domains and ranges as constraints makes it possible to guide the user in the process of providing information about instances. It is easy to compute the set of properties that can be applied to an instance, and then to ask the user to provide values for them. On the

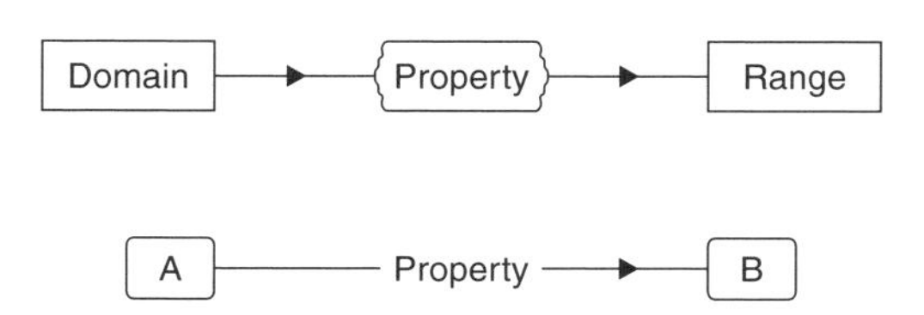

Figure 2.3 Treatment of domains and ranges

other hand, if domains and ranges are treated as axioms, any property can be applied to any instance, which makes it difficult to constrain the user's input.

- Similar problems occur when evolving the ontology. E.g., if B is removed from the extension of the concept RANGE, it can be computed that the PROPERTY between A and B must be removed. On the other hand, if domains and ranges are axioms, then it is not clear how to change the ontology so that it still makes sense.
- Treating domains and ranges as axioms introduces significant performance overhead in query answering. For example, to compute the extension of some concept, it is not sufficient to take the union of the extension of all subconcepts—one must examine a larger part of the instance pool to see which instances may be classified under the concept according to the domain and range axioms. Therefore, if only the constraint semantics is needed, the system will suffer from unnecessary performance overhead.

For ontologies that are consistent under the constraint semantics, interpretations under both definitions of the semantics match. This may be easily seen, since if ontology is not inconsistent, the negated conditions for instance membership can be moved from the left to the right side of the implication, by which the semantics becomes identical to the axiom semantics. Hence, the choice of semantics does not change the model theory or allowed entailments for (the common case of) ontologies consistent under constraint semantics.

Cardinalities

In our ontology modelling approach we treat cardinalities as constraints regulating the number of property instances that may be specified for instances of each concept. This is different from OWL and other description logic languages, where cardinalities are axioms specifying that instances with a particular number of property instances to some concept can be inferred to be instances of some concept. We find that constraining the number of property instances that are allowed for some instance is extremely useful for guiding the user in providing ontology instances. By knowing how many property instances can be provided for instances of some concept, the user can be asked to provide the appropriate number of values. Similar arguments as in the case of domain and range semantics apply here as well.

2.3.3 Example

Next we present an example OI model. A common problem in knowledge management systems is to model documents classified into various topic hierarchies. For each document we associate with it an author. Being an author is just a role of a person, and a single person may have other roles at the same time (e.g. he/she may be a researcher). This domain may be conceptualized as follows:

- Documents are modelled as instances of the DOCUMENT concept.
- Topics are modelled as instances of the TOPIC concept.
- There is a SUBTOPIC transitive property specifying that a topic is a subtopic of another topic.
- There is a HAS-AUTHOR property between a document and an author. Each author has a property HAS-WRITTEN that is inverse to HAS-AUTHOR.
- There is a HAS-TOPIC property between a document and a topic.

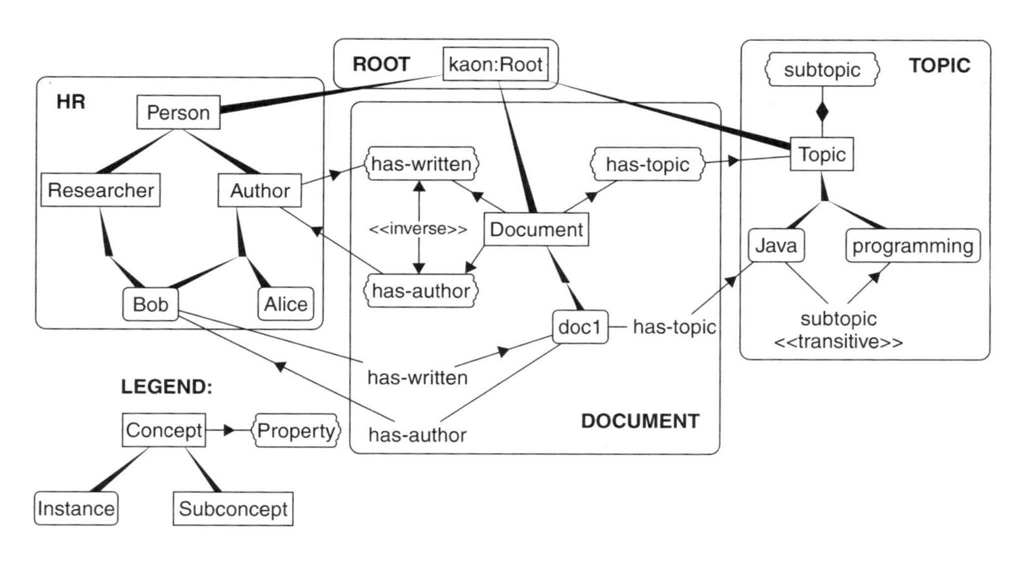

Figure 2.4 Example ontology

- Persons are modelled as instances of the Person concept.
- For each role there are subconcepts of Person. We present two: Researcher and Author.

The hierarchy of topics is a self-contained unit of information, so it is reasonable to assume there will be a need to reuse it. Therefore, it will be factored out into a separate OI model called TOPICS. Similarly, the personnel information will be separated into a human resources (HR) OI model, and the document information will be separated into a DOCUMENT OI model. Within a company it does not make sense to reuse just the ontology of HR—all information about people is centralized in one place. Therefore, the HR OI model will contain both ontology definitions as well as instances. Similar arguments may be made for other OI models. Figure 2.4 shows these OI models graphically.

The semantic interpretation of the example is as follows. All elements modelled as concepts may be interpreted as sets: Document is a set of all documents, Person is the set of all persons and Author is a subset of Person representing those persons that are authors as well. We may interpret Bob as a member of sets Person and Author and Document as a member of set kaon:Root. Topic is a set whose instances are individual topics, such as Java. Subtopic relationship is modelled using a transitive property. Inverse properties allow navigation in both directions.

2.4 Ontology Querying

Apart from representing domains with low conceptual gap, a way of querying ontologies is needed. Querying is closely related to inference—computing information which is not explicitly part of the model, but logically follows from the model. The query and inference capabilities available depend heavily on the ontology modelling approach. In this section we discuss some existing approaches to ontology querying and inference, and present our approach.

2.4.1 Approaches to Ontology Querying

We base our approach for ontology querying on the observation that a query can be seen as a function which, when applied to a data model, instantiates some results. As discussed in [22], it is beneficial if the source and target models of queries are the same. In such a way, queries can be composed functionally, giving the user a considerable degree of freedom.

In the case of a relational data model and relational algebra, this condition is fully satisfied, as the source and target models of algebra queries are both relational models, thus allowing the user to combine queries arbitrarily (e.g. through nesting). On the other hand, the relational model does not directly represent the semantics of modelled information and is thus not applicable to ontology modelling.

In other cases, e.g. in F-Logic [4], the domain of queries is the ontology model, but the target model is a list of variable bindings for which the query is satisfied in the model. We see several drawbacks to this approach. For one, F-Logic queries cannot be composed functionally, resulting in a more complex query language. Further, F-Logic queries destroy the semantics of the information. As an example let us consider an ontology consisting of people linked to the documents that they have written. A query selecting all persons with their corresponding documents, written in F-Logic as FORALL P,D $\leftarrow$ P:Person[hasWritten $\twoheadrightarrow$ D], may produce the result shown in Table 2.1.

We may observe that John has written two documents, but the query result reflects this fact by repeating the value John multiple times for P. The explicit information that there is an n:m relationship between persons and documents has been lost in the query process, and is now available only by analysing the intent of the query. This relationship may be reconstructed from the query result by collecting all values of D for which P is the same. In our view this makes the interpretation of the query external to the query itself, which does not follow the general desire to represent the semantics of information directly.

2.4.2 Conceptual Querying

In order to avoid the problems stated in the previous subsection, in our approach we focus on conceptual querying based on the observation that the basic ontology elements are concepts and properties, so to remain within the model, queries should generate concepts and properties as their results. This is a natural extension of the ontology model definition—concepts are sets of instances and queries are set of things that match some conditions. We may naturally unify

Table 2.1 Result of an example F-Logic query

P	D
John	Report1
John	Tutorial2
Bob	Calculation1

these two constructs and simply say that concepts can either be primitive, in which case they have explicitly defined extension (as described in Section 2.3), or can be defined intensionally, by specifying a condition constraining instance membership. In such spirit the enumeration of a concept's instances (primitive or otherwise) becomes the fundamental primitive for retrieving information from ontologies.

Our approach to querying ontologies is significantly influenced by description logics [23], a family of subsets of first-order logic. Central to the description logics paradigm is specifying concepts not only by enumerating individual instances, but by specifying necessary and/or sufficient conditions for instance membership. There are some differences, though. First, in description logics the focus is on inferring concept subsumption, i.e. deciding which concept is more specific than some other concept by examining only the definition of the concept. To obtain sound and complete subsumption inference procedure, the expressivity of the logics is sacrificed. In our case the focus is on efficient computation of a concept's extension, while the soundness and completeness of concept subsumption inference becomes of secondary importance. While description logics provide excellent ways of specifying structural conditions, they do not provide enough power for filtering. For example, in description logics it is not possible to form queries such as 'select all persons that are older then their spouses'.

Next we present our query approach in more detail. The queries may be concept or property expressions. Table 2.2 specifies the basic concept and property expressions, which can be combined into more complicated expressions. Enumerating a concept expression results in a set of member instances, whereas enumerating a property expression results in a set of pairs. The semantics of concept and property expressions is given in a model-theoretic way, building on the semantics presented in Section 2.3.2. We extend the function E^I to assign elements of the domain not only to entities, but to concept and property expressions as well.

Table 2.2 Query syntax and semantics

Expression	Semantics
c AND d	$C^I(E^I(c)) \cap C^I(E^I(d))$
c OR d	$C^I(E^I(c)) \cup C^I(E^I(d))$
NOT c	$\Delta^I \setminus C^I(E^I(c))$
SOME(p,c)	$\{x \mid \exists y \ \ (x, y) \in P^I(E^I(p)) \wedge y \in C^I(E^I(c))\}$
ALL(p,c)	$\{x \mid \exists y \ \ (x, y) \in P^I(E^I(p)) \Rightarrow y \in C^I(E^I(c))\}$
MUST_BE(p,c)	SOME(p,c) AND ALL(p,c)
{ i }	$\{E^I(i)\}$
FROM(p)	$\{x \mid \exists y \ \ (x, y) \in P^I(E^I(p))\}$
TO(p)	$\{y \mid \exists x \ \ (x, y) \in P^I(E^I(p))\}$
f(c)	$\{f(x) \mid x \in C^I(E^I(c))\}$
INVERSE(p)	$\{(y, x) \mid (x, y) \in P^I(E^I(p))\}$
CROSS(c,d)	$\{(x, y) \mid x \in C^I(E^I(c)) \wedge y \in C^I(E^I(d))\}$
p IN:1 c	$\{(x, y) \mid (x, y) \in P^I(E^I(p)) \wedge x \in C^I(E^I(c))\}$
p IN:2 c	$\{(x, y) \mid (x, y) \in P^I(E^I(p)) \wedge y \in C^I(E^I(c))\}$
p.r	$\{(x, z) \mid (x, y) \in P^I(E^I(p)) \wedge (y, z) \in P^I(E^I(r))\}$
f:1(p)	$\{(f(x), y) \mid (x, y) \in P^I(E^I(p))\}$
f:2(p)	$\{(x, f(y)) \mid (x, y) \in P^I(E^I(p))\}$

Some primitives from Table 2.2 may be expressed using other primitives. However, we chose to include them directly in the definition of our language because they benefit the user's understanding of the language. Primitives for property composition, concept cross product and application of functions to concept and property members are unique to our approach and typically do not appear in description logics. Next we give query examples of several queries (the characters [], <> and !! delimit the URIs of concepts, properties and instances, respectively).

The following is a typical description logic query which retrieves persons working on a research project:

```
[#Person] AND SOME(<#worksOn>,[#ResearchProject])
```

The following query allows easy navigation across properties and retrieves all persons working on a project sponsored by the EU:

```
[#Person] AND SOME(<#worksOn>.<#sponsored> {!#EU!})
```

The following query applies a function to the elements of a concept to retrieve the subconcepts of the PROJECT concept. An interesting point about this query is that it is supposed to return the set of concepts as a result. However, concepts can contain only instances, so a trick is applied. The query starts with a set of instances, interprets them as concepts (by traversing to the spanning concept), retrieves the `subconcepts` and returns the spanning instances of these concepts:

```
SUBCONCEPTS({!#Project!})
```

Schema queries can be easily chained. E.g. retrieving all concepts two levels under the PROJECT concept can be done in this way (the `SUBCONCEPTS^` function returns only direct subconcepts):

```
SUBCONCEPTS^(SUBCONCEPTS^({!#Project!}))
```

2.4.3 Navigation as a Complement for Queries

It is obvious that the presented query language itself cannot be used to express any query. For example, retrieving all documents with their authors and topics cannot be done in one query, because our queries can return only sets of items, not sets of tuples. However, this deficiency of the query language is compensated with the capability for navigation within the model. In our view the role of the query language is to provide entry points into the model, while retrieval of all ancillary information to each entry point is obtained using navigational primitives.

Retrieving all documents, their authors and topics can then be performed in the following way. First the list of relevant documents using the query language is obtained. Each document is then processed in a loop, where for each document the set of authors is retrieved by traversing the HASAUTHOR property and the set of topics is retrieved by traversing the HASTOPIC property. By doing that, the primitives for information retrieval follow closely the semantic structure of the ontology, so the semantics of the underlying information remains preserved.

However, there is a significant performance drawback in the specified approach. If there are n documents, then there will be $2 \times n + 1$ query requests issued (one for retrieving all documents and the two queries per document), which will clearly be a major cause of performance problems. Further, the order in which the information is retrieved is predefined by the order in which the navigational primitives are issued. This prevents the usage of various query optimization strategies that are considered to be the key factor for the success of relational databases systems. There seems to be a mismatch between the desire to retrieve as much information as needed on one side (to reduce communication overhead and to allow for optimizations) and to retrieve information using navigation on the underlying model (in order not to destroy the model's semantics).

To avoid such problems, we apply the following approach. Our queries can still be treated as concepts and return individuals or individual pairs at most. However, with each main query we allow specifying several additional queries. There queries are not part of the query result, but they are used as hints specifying which information is needed as well. For example, selecting all documents, their authors and topics can be done like this:

```
[#Document]
    [<#hasAuthor> IN:1 this]
    [<#hasTopic> IN:1 this]
```

The second and the third line specify that the HASAUTHOR and HASTOPIC properties with their first element equal to elements in the main query are of interest. Such additional queries can be treated as nested queries, and query unnesting [24] algorithms can be applied. Thus all necessary information can be retrieved, while allowing the database to apply any optimization strategy. The results of the query must be nested and can be accessed using navigational primitives.

2.5 Implementation

In this section we present how our ontology modelling and querying approach is applied to KAON [8]—a platform for developing and deploying semantics-driven enterprise applications. First we present KAON API—the central part of the system—and discuss its role within the overall KAON architecture. Next we discuss possible database schemata for storing ontologies.

2.5.1 KAON API within KAON Architecture

The manipulation of the ontologies in KAON is performed using KAON API—a set of interfaces offering access and manipulation of ontologies and instances. A UML view of the KAON API is shown in Figure 2.5.

The API closely follows the definitions presented in Section 2.3.1. For example, an OI model from definition 1 is represented as an OIModel object, which may include other OI models according to modularization constraints from definition 5. Ontology and instance pool objects are associated with each OI model. As per definition 2, an ontology consists of concepts (represented as Concept objects) and properties (represented as Property objects). A property has domain and range restrictions, as well as cardinality constraints. It may be an inverse of some other property, and may be marked as being transitive or symmetric. As per definition 3, an instance pool consists of instances (represented through Instance objects) that may be linked

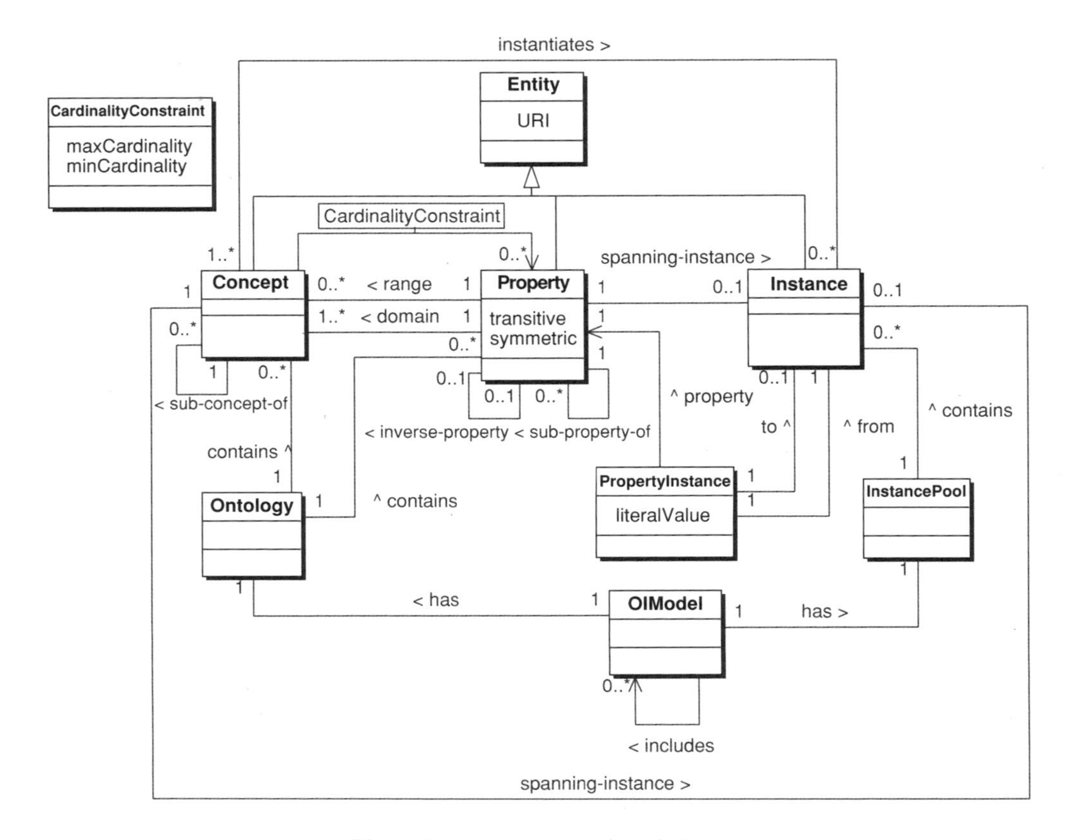

Figure 2.5 UML View of KAON API

with other instances through property instances (represented as PropertyInstance objects). Each concept or property may have an instance associated with it through a spanning object.

KAON API is embedded within the KAON conceptual architecture consisting of three layers as described in [25]. Building on top of KAON API, various tools have been realized within the Application and Services layer, such as OI modeler for ontology and metadata engineering and KAON Portal for (semi-)automatic generation of web portals from ontologies. These tools are responsible for providing the user interface.

KAON API is realized on top of the Data and Remote Services layer, which is responsible for realizing typical business-related requirements, such as persistence, reliability, transaction and concurrency support. The layer is realized within an EJB application server and uses relational databases for persistence. Apart from providing abstractions for accessing ontologies, KAON API also decouples actual sources of ontology data by offering different API implementations for various data sources. The following API implementations may be used:

- *Engineering Server*: an implementation of the KAON API that may be used for ontology engineering. This implementation provides efficient implementation of operations that are

common during ontology engineering, such as concept adding and removal in a transactional way. The schema for this API implementation is described in the following subsection.
- *Implementation for RDF repository access*: an implementation of KAON API based on RDF API (we reengineered Sergey Melnik's RDF API: http://www-db.stanford.edu/melnik/rdf/api.html) may be used for accessing RDF repositories. This implementation is primarily useful for accessing in-memory RDF models under local, autonomous operation mode. However, it may be used for accessing any RDF repository for which an RDF API implementation exists. KAON RDF Server is such a repository that enables persistence and management of RDF models, and is described in more detail in [26].
- *Implementation for accessing any database*: an implementation of KAON API may be used to lift existing databases to the ontology level. To achieve that, one must specify a set of mappings from some relational schema to the chosen ontology, according to principles described in [27]. E.g. it is possible to say that tuples of some relation make up a set of instances of some concept, and to map foreign key relationships into instance relationships. After translations are specified, a virtual OI model is generated, which will on each access translate the request into native database queries. In such a way the persistence of ontology information is obtained, while reusing well-known database mechanisms such as transactional integrity.

This architecture fulfils the requirements stated in Section 2.2. We consider it suitable for enterprise-wide application for the following reasons:

- Well-known technologies are used for ontology persistence and management, such as EJB servers and relational databases.
- These technologies already realize many of the needed requirements: transactions, concurrent processing and data integrity come 'for free'.
- Because of the structure of ontologies, the majority of queries may be executed by the underlying databases, thus ensuring scalability.

2.5.2 Persisting Ontologies

In this section we present our approach for persisting ontologies in engineering server. The name 'engineering server' comes from the fact that the persistence component is optimized for management of ontologies during the ontology engineering process. Ontology engineering is a cooperative consensus building process, during which many ontology modellers experiment with different modelling possibilities. As a result, addition, removal, merging and splitting of concepts are operations whose performance is most critical. Further, since several people are working on the same data set at once, it is important to support transactional ontology modification. At the same time, accessing large data sets efficiently is not so critical during ontology engineering. In order to support such requirements, a generic database schema, presented in Figure 2.6, is used. The schema is a straightforward translation of definitions 1, 2 and 3 into a relational model.

The schema is organized around the OIModelEntity table, whose single row represents a concept, instance and property attached to a single URI. This structure has been chosen due to the presence of meta-class modelling—by keeping all information within one table it is possible to access all information about an entity at once.

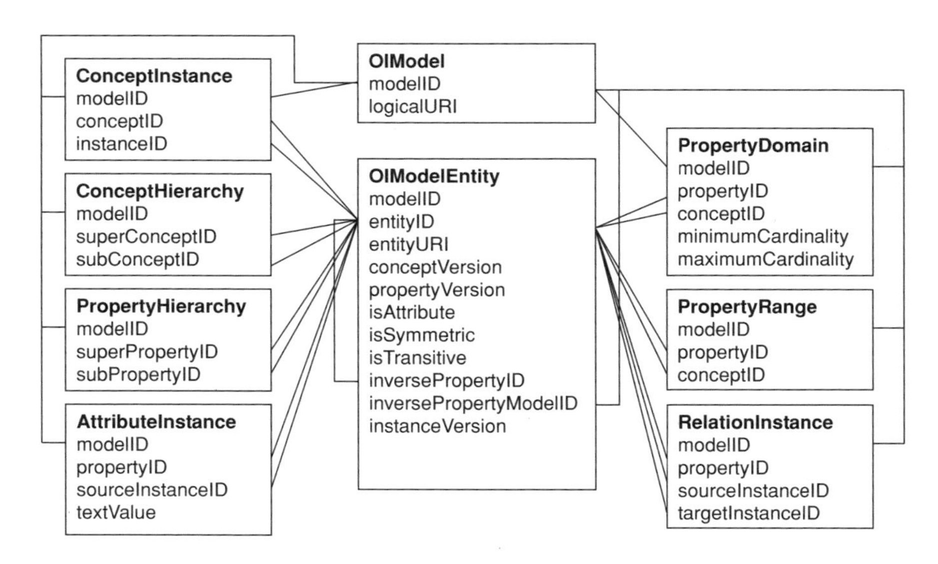

Figure 2.6 Generic KAON schema

Our schema design differs from the schema of RDF Query Language (RQL) [28] of the RDFSuite toolkit [29] in one significant point. In the RQL implementation, a unary table is assigned to each concept, where rows identify the concept's instances. Clearly, such a schema will offer much better performance, but has a significant drawback—adding (removing) a concept requires creating (removing) a table in the database. Such operations are quite expensive and may not be performed within a transaction. However, our goal is to design a system supporting users in cooperative ontology engineering, where creation and removal of concepts is quite common and must be transactional, whereas runtime performance is not that critical.

2.5.3 Evaluating Concept Queries

As mentioned in Section 2.4, efficient evaluation of extensions of intentionally defined concepts is of central importance in our approach for ontology querying. Most description logic reasoners (e.g. FaCT [30]) use tableaux reasoning [31] for proving whether an instance is a member of a concept. Such an inference procedure is needed because in description logics any concept can be instantiated. For example, in description logics it is possible to specify that i is an instance of the concept A OR B. In this case, it is not known whether i is in A or in B (or perhaps in both), but if A OR B is a subconcept of some other concept C, then it can be inferred that i is in C. To implement such inferences, a system providing disjunctive and incomplete reasoning is needed.

Our case is much simpler, since currently only primitive concepts can be instantiated. Hence, a statement from the previous paragraph cannot be made in the first place. Therefore, we may choose more traditional techniques for evaluating queries, which build on the rich experience of deductive database systems. Such evaluation strategies are much more efficient, and lend themselves well to implementation on top of relational database systems.

To evaluate queries we perform a translation of concept and property expressions into datalog programs, which we then evaluate bottom-up [32] using a semi-naive evaluation strategy. In order to restrict the amount of information retrieved from the extensional database only to relevant facts, we perform a generalized magic sets transformation of the program against the query asked [33]. Briefly, magic sets transformation simulates top-down through bottom-up computation, by applying sideways information passing, which propagates bindings of variables between predicates. Further, the magic program is extended with magic predicates containing information about the facts which are relevant to the query. Particular care has to be taken in case negation is used, since it is known that magic sets transformation of stratified programs can result in unstratified programs. In this case results from [34] are used to compute the well-founded model of the magic program, which has been proven to be equal to the minimal model of the original stratified program. By imposing some restrictions on the sideways information passing strategy, it is even possible to compute the well-founded model of a non-stratified datalog program using magic sets.

Extending the work from [35], Table 2.3 defines the transformation operator μ providing a transformation for concept and property expressions.

In the case that the engineering server is used as the underlying storage model for primitive concepts and properties, we represent each table of the server as an extensional database predicate. To capture the semantics of the model specified in Section 2.3.2, we use the default axiomatization as specified in Table 2.4.

Based on such an axiomatization, we complement the rules from Table 2.3 by the rules in Table 2.5, which define how to translate into datalog the references to primitive concepts and properties.

Table 2.3 Translating queries to datalog

$\mu[c \text{ AND } d](x) \equiv \mu[c](x), \mu[d](x)$
$\mu[c \text{ OR } d](x) \equiv p(x)$ with two additional rules $p(x) \leftarrow \mu[c](x).$, $p(x) \leftarrow \mu[d](x).$
$\mu[\text{NOT } c](x) \equiv \mu[\text{kaon:Root}](x), \neg\mu[c](x)$
$\mu[\text{SOME}(p, c)](x) \equiv \mu[p](x, y), \mu[c](y)$
$\mu[\text{ALL}(p, c)](x) \equiv \mu[\text{kaon:Root}](x), \neg p(x)$ with additional rule $p(x) \leftarrow \mu[p](x, y), \neg\mu[c](y).$
$\mu[\{i\}](x) \equiv x = i$
$\mu[\text{FROM}(p)](x) \equiv \mu[p](x, y)$
$\mu[\text{TO}(p)](x) \equiv \mu[p](y, x)$
$\mu[\text{f}(c)](x) \equiv \mu[c](y), f(x, y)$
$\mu[\text{INVERSE}(p)](x, y) \equiv \mu[p](y, x)$
$\mu[\text{CROSS}(c, d)](x, y) \equiv \mu[c](x), \mu[d](y)$
$\mu[p \text{ IN:1 } c](x, y) \equiv \mu[p](x, y), \mu[c](x)$
$\mu[p \text{ IN:2 } c](x, y) \equiv \mu[p](x, y), \mu[c](y)$
$\mu[p.r](x, y) \equiv \mu[p](x, z), \mu[r](z, y)$
$\mu[\text{f:1}(p)](x, y) \equiv \mu[p](z, y), f(x, z)$
$\mu[\text{f:2}(p)](x, y) \equiv \mu[p](x, z), f(y, z)$

Table 2.4 Default axiomatization of engineering server

ConceptHierarchy_t(X, Y) $\leftarrow$ ConceptHierarchy(X, Y).
ConceptHierarchy_t(X, Y) $\leftarrow$ ConceptHierarchy(X, Z), ConceptHierarchy_t(Z, Y).

PropertyHierarchy_t(X, Y) $\leftarrow$ PropertyHierarchy(X, Y).
PropertyHierarchy_t(X, Y) $\leftarrow$ PropertyHierarchy(X, Z), PropertyHierarchy_t(Z, Y).

ConceptInstance_a(C, I) $\leftarrow$ ConceptInstance(C, I).
ConceptInstance_a(C, I) $\leftarrow$ ConceptHierarchy_t(C, SC), ConceptInstance(SC, I).

AttributeInstance_a(P, S, V) $\leftarrow$ AttributeInstance(P, S, V).
AttributeInstance_a(P, S, V) $\leftarrow$ PropertyHierarchy_t(P, SP), AttributeInstance(SP, S, V).

RelationInstance_a(P, S, T) $\leftarrow$ RelationInstance(P, S, T).
RelationInstance_a(P, S, T) $\leftarrow$ PropertyHierarchy_t(P, SP), RelationInstance_a(SP, S, T).
RelationInstance_a(P, S, T) $\leftarrow$
    OIModelEntity($P, ATTR, SIM, TRANS, IP$),
    RelationInstance_a(IP, T, S).
RelationInstance_a(P, S, T) $\leftarrow$
    OIModelEntity($P, ATTR$, true, $TRANS, IP$),
    RelationInstance(P, T, S).
RelationInstance_a(P, S, T) $\leftarrow$
    OIModelEntity($P, ATTR, SIM$, true, IP),
    RelationInstance_a(P, S, M),
    RelationInstance_a(P, M, T).

PropertyInstance_a($P, S, NULL, V$) $\leftarrow$ AttributeInstance_a(P, S, V).
PropertyInstance_a($P, S, T, NULL$) $\leftarrow$ RelationInstance_a(P, S, T).

Table 2.5 Translating primitive concepts and properties to datalog

μ[concept-URI](x) $\equiv$ ConceptInstance_a(concept-URI, x)
μ[primitive-relation-URI](x, y) $\equiv$ RelationInstance_a(primitive-relation-URI, x, y)
μ[primitive-attribute-URI](x, y) $\equiv$ AttributeInstance_a(primitive-attribute-URI, x, y)

2.6 Related Work

In this section we discuss how our approach differs from other ontology and conceptual modelling approaches and tools available.

Entity-Relationship and Relational Modelling

In the database community entity-relationship (ER) modelling has established itself as the prevailing conceptual modelling paradigm. ER models consist of entity sets that define a set of attributes for each entity in the set and of relations of any arity between entities. Under recent extensions it is even possible to model subclassing between entities as well. ER modelling is just a tool for conceptual databases design. Before implementation these models are transformed into a logical model—nowadays this is the relational model. Evolving an implementing such models is more complex than if only one paradigm were used, since the conceptual and logical models must be kept in synchrony.

In our approach the logical model is hidden from the user. The users may use an ontology in a natural way, while enjoying the benefits of relational database technology for information storage and management.

RDF and RDFS

It has been argued by many (e.g. [20]) that the definition of RDFS is very confusing, because RDFS modelling primitives are used to define the RDFS language itself. Currently there is no specification for modularization of RDF models. The handling of lexical information is very messy—languages can be attached to the RDF model using XML attributes. Further, only 1:m relationships between lexical entries and ontology elements are possible.

RQL

Several query languages have been developed for RDF(S), with RDF Query Language (RQL) [28] being one of the most well known. RQL basically extends the functional approach of object query language (OQL) [36] with capabilities for querying schema as well as instances. This is achieved by creating a type system on top of the RDF model. Thus, each RDF object is given a certain type, and types for collections of objects of other types have been added. Each RQL query primitive can be viewed as a function which takes an object of a certain type and creates a new object of some type. To return multiple values, RQL introduces the tuple type as a sequence of values. Many RQL queries result in a collection of tuples, so they destroy the semantics, as discussed in Section 2.4.

UML

There have been proposals for using UML as an ontology modelling language (e.g. [37]). However, the reader may note that there are significant, but subtle, differences in standard object-oriented and ontology modelling. Classes of an object-oriented model are assigned responsibilities, which are encapsulated as methods. Often fictitious classes are introduced to encapsulate some responsibility (also known as pure fabrications [38]). On the other hand, ontologies do not contain methods so responsibility analysis is not important.

The fact that UML is targeted for modelling of software systems containing methods is far-reaching. For instance, an object cannot be an instance of more than one class and, if an object is created as an instance of some class, it is impossible for it to later become an instance of some other class—an object's class membership is determined at the point when the object is created.

In our modelling paradigm, however, these statements do not hold. Membership of an object in a class is often interpreted as statement that some unary predicate is true for this object. If properties of an object change as time passes, then the object should be reclassified. Membership of an object in different classes at the same time has the notion that some different aspects of that object are true simultaneously.

Frame-based languages

The application of the object-oriented modelling paradigm to ontology modelling has resulted in so called frame-based knowledge modelling languages, and F-logic [4] is one example. In F-logic, the knowledge is structured into classes having value slots. Frames may be instantiated, in which case slots are filled with values. F-logic introduces other advanced modelling constructs, such as expressing meta-statements about classes. Also, it is possible to define Horn-logic rules for inferring new information not explicitly present in the model.

Although F-Logic offers a comprehensive ontology model, the queries to F-Logic models result in the set of variable bindings, which are not within the model. Hence, the semantics of the data is obscured when the model is queried.

Description logics

A large body of research has been devoted to a subclass of logic-based languages, called description logics (a good overview is presented in [23]). In description logics an ontology consists of individuals (representing elements from the universe of discourse), roles (representing binary relations between individuals) and concepts (sets of elements from the universe of discourse). Concepts may either be primitive or specified through descriptions. Concept descriptions are generated by combining other concept definitions using constructors to specify constraints on the membership of individuals in the concept being defined. The primary inference procedure in description logics is that of subsumption—determining whether one concept description subsumes another one (whether all individuals of the subsumed concept must be individuals of the subsuming concept, regardless of how the instances are specified). Further, it is possible to enumerate all individuals of a concept. A large body of research has examined which concept constructors may be combined, while still allowing for a decidable and tractable inference procedure.

This led to a well-researched theory; however, as mentioned in [39], description logics have proven to be difficult to use due to the non-intuitive modelling style. Basically this is due to the mismatch with the predominant object-oriented way of thinking. For example, a common knowledge management problem is to model a set of documents that have topics chosen from a topic hierarchy. A typical solution is to create a DOCUMENT concept acting as a set of individual documents. However, modelling a topic hierarchy is not so straightforward, since it is not clear whether topics should be modelled as concepts or as individuals.

In object-oriented systems it is not possible to relate instances of classes with classes. Therefore, a possible approach is to model all topics as members of the concept TOPIC, and to introduce subtopic transitive relations between topic instances. To an experienced object-oriented modeller, this solution will be intuitive.

On the other hand, in description logic systems, since topics are arranged in a hierarchy, the preferred modelling solution is to arrange all topics in a concept hierarchy to rely on the subsumption semantics of description logics.[1] Thus, each topic will be a subtopic of its superconcepts. However, two problems arise:

- If topics are sets, what are the instances of this set? Most users think of topics as fixed entities, and not as (empty and abstract) sets.

[1] Thanks to Ian Horrocks for discussion on this topic.

- How to relate some document, e.g. d1, to a particular topic, e.g. t1? Documents will typically have a role has-topic that should be created between d1 and topic instance. But t1 is a concept, and there are no instances of t1. The solution exists, but is not intuitive—we do not specify exactly with which instance of t1 d1 is related, but say that it is related to 'some' instance of t1. In the syntax of CLASSIC [40] description logic system, this is expressed as

$$(createIndividual\ d1\ (some\ has\text{-}topic\ t1)).$$

Further, description logics typically do not support inclusion of lexical information in the model. Because of their focus on the sound and complete inference procedures, the power of filtering query results is limited. In description logics it is usually not possible to realize constraints—e.g. domains and ranges are there always treated as axioms (a proposal to overcome this limitation by using an epistemic operation has been put forward in [41] but to the best knowledge of the authors it has not been implemented in practice). Finally, to implement description logics fully, very advanced reasoning techniques, such as disjunctive and incomplete reasoning, are required. This hinders realization of systems that would fulfil the requirements from Section 2.2.

The appropriate query language for description logics is still under discussion. In [42] an approach for extending description logics with conjunctive querying has been presented. The approach is based on query roll-up, where a conjunctive query over description logics concept is transformed into an equivalent description logics Boolean query. However, as the authors mention, an efficient algorithm for evaluating such queries is still missing.

OIL

An important ontology modelling language is OIL [5]. This language has tried to combine the intuitive notions of frames, the clear semantics of description logics and the serialization syntax of RDF. It proposes a layered architecture. Core OIL defines constructs that equal in expressivity to RDF schema without the support for reification. Standard OIL contains constructs for ontology creation that are based on description logics. However, the syntax of the language hides the description logics background and presents a system that seems to have a more frame-based 'feel'. However, standard OIL does not support creation of instances. Instance OIL defines constructs for creation of instances, whereas heavy OIL is supposed to include the full power of description logics (it has not been defined yet).

Despite its apparent frame-based flavour, OIL is in fact a description logic system, thus our comments about description logics apply to OIL as well. OIL does support modularization of ontologies. However, it does not have a consistent strategy for management of lexical information.

RDFSuite

RDFSuite [29] is a set of tools for management of RDF information developed by the ICS-FORTH institute. It includes a validating parser for RDF, a database system for storing RDF and the RDF Query Language (RQL)—a functional query language for querying RDF repositories.

Our approach differs form the RDFSuite in several aspects. First, the ontology modelling paradigm explained in this paper is richer than that of the RDFSuite, as it includes symmetric, inverse and transitive properties, whereas the RDFSuite is limited to constructs of

RDF(S). Second, for manipulation of ontology data KAON API is used, which provides an object-oriented view of the ontology information, whereas RDFSuite relies on queries for accessing the information. The same API is used to isolate applications from actual data sources, thus allowing storage of ontology information not only in relational databases. Finally, the implementation of RQL does not support transactional updates for ontology engineering—creation of concepts involves creating tables, which cannot be executed within a transaction.

Sesame

Sesame [43] is an architecture for storing and querying RDF data developed by the Free University of Amsterdam and Aidministrator. It consists of an RDF parser, RQL query module and an administration module for managing RDF models. The differences between our approach and Sesame are similar to those between our approach and RDFSuite. Sesame also supports the RDF(S) modelling paradigm but does not provide the API for object-oriented access and modification of information.

Jena

Jena [44] is the set of tools for RDF manipulation developed by HP Labs. It implements a repository for RDF(S) and includes RDF Data Query Language (RDQL) implementation for querying. On top of that, it offers support for persistence of DAML ontologies, but with only very basic inferencing capabilities. An API for manipulation of raw RDF information is included, as well as for manipulation of DAML ontologies. However, there is no API for manipulation of RDF(S) ontologies. Further, RDQL query language is based on graph pattern matching, thus preventing the user from issuing recursive queries (e.g. for selecting all subconcepts of some concept).

2.7 Conclusion

In this chapter we present an approach for ontology modelling and querying, currently being developed within KAON. Our main motivation is to come up with an approach that can be used to build scalable enterprise-wide ontology-based applications using existing, well-established technologies.

In the chapter we argue that existing approaches for conceptual modelling lack some critical features, making them unsuitable for application within enterprise systems. From the technical point of view, these features include scalability, reliability and concurrency. From the conceptual modelling point of view, existing approaches lack the support for modularization and concept meta-modelling.

Based on the motivating usage scenarios, a set of requirements has been elicited. A mathematical definition of the ontology language has been provided, along with a denotational semantics. Next an approach for ontology querying adapting and extending the description logic paradigm has been presented. We have presented the current status of the implementation in the form of the KAON API—an API for management of ontologies and instances, along with the algorithms for evaluating ontology queries.

In future we shall focus on two important aspects of ontology modelling. First, we plan to investigate whether our approach for query execution can be extended to handle more description logics constructs. In particular we are interested in allowing the explicit instantiation of defined concepts while still being able to compute concept extensions efficiently. Second, in order to provide capabilities for modelling constraints and to increase the expressivity of our model, we are interested in finding ways to combine description logics with rules, but without significant performance degradation.

References

[1] T. Berners-Lee, J. Hendler, O. Lassila, The Semantic Web. *Scientific American*, **284**(5), 34–43, 2001.

[2] O. Lassila, R.R. Swick, Resource Description Framework (RDF) Model and Syntax Specification, http://www.w3.org/TR/REC-rdf-syntax/.

[3] D. Brickley, R.V. Guha, RDF Vocabulary Description Language 1.0: RDF Schema, http://www.w3.org/TR/rdf-schema/.

[4] M. Kifer, G. Lausen, J. Wu, Logical foundations of object-oriented and frame-based languages. *Journal of the ACM*, **42**, 741–843, 1995.

[5] D. Fensel, I. Horrocks, F. van Harmelen, S. Decker, M. Erdmann, M. Klein, OIL in a nutshell. In *Knowledge Acquisition, Modeling, and Management, Proceedings of the European Knowledge Acquisition Conference (EKAW-2000)*, October, pp. 1–16. Springer-Verlag, Berlin, 2000.

[6] P.F. Patel-Schneider, P. Hayes, I. Horrocks, F. van Harmelen, Web Ontology Language (OWL) Abstract Syntax and Semantics, http://www.w3.org/TR/owl-semantics/, November 2002.

[7] C. Davis, S. Jajodia, P. Ng, R. Yeh (eds), *Entity-Relationship Approach to Software Engineering: Proceedings of the 3rd International Conference on Entity-Relationship Approach*, Anahein, CA, 5–7, October. North-Holland, Amsterdam, 1983.

[8] B. Motik, A. Maedche, R. Volz, A Conceptual Modeling Approach for Semantics-driven Enterprise Applications. In *Proceedings of the 1st International Conference on Ontologies, Databases and Application of Semantics (ODBASE-2002)*, October 2002.

[9] M. Fowler, K. Scott, *UML Distilled: A Brief Guide to the Standard Object Modeling Language*, 2nd edn. Addison-Wesley, Reading, MA, 1999.

[10] A. Evans, A. Clark, *Foundations of the Unified Modeling Language*. Springer-Verlag, Berlin, 1998.

[11] G. Schreiber, Some challenge problems for the Web Ontology Language, http://www.cs.man.ac.uk/ horrocks/ OntoWeb/SIG/challenge-problems.pdf.

[12] A. Maedche, B. Motik, N. Silva, R. Volz, MAFRA—an ontology MApping FRAmework in the Context of the Semantic Web. In *Workshop on Ontology Transformation at ECAI-2002*, Lyon, France, July 2002.

[13] A. Maedche, B. Motik, L. Stojanovic, R. Studer, R. Volz, Managing multiple ontologies and ontology evolution in ontologging. In M.A. Musen, B. Neumann, R. Studer (eds) *Proceedings of the Conference on Intelligent Information Processing, World Computer Congress*, Montreal, Canada, 25–30 August 2002. Kluwer Academic, Boston, MA, 2002.

[14] A. Gangemi, N. Guarino, C. Masolo, A. Oltramari, Understanding top-level ontological distinctions. In *Proceedings of IJCAI 2001 Workshop on Ontologies and Information Sharing*, Seattle, WA, 6 August 2001.

[15] G. Schreiber, The Web is not well-formed. *IEEE Intelligent Systems*, **17**(2), 79–80, 2002.

[16] C.A. Welty, D.A. Ferrucci, What's in an instance? Technical report, RPI Computer Science, 1994.

[17] W. Chen, M. Kifer, D.S. Warren, HiLog: a foundation for higher-order logic programming. *Journal of Logic Programming*, **15**, 187–230, 1993.

[18] N.F. Noy, R.W. Fergerson, M.A. Musen, The knowledge model of Protege-2000: Combining interoperability and flexibility. In *2th International Conference on Knowledge Engineering and Knowledge Management (EKAW'2000)*, Juan-les-Pins, France, 2000.

[19] M. Fitting, *First-Order Logic and Automated Theorem Proving*, 2nd edn. Springer Verlag, Berlin, 1996.

[20] J. Pan and I. Horrocks, Metamodeling architecture of web ontology languages. In *Proceedings of the Semantic Web Working Symposium*, pp. 131–149, July 2001.

[21] P. Hayes, RDF model theory, http://www.w3.org/TR/rdf-mt/.

[22] R. Volz, D. Oberle, R. Studer, Views for light-weight web ontologies. In *Proceedings of ACM Symposium of Applied Computing (SAC)*, Melbourne, Florida, USA, March 2003.
[23] A. Borgida, Description logics are not just for the FLIGHTLESS-BIRDS: A new look at the utility and foundations of description logics. Technical Report DCS-TR-295, Department of Computer Science, Rutgers University, 1992.
[24] L. Fegaras, Query unnesting in object-oriented databases. In L. Haas, A. Tiwary (eds) *SIGMOD 1998, Proceedings ACM SIGMOD International Conference on Management of Data*, Seattle, WA, 1–4 June, pp. 49–60. ACM Press, New York, 1998.
[25] E. Bozak, M. Ehrig, S. Handschuh, A. Hotho, A. Maedche, B. Motik, D. Oberle, R. Studer, G. Stumme, Y. Sure, S. Staab, L. Stojanovic, N. Stojanovic, J. Tane, V. Zacharias, KAON—towards an infrastructure for semantics-based e-services. In K. Bauknecht, A. Min Tjoa, G. Quirchmayr (eds) *Proceedings of the 3rd International Conference on E-Commerce and Web Technologies, EC-Web 2002,* Aix-en-Provnce, France, 2–6 September 2002. Lecture Notes in Computer Science no. 2455 Springer-Verlag, Berling, 2002.
[26] R. Volz, B. Motik, A. Maedche, Poster at 1st International Semantic Web Conference ISWC, Sardinia, Italy, June 2002.
[27] N. Stojanovic, L. Stojanovic, R. Volz, A reverse engineering approach for migrating data-intensive web sites to the Semantic Web. In M.A. Musen, B. Neumann, R. Studer (eds) *Proceedings of the Conference on Intelligent Information Processing, World Computer Congress*, Montreal, Canada, 25–30 August 2002. Kluwer Academic, Boston, MA, 2002.
[28] G. Karvounarakis, S. Alexaki, V. Christophides, D. Plexousakis, M. Scholl, RQL: a declarative query language for RDF. In *11th International Conference on the WWW*, Hawaii, 2002.
[29] S. Alexaki, V. Christophides, G. Karvounarakis, D. Plexousakis, K. Tolle, The ICS-FORTH RDFSuite: managing voluminous RDF description bases. In S. Decker, D. Fensel, A. Sheth, S. Staab (eds) *Proceedings of the Second International Workshop on the Semantic Web—SemWeb '2001,* Hong Kong, 1 May 2001. CEUR Workshop Proceedings, vol. 40, pp. 1–13, 2001. Available at http://sunsite.informatic.rwth-aachen.de/Publications/CEUR-WS/Vol-40.
[30] I. Horrocks, FaCT and iFaCT. In *Proceedings of the International Workshop on Description Logics (DL'99),* Linköping, Sweden, 30 July–1 August 1999.
[31] F. Baader, U. Sattler, Tableau algorithms for description logics. In R. Dyckhoff (ed.) *Proceedings of the International Conference on Automated Reasoning with Tableaux and Related Methods (Tableaux 2000)*, St Andrews, UK, vol. 1847, pp. 1–18, Springer-Verlag, Berlin, 2000.
[32] S. Abiteboul, R. Hull, V. Vianu, *Foundations of Databases*. Addison-Wesley, Reading, MA, 1995.
[33] C. Beeri, R. Ramakrishnan. On the power of magic. In M.Y. Vardi (ed.) Proceedings of the Sixth ACM SIGACT-SIGMOD-SIGART Symposium on Principles of Database Systems, pp. 269–284. ACM Press, New York, 1987.
[34] D.B. Kemp, D. Srivastava, P.J. Stuckey. Magic sets and bottom-up evaluation of well-founded models. In V. Saraswat, K. Ueda (eds) *Proceedings of the 1991 International Symposium on Logic Programming*, San Diego, CA, pp. 337–354. MIT Press, Cambridge, MA, 1991.
[35] B. Grossof, I. Horrocks, R. Volz, S. Decker, Description logic programs: combining logic programs with description logic. In *Proceedings of WWW 2003*, Budapest, Hungary, May 2003.
[36] R.G.G. Cattell, D.K. Barry, R. Catell, M. Berler, J. Eastman, D. Jordan, C. Russell, O. Schadow, T. Stanienda, F. Velez, *The Object Data Standard: ODMG 3.0*. Morgan Kaufmann, San Francisco, CA, 2000.
[37] S. Cranefield, M. Purvis. UML as an ontology modelling language. In T.L. Dean (ed.) *Proceedings of the Workshop on Intelligent Information Integration, 16th International Joint Conference on Artificial Intelligence (IJCAI-99)*, Stockholm, Sweden, 31 July–6 August 1999. Morgan Kaufmann, San Francisco, CA, 1999.
[38] C. Larman, *Applying UML and Patterns: An Introduction to Object-Oriented Analysis and Design and the Unified Process*, 2nd edn. Prentice Hall, Upper Saddle River, NJ, 2001.
[39] S. Bechhofer, C. Goble, I. Horrocks, DAML+OIL is not enough. In *Proceedings of SWWS '1: The First Semantic Web Working Symposium,* Stanford University, CA, 30 July–1 August. Available at http://potato.cs.man.ac.uk/papers/not-enough.pdf.
[40] D.L. McGuinness, J.R. Wright, An industrial strength description logic-based configurator platform. *IEEE Intelligent Systems*, **13**(4), 69–77, 1998.
[41] F.M. Donini, M. Lenzerini, D. Nardi, A. Schaerf, W. Nutt, Adding epistemic operators to concept languages. In B. Nebel, C. Rich, W.R. Swartout (eds) *Proceedings of the Third International Conference on the Principles of Knowledge Representation and Reasoning (KR'92)*, pp. 342–353. Morgan Kaufmann, San Francisco, CA, 1992.

[42] I. Horrocks, S. Tessaris, Querying the semantic web: a formal approach. In I. Horrocks, J.A. Hendler (eds) *Proceedings of the 2002 International Semantic Web Conference (ISWC 2002)*, Sardinia, Italy, 9–12 June 2002. Springer-Verlag, Berlin, 2002.

[43] J. Broekstra, A. Kampman, F. van Harmelen, Sesame: an architecture for storing and querying RDF data and schema information. In D. Fensel, J. Hendler, H. Lieberman, and W. Wahlster (eds) *Semantics for the WWW.* MIT Press, Cambridge, MA, 2001.

[44] B. McBride, Jena: implementing the RDF model and syntax specification, http://www-uk.hpl.hp.com/people/bwm/papers/20001221-paper/.

3

Adding Multimedia to the Semantic Web: Building and Applying an MPEG-7 Ontology

Jane Hunter

3.1 Introduction

Audiovisual resources in the form of still pictures, graphics, 3D models, audio, speech and video will play an increasingly pervasive role in our lives, and there will be a growing need to enable computational interpretation and processing of such resources. Forms of representation that will allow some degree of machine interpretation of the meaning of audiovisual information will be necessary. The goal of MPEG-7 [1] is to support such requirements by providing a rich set of standardized tools to enable the generation of audiovisual descriptions which can be understood by machines as well as humans, to enable fast, efficient retrieval from digital archives (pull applications) as well as filtering of streamed audiovisual broadcasts on the Internet (push applications). The main elements of the MPEG-7 standard are:

- Descriptors (D): representations of Features, that define the syntax and the semantics of each feature representation.
- Description Schemes (DS), that specify the structure and semantics of the relationships between their components. These components may be both Descriptors and Description Schemes.
- A Description Definition Language (DDL) to allow the creation of new Description Schemes and, possibly, Descriptors and to allow the extension and modification of existing Description Schemes.
- System tools, to support multiplexing of descriptions, synchronization of descriptions with content, transmission mechanisms, coded representations (both textual and binary formats) for efficient storage and transmission, management and protection of intellectual property in MPEG-7 descriptions etc.

Multimedia Content and the Semantic Web Edited by Giorgos Stamou and Stefanos Kollias

XML Schema language was chosen as the DDL [5] for specifying MPEG-7 descriptors and description schemes because of its ability to express the syntactic, structural, cardinality and datatyping constraints required by MPEG-7 and because it also provides the necessary mechanisms for extending and refining existing DSs and Ds. However, there is a need for a machine-understandable representation of the semantics associated with MPEG-7 DSs and Ds to enable the interoperability and integration of MPEG-7 descriptions with metadata descriptions from other domains. New metadata initiatives such as TV-Anytime [6], MPEG-21 [7] and NewsML [8], and communities such as the museum, educational, medical and geospatial communities want to combine MPEG-7 multimedia descriptions with new and existing metadata standards for simple resource discovery (Dublin Core [9]), rights management (INDECS [10]), geospatial (FGDC [11]), educational (GEM [12], IEEE LOM [13]) and museum (CIDOC CRM [14]) content, to satisfy their domain-specific requirements. In order to do this, there needs to be a common understanding of the semantic relationships between metadata terms from different domains.

The Web Ontology Working group [15] of the W3C has recently published the Web Ontology Language (OWL) as a Candidate Recommendation [16]. In this chapter, we first describe an MPEG-7 ontology which we have developed using OWL, to define the semantics of MPEG-7 Descriptors and Description Schemes. Second, we describe how the generated MPEG-7 ontology can be utilized and understood by other domains on the Semantic Web to increase the potential reuse, sharing and exchange of multimedia content over the Internet. In particular, we describe one extremely valuable application of the MPEG-7 ontology—the inferencing of domain-specific semantic descriptions of multimedia content from combinations of low-level automatically extracted features.

The structure of this chapter is as follows. In Section 3.2 we describe the methodology, problems encountered and results of building an OWL representation of the MPEG-7 ontology. In Section 3.3 we describe how the OWL MPEG-7 ontology can be used to infer high-level domain-specific semantic descriptions of multimedia content from low-level, automatically extracted (MPEG-7) features, using predefined inferencing rules. In Section 3.4 we describe how the resulting semantic descriptions can enable sophisticated semantic querying of multimedia resources in terms familiar to the user's domain. Section 3.5 concludes with a discussion of this work and our plans for the future. The overriding objective of this work is to ensure that the information and knowledge within the multimedia content has a much greater chance of being discovered and exploited by services, agents and applications on the web.

3.2 Building an MPEG-7 Ontology

During the early development stages of MPEG-7, Unified Modelling Language (UML) [17] was used to model the entities, properties and relationships (description schemes and descriptors) which comprised MPEG-7. However, the massive size of the specification (the Multimedia Description Schemes specification [18] is almost 800 pages and that is only one out of seven parts), combined with the belief that the UML models were a development tool which duplicated information in the XML Schemas, led to the decision to drop them from the final specifications. Although the lack of an existing data model hinders the development of an MPEG-7 ontology, it also means that the generated ontology will be even more valuable, providing both a data model and a definition of the semantics of MPEG-7 terms and the semantic relationships between them. Building the ontology should also highlight any inconsistencies, duplication or ambiguities which exist across the large number of MPEG-7 description schemes and descriptors. Without a data model to build on, the class and property hierarchies and semantic definitions had to be derived through reverse engineering of the existing XML Schema definitions, together with interpretation of the English-text semantic descriptions. To simplify the process, we used a core subset of the MPEG-7 specification together with a top-down approach to generate the ontology described here. An additional very helpful mechanism for determining the data model was to generate the Document Object Model (DOM) for the XML Schema (using XML Spy) and to use this to assist with the generation of the class and property hierarchies. Hence, our approach was to first determine the basic multimedia entities (classes) and their hierarchies from the Multimedia Description Scheme basic entities [17]. This process is described in Section 3.2.1. Next, the structural hierarchies were determined from the Segment Description Schemes (Section 3.2.2). Section 3.2.3 describes the OWL representations of the MPEG-7 visual and audio descriptors defined in [19] and [20] respectively.

3.2.1 Top-level MPEG-7 Multimedia Entities

The Multimedia Content entities are described in section 4.4 of the MDS [18]. The OWL class hierarchy corresponding to these basic entities is illustrated in Figure 3.1 and Box 3.1.

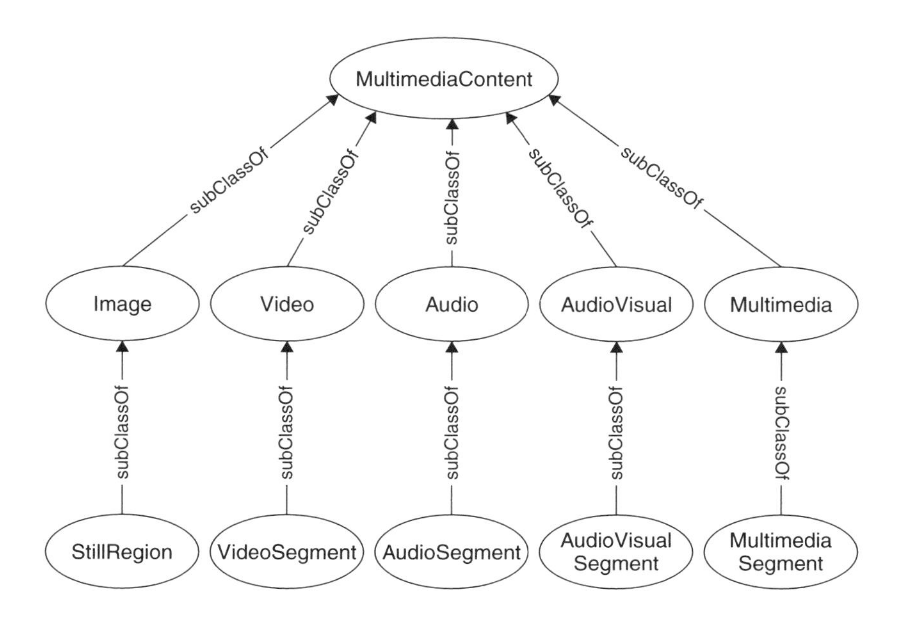

Figure 3.1 Top-level class hierarchy of MPEG-7 multimedia entities

Box 3.1

```
<?xml version='1.0'?>
<!DOCTYPE rdf:RDF [
   <!ENTITY mpeg7 "http://www.mpeg.org/WD2003/mpeg7#" >
   <!ENTITY owl "http://www.w3.org/2002/07/owl#" >
   <!ENTITY xsd "http://www.w3.org/2001/XMLSchema#" > ]>
<rdf:RDF
 xmlns = "http://www.mpeg.org/WD2003/mpeg7#"
 xmlns:mpeg7 = "http://www.mpeg.org/WD2003/mpeg7#"
 xml:base = "http://www.mpeg.org/WD2003/mpeg7#"
 xmlns:owl = "http://www.w3.org/2002/07/owl#"
 xmlns:rdf = "http://www.w3.org/1999/02/22-rdf-syntax-ns#"
 xmlns:rdfs= "http://www.w3.org/2000/01/rdf-schema#"
 xmlns:xsd = "http://www.w3.org/2001/XMLSchema#">
 <owl:Ontology rdf:about="">
  <rdfs:comment>An OWL ontology for MPEG-7</rdfs:comment>
  <rdfs:label>MPEG-7 Ontology</rdfs:label>
 </owl:Ontology>
<owl:Class rdf:ID="MultimediaContent">
  <rdfs:label>MultimediaContent</rdfs:label>
  <rdfs:comment>The class of multimedia data</rdfs:comment>
  <rdfs:subClassOf rdf:resource="http://www.w3.org/2000/01/
rdf-schema#Resource"/>
</owl:Class>
<owl:Class rdf:ID="Image">
  <rdfs:label>Image</rdfs:label>
  <rdfs:comment>The class of images</rdfs:comment>
  <rdfs:subClassOf rdf:resource="#MultimediaContent"/>
</owl:Class>
<owl:Class rdf:ID="Video">
  <rdfs:label>Video</rdfs:label>
  <rdfs:comment>The class of videos</rdfs:comment>
  <rdfs:subClassOf rdf:resource="#MultimediaContent"/>
</owl:Class>
<owl:Class rdf:ID="Audio">
  <rdfs:label>Audio</rdfs:label>
  <rdfs:comment>The class of audio resources</rdfs:comment>
  <rdfs:subClassOf rdf:resource="#MultimediaContent"/>
</owl:Class>
<owl:Class rdf:ID="AudioVisual">
  <rdfs:label>AudioVisual</rdfs:label>
  <rdfs:comment>The class of audiovisual resources</rdfs:comment>
  <rdfs:subClassOf rdf:resource="#MultimediaContent"/>
</owl:Class>
<owl:Class rdf:ID="Multimedia">
  <rdfs:label>Multimedia</rdfs:label>
  <rdfs:comment>The class of multimedia resources</rdfs:comment>
  <rdfs:subClassOf rdf:resource="#MultimediaContent"/>
</owl:Class>
</rdf:RDF>
```

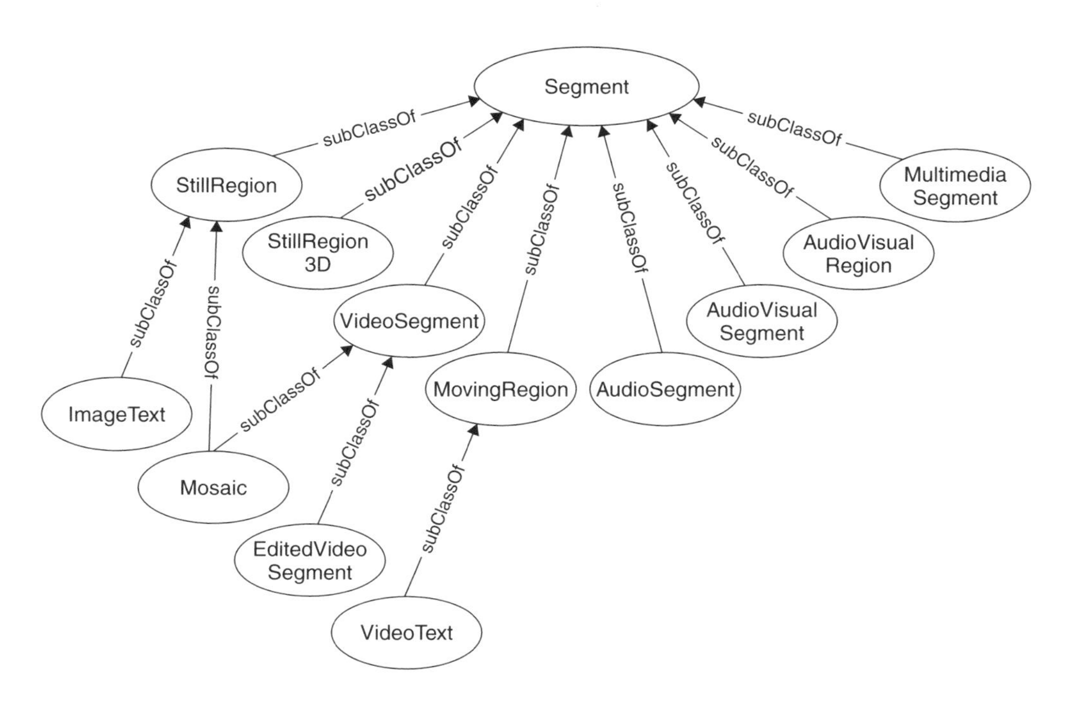

Figure 3.2 Segment class hierarchy

3.2.2 MPEG-7 Multimedia Segments and Hierarchical Structures

Due to the continuous nature of multimedia, each multimedia object can be segmented into a number of spatial, temporal, spatio-temporal or mediasource segments, depending on its media type. Figure 3.2 illustrates the multimedia segment types in the segment class hierarchy.

The OWL representation corresponding to Figure 3.2 is in Appendix A.1. The relationship of these segment types to multimedia entities is illustrated in Figure 3.3. Multimedia resources can be segmented or decomposed into sub-segments through four types of decomposition:

- Spatial decomposition, e.g. spatial regions within an image.
- Temporal decomposition, e.g. temporal video segments within a video.
- Spatio-temporal decomposition, e.g. moving regions within a video.
- Mediasource decomposition, e.g. the different tracks within an audio file or the different media objects within a SMIL presentation.

The OWL representation for the SegmentDecomposition property hierarchy is given in Appendix A.2.

If we consider the decomposition of a VideoSegment, then in order to explicitly define the allowable types of decomposition of VideoSegments and the possible types of segments which result from the decomposition, then a further level of sub-properties is required. OWL [21] permits multiple range statements but appears to interpret the resulting range to be the *intersection* of these classes. In this case, we want to specify that the range will be an instance

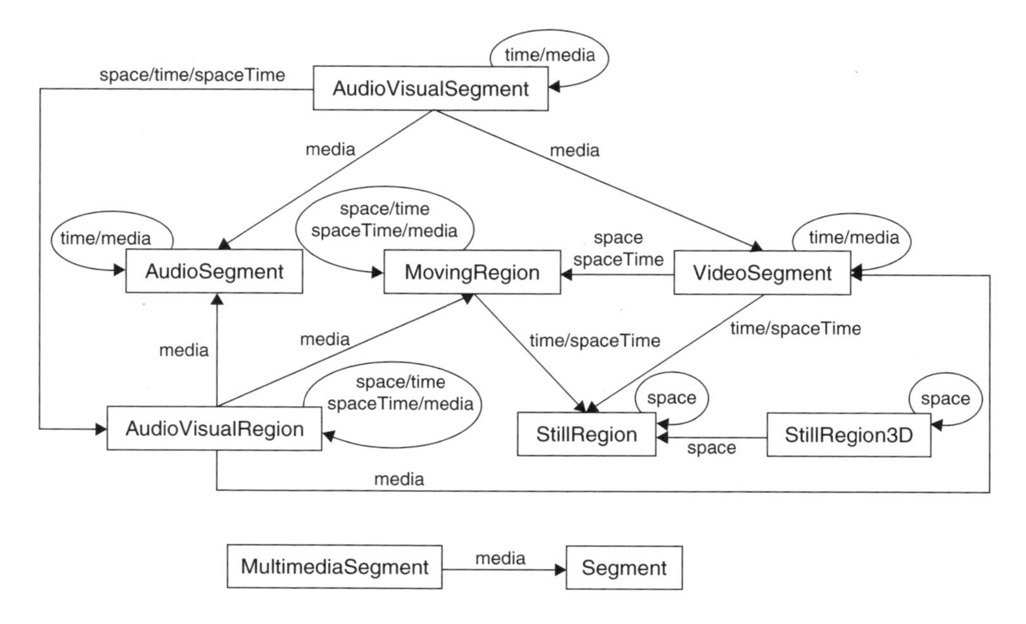

Figure 3.3 Valid types of segment decomposition

from the *union* of the two classes (VideoSegment and Moving Region). In order to do this we could use `owl:unionOf` to define a class which is the union of these two classes and then specify this new class as the range. This is illustrated in Appendix A.3.

Also associated with the segment classes are the spatialMask, temporalMask, spatiotemporalMask and mediaspaceMask properties, which define the location of the segment relative to its containing media object. Their semantic definitions and class associations are fully described within the complete MPEG-7 OWL ontology, which can be found at http://metadata.net/sunago/fusion/mpeg7.owl.

3.2.3 Low-Level Visual and Audio Descriptors

Examples of widely used visual features or properties which are applicable to the visual entities (Image, Video, AudioVisual, StillRegion, MovingRegion, VideoSegment) include:

- Colour
- Texture
- Motion
- Shape.

Each of these features can be represented by a choice of descriptors. Table 3.1 illustrates the relationship between MPEG-7 descriptors and visual features. Precise details of the structure and semantics of these visual descriptors are provided in [19].

Table 3.1 Visual features and their corresponding descriptors

Type	Feature	Descriptors
Visual	Color	DominantColor ScalableColor ColorLayout ColorStructure GoFGoPColor (extension of ColorStructure)
	Texture	HomogeneousTexture TextureBrowsing EdgeHistogram
	Shape	RegionShape ContourShape Shape3D
	Motion	CameraMotion MotionTrajectory ParametricMotion MotionActivity

Similarly there is a set of audio features are applicable to MPEG-7 entities that contain audio (Video, AudiVisual, Audio, AudioSegment). Each of the following audio features can be represented by one or more audio descriptors:

- Silence
- Timbre
- Speech
- Melody.

The XML Schema specifications of the audio descriptors are described in detail in [20]. Table 3.2 illustrates the audio descriptors corresponding to each audio feature.

Table 3.2 Audio features and their corresponding descriptors

Type	Feature	Descriptors
Audio	Silence	Silence
	Timbre	InstrumentTimbre HarmonicInstrumentTimbre PercussiveInstrumentTimbre
	Speech	Phoneme Articulation Language
	MusicalStructure	MelodicContour Rhythm
	SoundEffects	Reverberation, Pitch, Contour, Noise

Table 3.3 Relationships between segment types and audio and visual descriptors

Feature	Video Segment	Still Region	Moving Region	Audio Segment
Time	X	—	X	X
Shape	—	X	X	—
Color	X	X	X	—
Texture	—	X	—	—
Motion	X	—	X	—
Audio	X	—	—	X

Only certain low-level visual and audio descriptors are applicable to each segment type. Table 3.3 illustrates the association of visual and audio descriptors to different segment types. RDF Schema must be able to specify the constraints on these property-to-entity relationships.

Figure 3.4 and the code in Box 3.2 illustrate how OWL is able express these constraints through the domain and range values on the property definition of the color descriptor.

Figure 3.5 illustrates the class and property relationships for the MPEG-7 visual descriptors: Shape, Texture, Motion and Locator. Figures 3.6–3.8 show the class/property diagrams for the RegionLocator, SpatioTemporalLocator and TemporalInterpolationVisualDescriptor.

We will not describe the OWL representations for these descriptors in detail here, but the complete OWL representation of the MPEG-7 ontology is available at http://metadata.net/sunago/fusion/mpeg7.owl.

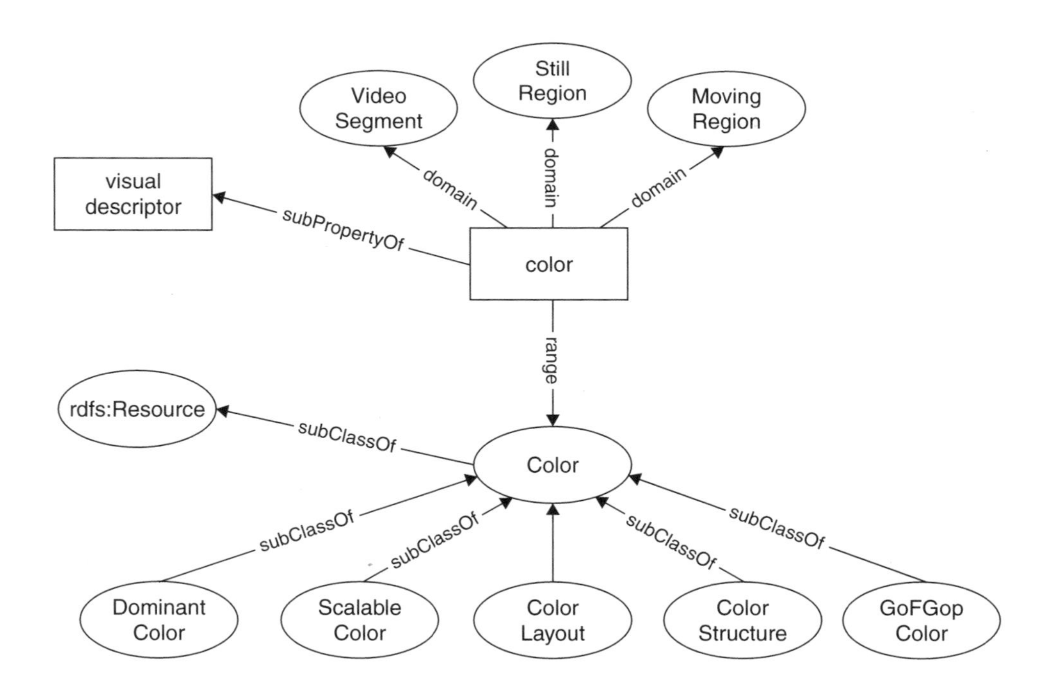

Figure 3.4 The Color visual descriptor

Box 3.2

```
<owl:Class rdf:ID="Color">
  <rdfs:label>Color</rdfs:label>
  <rdfs:comment>Color of a visual resource</rdfs:comment>
  <rdfs:subClassOf rdf:resource="http://www.w3.org/2000/01/
rdf-schema#Resource"/>
</owl:Class>
<owl:Class rdf:ID="DominantColor">
  <rdfs:label>DominantColor</rdfs:label>
  <rdfs:comment>The set of dominant colors in an
arbitrarily-shaped region.</rdfs:comment>
  <rdfs:subClassOf rdf:resource="#Color"/>
</owl:Class>
<owl:Class rdf:ID="ScalableColor">
  <rdfs:label>ScalableColor</rdfs:label>
  <rdfs:comment>Color histogram in the HSV color space.
  </rdfs:comment>
  <rdfs:subClassOf rdf:resource="#Color"/>
</owl:Class>
<owl:Class rdf:ID="ColorLayout">
  <rdfs:label>ColorLayout</rdfs:label>
  <rdfs:comment>Spatial distribution of colors.</rdfs:comment>
  <rdfs:subClassOf rdf:resource="#Color"/>
</owl:Class>
<owl:Class rdf:ID="ColorStructure">
  <rdfs:label>ColorStructure</rdfs:label>
  <rdfs:comment>Describes color content and the structure
of this content.</rdfs:comment>
  <rdfs:subClassOf rdf:resource="#Color"/>
</owl:Class>
<owl:Class rdf:ID="GoFGoPColor">
  <rdfs:label>GoFGoPColor</rdfs:label>
  <rdfs:comment>Group of frames/pictures color descriptor
.</rdfs:comment>
  <rdfs:subClassOf rdf:resource="#ScalableColor"/>
</owl:Class>
<rdf:Property rdf:ID="color">
  <rdfs:label>color</rdfs:label>
  <rdfs:comment>The color descriptor is applicable to video
segments, still regions and moving regions.</rdfs:comment>
  <rdfs:subPropertyOf rdf:resource="#visualDescriptor"/>
  <rdfs:domain rdf:resource="#VideoSegment"/>
  <rdfs:domain rdf:resource="#StillRegion"/>
  <rdfs:domain rdf:resource="#MovingRegion"/>
  <rdfs:range rdf:resource="#Color"/>
</rdf:Property>
```

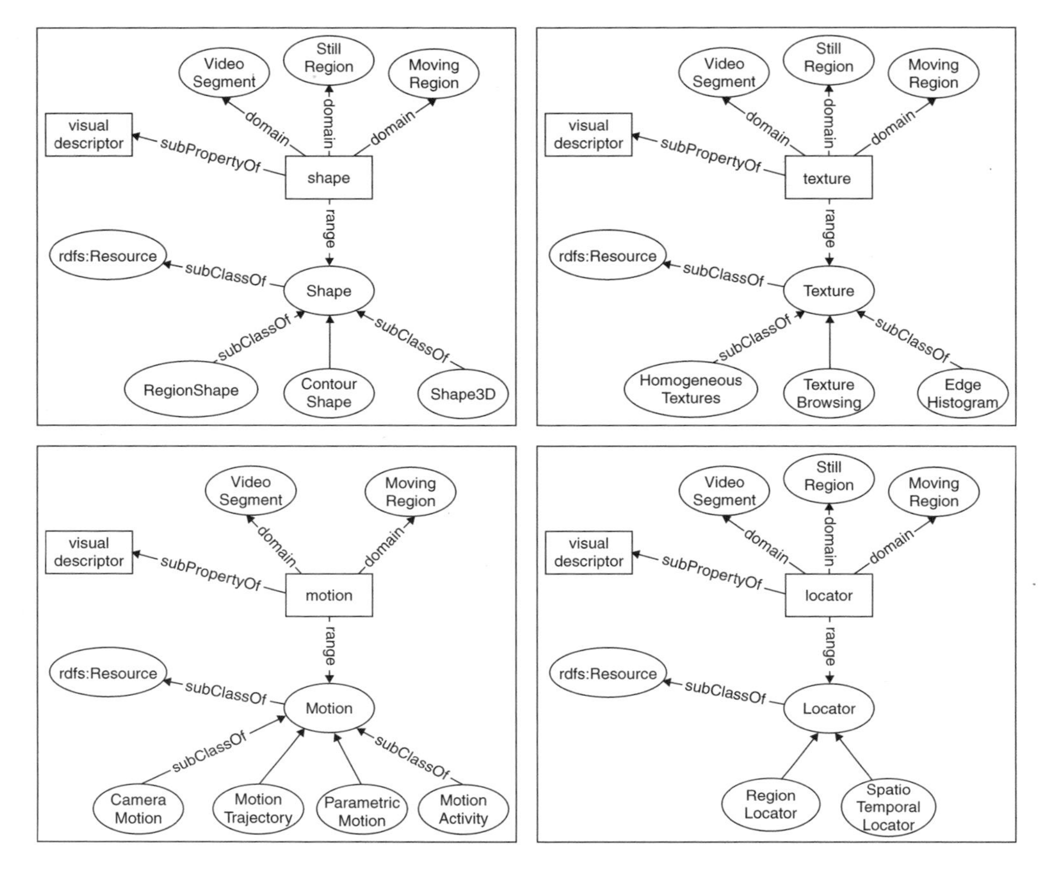

Figure 3.5 The Shape, Texture, Motion and Locator visual descriptors

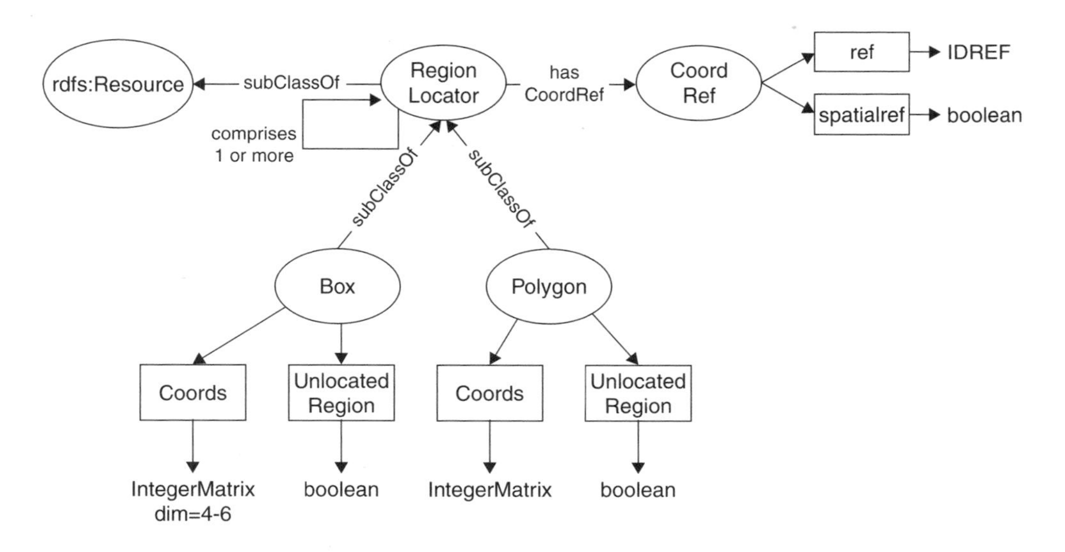

Figure 3.6 The RegionLocator visual descriptor

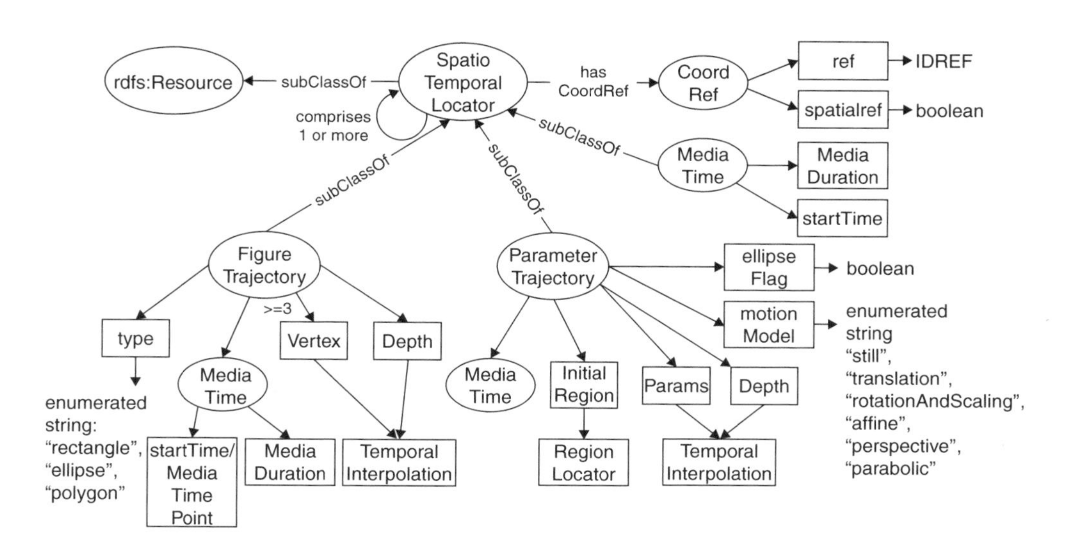

Figure 3.7 The SpatioTemporalLocator visual descriptor

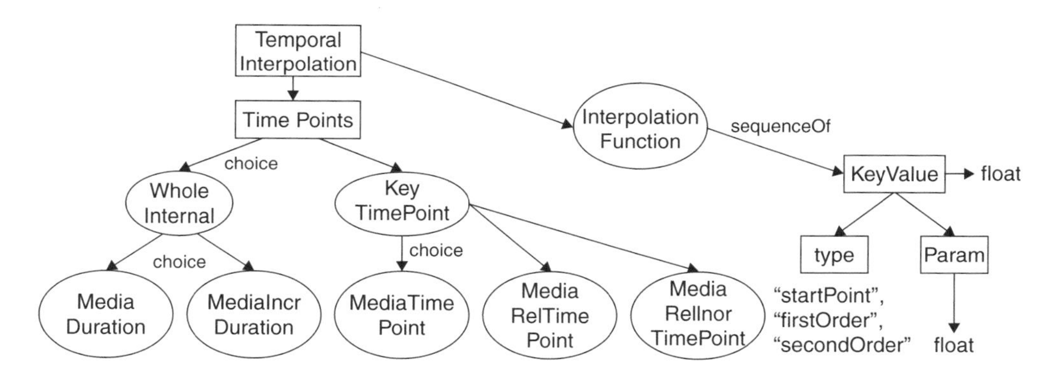

Figure 3.8 The TemporalInterpolation visual descriptor

The next section describes research we are currently undertaking which demonstrates how the MPEG-7 ontology can be usefully applied to facilitate the generation of semantic descriptions of multimedia content.

3.3 Inferring Semantic Descriptions of Multimedia Content

To enable multimedia content to be discovered and exploited by services, agents and applications on the Semantic Web, it needs to be described semantically. Generating descriptions of multimedia content is inherently problematic because of the volume and complexity of the data, its multidimensional nature and the potentially high subjectivity of human-generated descriptions. Significant progress has been made in recent years on automatic segmentation or structuring of multimedia content and the recognition of low-level features within such content. However, comparatively little progress has been made on machine-generation of semantic descriptions of audiovisual information. Within this chapter we describe our recent and ongoing research efforts within the FUSION project at DSTC [22], which attempt to combine standards for multimedia content description (MPEG-7) with recent developments in Semantic Web technologies, to develop systems which maximize the potential knowledge that can be mined from large heterogeneous multimedia information sets on the Internet. More specifically, we describe our research into the inferencing of high-level domain-specific semantic descriptions of multimedia content from low-level, automatically extracted (MPEG-7) [6] features, using ontologies and predefined inferencing rules. Such semantic descriptions enable sophisticated semantic querying of multimedia resources in terms familiar to the user's domain and ensure that the information and knowledge within the multimedia content has a much greater chance of being discovered and exploited by services, agents and applications on the web.

Within the scope of the FUSION application we are interested in generating XML descriptions of microscopy images of cross-sections of fuel cells to enable their analysis and optimization by scientists. For this application we use an application profile which draws on and combines two existing metadata standards (MPEG-7 and OME) with a third metadata schema developed specifically for this domain (FUSION):

- MPEG-7 (Multimedia Content Description Interface): for describing the low-level features and encoding format of images;
- OME (Open Microscopy Environment): a standard for describing the source of microscopy data (this overlaps with `mpeg7:CreationInfoDS`);
- FUSION: provides the metadata description of a fuel cell as required by fuel cell scientists or analysts. This includes descriptions of the thickness, surface area, density, porosity and conductivity of the anode, cathode and electrolyte which compose the fuel cell.

The MPEG-7 descriptions are extracted automatically by applying image recognition software to the images. The OME metadata is also generated automatically by manipulating the output from the software provided with the microscope. The FUSION description will be generated by defining and applying inferencing rules which will infer the anode, cathode and electrolyte components and their attributes from the automatically extracted low-level MPEG-7 features (regions, colour, shape, texture).

In order to enable semantic interoperability between the MPEG-7, OME and FUSION metadata vocabularies, and to define the semantic and inferencing relationships between the terms in these different schemas, we needed to develop ontologies for each of these vocabularies and merge or harmonize them. We used the top-level or core ABC ontology [14] developed within the Harmony project to do this. The ABC ontology provides a global and extensible model that expresses the basic concepts that are common across a variety of domains and provides the basis for specialization into domain-specific concepts and vocabularies. Figure 3.9 illustrates this process at a very high level.

3.3.1 Linking Ontologies to XML data

Once we have defined XML Schemas for the metadata descriptions and the corresponding ontologies in OWL, we need a way of linking the semantic definitions in the ontologies to the corresponding elements in the schemas or instances. In an earlier paper [23] we described how

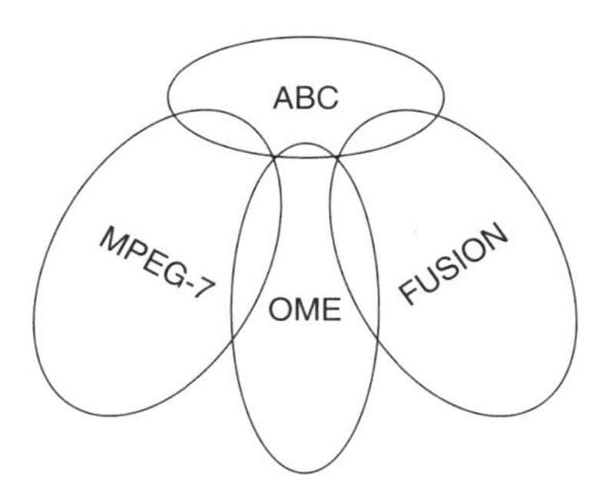

Figure 3.9 Ontology harmonization using ABC

the `xx:semantics` attribute in XML Schema documents can be used to link element definitions to an ontology term, thereby linking the semantic definitions to the preferred encoding or XML representation. The advantages of this approach are the separation of encodings from semantics and the consequent freedom to structure and name elements in XML to suit the application or user community without altering the meaning of the term. Appendix E contains the XML Schema for a fuel cell description and illustrates the use of the `xx:semantics` attribute to link this Schema to its corresponding OWL ontology. The code in Box 3.3 is a brief excerpt from this.

Box 3.3

```
<?xml version="1.0" encoding="UTF-8"?>
<xs:schema xmnls="http://www.w3.org/2001/XMLSchema"
xmnls:xs="http://www.w3.org/2001/XMLSchema"
xmlns:xx="http://www.example.org/XMLRDFSchemaBridge"
targetNamespace="http://metadata.net/fusion/FUSION">

   <xs:annotation>
     <xs:documentation>Draft XML Schema for FUSION
</xs:documentation>
   </xs:annotation>

   <xs:element name="fuelcell" type="fuelcell"/>
   <xs:complexType name="fuelcell"
     xx:semantics="http://metadata.net/fusion/FUSION#fuelcell">
     <xs:sequence>
        <xs:element name="anode" type="anodeType"/>
        <xs:element name="electrolyte" type="electrolyteType"/>
        <xs:element name="cathode" ref="cathodeType"/>
     </xs:sequence>
     <xs:attribute name="id" use="required"/>
   </xs:complexType>
...
</xs:schema>
```

3.3.2 Automatic Feature Extraction

Automatic feature extraction for images is a significant research area, and a wide variety of tools and mechanisms exist for analysing scientific and other types of images (e.g. see [24]). Within the scope of this project, we are not particularly interested in the underlying method by which these programs analyse images but we are interested in the outputs or features that such programs produce and whether they can be used by our domain experts or inferencing engines to infer the occurrence of higher-level semantic concepts, e.g. regions or objects of a particular colour, texture and shape. Although we have designed our system so that it can invoke automatic image analysis tools that have been set up as web services, we have chosen initially to use MATLAB [25] because it is a popular and powerful tool that is currently being widely used by microscopists and scientists for analysing scientific images. In an ideal system, SOAP [26] and WSDL [27] would be employed to make the automatic feature extraction tools available as web services. This would allow greater choice and flexibility as more advanced multimedia extraction and analysis tools become available.

MATLAB is capable of producing a large amount of low-level data about the features and objects within an image, e.g. area, mean density, standard deviation density, perimeter/length, X/Y centre, integrated density etc. The data produced by MATLAB needs to be mapped to the image analysis (MPEG-7) description and related (using the inferencing rules) to the high-level semantic terms in the FUSION description. In particular, we want to map regions extracted from the images to the fuel cell components such as the anode, cathode, electrolyte, catalyst and substrate. This process is described in detail in the next section.

3.3.3 Semantic Inferencing Rules

Rule Description Techniques

The most popular markup language for rules and the current most likely candidate for adoption by the Semantic Web community is RuleML [12,28], the Rule Markup Language. Developed by a consortium of industry and academic partners called the Rule Markup Initiative, RuleML provides a markup language which allows for descriptions of rules of the form `if` *x*`(`*a*`,` *b*`)` `[and|or]` *y*`(`*b*`,` *c*`)` `then` *z*`(`*b*`,` *d*`)` and facts of the form *x*`(`*a*`,` *b*`)` in XML format. Some translation and tool support is available for RuleML. A Java reasoning engine called Mandarax [29] is available, which allows querying and application of RuleML rules and facts. Other possible rule standards are the eXtensible Rule Markup Language (XRML) [30] and the Description Logic Markup Language (DLML) [31], neither of which is as well developed or widely supported as RuleML.

In addition to rules defining relationships based on purely semantic lines, our particular application also requires rules to describe mathematical relationships within and between ontology terms and data elements. This is a common requirement for multimedia data associated with scientific applications. For example, an electrode contains catalyst particles, and data on their average size is required—this information can be calculated from the size recorded for each of the catalyst particles using a mathematical formula. MathML [32], the W3C's maths markup language, provides a comprehensive method for describing both presentation and content of mathematical formulas. While MathML also has set theory and Boolean logic operators and could therefore also be used to define logic rules, it lacks the reasoning engine support of RuleML and can only (natively) define relationships such as equals, greater than etc. and cannot describe higher-level semantic relationships such as `depicts`. Box 3.4 illustrates a example of MathML describing a simple formula.

Box 3.4

```
<!-- The mean width of an anode is the mean of the difference
between y1 and y2 coords of the object found in the media. -->
<mathml:reln><mathml:eq/>
   <mathml:ci>/anode/width/mean</mathml:ci>
   <mathml:ci>
     <mathml:mean>
        <mathml:reln><mathml:minus/>
           <mathml:ci>//mpeg7/object/region/coords/y2</mathml:ci>
           <mathml:ci>//mpeg7/object/region/coords/y1</mathml:ci>
        </mathml:reln>
     </mathml:mean>
   </mathml:ci>
</mathml:reln>
```

The FUSION project accesses data from a broad range of sources (manufacturing, performance, microscopy and image analysis data) that are related through merged ontologies. Hence FUSION requires both semantic (ABC, MPEG-7, FUSION) and mathematical relationships to be defined. Therefore both RuleML (semantics) and MathML (mathematics) are required to achieve integration of comprehensive, complete data sets with a minimum of user effort. In addition, the variables used in the rules and formulas may be in different documents and can be either abstract (ontological terms) or specific (XML elements). We use XPath [33] to identify and retrieve the specific variable values or elements from the XML instances and the `xx:semantics` attribute to link this back to the ontological definitions.

To apply the defined rules set and infer high-level semantic annotations, we can either use pre-existing engines (e.g. Mandarax for RuleML with a wrapper for formatting data input) or write our own application. At this stage we have decided to use Mandarax, but because of the tenuous interface between OWL ontologies and RuleML it may be necessary to develop extensions or wrappers to Mandarax to incorporate knowledge implicitly defined with the existing OWL ontologies (e.g. subsumption relationships or transitive, symmetric or inverse relationships).

A Simple Semantic Inferencing Example

Suppose the user is interested in verifying the theory that small catalyst particle size affects the performance of fuel cells. The system does not explicitly record such information. However, by using rules to infer semantic descriptions from the features extracted from the microscopy images plus the implicit relationships between terms defined in the harmonised ontologies, this information can be derived.

Box 3.5

```
mpeg7:depicts (mpeg7:StillRegion@ID, fusion:Catalyst)
        IF mpeg7:DominantColor > (112,112,112)
        AND mpeg7:HomogeneousTexture == ... ..
        AND mpeg7:RegionShape is circle
        AND ome:magnification >= 40.
```

Appendix B contains an MPEG-7 XML description generated from applying MATLAB image analysis to a microscopy image and mapping the output to MPEG-7. Using RuleML to express a simple rule such as the one described in Box 3.5, it can be inferred that the feature in the image being referred to, http://metadata.net/FUSION/media/image.jp2, depicts a Catalyst as defined in the FUSION fuel cell ontology.

Box 3.6 illustrates the complete RuleML representation of this example. In this example, the relationship being described is `mpeg7:depicts` and the variables are identified as `mpeg7:Color`, `mpeg7:Texture` and `mpeg7:Shape`. The `mpeg7` ontology provides the additional knowledge that `mpeg7:DominantColor`, `mpeg7:ScalableColor`, `mpeg7:ColorLayout` etc. are all `rdfs:subClassOf` `mpeg7:Color`. To find elements within the data which contain information about colour, the `xx:semantics` attribute in the XML Schema is used to link to the ontology and discover those tags which are relevant to `mpeg7:color` and to extract their values in order to apply the rule.

Box 3.6

```
<?xml version="1.0" encoding="UTF-8"?>
<rulebase xmlns="http://www.ruleml.org/"
          xmlns:xsi="http://www.w3.org/2001/XMLSchema-instance"
          xmlns:mpeg7="urn:mpeg:mpeg7:schema:2001"
          xmlns:mathml="http://www.w3.org/1998/Math/MathML"
          xmlns:xml="http://www.w3.org/XML/1998/namespace"
          xmlns:fusion="http://metadata.net/FUSION">
   <imp>
<!-- A StillRegion depicts Catalyst if the colour value is greater
than (112, 112, 112) and the shape is logically equivalent to a
circle, the texture is . and the magnification at which the
image was taken is greater than or equal to 40 -->
     <_head>
        <atom>
           <_opr>
              <rel>mpeg7:depicts</rel>
           </_opr>
           <var>mepg7:StillRegion@ID</var>
           <var>fusion:Catalyst</var>
        </atom>
     </_head>
     <_body>
```

```
        <and>
            <atom>
                <_opr>
                    <rel>mathml:gt</rel>
                </_opr>
                <ind>mpeg7:Color</ind>
                <ind>(113,113,113)</ind>
            </atom>
            <atom>
                <_opr>
                    <rel>mathml:equivalent</rel>
                </_opr>
                <ind>mpeg7:Shape</ind>
                <ind>circle</ind>
            </atom>
            <atom>
                <_opr>
                    <rel>mathml:lt</rel>
                </_opr>
                <ind>mpeg7:Texture</ind>
                <ind>... .</ind>
            </atom>
            <atom>
                <_opr>
                    <rel>mathml:geq</rel>
                </_opr>
                <ind>ome:magnification</ind>
                <ind>40.0</ind>
            </atom>
        </and>
    </_body>
  </imp>
</rulebase>
```

Once the features which depict catalysts have been identified, the second step is to determine whether or not they qualify as 'small'. The user defines `small` to be catalysts of size less than 50^2 microns. The FUSION ontology defines `area` to be a sub-class of `size`. By comparing the `area` value for the catalyst features with the required value, fuel cells with 'small' catalyst particles can be discovered. The fuel cell ID associated with the microscopy information can be used to retrieve the performance data for each of these fuel cells. By appropriately presenting the images and performance data in parallel, the user will be able to determine whether his/her hypothesis appears reasonable and worthy of further exploration.

The fuzzy nature of image indexing and retrieval means that users do not require exact matches between colour, texture or shape but only require similarity-based search and retrieval. In this first prototype, we have implemented this by including a default threshold of 10% (which can be modified) in matching instances against the values specified in the rules. However, we are planning to develop a graphical rule editing tool which supports the definition of rules (in RuleML) using QBE-type tools, e.g. a StillRegion of an image is a cathode if its colour is like this, its texture is like this and its shape is like this.

This simple example shows how, by using inferencing rules to assign semantic meaning to image features and ontologies to define further semantic relationships, information not explicitly recorded by the system can be mined and used to enable more sophisticated search and retrieval. Such inferencing may be done on-the-fly or in advance and stored in a knowledge repository (e.g. a relational or XML database) to enable more efficient querying and retrieval. In our system we are carrying out the inferencing in advance and then storing and indexing the inferred information in a database. At this stage we are using MySQL but we will shortly be evaluating the Tamino XML [34] database.

3.4 Semantic Querying and Presentation

The high-level semantic descriptions that have been inferred, together with the knowledge that is both explicitly and implicitly defined within harmonized ontologies (ABC, MPEG-7, FUSION, OME), enable more sophisticated querying to be performed in the natural language of the community that is attempting to mine knowledge from the multimedia content.

For example, consider the following, which is typical of the kind of queries carried out by fuel cell analysts: 'How does the mean catalyst size in electrodes of width < 20 microns affect electrode conductivity?'.

In order to answer this query we need to:

1. find all electrodes of width < 20 microns;
2. find performance/conductivity data for those fuel cells;
3. find the mean catalyst size for those electrodes.

These three steps are described in detail below.

Step 1: Find all electrodes with width < 20 microns

- The image regions which represent electrodes are determined by applying inferencing rules to the fuel cell images.
- The width is calculated from the `feature.length` property value for image regions that are electrodes. This may be difficult due to different magnifications of the microscopy images, and the need for conversions between units, e.g. 16 500 nm = 16.5 microns, which is < 20 microns and fits the requirements.
- The `image.xml` files are retrieved for images which contain electrodes of width < 20 microns.
- The `image.xml` files are linked to the `microscopy.xml` files which contain the Fuel Cell ID.

Step 2: Find performance/conductivity data for those fuel cells

- The Fuel Cell ID is used to access the related `performance.xml` data that records the AC impedance test results, which describe electrode conductivity.

Step 3: Find the mean catalyst size for those electrodes

- Using the Fuel Cell ID, all related images can be retrieved.
- Using the rule which describes which features are catalysts, their area (which is a sub-class of `size`; this is known from the ontology) can be derived.
- Using the MathML encoded rule for calculating `mean`, the mean value for all of the returned catalyst sizes for each electrode can be calculated.

The final step is to present all of the relevant related retrieved data in an integrated, synchronized and coherent presentation which will enable scientists to detect trends or patterns or visualize the results of hypotheses that would not be possible through traditional search interfaces. An earlier research prototype, developed through a collaboration between DSTC and CWI [35], will be refined specifically to provide the presentation engine for this project. The system dynamically generates multimedia presentations in the Synchronized Multimedia Integration Language (SMIL) 2.0 [36] format by aggregating retrieved results sets based on the semantic relationships between them (which have been inferred by applying predefined inferencing rules to the associated metadata) (Figure 3.10). CWI's Cuypers system [37] is used to apply constraints such as page size, bandwidth, user preferences etc., to adapt the presentation components to the user's device capabilities and platform. In the earlier OAI-based prototype [8], we hardwired the mappings from semantic relationships between retrieved digital objects to spatio-temporal relationships in the presentations. For the FUSION project, we plan to develop a graphical user interface which allows the fuel cell experts to interactively map semantic relationships to their preferred spatio-temporal presentation modes. For example, sequences of images, in increasing order of magnification, may be displayed either as a slide show or tiled horizontally or vertically and either to the left or right of corresponding performance

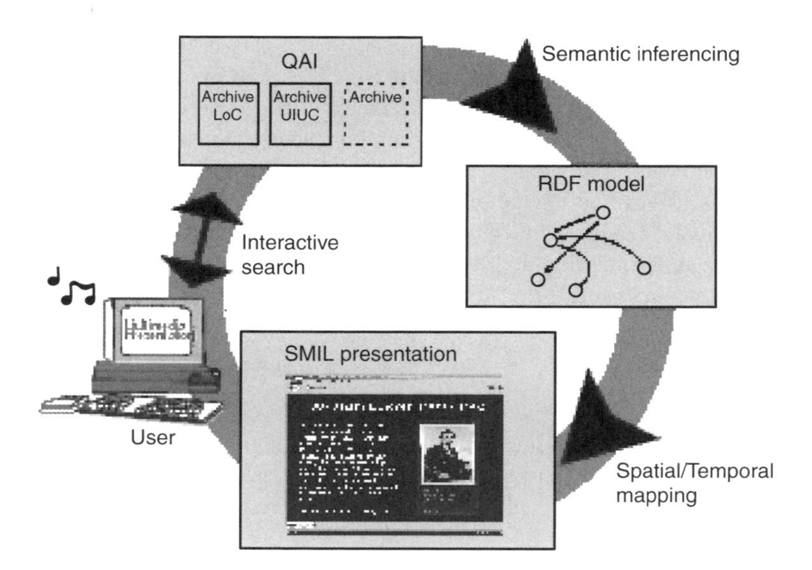

Figure 3.10 The automatic SMIL presentation generator

data. The domain experts can interactively define and modify their preferred modes of layout and presentation and their preferences will be reflected in the results of their next search.

For example, the SMIL visualization tool will be useful in enabling a better understanding of how fuel cells degrade over time. Sequences of micrographic images retrieved from identically manufactured fuel cells after increasing duration of usage could be displayed as animations, synchronized in parallel with the display of corresponding performance data. When SMIL presentations reveal new information or previously unrecognized patterns or trends, users will be able to save and annotate them, for later retrieval, as evidence to support new hypotheses or theories.

3.5 Conclusions

In this chapter, we began by describing the reasons why a machine-understandable MPEG-7 ontology is required if the knowledge and information within multimedia content is going to be exploited to its full potential. We then described the methodology, problems encountered and results of building an OWL representation of the MPEG-7 vocabulary. Finally, we described how we are using the resulting MPEG-7 ontology within an eScience project to enable the automatic generation of semantic descriptions of microscopic images and the development of more sophisticated knowledge mining and information integration services.

Although the inferencing work which we have described in this chapter is still at a relatively preliminary stage, we have defined and developed an overall framework, based on Semantic Web technologies, that will enable the semantic indexing and retrieval of multimedia content, and more specifically scientific image content. We believe that the proposed application of Semantic Web technologies will enhance prior research which has already demonstrated that semantic indexing of multimedia is improved by combining a variety of low-level audiovisual cues automatically extracted from the content. Employing formal languages (XML Schema, OWL), standardized metadata schemas (MPEG-7, OME), ontologies and inferencing rules to deduce high-level semantic descriptions from low-level features will expedite the semantic indexing process for multimedia content, enhance its potential interoperability and hence enhance its ability to be discovered, processed, filtered, exchanged and combined by people, machines, services and agents on the web. Furthermore, the proposed architecture decouples the automatic feature extraction tools from the domain-specific inferencing rules, ontologies, and semantic querying and presentation engines, thus maximizing the system's flexibility, scalability and extensibility so that technological advances in the future can easily be incorporated.

In the future we plan to:

- develop a graphical rule editing tool which supports the definition of rules (in RuleML) using Query-By-Example (QBE)-type tools, e.g. a StillRegion of an image is a cathode if its colour is like this, its texture is like this and its shape is like this;
- investigate the integration of sample database sets manually annotated by domain experts, which can be used to improve the initial rule definitions, i.e. implement a rule-based system which learns and adapts based on users' manual annotations and feedback;

- investigate and compare alternative database or repository options for storing the data, metadata and images, e.g. D-Space, Tamino or a traditional relational database such as Oracle or MySQL;
- further develop the system as a web services architecture by enabling both the automatic feature extraction tools and the sets of inferencing rules and Mandarax reasoning engine to be accessed or discovered as web services and made available to all members of a particular community. Such an architecture enables collaborative development and sharing of rules for the benefit of the whole community but still allows the researchers to retain their (copyright) images and data;
- develop a user interface to the SMIL presentation/visualization engine which enables the users to define their preferred mappings from semantic relationships between atomic multimedia resources to spatio-temporal presentation modes;
- continue the implementation of a prototype knowledge management system for the fuel cell community and carry out detailed evaluation and usability studies and system refinements.

References

[1] J. Martinez, Overview of the MPEG-7 Standard (version 5.0), ISO/IEC JTC1/SC29/WG11 N4031, Singapore, March 2001. Available at http://www.cselt.it/mpeg/standards/mpeg-7/mpeg-7.htm.

[2] XML Schema Part 0: Primer, W3C Recommendation, 2 May 2001. Available at http://www.w3.org/TR/xmlschema-0.

[3] XML Schema Part 1: Structures, W3C Recommendation, 2 May 2001. Available at http://www.w3.org/TR/xmlschema-1/.

[4] XML Schema Part 2: Datatypes, W3C Recommendation, 2 May 2001. Available at http://www.w3.org/TR/xmlschema-2/.

[5] ISO/IEC 15938-2 FCD Information Technology—Multimedia Content Description Interface—Part 2: Description Definition Language, March 2001, Singapore.

[6] TV-Anytime Forum, http://www.tv-anytime.org/.

[7] MPEG-21 Multimedia Framework, http://www.cselt.it/mpeg/public/mpeg-21_pdtr.zip.

[8] NewsML, http://www.newsml.org/.

[9] Dublin Core Metadata Element Set, Version 1.1, 2 July 1999. Available at http://www.purl.org/dc/documents/rec-dces-19990702.htm.

[10] G. Rust, M. Bide, The indecs Metadata Schema Building Blocks. *Indecs Metadata Model*, November, 1999. Available at http://www.indecs.org/results/model.htm.

[11] Content Standard for Digital Geospatial Metadata (CSDGM), http://www.fgdc.gov/metadata/contstan.html.

[12] The Gateway to Educational Materials, http://www.the gateway.org.

[13] IEEE Learning Technology Standards Committee's Learning Object Meta-data Working Group. Version 3.5 Learning Object Meta-data Scheme.

[14] ICOM/CIDOC Documentation Standards Group, Revised Definition of the CIDOC Conceptual Reference Model, September 1999. Available at http://www.geneva-city.ch:80/musinfo/cidoc/oomodel.

[15] W3C, Web-Ontology (WebOnt) Working Group, http://www.w3.org/2001/sw/WebOnt/.

[16] W3C, OWL Web Ontology Language Overview, W3C Candidate Recommendation, 18 August 2003, edited by Deborah L. McGuinness and Frank van Harmelen. Available at http://www.w3.org/TR/owl-features-20030818.

[17] UML Resource Center, http://www.rational.com/uml/index.jsp.

[18] ISO/IEC 15938-5 FCD Information Technology—Multimedia Content Description Interface—Part 5: Multimedia Description Schemes, March 2001, Singapore.

[19] ISO/IEC 15938-3 FCD Information Technology—Multimedia Content Description Interface—Part 3: Visual, March 2001, Singapore.

[20] ISO/IEC 15938-4 FCD Information Technology—Multimedia Content Description Interface—Part 4: Audio, March 2001, Singapore.

[21] DAML +OIL, March 2001, http://www.daml.org/2001/03/daml+Oil-index.

[22] RDF Schema Specification 1.0, W3C Candidate Recommendation, 27 March 2000. Available at http://www.w3.org/TR/rdf-schema/.
[23] Open Archives Initiative (OAI), http://www.openarchives.org/.
[24] S. Little, J. Geurts, J. Hunter, Dynamic generation of intelligent multimedia presentations through semantic inferencing. In M. Agosti, C. Thanos (eds) *Research and Advances Technology for Digital Technology: 6th European Conference, ECDL 2002*, Rome, Italy, 16–18 September 2002, pp. 158–175. Lecture Notes in Computer Science no. 2458. Springer-Verlag, Heidelberg, 2002.
[25] E. Miller, The W3C Semantic Web activity, presented at the International Semantic Web Workshop, 30–31 July, 2001, Stanford University, California.
[26] W3C, XML Schema Part 0: Primer, W3C Recommendation, 2 May 2001, edited by David C. Fallside. Available at http://www.w3.org/TR/xmlschema-0/.
[27] W3C, XML Schema Part 1: Structures, W3C Recommendation, 2 May 2001, edited by Henry S. Thompson, David Beech, Murray Maloney and Noah Mendelsohn. Available at http://www.w3.org/TR/xmlschema-1.
[28] W3C, XML Schema Part 2: Datatypes, W3C Recommendation, 2 May 2001, edited by Paul V. Biron and Ashok Malhotra. Available at http://www.w3.org/TR/xmlschema-2/.
[29] O. Lassila, R.R Swick, Resource Description Framework (RDF) Model and Syntax Specification, W3C Recommendation, 22 February 1999. Available at http://www.w3.org/TR/REC-rdfsyntax/.
[30] Open Microscopy Environment (OME), http://www.openmicroscopy.org/.
[31] C. Lagoze, J. Hunter, The ABC ontology and model. *Journal of Digital Information*, **2**(2), 2001.
[32] Y. Deng, B.S. Manjunath, Content-based search of video using color, texture and motion. In *Proceedings of the IEEE International Conference on Image Processing*, vol. 2, pp. 13–16, Santa Barbara, CA, October 1997.
[33] S. Marchand-Maillet, Content-based video retrieval: an overview, technical report no. 00.06, CUI—University of Geneva, Geneva, Switzerland, 2000.
[34] M.R. Naphade, I. Kozintsev, T.S. Huang, K. Ramchandran, A factor graph framework for semantic indexing and retrieval in video. In *IEEE Workshop on Content-based Access of Image and Video Libraries (CBAIVL'00)*, Hilton Head, South Carolina, 16 June 2000.
[35] S.F. Chang, W. Chen, H. Sundaram, Semantic visual templates: linking visual features to semantics. In *IEEE International Conference on Image Processing (ICIP '98)*, Chicago, IL, pp. 531–535, 1998.
[36] B. Adams, G. Iyengar, C. Lin, M. Naphade, C. Neti, H. Nock, J. Smith, Semantic indexing of multimedia content using visual, audio, and text cues. *EURASIP Journal on Applied Signal Processing*, **2003**(2), 170–185, 2003.
[37] R. Zhao, W.I. Grosky, Negotiating the semantic gap: from feature maps to semantic landscapes. *Pattern Recognition*, **35**(3), 51–58, 2002.

Appendix A

Appendix A.1 MPEG-7 Segment Class Hierarchy

```
<owl:Class rdf:ID="Segment">
  <rdfs:label>Image</rdfs:label>
  <rdfs:comment>The class of images</rdfs:comment>
  <rdfs:subClassOf rdf:resource="#MultimediaContent"/>
</owl:Class>
<owl:Class rdf:ID="StillRegion">
  <rdfs:label>StillRegion</rdfs:label>
  <rdfs:comment>2D spatial regions of an image or video frame
</rdfs:comment>
  <rdfs:subClassOf rdf:resource="#Segment"/>
```

```
    <rdfs:subClassOf rdf:resource="#Image"/>
  </owl:Class>
  <owl:Class rdf:ID="ImageText">
    <rdfs:label>ImageText</rdfs:label>
    <rdfs:comment>Spatial regions of an image or video framethat
  correspond to text or captions</rdfs:comment>
    <rdfs:subClassOf rdf:resource="#StillRegion"/>
  </owl:Class>
  <owl:Class rdf:ID="StillRegion3D">
    <rdfs:label>StillRegion3D</rdfs:label>
    <rdfs:comment>3D spatial regions of a 3D image</rdfs:comment>
    <rdfs:subClassOf rdf:resource="#Segment"/>
    <rdfs:subClassOf rdf:resource="#Image"/>
  </owl:Class>
  <owl:Class rdf:ID="VideoSegment">
    <rdfs:label>VideoSegment</rdfs:label>
    <rdfs:comment>Temporal intervals or segments of video data
  </rdfs:comment>
    <rdfs:subClassOf rdf:resource="#Segment"/>
    <rdfs:subClassOf rdf:resource="#Video"/>
  </owl:Class>
  <owl:Class rdf:ID="MovingRegion">
    <rdfs:label>MovingRegion</rdfs:label>
    <rdfs:comment>2D spatio-temporal regions of video data
  </rdfs:comment>
    <rdfs:subClassOf rdf:resource="#Segment"/>
  </owl:Class>
  <owl:Class rdf:ID="VideoText">
    <rdfs:label>VideoText</rdfs:label>
    <rdfs:comment>Spatio-temporal regions of video data that
  correspond to text or captions</rdfs:comment>
    <rdfs:subClassOf rdf:resource="#MovingRegion"/>
  </owl:Class>
  <owl:Class rdf:ID="AudioSegment">
    <rdfs:label>AudioSegment</rdfs:label>
    <rdfs:comment>Temporal intervals or segments of audio data
  </rdfs:comment>
    <rdfs:subClassOf rdf:resource="#Segment"/>
    <rdfs:subClassOf rdf:resource="#Audio"/>
  </owl:Class>
  <owl:Class rdf:ID="AudioVisualSegment">
    <rdfs:label>AudioVisualSegment</rdfs:label>
    <rdfs:comment>Temporal intervals or segments of audiovisual
  data</rdfs:comment>
    <rdfs:subClassOf rdf:resource="#Segment"/>
    <rdfs:subClassOf rdf:resource="#AudioVisual"/>
  </owl:Class>
  <owl:Class rdf:ID="AudioVisualRegion">
    <rdfs:label>AudioVisualRegion</rdfs:label>
```

```
  <rdfs:comment>Arbitrary spatio-temporal segments of AV data
</rdfs:comment>
  <rdfs:subClassOf rdf:resource="#Segment"/>
</owl:Class>
<owl:Class rdf:ID="MultimediaSegment">
  <rdfs:label>MultimediaSegment</rdfs:label>
  <rdfs:comment>Segment of a composite multimedia presentation
</rdfs:comment>
  <rdfs:subClassOf rdf:resource="#Multimedia"/>
  <rdfs:subClassOf rdf:resource="#Segment"/>
</owl:Class>
```

Appendix A.2 MPEG-7 Decomposition Property Hierarchy

```
<rdf:Property rdf:ID="decomposition">
   <rdfs:label>decomposition of a segment</rdfs:label>
   <rdfs:domain rdf:resource="#MultimediaContent"/>
   <rdfs:range rdf:resource="#Segment"/>
</rdf:Property>
<rdf:Property rdf:ID="spatial_decomposition">
   <rdfs:label>spatial decomposition of a segment</rdfs:label>
   <rdfs:subPropertyOf rdf:resource="#decomposition"/>
   <rdfs:domain rdf:resource="#MultimediaContent"/>
   <rdfs:range rdf:resource="#Segment"/>
</rdf:Property>
<rdf:Property rdf:ID="temporal_decomposition">
   <rdfs:label>temporal decomposition of a segment</rdfs:label>
   <rdfs:subPropertyOf rdf:resource="#decomposition"/>
   <rdfs:domain rdf:resource="#MultimediaContent"/>
   <rdfs:range rdf:resource="#Segment"/>
</rdf:Property>
<rdf:Property rdf:ID="spatio-temporal_decomposition">
   <rdfs:label>spatio-temporal decomposition of a segment
   </rdfs:label>
   <rdfs:subPropertyOf rdf:resource="#decomposition"/>
   <rdfs:domain rdf:resource="#MultimediaContent"/>
   <rdfs:range rdf:resource="#Segment"/>
</rdf:Property>
<rdf:Property rdf:ID="mediaSource_decomposition">
   <rdfs:label>media source decomposition of a segment
   </rdfs:label>
   <rdfs:subPropertyOf rdf:resource="#decomposition"/>
   <rdfs:domain rdf:resource="#MultimediaContent"/>
   <rdfs:range rdf:resource="#Segment"/>
</rdf:Property>
```

Appendix A.3 VideoSegment Decomposition Property Hierarchy

```
<rdf:Property rdf:ID="videoSegment_spatial_decomposition">
   <rdfs:label>spatial decomposition of a video segment
</rdfs:label>
   <rdfs:subPropertyOf rdf:resource="#spatial_decomposition"/>
   <rdfs:domain rdf:resource="#VideoSegment"/>
   <rdfs:range rdf:resource="#MovingRegion"/>
</rdf:Property>
<rdf:Property rdf:ID="videoSegment_temporal_decomposition">
   <rdfs:label>temporal decomposition of a video segment
</rdfs:label>
   <rdfs:subPropertyOf rdf:resource="#temporal_decomposition"/>
   <rdfs:domain rdf:resource="#VideoSegment"/>
   <rdfs:range rdf:resource="#VideoSegment"/>
   <rdfs:range rdf:resource="#StillRegion"/>
</rdf:Property>
<rdf:Property rdf:ID="videoSegment_spatio-temporal
_decomposition">
   <rdfs:label>spatio-temporal decomposition of a video segment
</rdfs:label>
   <rdfs:subPropertyOf rdf:resource="#spatio-temporal
_decomposition"/>
   <rdfs:domain rdf:resource="#VideoSegment"/>
   <rdfs:range rdf:resource="#MovingRegion"/>
   <rdfs:range rdf:resource="#StillRegion"/>
</rdf:Property>
<rdf:Property rdf:ID="videoSegment_mediaSource_decomposition">
   <rdfs:label>media source decomposition of a video segment
</rdfs:label>
   <rdfs:subPropertyOf rdf:resource="#mediaSource_decomposition"/>
   <rdfs:domain rdf:resource="#VideoSegment"/>
   <rdfs:range rdf:resource="#VideoSegment"/>
</rdf:Property>
```

Appendix B MPEG-7 Description of a Fuel Cell

```
<?xml version="1.0" encoding="iso-8859-1"?>
<Mpeg7 xmlns="urn:mpeg:mpeg7:schema:2001"
      xmlns:xsi="http://www.w3.org/2001/XMLSchema-instance"
      xmlns:mpeg7="urn:mpeg:mpeg7:schema:2001"
      xmlns:xml="http://www.w3.org/XML/1998/namespace"
      xsi:schemaLocation="urn:mpeg:mpeg7:schema:2001
.\Mpeg7-2001.xsd">
```

```
   <Description xsi:type="ContentEntityType">
      <MultimediaContent xsi:type="ImageType">
         <Image>
            <MediaLocator>
               <MediaUri>http://metadata.net/FUSION/
media/image012.jp2
               </MediaUri>
            </MediaLocator>
            <CreationInformation>
               <CreationUri>http://metadata.net/FUSION/
microscopy.xml
               </CreationUri>
            </CreationInformation>
            <MediaInformation>
               <MediaProfile>
                  <MediaFormat>
                     <Content href="urn:mpeg:mpeg7:cs:ContentCS
:2001:2">
                        <Name>visual</Name>
                     </Content>
                     <FileFormat href="urn:mpeg:mpeg7:cs
:FileFormatCS:2001:3">
                        <Name xml:lang="en">JPEG2000</Name>
                     </FileFormat>
                     <FileSize>10483</FileSize>
                     <VisualCoding>
                        <Format href="urn:mpeg:mpeg7:cs
:VisualCodingFormatCS:2001:1"  colorDomain="binary">
                           <Name xml:lang="en">JPEG2000</Name>
                        </Format>
                        <Pixel aspectRatio="0.75" bitsPer="8"/>
                        <Frame height="288" width="352"
rate="25"/>
                     </VisualCoding>
                  </MediaFormat>
               </MediaProfile>
            </MediaInformation>
            <Object id="object1">
               <Mask xsi:type="SpatialMaskType">
                  <SubRegion>
                     <Polygon>
                        <Coords mpeg7:dim="2 5"> 5 25 10 20 15
15 10 10 5 15</Coords>
                     </Polygon>
                  </SubRegion>
               </Mask>
               <VisualDescriptor xsi:type="ScalableColorType"
numOfCoeff="16" numOfBitplanesDiscarded="0">
                  <Coeff> 1 2 3 4 5 6 7 8 9 0 1 2 3 4 5 6
</Coeff>
```

```
                </VisualDescriptor>
                <VisualDescriptor xsi:type="RegionShapeType">
                   <MagnitudeOfART>0 0 0 0 0 0 0 0 0 0 0 0 0 0 0
0 0 0 0 0 0 0 0 0 0 0 0 0 0 0 0 0 0 0 0 </MagnitudeOfART>
                </VisualDescriptor>
                <VisualDescriptor xsi:type=
"HomogenousTextureType">
                   ...
                </VisualDescriptor>
             </Object>
          </Image>
       </MultimediaContent>
    </Description>
</Mpeg7>
```

Appendix C OME Description of Fuel Cell Image

```
<?xml version="1.0" encoding="UTF-8"?>
<OME xmlns="http://www.openmicroscopy.org/XMLschemas/OME/FC/
ome.xsd"
xmlns:Bin="http://.../XMLschemas/BinaryFile/RC1/BinaryFile.xsd"
xmlns:xsi="http://www.w3.org/2001/XMLSchema-instance"
xsi:schemaLocation="http://...ome.../XMLschemas/OME/FC/ome.xsd
                    http://...ome.../XMLschemas/OME/FC/ome.xsd
                    http://...ome.../XMLschemas/STD/RC2/STD.xsd">
...
   <Instrument ID="urn...Instrument:123456">
      <Microscope Manufacturer="Zeiss" Model="foo"
SerialNumber="bar" Type="Upright"/>
      <LightSource ID="urn...LightSource:123456"
Manufacturer="Olympus" Model="WMD Laser" SerialNumber="1...4">
         <Laser Type="Semiconductor" Medium="GaAs">
            <Pump ID="urn...LightSource:123789"/>
         </Laser>
      </LightSource>
      <LightSource ID="urn...LightSource:123123" Manufacturer=
"Olympus" Model="Realy Bright Lite" SerialNumber="1jhf16">
         <Arc Type="Hg"/>
      </LightSource>
      <Detector ID="urn...Detector:123456" Type="CCD"
Manufacturer="Kodak" Model="Instamatic"
SerialNumber="frf8u198"/>
      <Objective ID="urn...Objective:123456" Manufacturer=
"Olympus" Model="SPlanL" SerialNumber="456anxcoas123">
```

```
          <LensNA>2.4</LensNA>
          <Magnification>40.0</Magnification>
       </Objective>
       <Filter ID="urn...Filter:123456">
          <FilterSet Manufacturer="Omega" Model="SuperGFP"
LotNumber="123LJKHG123"/>
       </Filter>
       <OTF ID="urn...OTF:123456" PixelType="int8"
OpticalAxisAvrg="true" SizeX="512" SizeY="512">
          <ObjectiveRef ID="urn...Objective:123456"/>
          <FilterRef ID="urn...Filter:123456"/>
          <Bin:External Compression="bzip2" SHA1="012...456"
href="OTF123.otf"/>
       </OTF>
    </Instrument>
...
</OME>
```

Appendix D FUSION Description of a Fuel Cell Image

```
<?xml version="1.0" encoding="UTF-8"?>
<fuelcell xmlns="http://metadata.net/fusion/FUSION">
   <anode>
      <density context="fusion:catalyst" unit="per micron
squared">42
      </density>
      <porosity unit="microns">42</porosity>
      <surfaceArea context="fusion:fuelcell" unit="microns
squared">42
      </surfaceArea>
      <surfaceArea context="fusion:electrolyte" unit="microns
squared">42
      </surfaceArea>
      <width>
         <mean unit="microns">42</mean>
         <sdev unit="microns">8</sdev>
      </width>
      <catalyst id="1">
         <density context="fusion:activeMetal" unit="per micron
squared">42
         </density>
         <nearestNeighbour degree="1" unit="">42
</nearestNeighbour>
         <shape>circle</shape>
         <size unit="microns">42</size>
         <surfaceArea unit="microns squared">42</surfaceArea>
```

```
            <size context="fusion:activeMetal">
               <mean unit="microns">42</mean>
               <sdev unit="microns">8</sdev>
            </size>
            <activeMetal id="1">
               <nearestNeighbour degree="1" unit="microns">42
 </nearestNeighbour>
               <shape>circle</shape>
               <size unit="microns">42</size>
               <surfaceArea unit="microns squared">42</surfaceArea>
            </activeMetal>
            <catalystSupport id="1">
               <composition>carbon</composition>
               <porosity unit="percentage">42</porosity>
               <size unit="microns">42</size>
            </catalystSupport>
         </catalyst>
      </anode>
      <electrolyte>
         <conductivity unit="">42</conductivity>
         <composition>nafion</composition>
         <porosity unit="percentage">42</porosity>
         <width>
            <mean unit="microns">42</mean>
            <sdev unit="microns">8</sdev>
         </width>
      </electrolyte>
      <cathode>same as anode</cathode>
</fuelcell>
```

Appendix E XML Schema for FUSION

```
<?xml version="1.0" encoding="UTF-8"?>
<xs:schema xmnls="http://www.w3.org/2001/XMLSchema"
xmnls:xs="http://www.w3.org/2001/XMLSchema"
xmlns:xx="http://www.example.org/XMLRDFSchemaBridge"
targetNamespace="http://metadata.net/fusion/FUSION">
   <xs:annotation>
      <xs:documentation>Draft XML Schema for FUSION
</xs:documentation>
   </xs:annotation>
   <xs:element name="fuelcell" type="fuelcell"/>
   <xs:complexType name="fuelcell" xx:semantics="http:
//metadata.net/fusion/FUSION#fuelcell">
      <xs:sequence>
```

```
        <xs:element name="anode" type="anodeType"/>
        <xs:element name="electrolyte" type="electrolyteType"/>
        <xs:element name="cathode" ref="cathodeType"/>
     </xs:sequence>
     <xs:attribute name="id" use="required"/>
  </xs:complexType>
  <xs:complexType name="anodeType" xx:semantics="http:
//metadata.net/fusion/FUSION#anode">
     <xs:sequence>
        <xs:element name="density" type="densityType"/>
        <xs:element name="porosity" type="porosityType"/>
        <xs:element name="surfaceArea" type="surfaceAreaType"/>
        <xs:element name="width" type="widthType"/>
        <xs:element name="catalyst" type="catalystType"/>
     </xs:sequence>
  </xs:complexType>
  <xs:complexType name="catalystType" xx:semantics="http:
//metadata.net/fusion/FUSION#catalyst">
     <xs:sequence>
        <xs:element name="density" type="densityType"/>
        <xs:element name="nearestNeighbour"
type="nearestNeighbourType"/>
        <xs:element name="shape" type="shapeType"/>
        <xs:element name="size" type="sizeType"/>
        <xs:element name="surfaceArea" type="surfaceAreaType"/>
        <xs:element name="activeMetal" type="activeMetalType"/>
        <xs:element name="catalystSupport"
type="catalystSupportType"/>
     </xs:sequence>
     <xs:attribute name="id" use="required"/>
  </xs:complexType>
  <xs:complexType name="activeMetalType">
     <xs:sequence>
        <xs:element name="nearestNeighbour"
type="nearestNeighbourType"/>
        <xs:element name="shape" type="shapeType"/>
        <xs:element name="size" type="sizeType"/>
        <xs:element name="surfaceArea" type="surfaceAreaType"/>
     </xs:sequence>
     <xs:attribute name="id" use="required"/>
  </xs:complexType>
  <xs:complexType name="catalystSupportType">
     <xs:sequence>
        <xs:element name="composition" type="compositionType"/>
        <xs:element name="porosity" type="porosityType"/>
        <xs:element name="size" type="sizeType"/>
     </xs:sequence>
  </xs:complexType>
  <xs:complexType name="electrolyteType" xx:semantics="http:
```

```
//metadata.net/fusion/FUSION#electrolyte">
        <xs:sequence>
            <xs:element name="conductivity" type="conductivityType"/>
            <xs:element name="composition" type="compositionType"/>
            <xs:element name="porosity" type="porosityType"/>
            <xs:element name="width" type="widthType"/>
        </xs:sequence>
    </xs:complexType>
    <xs:complexType name="cathodeType" xx:semantics="http:
//metadata.net/fusion/FUSION#cathode">
        <xs:sequence>
            <xs:element name="density" type="densityType"/>
            <xs:element name="porosity" type="porosityType"/>
            <xs:element name="surfaceArea" type="surfaceAreaType"/>
            <xs:element name="width" type="widthType"/>
            <xs:element name="catalyst" type="catalystType"/>
        </xs:sequence>
    </xs:complexType>
    <xs:simpleType name="compositionType" xx:semantics="http:
//metadata.net/fusion/FUSION#composition">
        <restriction base="string"/>
    </xs:simpleType>
    <xs:complexType name="conductivityType" xx:semantics="http:
//metadata.net/fusion/FUSION#conductivity">
        <xs:attribute name="unit" type="xs:string"/>
    </xs:complexType>
    <xs:complexType name="densityType" xx:semantics="http:
//metadata.net/fusion/FUSION#density">
        <xs:attribute name="context" type="xs:anyURI"
use="required"/>
        <xs:attribute name="unit" type="xs:string"/>
    </xs:complexType>
    <xs:complexType name="nearestNeighbourType" xx:semantics=
"http://metadata.net/fusion/FUSION#nearestNeighbour">
        <xs:attribute name="degree" type="xs:integer"
use="required"/>
        <xs:attribute name="unit" type="xs:string"/>
    </xs:complexType>
    <xs:complexType name="porosityType" xx:semantics="http:
//metadata.net/fusion/FUSION#porosity">
        <xs:attribute name="unit" type="xs:string"/>
    </xs:complexType>
    <xs:simpleType name="shapeType" xx:semantics="http:
//metadata.net/fusion/FUSION#shape">
        <restriction base="string"/>
    </xs:simpleType>
    <xs:complexType name="sizeType" xx:semantics="http:
//metadata.net/fusion/FUSION#size">
        <xs:sequence>
            <xs:element name="mean" type="meanType"/>
```

```
            <xs:element name="sdev" type="sdevType"/>
        </xs:sequence>
        <xs:attribute name="unit" type="xs:string"/>
    </xs:complexType>
    <xs:complexType name="surfaceAreaType" xx:semantics="http:
//metadata.net/fusion/FUSION#surfaceArea">
        <xs:attribute name="context" type="xs:anyURI"
use="required"/>
        <xs:attribute name="unit" type="xs:string"/>
    </xs:complexType>
    <xs:complexType name="widthType" xx:semantics="http:
//metadata.net/fusion/FUSION#width">
        <xs:sequence>
            <xs:element name="mean" type="meanType"/>
            <xs:element name="sdev" type="sdevType"/>
        </xs:sequence>
    </xs:complexType>
    <xs:complexType name="meanType" xx:semantics="http:
//metadata.net/fusion/FUSION#mean">
        <xs:attribute name="unit" type="xs:string"/>
    </xs:complexType>
    <xs:complexType name="sdevType" xx:semantics="http:
//metadata.net/fusion/FUSION#sdev">
        <xs:attribute name="unit" type="xs:string"/>
    </xs:complexType>
</xs:schema>
```

4

A Fuzzy Knowledge-Based System for Multimedia Applications

Vassilis Tzouvaras, Giorgos Stamou and Stefanos Kollias

4.1 Introduction

Currently, computers are changing from single isolated devices into entry points to the global network of information exchange and business transactions called the World Wide Web (WWW). For this reason, support data, information and knowledge exchange have become key issues in current computer technology. The Semantic Web (SW) will provide intelligent access to heterogeneous, distributed information.

In parallel to the SW advancements, the representation and management of uncertainty, imprecision and vague knowledge that exists in real-life applications has received considerable attention in the artificial intelligence (AI) community in an attempt to extend existing knowledge representation systems to deal with the imperfect nature of real-world information. Furthermore, a lot of work has been carried out on the development of reasoning engines that can interpret imprecise knowledge [1].

New applications in the context of the SW and AI have renewed interest in the creation of knowledge-based systems that can represent and interpret imprecise and vague information. A knowledge-based system, as the name suggests, is a computer-based system that emulates the reasoning process of a human expert within a specific domain of knowledge. Knowledge-based systems are primarily built for the purpose of making the experience, understanding and problem solving capabilities of the expert in a particular subject area available to the non-expert in this area. In addition, they may be designed for various specific activities, such as consulting, diagnosis, learning, decision support, design, planning and research.

A typical architecture of a knowledge-based system consists of a knowledge base and an inference engine. These two units, together with some interface for communicating with the user, form the minimal configuration that is a called a knowledge-based system.

Multimedia Content and the Semantic Web Edited by Giorgos Stamou and Stefanos Kollias

The proposed knowledge base has a hybrid structure. It consists of the general knowledge pertaining to the domain of interest and the propositional rules, which are used to infer implicit knowledge.

For the representation of the general knowledge, we used Description Logics (DLs). In the past decade a substantial amount of work has been carried out in the context of DLs. DLs are logical reconstructions of the so-called frame-based knowledge representation languages, with the aim of providing a simple, well-established Tarski-style [2] declarative semantics to capture the meaning of the most popular features of structured representation of knowledge. An important point is that DLs are considered to be attractive in knowledge-based systems because they are a good compromise between power and computational complexity.

Experience in using DLs in applications has shown that in many cases we would like to extend the representational and reasoning capabilities of them. In particular, the use of DLs in the context of multimedia points out the necessity of extending DLs with capabilities which allow the treatment of the inherent imprecision in multimedia object representation, retrieval and detection [1]. In fact, classical DLs are insufficient for describing multimedia situations since retrieval matching and detection are not usually true or false situations.

Along with the general knowledge, the knowledge base consists of the propositional rules. These rules are in the form '*IF A then B*', where *A* and *B* are called logic variables. Propositional rules can be represented using propositional logic. The main concern of propositional logics (PLs) is the study of rules by which new logic variables can be produced as functions of some given logic variables. The difference between DLs and PLs is that PLs are not concerned with the internal structure of the propositions that the logic variables represent.

The inference engine of a knowledge-based system operates on the knowledge base. Its operation is to generate knowledge that is not explicitly recorded in the knowledge base. Since we have a hybrid knowledge base, we perform two types of inferencing. The inference engine applied on the general knowledge is based on the tableaux calculus [1]. For the interpretation of the propositional rules, there exist two approaches. The first is data driven and is exemplified by the generalized modus ponens tautology. In this case available data are supplied to the knowledge-based system, which then uses them to evaluate relevant propositional rules and draw all possible conclusions. An alternative method of evaluation is goal driven. It is exemplified by the generalized modus tollens tautology of logical inference. Here the system searches for data specified in the If-clauses of inference rules that will lead to the objective. These data are found either in the knowledge base, in the Then-clauses of other propositional rules or by querying the user.

To this end, in this chapter, we propose an extended version of DLs with fuzzy capabilities, for the representation of the general knowledge. Moreover, we propose a fuzzy extension of propositional rules using the generalized fuzzy modus ponens tautology. Finally, a hybrid inference engine that is capable of interpreting fuzzy propositional rules is proposed [3]. This inference engine is supported by an adaptation algorithm that can learn and adapt the weights of the rules using predefined input–output data.

The proposed extensions have been tested in a facial expression recognition study. We have constructed fuzzy propositional rules using the Facial Expression Parameters (FAPs) defined in the MPEG-4 standard [4]. The continuity of emotion space, the uncertainty involved in the feature estimation process and the required ability of the system to use prior knowledge, while being capable of adapting its behaviour to its users' characteristics, make appropriate the use of the proposed fuzzy extensions and inference engine. In addition, the fuzzy nature of the system is appropriate for multimedia information retrieval since it allows the treatment of inherent imprecision in multimedia object representation and retrieval. Classical multimedia systems are insufficient for describing multimedia retrieval, matching and detecting situations.

The structure of the chapter is as follows. Section 4.2 outlines the structure of the knowledge base. The basic concept of DLs and PLs along with their respective fuzzy extensions are illustrated. Section 4.3 presents the inference engine of the propositional rules and its adaptation algorithm. In Section 4.4 the proposed fuzzy knowledge-based system is demonstrated through a facial expression recognition study.

4.2 Knowledge Base Formalization

The structure of the knowledge base is hybrid. It consists of the general knowledge and the propositional rules. The general knowledge contains concepts and relations pertaining to the domain of interest. The propositional rules are constructed using concepts and relations defined in the general knowledge, and are used to infer higher-level knowledge not explicitly recorded. For the formalization of the general knowledge we use DLs. DLs are chosen since they are a good compromise between power and computational complexity. For the formalization of the inference rules we use PLs. For both DLs and PLs we propose some fuzzy extensions, which are necessary for multimedia applications since multimedia information is vague and uncertain.

4.2.1 Fuzzy extension in DLs

Many researchers in the past have been involved with the use of fuzzy set theory to extend the DLs in order to deal with imprecision [1,2,5].

The specific DL we will extend and propose some fuzzy extension for is *ALC*, a significant representative of DLs. First, we will introduce the basic concepts of *ALC*. Second, we will present our fuzzy extension of *ALC*. In this chapter, we will generalize work done by other researchers [1] in the field of fuzzy DLs.

4.2.2 Classical DLs

Concepts are expressions that collect the properties, described by means of roles, of a set of individuals. From a first order logic (FOL) point of view, concepts can be seen as unary predicates, whereas roles are interpreted as binary predicates.

A concept denoted by C or D, in the language *ALC*, is build out of primitive concepts according to the following syntax rules:

$$
\begin{aligned}
C,\ D \rightarrow\ & \top \mid (top\ concept) \\
& \bot \mid (bottom\ concept) \\
& A \mid (primitive\ concpet) \\
& C \sqcap D \mid (concept\ conjuction) \\
& C \sqcup D \mid (concept\ disjunction) \\
& \neg C \mid (concept\ negation) \\
& \forall R.C \mid (universal\ quantification) \\
& \exists R.C \mid (existential\ quantification)
\end{aligned}
$$

Dls have a clean model-theoretic semantics, based on the notion of interpretation. An interpretation I is a pair $I = (\Delta^I, \cdot^I)$ consisting of a non-empty set Δ^I and an interpretation function $\cdot^I$ mapping different individuals into different elements of Δ^I, primitive concepts into subsets of Δ^I and primitive roles into subsets of $\Delta^I \times \Delta^I$. The interpretation of complex concepts is defined as:

$$
\begin{aligned}
&\top^I = \Delta^I \\
&\bot^I = 0 \\
&(C \sqcap D)^I = C^I \cap D^I \\
&(C \sqcup D)^I = C^I \cup D^I \\
&(\neg C)^I = \Delta^I \backslash C^I \\
&(\forall R.C)^I = \{a \in \Delta^I | \forall b.(a, b) \in R^I \to b \in C^I\} \\
&(\exists R.C)^I = \{a \in \Delta^I | \exists b.(a, b) \in R^I\}.
\end{aligned}
$$

A knowledge base in DLs comprises two components, the *TBox* and the *ABox*. The *TBox* introduces the terminology of an application domain, while the *ABox* contains assertions about named individuals in terms of this vocabulary.

An assertion a is an expression of type $a : C$,which means that a is an instance of C, or an expression of type $(a, b) : R$, which mean that a, b are instances of R. An interpretation I satisfies $a : C$ iff $a^I \in C^I$and $(a, b) : R$ iff $(a^I, b^I) \in R^I$.

In the most general case, terminological axioms have the form

$$C \sqsubseteq D \quad (R \sqsubseteq S) \qquad \text{or} \qquad C \equiv D \quad (R \equiv S)$$

where C, D are concepts and R, S are roles. Axioms of the first kind are called inclusions, while those of the second kind are called equalities. An interpretation I satisfies an inclusion $C \sqsubseteq D$ if $C^I \subseteq D^I$, and it satisfies an equality if $C^I = D^I$. If T is a set of axioms, then I satisfies Tiff Isatisfies each element of T. If I satisfies an axiom, then we say that it is a model of this axiom. Two axioms or two sets of axioms are equivalent if they have the same models.

4.2.3 Fuzzy Extension

The concepts in DLs are interpreted as crisp sets, i.e. an individual either belongs to the set or not. However, many real-life concepts are vague in the sense that they do not have precisely defined membership criteria. The main idea underlying the fuzzy extensions of DLs proposed in [2, 5] is to leave the syntax as it is, but to use fuzzy logic for defining the semantics. Our fuzzy extension generalizes the work done in [1]. Straccia's work is based on Zadeh's work on fuzzy sets. A fuzzy set S with respect to a universe U is characterized by a membership function $\mu_s : U \to [0, 1]$, assigning an S-membership degree, $\mu_s(u)$, to each element u in U. $\mu_s(u)$ gives us an estimation of the belonging of u to S. Typically, if $\mu_s(u) = 1$ then u definitely belongs to S, while $\mu_s(u) = 0.8$ means that u is likely to be an element of S. The membership function has to satisfy three restrictions. The minimum, the maximum and the complement operators, called the standard fuzzy operations, perform in precisely the same way as the corresponding operations for classical sets when the range of membership grades is restricted to the set $\{0, 1\}$. That is, the standard fuzzy operations are generalizations of the corresponding classical set operations. However, for each of the three operators, there exists a broad class of functions whose members qualify as generalizations of the classical operation as well. Straccia proposed the standard fuzzy operators. In our extension triangular norms and conorms operators are considered as fuzzy intersections and fuzzy unions, respectively. Also, the general fuzzy complement is considered instead of the standard fuzzy complement $(1 - a)$. Each of the classes of the operators is characterized by properly justified axioms. A t-norm (triangular norm) is a function t: $[0, 1] \times [0, 1] \rightarrow [0, 1]$ satisfying for any $a, b, d \in [0, 1]$ the following four conditions:

1. $t(a, 1) = a$ and $t(a, 0) = 0$
2. $b \leq d$ implies $t(a, b) \leq t(a, d)$
3. $t(a, b) = t(b, a)$
4. $t(a, t(b, d)) = t(t(a, b), d)$

Moreover, it is called *Archimedean* iff:

1. t is a continuous function
2. $t(a, a) < a, \forall a \in (0, 1)$ (idempotent).

An s-norm is a function $\sigma : [0, 1] \times [0, 1] \rightarrow [0, 1]$ satisfying for any $a, b, d \in [0, 1]$2–4 of the t-norm conditions and a dual boundary condition:

$$s(a, 0) = 0 \quad \text{and} \quad s(a, 1) = 1$$

Having defined the fuzzy set intersection and union, we can generalize the three main set restrictions that can be applied in a membership function. This generalization will help us later in this section to define in a better way the fuzzy existential and universal quantification operators. t-norms and s-norms are the functions that qualify as fuzzy intersections and fuzzy unions, respectively:

$$\mu(s_1 \cap s_2)(u) = t(\mu s_1(u), \mu s_2(u))$$
$$\mu(s_1 \cup s_2)(u) = \sigma(\mu s_1(u), \mu s_2(u))$$
$$\mu \overline{s_1}(u) = c(\mu s_1(u)),$$

where t is a t-norm, σ is an s-norm and c is a fuzzy complement. The choice of the min and max operators as the set of intersection and union, respectively, as unique possible choice has been made by Bellman and Giertz [6]. However, this is contradictory to Zadeh's work on fuzzy union and intersection operators [7].

Four examples of t-norms are (each defined for all $a, b \in [0, 1]$):

standard intersection: $t(a, b) = \min(a, b)$
algebraic product: $t(a, b) = a \cdot b$
bounded difference: $t(a, b) = \max(0, a + b - 1)$
Yager: $1 - \min\{1, [(1 - a)^w + (1 - b)^w]^{\frac{1}{w}}\} \quad w > 0$

The corresponding dual s-norms are (each defined for all $a, b \in [0, 1]$):

standard union: $\sigma(a, b) = \max(a, b)$
algebraic sum: $\sigma(a, b) = a + b - a \cdot b$
bounded sum: $\sigma(a, b) = \min(1, a + b)$
Yager: $\min[1, (a^w + b^w)^{\frac{1}{w}}] \quad w > 0$

An example of fuzzy complement is defined by ($a \in [0, 1]$):

$$c_w(a) = (1 - a^w)^{1/w}$$

where $w \in (0, \infty)$, and is known as the Yager complement. When $w = 1$, this function becomes the classical fuzzy complement of $c(a) = 1 - a$.

According to Straccia's fuzzy DLs and our generalization using t-norms and s-norms a concept is interpreted as a fuzzy set rather than a classical set and thus concepts becomes imprecise.

The fuzzy DLs have a clean model-theoretic semantics based on the notion of the fuzzy interpretation. A fuzzy interpretation is defined using the classical definition. It is a pair $I = (\Delta^I, \cdot^I)$ consisting of a non-empty set Δ^I and an interpretation mapping function $\cdot^I$. Thus, an interpretation now assigns fuzzy sets to concepts and roles as:

$$\begin{aligned} f\ concept\ A^I &: \Delta^I \to [0, 1] \\ f\ role\ R^I &: \Delta^I \times \Delta^I \to [0, 1] \end{aligned}$$

The interpretation of the Boolean operators and the quantifiers must then be extended from $\{0, 1\}$ to the interval [0, 1]. Therefore if A is a concept then A^I will be interpreted as the membership degree function of the fuzzy concept A. For example, if $a \in \Delta^I$ then $A^I(a)$ gives us the degree to which the object a belongs to the fuzzy concept A, i.e. $A^I(a) = 0.8$. Additionally, the interpretation function $\cdot^I$ must satisfy the following conditions for all $a \in \Delta^I$:

$$\begin{aligned}
&\top^I(a) = 1 \\
&\bot^I(a) = 0 \\
&(C \sqcap D)^I(a) = t(C^I(a), D^I(a)) \\
&(C \sqcup D)^I(a) = \sigma(C^I(a), D^I(a)) \\
&(\neg C)^I(a) = c(C^I(a)) \\
&(\forall R.C)^I(a) = \inf_{a' \in \Delta^I} \sigma(c(R^I(a, a')), C^I(a')) \\
&(\exists R.C)^I(a) = \sup_{a' \in \Delta^I} t(R^I(a, a'), C^I(a'))
\end{aligned}$$

Note that $\forall R.C$ corresponds to the FOL formula $\forall x.(\neg R(x, y) \vee C(y))$, and $\exists R.C$ corresponds to the FOL formula $\exists x.(R(x, y) \wedge C(y))$.

The usual interpretations of conjuction as minimum, disjunction as maximum, negation as $(1 - a)$, universal quantifier as infimum and existential quantifier as supremum are defined in [1,2,5]. In this chapter, we consider the maximum as s-norm, the minimum as t-norm and $(1 - a)$ as the fuzzy complement. Fuzzy norms can change the behaviour of the knowledge base. It is obvious that the semantics of a sentence that uses the minimum t-norm are different from one that uses the product t-norm.

Two concepts C and D are said to be equivalent when $C^I = D^I$ for all interpretations I. In classical sets, the operations of intersection and union are dual with respect to the complement. As for the classical sets, dual relationships between concepts hold, e.g.:

$$(C \sqcap D) \cong \neg(\neg C \sqcap D) \text{ and } (\forall R.C) \cong \neg(\exists R.\neg C)$$

However, only some combinations of t-norms, t-conorms and fuzzy complements can satisfy the duality. A t-norm t and a t-conorm σ are dual with respect to the standard complement c iff:

$$c(t(a, b)) = \sigma(c(a), c(b)$$

and

$$c(\sigma(a, b)) = t(c(a), c(b))$$

where these equations describe the De Morgal law for fuzzy sets [8].

The fuzzy DLs knowledge base, like the classical DLs, consists of fuzzy assertions and fuzzy terminologies. A fuzzy assertion $a \leq k$, where $k \in [0, 1]$ constraints the truth value of a to be less or equal to k. An interpretation I satisfies a fuzzy assertion $a : C \leq k$ iff $C^I(a^I) \leq k$. A fuzzy terminological axiom, like the classical case, is either a concept equality or a concept inclusion. A fuzzy interpretation I satisfies a fuzzy concept equality $C\hat{o}D$ iff $C^I \leq D^I$ and a fuzzy concept inclusion $C = D$ iff $C^I = D^I$.

4.2.4 *Propositional Logics*

Logic is the study of the methods and principles of reasoning in all its possible forms. Classical logic deals with propositions that are required to be either true or false. Each proposition has its opposite, which is usually called a negation of the proposition. A proposition and its negation are required to assume opposite truth values.

Propositional logic deals with combinations of variables that stand for arbitrary propositions. These variables are usually called logic variables (or propositional variables). As each variable stands for a hypothetical proposition, it may assume either of the two truth values; the variable is not committed to either truth value unless a particular proposition is substituted for it.

One of the main concerns of the propositional logic is the study of rules by which new logic variables can be produced as functions of some given logic variables. It is not concerned with the internal structure of the proposition that the logic variables represent.

Assume that n logic variables $u_1, u_2, \ldots, u_n$ are given. A new logic variable can then be defined by a function that assigns a particular truth value to the new variable for each combination of the truth values of the given variables. This function is usually called a logic function. Logic functions of one or two variables are usually called logic operations. The key issue of propositional logic is the expression of all the logic functions of n variables ($n \in N$), the number of which grows extremely rapidly with increasing values of n with the aid of a small number of simple logic functions. These simple functions are preferably logic operation of one or two variables, which are called logic primitives.

Two of the many sets of primitives have been predominant in propositional logic: (i) negation, conjunction and disjunction; and (ii) negation and implication. By combining, for example, negations, conjunctions and disjunctions in appropriate algebraic expressions, referred to as logic formulas, we can form any other logic function.

When a variable represented by a logic formula is always true regardless of the truth values assigned to the variables participating in the formula, it is called a tautology; when it is always false, it is called a contradiction.

Various forms of tautologies can be used for making deductive inferences [8]. They are referred to as inference rules. Examples of some tautologies frequently used as inference rules are:

$(a \wedge (a \Rightarrow b)) \Rightarrow b$	modus ponens
$(\bar{b} \wedge (a \Rightarrow b)) \Rightarrow \bar{a}$	modus tollens
$((a \Rightarrow b) \wedge (b \Rightarrow c)) \Rightarrow (a \Rightarrow c)$	hypothetical sylogism

In this chapter we use the modus ponens tautology, in its fuzzy extension, to construct rules, which states that given two true propositions, a and $a \Rightarrow b$, the truth of the proposition b may be inferred. Every tautology remains a tautology when any of its variables is replaced with any arbitrary logic formula.

4.2.5 Fuzzy Propositions

The fundamental difference between classical propositions and fuzzy propositions is in the range of their truth values. While each classical proposition is required to be either true or false, the truth or falsity of fuzzy propositions is a matter of degree. Assuming that truth and falsity are expressed by values 0 and 1, respectively, the degree of truth of each fuzzy proposition is expressed by a number in the unit interval [0, 1]. There are various fuzzy propositions, which are classified into the following four types.

1. Unconditional and unqualified propositions
2. Unconditional and qualified propositions
3. Conditional and unqualified propositions
4. Conditional and qualified propositions.

In this chapter we will discuss the interpretation of the third type, conditional and unqualified propositions.

Propositions p of the conditional and unqualified type are expressed by the canonical form:

$$p : \text{If } X \text{ is } A, \quad \text{then } Y \text{ is } B$$

where X, Y are variables whose values are in the sets X, Y, respectively, and A, B are fuzzy sets on X, Y, respectively. These propositions may also be viewed as propositions of the form:

$$\langle X, Y \rangle \text{ is } R$$

where R is a fuzzy set on $X \times Y$ that is determined for each $x \in X$ and each $y \in Y$ by the formula:

$$R(x, y) = \ell[A(x), B(y)] \tag{4.1}$$

where ℓ denotes a binary operation on [0, 1] representing a fuzzy implication, and R expresses the relationship between the variables X and Y involved in the given fuzzy proposition. For each $x \in X$and each $y \in Y$, the membership grade $R(x, y)$ represents the truth value of the proposition:

$$p_{xy} : \text{If } X = x, \text{then } Y = y$$

Now, the truth values of the propositions $X = x$ and $Y = y$ are expressed by the membership grades $A(x)$ and $B(y)$, respectively. Consequently, the truth value of the proposition p_{xy}, given by $R(x, y)$, involves a fuzzy implication in which $A(x)$ is the truth value of the antecedent and $B(y)$ is the truth value of the consequent.

4.2.6 Compositional Rules of Inference

Assume that R is a fuzzy relation on $X \times Y$ and A', B' are fuzzy sets on X and Y, respectively. Then if R and A' are given, we can obtain B' by the equation:

$$B'(y) = \sup_{x \in x} t[A'(x), R(x, y)] \tag{4.2}$$

for all $y \in Y$. This equation, which can also be written in the matrix form as:

$$B' = A' \circ R \tag{4.3}$$

is called the compositional rule of inference. This procedure is called the generalized fuzzy modus ponens.

The fuzzy relation employed in equation (4.2) is usually not given directly, but in some form. In the case that the relation is embedded in a single conditional fuzzy proposition, then it is determined using the fuzzy implication operator, equation (4.1). A more general case, in which the relation emerges from several conditional fuzzy propositions, is as follows:

Rule 1: If X is A_1, then Y is B_1
Rule 2: If X is A_2, then $Y$is B_2
⋮
Rule n: If X is A_n, then Y is B_n

In the multi-conditional case R is determined by solving appropriate fuzzy relational equations as explained in the following section.

4.3 Fuzzy Propositional Rules Inference Engine

As previously described, any conditional (If-Then) fuzzy proposition can be expressed in terms of a fuzzy relation R between the two variables involved. One way to determine R is using the fuzzy implication, which operates on fuzzy sets involved in the fuzzy proposition. However, the problem of determining R for a given conditional fuzzy proposition can be detached from fuzzy implications, and R can be determined using fuzzy relational equations.

As described, the equation to be solved for fuzzy modus ponens has the form:

$$B = A \circ^{t} R, \tag{4.4}$$

where A and B are given fuzzy sets that represent, respectively, the If-part and the Then-part in the conditional fuzzy proposition involved and t is a t-norm. It will be proved in the following section that equation (4.4) is solvable for R if $A \circ^{\omega_t} B$ is a solution, where:

$$\omega_t(a, b) = \sup\{x \in [0, 1] | t(a, x) \leq b\} \tag{4.5}$$

for every $a, b \in [0, 1]$ and a continuous t-norm t.

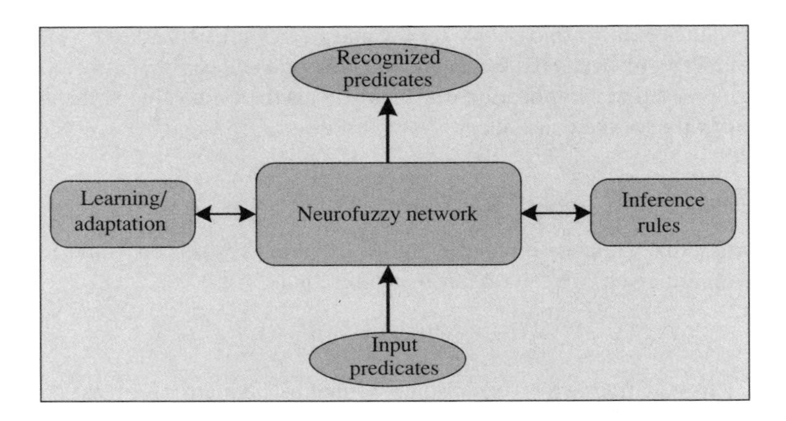

Figure 4.1 The neurofuzzy architecture

In the following section we present a complete algorithm for solving fuzzy relational equations for the interpretation of inference rules in the respective fuzzy extension of propositional logics. The proposed interpretation algorithm is realized using a hybrid neurofuzzy architecture (Figure 4.1).

Fuzzy systems are numerical model-free estimators. While neural networks encode sampled information in a parallel distributed framework, fuzzy systems encode structured, empirical (heuristic) or linguistic knowledge in a similar numerical framework [9]. Although they can describe the operation of the system in natural language with the aid of human-like If-Then rules, they do not provide the highly desired characteristics of learning and adaptation. The use of neural networks in order to realize the key concepts of a fuzzy logic system enriches the system with the ability of learning and improves the sub-symbolic to symbolic mapping [10].

The proposed neurofuzzy network is supported by an adaptation algorithm. This algorithm uses predefined input–output data to (i) initiate and (ii) adapt the weights of the fuzzy propositional rules. It is important to state that using the adaptation algorithm we are not altering the knowledge or generating new knowledge, but we refine the existed knowledge to achieve the optimal behaviour of the knowledge-based system. Finally, using the adaptation algorithm we incorporate uncertainty by using degrees of confidence. The degree of confidence measures the belief of the existence of the specific concept or relation, since real-life applications involve uncertainty and fuzzy hypotheses.

4.3.1 Neurofuzzy Network

Let $y = [y_1, y_2, \ldots, y_m]$ denote a fuzzy set defined on the set of output predicates, the truth of which will be examined. Actually, each y_i represents the degree to which the ith output fuzzy predicate is satisfied. The input of the proposed neurofuzzy network is a fuzzy set $x = [x_1, x_2, \ldots, x_n]$ defined on the set of the input predicates, with each x_i representing the degree to which the ith input predicate is detected. The proposed network represents the association $f\colon X \to Y$, which is the knowledge of the system, in a neurofuzzy structure. After the evaluation of the input predicates, some output predicates represented in the knowledge of the system can be recognized with the aid of fuzzy systems reasoning [8]. One of the widely used ways of constructing fuzzy inference systems is the method of approximate reasoning, which can be implemented on the basis of compositional rule of inference [8]. The need for results with theoretical soundness leads to the representation of fuzzy inference systems on the basis of generalized sup-t-norm compositions [9, 11].

The class of t-norms has been studied by many researchers [10–12]. Using the definition $\omega_t$in equation (4.5) two additional operators $\widehat{\omega}_t, \breve{\omega}_t : [0,1] \times [0,1] \rightarrow [0,1]$ are defined by the following relations:

$$\widehat{\omega}_t \quad (a,b) = \begin{cases} 1 & a < b \\ a \widehat{\otimes}^t b & a \geq b \end{cases}$$

$$\breve{\omega}_t(a,b) = \begin{cases} 0 & a < b \\ a \breve{\otimes}^t b & a \geq b \end{cases}$$

where $a \widehat{\otimes}^t b = \sup\{x \in [0,1] : t(a,x) = b\}$, $a \breve{\otimes}^t b = \inf\{x \in [0,1] : t(a,x) = b\}$.

With the aid of the above operators, compositions of fuzzy relations can be defined. These compositions are used in order to construct fuzzy relational equations, and represent the rule-based symbolic knowledge with the aid of fuzzy inference [13].

Let X, Z, Y be three discrete crisp sets with cardinalities n, l and m, respectively, and $A(X, Z)$, $B(Z, Y)$ be two binary fuzzy relations. The definitions of sup-t and inf-$\widehat{\omega}_t$ compositions are given by:

$$(A \circ^t B)(i,j) = \sup_{k \in N_l} t\{A(i,k), B(k,j)\}, i \in N_n, j \in N_m \tag{4.6}$$

$$(A \circ^{\widehat{\omega}_t} B)(i,j) = \inf_{k \in N_l} \widehat{\omega}_t\{A(i,k), B(k,j)\}, i \in N_n, j \in N_m \tag{4.7}$$

Let us now proceed to a more detailed description of the proposed neurofuzzy architecture. It consists of two layers of compositional neurons, which are extensions of the conventional neurons (Figure 4.2) [13]. While the operation of the conventional neuron is described by the equation:

$$y = \alpha\left(\sum_{i=1}^{n} w_i x_i + \vartheta\right) \tag{4.8}$$

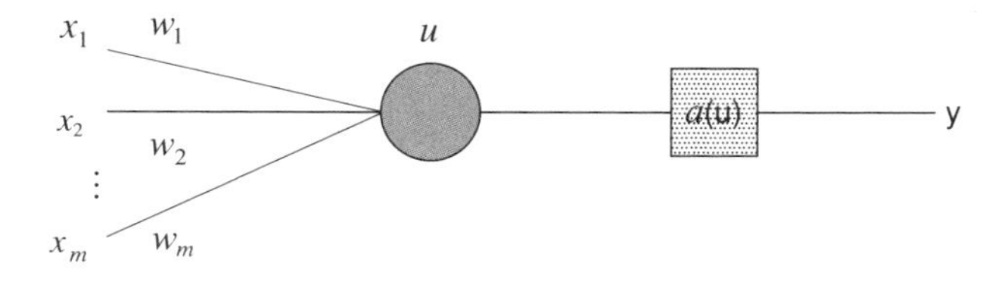

Figure 4.2 The structure of a compositional neuron

where α is nonlinearity, ϑ is threshold and w_i are the weights, the operation of the sup-t compositional neuron is described by the equation:

$$y = a'\left\{ \sup_{j \in N_n} t\,(x_i, w_i) \right\} \tag{4.9}$$

where t is a t-norm and α is the following activation function:

$$a'(z) = \begin{cases} 0 & x \in (-\infty, 0) \\ x & x \in [0, 1] \\ 1 & x \in (0, +\infty) \end{cases}$$

A second type of compositional neuron is constructed using the $\widehat{\omega}_t$ operation. The neuron equation is given by:

$$y = a'\left\{ \inf_{j \in N_n} \widehat{\omega}_t\,(x_i, w_i) \right\} \tag{4.10}$$

The proposed architecture is a two-layer neural network of compositional neurons (Figure 4.3). The first layer consists of the inf-$\widehat{\omega}_t$ neurons and the second layer consists of the sup-t neurons. The system takes as input predicates, and gives to the output the recognized

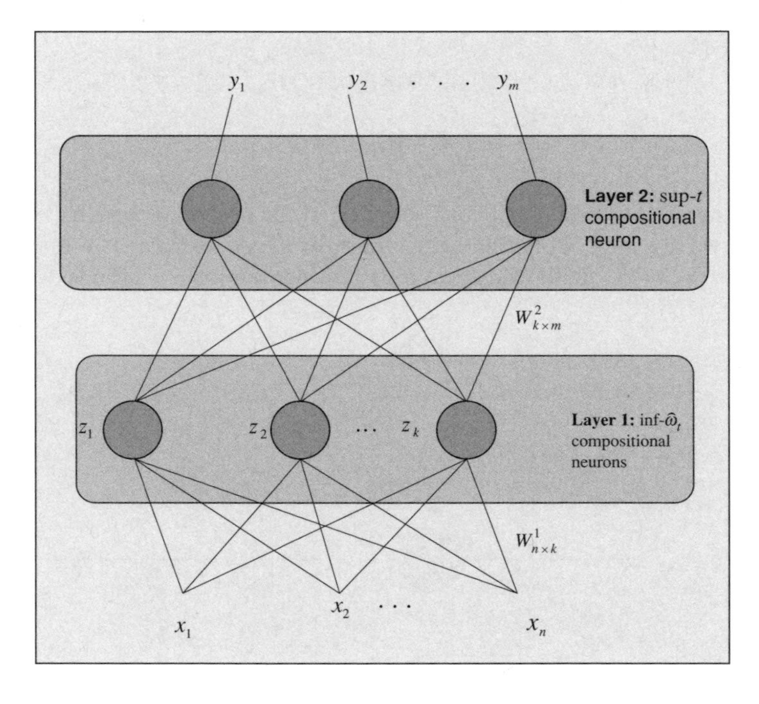

Figure 4.3 The neurofuzzy network

output predicates. The first layer computes the antecedents of the mapping rules, while the second implements the fuzzy reasoning using the fuzzy modus ponens schema.

The rules are used to initialize the neurofuzzy network (giving its initial structure and weights). During the learning process the number of neurons in the hidden layer and the weights of the two layers may change with the aid of learning with the objective of error minimization. The learning algorithm that supports the above network is applied in each layer independently. During the learning process, the weight matrices are adapted in order to approximate the solution of the fuzzy relational equation describing the association of the input with the output. Using a traditional minimization algorithm (for example the steepest descent), we cannot take advantage of the specific character of the problem. The algorithm that we use is based on a more sophisticated credit assignment that 'blames' the neurons of the network using knowledge about the topographic structure of the solution of the fuzzy relation equation [13]. After the learning process, the network keeps its transparent structure and the new knowledge represented in it can be extracted in the form of mapping If-Then rules.

4.3.2 Learning Operation

In the process of knowledge adaptation, the If-Then rules are inserted into the proposed neurofuzzy system. This refers to automatically transforming the structured knowledg provided by the knowledge base in order to perform the following:

1. Define the required input predicates as *input predicate(1), input predicate(2),..., input predicate(n)*. The input predicates will define the set $X = \{x_1, x_2, \ldots, x_m\}$.
2. Define the required output predicates as *output predicate(1), output predicate(2),..., output predicate(n)*. The output predicates will define the set $Y = \{y_1, y_2, \ldots, y_m\}$.
3. Insert the a priori knowledge given in If-Then rules of the form 'if *input predicate(1)* and *input predicate(2)* and ... then *output predicate(5)*' into the neurofuzzy structural elements (the weights of the neurofuzzy system). The number of different antecedents (If-parts of the rules) defines the set $Z = \{z_1, z_2, \ldots, z_l\}$. The predicates could be associated with confidence levels in order to produce the antecedents; this means that the antecedents could have the form (*input predicate(1), input predicate(2)*, 0.7, 0.9), with the 0.7 and 0.9 values corresponding to confidence levels. The above degrees are used in order to define the weights $\mathbf{W}^1_{ij}, i \in N_n, j \in N_1$ of the first layer. Furthermore, the consequences could also be associated with confidence levels, i.e. 'if *input predicate(1)* and *input predicate(2)* and ... then *output predicate(5)* with confidence 0.7'. These values are used in order to define the weights $\mathbf{W}^2_{ij}, i \in N_l, j \in N_m$ of the second layer.

The knowledge refinement provided by the proposed neurofuzzy system will now be described. Let $X = \{x_1, x_2, \ldots, x_m\}$ and $Y = \{y_1, y_2, \ldots, y_m\}$ be the input and output predicate sets, respectively, and let also $R = \{r_1, r_2, \ldots, r_p\}$ be the set of rules describing the knowledge of the system. The set of antecedents of the rules is denoted by $Z = \{z_1, z_2, \ldots, z_l\}$ (see the structure of the neurofuzzy system given in Figure 4.3). Suppose now that a set of input–output data $D = \{(A_i, B_i), i \in N_q\}$, where and $B_i \in F(Y)$ ($F(*)$ is the set of fuzzy sets defined on $*$), is given sequentially and randomly to the system (some of them are allowed to reiterate before the first appearance of some others). The data sequence is described as $(A^{(q)}, B^{(q)}), q \in N$, where

$(A^{(i)}, B^{(i)}) \in D$. The problem that arises is finding the new weight matrices $\mathbf{W}^1_{ij}$, $i \in N_n$, $j \in N_l$ and $\mathbf{W}^2_{ij}$, $i \in N_l$, $j \in N_m$ for which the following error is minimized:

$$\varepsilon = \sum_{i \in N_q} \| B_i - Y^i \| \tag{4.11}$$

where y^i, $i \in N_q$ is the output of the network when the input A_i is given. The process of the minimization of the above error is based on the resolution of the following fuzzy relational equations:

$$\mathbf{W}^1 \circ^{\widehat{\omega}_t} \mathbf{A} = \mathbf{Z} \tag{4.12}$$

$$\mathbf{Z}^1 \circ^t \mathbf{W}^2 = \mathbf{B} \tag{4.13}$$

where t is a continuous t-norm and $\mathbf{Z}$ is the set of antecedents fired when the input $\mathbf{A}$ is given to the network.

For the resolution of the above problem the adaptation process changes the weight matrices $\mathbf{W}^1$ and $\mathbf{W}^2$ in order to approximate a solution of the above fuzzy relational equations. During its operation the proposed network can generalize in a way that is inspired from the theory of fuzzy systems and the generalized modus ponens. Let us here describe the adaptation of the weights of the second layer (the adaptation of the first layer is similar). The proposed algorithm converges independently for each neuron. For simplicity and without loss of generality, let us consider only the single neuron case. The response of the neuron $f^{(k)}$ at time k is given by:

$$f^{(k)} = \sup_{i \in N_l} t\left(z_i^{(k)}, w_i^{(k)}\right) \tag{4.14}$$

where $w_i^{(k)}$ are the weights of the neuron and $z_i^{(k)}$ the input, at time k. The desired output at time k is $B_i^{(k)}$. The algorithm works in the following way. The weights are initialized as $w_i^{(0)}$, $i \in N_l$. The input $z^{(k)}$ and the desired output $B^{(k)}$ are processed, the response of the network $f^{(k)}$ is computed and the weight is updated accordingly (online variant of learning):

$$w_i^{(k+1)} = w_i^{(k)} + \Delta w_i^{(k)}$$

$$\Delta w_i^{(k)} = \eta l_s$$

$$l_s = \begin{cases} \eta_1\left(\breve{\omega}_t\left(z_i^{(k)}, B^{(k)}\right) - w_i^{(k)}\right), & \text{if} \quad w_i^{(k)} < \breve{\omega}_t\left(z_i^{(k)}, B^{(k)}\right) \\ \eta_2\left(w_i^{(k)} - \widehat{\omega}_t\left(z_i^{(k)}, b^{(k)}\right)\right), & \text{if} \quad w_i^{(k)} > \widehat{\omega}_t\left(z_i^{(k)}, B^{(k)}\right) \end{cases}$$

where η, η_1, η_2 are the learning rates. The adaptation is activated only if $\left|\varepsilon(B^{(k)}, y^{(k)})\right| > \varepsilon_c$, where ε_c is an error constant.

If the t-norm is Archimedean, then the learning signal is computed as:

$$l_s = \left(\widehat{\omega}_t\left(z_i^{(k)}, b^{(k)}\right) - w_i^{(k)}\right), \text{ if } z_i^{(k)} \geq b^{(k)} \text{ and } z_i^{(k)} \neq 0, \text{ else } l_s = 0$$

With the aid of the above learning process (and similar for the first layer, since the operator $\widehat{\omega}_t$ is also used in order to solve the fuzzy relational equation of the first layer [8]), the network approximates the solutions of the fuzzy relational equations given above and thus minimizes the error.

4.4 Demonstration

In order to demonstrate the applicability of the proposed extensions, we illustrate an experimental study dealing with facial expression recognition. The basic motivation for examining this particular application stems from several studies for facial expression recognition that are based on image/video features. In this experimental study we employ MPEG-4, through the use of the Facial Animation Parameters (FAPs) that use intermediate states to characterize facial expressions [4]. Intermediate states refer to the fact that no low-level image/video features (pixel values, motion vectors, colour histograms) are used directly for modelling the expressions. FAPs in real images and video sequences are defined through the movement of some points that lie in the facial area and are able to be automatically detected. Quantitative description of FAPs based on particular Facial Points (FPs), which correspond to the movement of protuberant facial points, provides the means of bridging the gap between image analysis and expression recognition. In this chapter we are not particularly interested in the extraction of image/video features and the calculation of the FAPs. We employ FAPs to construct rules and knowledge that formally represent the facial expressions recognition domain. The output of the algorithm [4] that calculates the FAPs values of an image is stored in an XML file. The list of FAPs that are used is shown in Table 4.1 and the list of the possible facial expressions to be recognized in Table 4.2. Figure 4.4 shows the distances that define FAPs and FPs. The fuzzy nature of the facial expression recognition task, in combination with the use of structured knowledge that can be implemented in a machine processable way, make suitable the use of the proposed knowledge-based system.

Table 4.1 The set of FAPs

Nr	FAP name	D
F1	open_jaw	d_2
F2	lower_top_midlip	d_1
F3	raise_bottom_midlip	d_3
F4	widening_mouth	d_4
F5	close_left_eye	d_6
F6	close_right_eye	d_5
F7	raise_left_inner_eyebrow	d_8
F8	raise_right_inner_eyebrow	d_7
F9	raise_left_medium_eyebrow	d_{10}
F10	raise_right_medium_eyebrow	d_9
F11	raise_left_outer_eyebrow	d_{11}
F12	raise_right_outer_eyebrow	d_{13}
F13	squeeze_left_eyebrow	d_{12}
F14	squeeze_right_eyebrow	d_{15}
F15	wrinkles_between_eyebrows	d_{23}
F16	raise_left_outer_cornerlip	d_{22}
F17	raise_right_outer_cornerlip	d_{11}

Table 4.2 The set of expressions

Nr	Expression name
E1	Joy
E2	Sadness
E3	Anger
E4	Fear
E5	Disgust
E6	Surprise

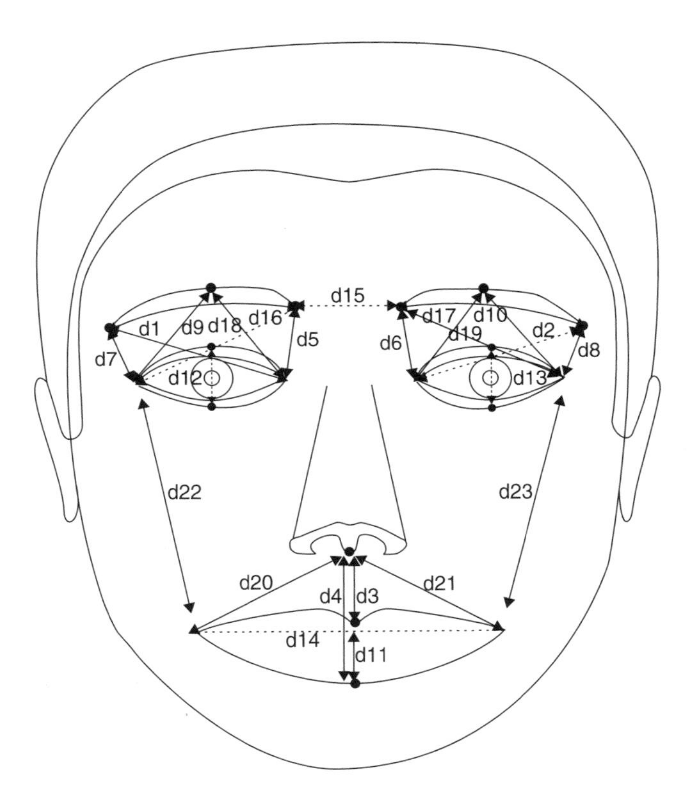

Figure 4.4 The distances that define FAPs and FPs

4.4.1 Knowledge-based System Implementation

In recent years significant process has been made in the area of knowledge management. The SW initiative has increased the number of researchers involved in knowledge management techniques and web language technologies. The new advancements facilitate the development of knowledge-based systems capable of performing inferencing using knowledge expressed in machine understandable and processable languages.

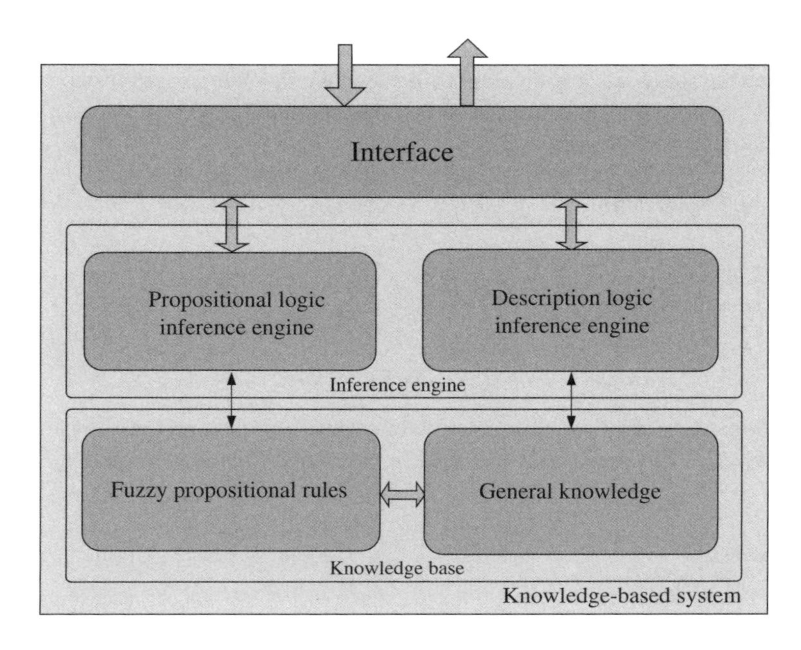

Figure 4.5 The knowledge-based system architecture

The architecture of the proposed knowledge-based system is depicted by the block diagram in Figure 4.5. It consists of the knowledge base, the inference engines and an interface for communicating with the user or the application. The structure of the knowledge base, as previously explained, is hybrid. It consists of the fuzzy propositional rules and the general knowledge.

Two types of inferencing are performed on the knowledge base. The inference engines function independently. The inference rules are interpreted using the proposed inference algorithm. This inference algorithm operates on a series of fuzzy propositional rules and makes fuzzy inferencing. For the general knowledge, we use a classical DL reasoner, based on the tableaux calculus. Unfortunately, a drawback of the tableaux calculus in [8] is that any system that would like to implement this fuzzy logic extension has to be worked out from scratch. Complete algorithms are proposed in [1,2] for solving these inference problems in the respective fuzzy extension of *ALC* but neither of them is implemented. However, even if there was an inference engine implemented capable of using the fuzzy extensions proposed in [1,2], it would cover only one specific case, and not the whole set of t-norms. For that reason, the proposed fuzzy DL extension has not been used in this demonstration of the knowledge-based system.

4.4.2 General Knowledge

For the implementation of the general knowledge we have used a machine processable language. Ontology Web Language (OWL) implements the described DLs operators. More specifically, OWL DL [14] has been used for the construction of the general knowledge ontology. For editing the ontology we have used Construct ontology editor [14] and the reasoning is performed using the Cerebra inference engine [14].

In order to create the knowledge base we must first specify the domain of interest. In this experimental study we are dealing with the domain of facial expression recognition. Therefore, an ontology has been produced for representing knowledge concerning MPEG-4 facial information and expressions.

The general knowledge has been formally defined based on the described DLs operators. A small set of the formal definitions of the knowledge is shown below:

$$\begin{aligned}
&person \equiv body \sqcap face\\
&face \equiv mouth \sqcap nose \sqcap ears \sqcap eyelids \sqcap eyes\\
&facialExpressions \equiv sadness \sqcap joy \sqcap mad\\
&joyPerson \equiv person \sqcap \exists\ hasFacialExpression.joy\\
&neutral \equiv face \sqcap \neg facialExpression
\end{aligned}$$

Figure 4.6 shows the graphical representation of a sample of the ontology constructed for the general knowledge. The OWL DL code of the presented sample is presented in Box 4.1.

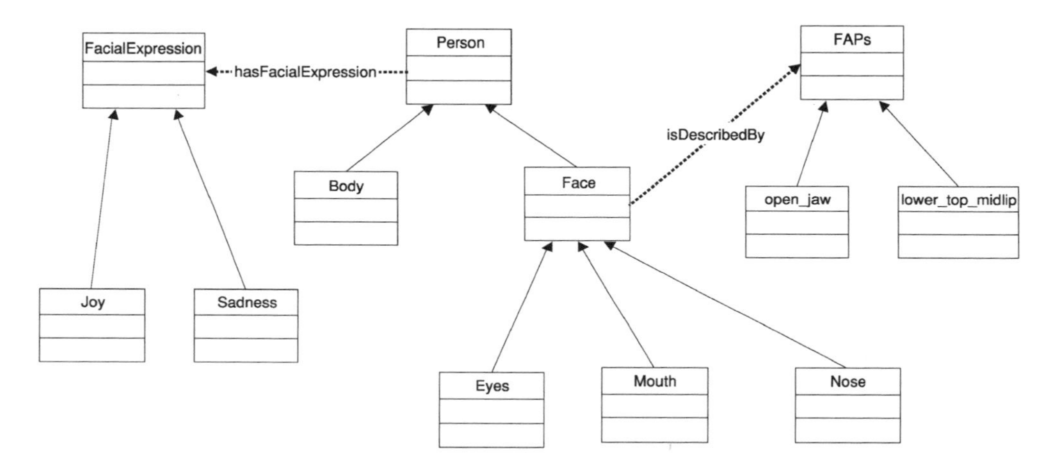

Figure 4.6 A sample of the Facial Expression ontology

Box 4.1

```
<?xml version="1.0" encoding="UTF-8"?>
<rdf:RDF
xmlns:rdf="http://www.w3.org/1999/02/22-rdf-syntax-ns#"
xmlns:owl="http://www.w3.org/2002/07/owl#"
xmlns:xsd="http://www.w3.org/2001/XMLSchema#"
xmlns:rdfs="http://www.w3.org/2000/01/rdf-schema#"
xmlns:dc="http://purl.org/dc/elements/1.1/"
xmlns:ni="http://www.networkinference.com/2003/03/owl#"
xml:base="http://www.image.ntua.gr/~tzouvaras/
FacialExpreesionOntology#"
```

```
xmlns="http://www.image.ntua.gr/~tzouvaras/
FacialExpreesionOntology#"
xmlns:FacialExpreesionOntology=
"http://www.image.ntua.gr/~tzouvaras/
FacialExpreesionOntology#">
<owl:Ontology rdf:about="">
<owl:versionInfo>
1.0</owl:versionInfo>
</owl:Ontology>
<owl:AnnotationProperty rdf:about="http://www.networkinference
.com/2003/03/owl#datatypeImports"/>
<owl:Class rdf:about="http://www.w3.org/2002/07/owl#Thing"/>
<owl:Class rdf:ID="Person">
<rdfs:subClassOf>
<owl:Restriction>
<owl:onProperty rdf:resource="#hasFacialExpression"/>
<owl:allValuesFrom rdf:resource="#FacialExpression"/>
</owl:Restriction>
</rdfs:subClassOf>
</owl:Class>
<owl:Class rdf:ID="Body">
<rdfs:subClassOf rdf:resource="#Person"/>
</owl:Class>
<owl:Class rdf:ID="Face">
<rdfs:subClassOf>
<owl:Restriction>
<owl:onProperty rdf:resource="#isDescribedBy"/>
<owl:allValuesFrom rdf:resource="#FAPs"/>
</owl:Restriction>
</rdfs:subClassOf>
<rdfs:subClassOf rdf:resource="#Person"/>
</owl:Class>
. . . </rdf:RDF>
```

4.4.3 Fuzzy Propositional Rules

There are many languages for expressing rules. We have selected RuleML [15] since it is the most popular and most likely candidate for adoption by the Semantic Web community. It provides a markup language which allows descriptions of the form of an implication between an antecedent (body) and a consequent (head). The intended meaning can be read as: whenever the conditions specified in the antecedent hold, then the conditions specified in the consequent must also hold. Both the antecedent (body) and consequent (head) consist of zero or more atoms. An empty antecedent is treated as trivially true, so the consequent must also be satisfied by every interpretation. An empty consequent is treated as trivially false, so the antecedent must also not be satisfied by any interpretation.

More specifically, we have used the RuleML Lite [15] concrete syntax to implement the propositional rules. In this syntax, all predicates are unary or binary in order to be compatible with OWL DL. In RuleML we can also incorporate uncertainty degrees, which are very significant in order to represent fuzzy propositional rules. The incorporated uncertainty degrees constitute absolute meta-knowledge about a rule.

For the creation of the rules, we used the defined FAPs as predicates. An example of a rule is shown below:

$$\begin{aligned} RULE:\ & IF\ open\_jaw(0.8)\ AND\ lower\_top\_midlip(0.5) \\ & AND\ squeeze\_left\_eyebrow(0.9) \\ & AND\ wrinkles\_between\_eyebrows(0.75) \\ & THEN\ joy(0.7) \end{aligned}$$

The Rule-ML representation of this example is presented in Box 4.2. This example presents a rule that depicts the expression Joy.

Box 4.2

```
<?xml version="1.0" encoding="UTF-8"?>
<rulebase xmlns="http://www.ruleml.org/"
xmlns:xsi="http://www.w3.org.2001/xmlSchema-insance"
xmlns:ermis="http://www.image.ntua.gr/~tzouvaras/
FacialExpressionOntology"
<imp>
  <_head>
     <atom>
        <_opr>
          <Cterm>ermis:joy</Cterm>
      </_opr>
      <var>ermis:_face</var>
      <ind>0.7</ind>
   </atom>
 </_head>
 <body>
    <and>
      <atom>
        <_opr>
          <Cterm>ermis:open_jaw </Cterm>
        </_opr>
           <var>ermis:_face</var>
           <ind>0.8</ind>
        </atom>
        <atom>
         <_opr>
           <Cterm>ermis:lower_top_midlip </Cterm>
         </_opr>
          <var>ermis:_face</var>
          <ind>0.5</ind>
      </atom>
      <atom>
       <_opr>
         <Cterm>ermis:squeeze_left_eyebrow </Cterm>
```

```
        </_opr>
          <var>ermis:_face</var>
          <ind>0.9</ind>
        </atom>
        <atom>
         <_opr>
        <Cterm>ermis:wrinkles_between_eyebrows </Cterm>
         </_opr>
           <var>ermis:_face</var>
           <ind>0.7</ind>
       </atom>
      </and>
     </body>
    </imp>
  </rulebase>
```

Regarding the interpretation of the propositional rules, there are available implemented inference engines [16]. However, they have a major drawback. They cannot perform inferencing in vague and uncertain knowledge. The proposed inference engine has two major advantages. It can incorporate the uncertainty of using fuzzy implication operators, and it has the ability to automatically refine the rules, using key concepts of artificial neural networks.

In this experimental study we used XPath to identify and retrieve the values that are needed in the propositional rules. These variable values are retrieved from the XML instances, which is the output of the software that calculates that FAPs values of an image.

4.4.4 Experiment Setup

In the current study we used the EKMAN database [17], which contains 300 images. The set of propositional rules that was used was constructed based on [4] and is shown in Table 4.3.

The inferencing may be done on-the-fly or in advance. In this experimental study the inferencing was done in advance and the output saved in an XML file. In this way we enable more sophisticated querying to be performed. For example, the user can ask for an image that combines the happy expression with confidence value (0.75) and the normal expression with confidence (0.2). Or an expert can ask for images that have activated specific FAPs (open_jaw(0.4), closed_left_eye(0.9), ...). The fuzzy extensions of DLs and propositional rules are suitable, generally, for multimedia applications since multimedia information is vague and uncertain.

4.4.5 Adaptation of the Rules

In this section we present some simulation results illustrating the operation of the proposed adaptation algorithm. Also, we demonstrate the performance of the proposed neurofuzzy network before and after the adaptation procedure. It operates on the meta-knowledge of the

Table 4.3 The fuzzy propositional rules used in the experiment along with weight, representing the degree of confidence

Nr	FAP	Expression	Antecedent	Consequent
1	F1+F5+F9+F11	E1	0.5, 0.7, 0.6, 0.9	0.8
2	F1+F3+F9+F10+F15	E1	0.6, 0.6, 0.7, 1, 0.4	0.5
3	F1+F4+F7+F9+F10+F14	E1	1, 0.7, 0.7, 1, 0.4, 0.8	0.7
4	F3+F5+F7+9+F15	E1	0.8, 0.6, 0.8, 1, 0.5	0.8
5	F4+F8+F7+F9+F14+F17	E2	1, 0.9, 0.7, 0.6, 0.4, 0.6	0.9
6	F5+F7+F8+F14+F15	E2	0.7, 0.8, 0.7, 1, 0.4	0.6
7	F1+F2+F5+F14+F16	E2	0.7, 0.6, 0.7, 1, 0.7	0.9
8	F7+F8+F9+F13+F16+F17	E3	0.6, 0.7, 0.7, 1, 0.4, 1	0.5
9	F6+F9+F10+F12+F15	E3	0.6, 0.8, 0.7, 1, 0.4	0.8
10	F5+F8+F10+F12+F6	E3	1, 1, 0.7, 1, 0.4	0.8
11	F2+F6+F9+F12+F16	E4	0.7, 0.6, 0.7, 1, 0.4	0.5
12	F2+F4+F9+F11+F15+F17	E4	0.6, 0.6, 0.7, 1, 0.7, 0.4	0.6
13	F7+F6+F8+F14+F15	E4	0.6, 0.9, 0.9, 1, 0.4	0.6
14	F6+F7+F8+F13+F15	E5	0.9, 0.4, 0.3, 0.8, 0.8	0.9
15	F8+F9+F11+F12+F17	E5	0.9, 0.6, 0.7, 1, 0.4	0.9
16	F4+F8+F11+F15+F17	E5	0.8, 0.6, 0.7, 0.7, 0.4	0.7
17	F6+F7+F12+F15+F17	E6	0.6, 0.7, 0.9, 0.6, 0, 1	0.6
18	F8+F9+F12+F14+F17	E6	0.6, 0.6, 1, 0.3, 0.8	0.7
19	F1+F8+F9+F14+F16+F17	E6	0.8, 0.4, 0.9, 1, 0.5, 0.4	0.8
20	F5+F7+F13+F14+F15	E6	0.6, 0.9, 0.7, 0.8, 0.8	0.9

rules, the weights. It does not alter the meaning of the knowledge but adapts the existed knowledge according to predefined input–output data, which are provided by the experts. The results of the algorithm are new adapted weights that operate better in the working context. The new weights are replacing the old weights in the knowledge base. Every time the context is changed the weights must be adapted since the proposed propositional rules are context sensitive.

For the system's adaptation demonstration, we consider a small set of the described inputs, outputs and propositional rules. We have 10 input predicates, $X = \{x_1, x_2, \ldots, x_{10}\}$, 8 antecedents, $Z = \{z_1, z_2, \ldots, z_8\}$, and 3 output predicates, $Y = \{y_1, y_2, y_3\}$. The rules of the network $R = \{R_1, R_2, \ldots, R_{10}\}$ are shown in Table 4.4. The four data that are used to adapt the weights of the network are:

$$
\begin{array}{l}
([.9\ \ .6\ \ .7\ \ .7\ \ 0\ \ 0\ \ 0\ \ 0\ \ 0\ \ 0],[.8\ \ 0\ \ 0])\\
([0\ \ 0\ \ 0\ \ .6\ \ .5\ \ .9\ \ 0\ \ 0\ \ 0\ \ 0],[0\ \ 0\ \ .7])\\
([.7\ \ .9\ \ 0\ \ 0\ \ 0\ \ 0\ \ 0\ \ 0\ \ 0\ \ .8],[.7\ \ 0\ \ 0])\\
([0\ \ 0\ \ 0\ \ .5\ \ 0\ \ 0\ \ .9\ \ 0\ \ 0\ \ .7],[0\ \ .8\ \ 0])
\end{array}
$$

The weights of the two layers are initialized using the rules shown in Table 4.4. The antecedent part is used for $\mathbf{W}^1$ and the output (consequence) is used for $\mathbf{W}^2$. The two matrices are shown below:

Table 4.4 The sample set of rules of the system

R	Antecedent	Output
1	$x_1 + x_2 + x_3 + x_4$	y_1
2	$x_1 + x_3 + x_4 + x_8$	y_2
3	$x_2 + x_3 + x_4 + x_9$	y_2
4	$x_5 + x_6 + x_7$	y_3
5	$x_4 + x_7 + x_{10}$	y_2
6	$x_3 + x_5 + x_8 + x_9$	y_1
7	$x_1 + x_2 + x_{10}$	y_1
8	$x_4 + x_6 + x_7$	y_3

$$W^1 = \begin{bmatrix} 1 & 1 & 1 & 1 & 0 & 0 & 0 & 0 & 0 & 0 \\ 1 & 0 & 1 & 1 & 0 & 0 & 0 & 1 & 0 & 0 \\ 0 & 1 & 1 & 1 & 0 & 0 & 0 & 0 & 1 & 0 \\ 0 & 0 & 0 & 0 & 1 & 1 & 1 & 0 & 0 & 0 \\ 0 & 0 & 0 & 1 & 0 & 0 & 1 & 0 & 0 & 1 \\ 0 & 0 & 1 & 0 & 1 & 0 & 0 & 1 & 1 & 0 \\ 1 & 1 & 0 & 0 & 0 & 0 & 0 & 0 & 0 & 1 \\ 0 & 0 & 0 & 1 & 1 & 1 & 0 & 0 & 0 & 0 \end{bmatrix} \qquad W^2 = \begin{bmatrix} 0 & 0 & 1 \\ 0 & 1 & 0 \\ 0 & 1 & 0 \\ 0 & 0 & 1 \\ 0 & 1 & 0 \\ 1 & 0 & 0 \\ 1 & 0 & 0 \\ 0 & 0 & 1 \end{bmatrix}$$

We have used the Yager t-norm [8] with parameter value $p = 2$, and learning rate $\eta = 1$. The neurons are adapted independently, and after 10 iterations the weights of the two layers are given by:

$$W^1 = \begin{bmatrix} .8 & .75 & .65 & 1 & 0 & 0 & 0 & 0 & 0 & 0 \\ 1 & 0 & 1 & 1 & 0 & 0 & 0 & .7 & 0 & 0 \\ 0 & .65 & 1 & 1 & 0 & 0 & 0 & 0 & 1 & 0 \\ 0 & 0 & 0 & 0 & .76 & 1 & 1 & 0 & 0 & 0 \\ 0 & 0 & 0 & .5 & 0 & 0 & 1 & 0 & 0 & 1 \\ 0 & 0 & 1 & 0 & 1 & 0 & 0 & .7 & 1 & 0 \\ 1 & .8 & 0 & 0 & 0 & 0 & 0 & 0 & 0 & 1 \\ 0 & 0 & 0 & .67 & 1 & .9 & 0 & 0 & 0 & 0 \end{bmatrix} \qquad W^2 = \begin{bmatrix} 0 & 0 & .9 \\ 0 & 1 & 0 \\ 0 & .75 & 0 \\ 0 & 0 & 1 \\ 0 & 1 & 0 \\ .8 & 0 & 0 \\ 1 & 0 & 0 \\ 0 & 0 & 1 \end{bmatrix}$$

We observe that the adaptation procedure has refined the weight of the rules and consequently the semantics of the knowledge base. The error of the system, as shown in Figure 4.7, became zero. In Figure 4.7 the error performance is illustrated, using learning rates of 0.5 and 1. It can be seen that the structure of the knowledge is not changed. The algorithm only adapts the existed knowledge and is not generating new knowledge. The new weights are replacing the old weights in the knowledge base.

4.5 Conclusion and Future Work

In this chapter, we have presented a fuzzy extension in DLs based on fuzzy sets and triangular norm and a fuzzy extension of PLs based on compositional rule of inference for

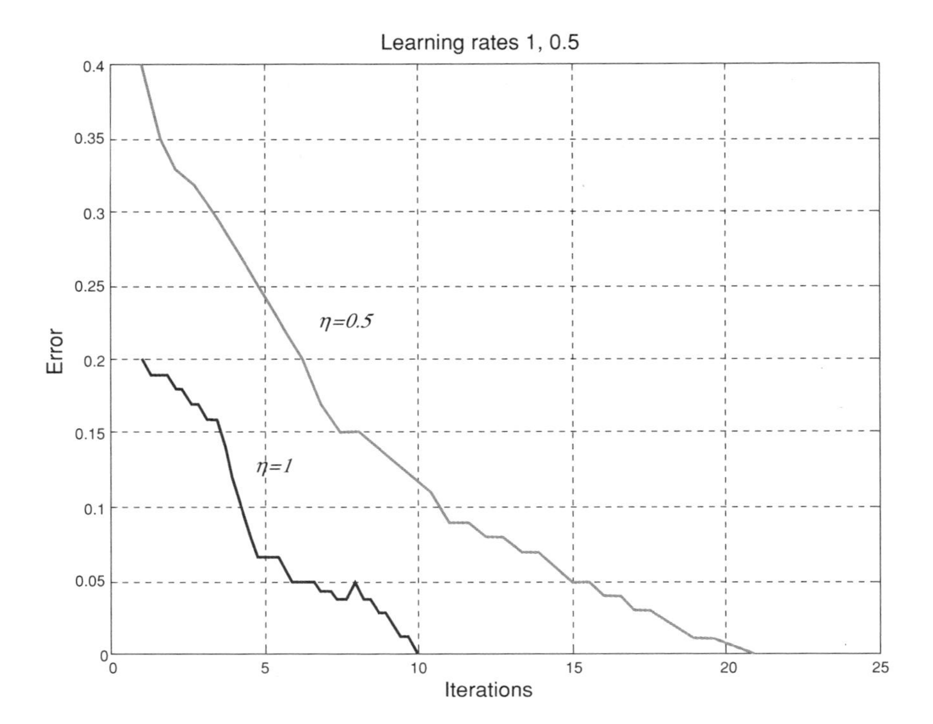

Figure 4.7 The error of the neural network during the adaptation process

constructing fuzzy inference rules of the form *If A and B Then C*. A hybrid neurofuzzy network that is capable of performing inferencing on fuzzy propositional rules is also proposed. This inference engine is supported by an adaptation algorithm that refines the weights of the rules according to predefined input–output data. The proposed fuzzy extensions are suitable for multimedia applications since multimedia information incorporates uncertainty and vague knowledge. In particular, the use of DLs and inference rules in the context of multimedia point out the necessity of extending them with fuzzy capabilities which allow the treatment of the inherent imprecision in multimedia object representation, retrieval, matching and detection.

The proposed extensions have been tested in a facial expression recognition study using fuzzy propositional rules created through the use of the FAPs defined in the MPEG-4 standard. The continuity of emotion space, the uncertainty involved in the feature estimation process and the required ability of the system to use prior knowledge, while being capable of adapting its behaviour to its users' characteristics, make appropriate the use of the proposed fuzzy extensions and inference engine.

Finally, our plans are to extend the existed tableaux algorithm to support the proposed fuzzy extensions of DLs. Complete algorithms for solving a set of the fuzzy extensions are defined in [1,2]. Our premise is to extend the fuzzy tableaux algorithm to support the whole set of t-norms and s-norm operators.

References

[1] U. Straccia, Reasoning within fuzzy description logics. *Journal of Artificial Intelligence Research*, **14**, 137–166, 2001.
[2] J. Yen, Generalising term subsumption languages to fuzzy logic. In *Joint Conference on Artificial Intelligence (IJCAI-91)*, Sydney, Australia, pp. 472–477, 1991.
[3] V. Tzouvaras, G. Stamou, S. Kollias Knowledge refinement using fuzzy compositional neural networks. In *ICANN 2003,* Istanbul, June 2003, pp. 933–942. Lecture Notes in Artificial Intelligence, Springer-Verlag, Berlin, 2003.
[4] A. Raouzaiou, N. Tsapatsoulis, V. Tzouvaras, G. Stamou, S. Kollias, A hybrid intelligence system for facial expression recognition. In Eunite 2002, *European Symposium on Intelligent Technologies, Hybrid Systems and their implementation on Smart Adaptive Systems*, Algarve, Portugal, pp. 1109–1116. Lecture Notes in Computer Science, Springer-Verlag, Heidelberg, 2003.
[5] C. Tresp, R. Monitor, A description logic for vague knowledge. In *Proceedings of the 13th European Conference on Artificial Intelligence (ECAI-98)*, Brighton, UK, 1998.
[6] R. Bellman, M. Giertz, On the analytic formalism of the theory of fuzzy sets. *Information Sciences*, **5**, 149–156, 1973.
[7] L.A. Zadeh, Fuzzy sets. *Information Control,* **8**(3), 338–353, 1976.
[8] G. Klir, Bo Yuan, *Fuzzy Sets and Fuzzy Logic: Theory and Applications*. Englewood Cliffs, NJ, Prentice Hall, 1995.
[9] B. Kosko, *Neural Networks and Fuzzy Systems: A Dynamical Approach to Machine Intelligence*. Englewood Cliffs, NJ, Prentice Hall, 1992.
[10] C.-T. Lin, C.S. Lee, *Neural Fuzzy Systems: A Neuro-Fuzzy Synergism to Intelligent Systems*, Englewood Cliffs, NJ, Prentice Hall, 1995.
[11] S. Jenei, On Archimedean triangular norms. *Fuzzy Sets and Systems*, **99**, 179–186, 1998.
[12] K. Hirota, W. Pedrycz, Solving fuzzy relational equations through logical filtering. *Fuzzy Sets and Systems*, **81**, 355–363, 1996.
[13] G.B. Stamou, S.G. Tzafestas, Neural fuzzy relational systems with a new learning algorithm. *Mathematics and Computers in Simulation,* **51**(314), 301–304, 2000.
[14] Web Ontology Language (OWL) guide, W3C recommendation, http://www.w3./TR/owl-guide.
[15] The Rule MarkUp Language Initiative (RuleML), http://www.dfki.uni.kl.de/ruleml.
[16] The Mandarax project, http://www.mandarax.org.
[17] P. Ekman, W. Friesen, *The Facial Action Coding System*. Consulting Psychologists Press, San Francisco, CA, 1978.

Part Two

Multimedia Content Analysis

5

Structure Identification in an Audiovisual Document

Philippe Joly

5.1 Introduction

This chapter gathers different works mainly published in [1] considering shot segmentation, in [2] considering results evaluation and in [3] considering editing work description. The last part is dedicated to macrosegmentation, with the presentation of a set of tools that have been developed in order to detect sequences or 'highlights', or to analyse the temporal structure in some well-defined context [4].

If we try to do a classification of all the different segmentation tools that have been developed to date with regard to the kind of results produced, we may first consider shot segmentation algorithms. This is obviously the larger family of segmentation tools. Information about shot boundaries, generated during the editing process, could be kept in a digital format along with the final document. Unfortunately, this information is lost in most cases, and must be extracted by automatic tools. As we will see, all the effects, and their parameters, which are produced during the editing step are not always taken into account by those automatic tools, and some efforts in the research field are still required to overcome these limitations.

Nevertheless, shots are considered as the basic unit for automatic structure analysis tools. But we can define a second class of segmentation, sometimes called 'microsegmentation', aimed at defining units which are generally smaller than shots. Microsegmentation tools can be based on event detection (e.g. a character or an object entering the visual scene, music starting on the soundtrack). They may also rely on the evolution of low-level feature values in order to identify homogeneous segments (e.g. global motion direction or magnitude, dominant colours). Actually, most traditional low-level features can be used to perform a microsegmentation of almost any video contents. Results have more or less significance with regard to the content.

The last class of segmentation is addressing the problem of the detection of larger units than shots such as 'sequences' or 'chapters'. Works on that topic must perform a deeper analysis with regard to the semantic content, trying to detect events or to track some significant

Multimedia Content and the Semantic Web Edited by Giorgos Stamou and Stefanos Kollias

elements in the content. To achieve this, these tools usually involve a large set of tools which can be individually used for microsegmentation purposes. In order to combine their results to identify large temporal structures, some stochastic approaches are usually involved such as hidden Markov models. But macrosegmentation seems to be really efficient today only on well-formatted documents (such as sport events and TV news programmes), as will be shown in the last part of this chapter.

5.2 Shot Segmentation

Segmentation into shots is one of the basic processes that must be applied to edited videos in order to extract information on the document structure. This information is useful for some post-processing operations, to edit abstracts, for automatic document classification etc.

Many syntactic elements of a video document are introduced during the editing step at the end of the production process. Those elements are of several kinds. They can be transitions, subtitles or intertitles, logos, audio jingles or additional noises etc. Detecting transitions is a major issue in digital video processing because those effects are used to separate two different contents, so knowledge of their location is used, for example:

- to improve compression rates;
- to apply some other automatic analysis (such as motion analysis) only on shots which can be then considered as continuous temporal units and in most of the cases as homogeneous content;
- to build an index on the temporal dimension.

But if knowledge of transition locations can be considered as critical for some applications, the extraction of this knowledge is often based on simplistic models of those transitions. There are at least two good reasons to use those simplistic models: first, they can be easily exploited in some very fast algorithms, which are usually expected to be able to process a video stream faster than 'real time' (ie. at a higher rate than 25 or 30 frames per second); the second reason is that they are sufficient to obtain very good results on most video content. In fact, the most common transition effect used to link two shots is the cut (the two shots are simply pasted one after the other). This effect can be easily characterized by a high decorrelation of pixel values before and after the transition. As cuts usually represent more than 90% of the transitions involved in edited video documents, and as these effects can be detected with a reliability that is generally higher than 90% shot segmentation is often too quickly considered as a solved problem today.

When considering more precisely syntactic elements which are added during the editing step, we can find two other kinds of effects: gradual transitions and composition effects. Some propositions have been made to automatically localize and identify some gradual transitions (mainly fades and wipes). In the general case, the reliability of their detection is not as good as for cuts, and some improvements should still be made in that field. Gradual transitions are generally used to highlight major changes in the video content and can play an important role in macrosegmentation into sequences [5].

Composition effects are used to create or to delete a sub-region of the screen where a second video content is played (inside the current video frame). These effects are really characteristic of specific video documents such as trailers, talk shows or TV news (when a distant speaker appears inside a small rectangle at the side of the screen), TV games (to show two players simultaneously, for example) etc.

Gradual transitions and composition effects are not as frequent as cuts in video documents, but their presence is more significant of a specific content. Cuts can be considered as elementary transitions; gradual transitions and composition effects are always intentionally used instead of a cut to highlight some content, to create a rhythm, and to temporally and spatially structure the visual information. They are a major clue for high-level indexing tools.

Concerning detection of cuts and fades, many works have been developed to propose shot segmentation algorithms. The very first propositions were based on the detection of a local high variation of a measure made on consecutive pixel values which was supposed to highlight the presence of a cut (such as variation of pixel intensity [6], variation of statistical local values [7], colour histograms [8] etc). Results of other digital processing appear to be also interesting for shot segmentation such as the evolution of the number of detected edges [9] or of a motion estimator [10].

Only a few papers propose an algorithm able to handle also gradual transition effects like wipes. The detection of those effects requires analysis of the spatial progression of the variation due to the effect. Some hypotheses are usually made on some features of the effect (such as the linearity of the progression, the allowed orientations for such an effect etc.) [11].

First proposals made for fade detection were generally based on one of the two following rules. In some cases, it has been considered that the sum of the small variations observed on particular measures during the effect was equal to the high variation observed on the same measures when a cut occurs (see [7] for example). A more general approach consists in considering that the variation between the frame before the effect and the frame after must be equal to that obtained between frames before and after a cut. This second approach allows detection of any kind of gradual transition effect, but is extremely dependent on the presence of camera motion.

Among the most convincing algorithms to detect gradual transition effects, we can mention [12] where given labels are attributed to pixels considering their evolution during several frames. A transition effect is detected when the proportion of associated labels exceeds a given threshold. We also have to cite proposals made to automatically detect transitions on compressed video data, such as [13], where DC coefficients are analysed because of their characteristic evolution during a fade. There is today a rather great number of different segmentation algorithms into shots. Whereas only a few works have been made in order to define a common protocol to compare those methods with each other [14], we can find several papers describing different segmentation tools and comparing their results [15,16].

To go further in the editing work analysis, we have to take into account composition effects. The temporal localization of those effects can be used to classify shots or documents in which they appear. Furthermore, the spatial localization can be use to refine results of shot segmentation algorithms (as will be explained in Sections 5.4 and 5.5).

For the spatial analysis, we have to introduce the concept of editing regions. These regions are places on the screen where a single video flow is sent during the composition. There may be several editing regions on a single frame (Figure 5.1).

We can distinguish two types of composition effects. We classify them according to whether they introduce or delete an editing region.

Once the editing region is spatially localized, one may consider that the video flow shown in this region can be the result of a previous editing work. So, some transition effects can appear in the editing region. Let us call them internal transition effects. When the size of the

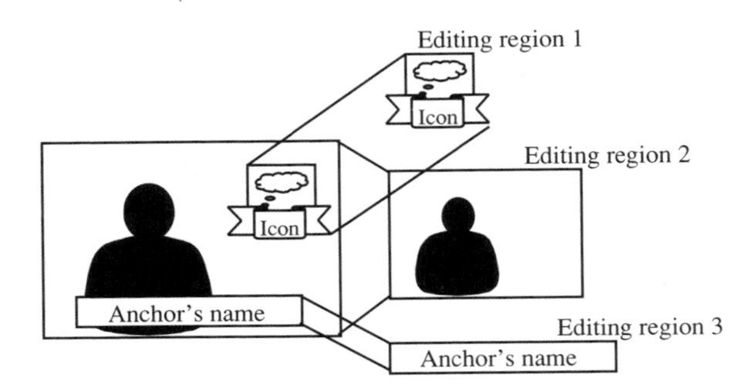

Figure 5.1 Typical editing regions for an anchor shot in a TV news programme

editing region is important, most of the segmentation algorithms into shots confuse these effects and global transitions. We could propose a recursive definition of composition and transition effects: the video in an editing region introduced by a composition effect may contain internal transitions and compositions, which introduce (or delete) new sub-editing-regions, etc. In fact, the presence of editing regions inside editing regions is not really critical for analytic purposes (which would not have been the case for synthetic ones), and so three video editing levels can be kept for an accurate analysis of the video structure:

- the higher one corresponds to global transitions which occurs on the whole screen;
- the second corresponds to the introduction or the suppression of editing regions by composition effects;
- the last one corresponds to internal transitions appearing inside an editing region.

A first algorithm has been developed to automatically identify editing levels on the basis of some previous results obtained after a first shot segmentation [17], where the main ideas are:

- Considering a statistical model of the variation during a cut, a fade and a sequence without transition effects, theoretical thresholds can be defined for a pixel-to-pixel difference algorithm.
- Effects due to object or camera motion are reduced by applying temporal filters on successive estimations of the variation. This approach produces results which are not as good as those obtained with a motion-compensated approach but highly reduces the computational cost.
- Illumination changes are taken into account mainly with histogram equalization.
- Fade detection also involves an algorithm quite similar to the label method explained in [12].

On the base of this algorithm, cut detection and fade detection are performed in an independent way. Detection thresholds are first reduced in order to detect global effects as well as local effects. Once an effect is detected, the region of activity (where the effect occurs) is localized in order to identify its potential editing level. Then, the validity of the first detection can be checked on the base of the size of this region (detection thresholds can then be precisely fixed). Values corresponding to effects associated to a lower value than the new computed thresholds are rejected (Figure 5.2).

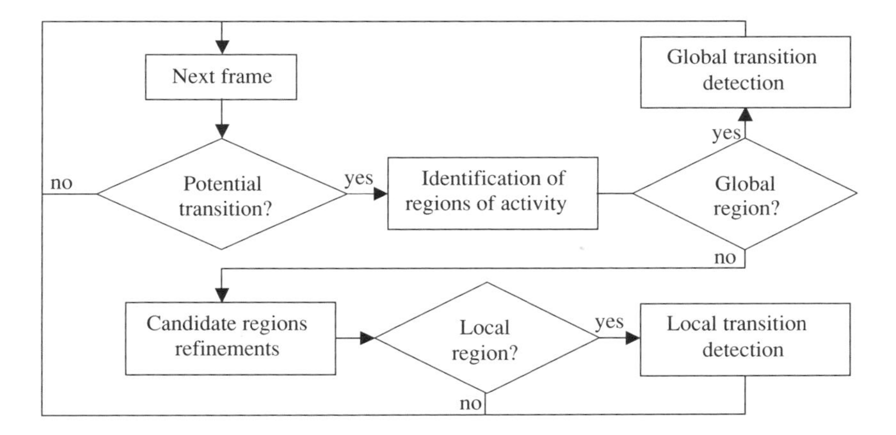

Figure 5.2 Editing level detection and classification

The region of activity is determined in two steps:

- A binary image is computed where only pixels that are detected as affected by the effect are kept (Figure 5.3). Some morphological operators are then applied in order to reduce noise and to agglomerate pixels into regions. If the resulting regions are few in number and are covering nearly all of the original frame, the potential effect is classified as a global one. Then thresholds used to detect global effects are applied to finally decide whether or not a global transition occurred.
- If the resulting regions are too small, only those with a compacity measure higher than a given threshold, and a similar region at the same place a few frames before or a few frames after, are kept. This last condition expresses the fact that there are always at least two editing operations required to produce an effect in a local area (creation, deletion and sometimes internal transitions). Thresholds used to detect effects are then proportionally adapted to the size of the region of activity to finally determine the presence of a transition. The decision to classify as a composition or as an internal effect is taken while considering that an internal transition can only occur in an existing editing region. In that sense, edges of such a region before the effect are searched using a chamfer matching algorithm. If similar edges to those of the considered region can be found, then the effect is classified as an internal transition; otherwise it is classified as a composition one.

Figure 5.3 Localization of a region of activity leading to a composition effect classification

Table 5.1 Global transition detection results

Editing level classification	Global cuts		Global fades	
	Recall	Precision	Recall	Precision
None	82.7	79.6	25.5	21.91
Present	89.8	98.0	28.5	42.0

Experiments have been made on a video corpus of 257 556 frames containing 2296 global cuts and 191 global fades. The evaluation was made on the base of a set of rules and measures defined to compare segmentation-into-shot algorithms (see Section 5.3). False detections correspond to detected effects, which do not actually exist in the video (including local effects, which must not be accepted as a global transition). An actual global effect classified as a local one is considered as a missed detection, as well as actual global transitions which are not detected. Furthermore, some very strict rules have been followed on the localization and on the duration to agree on the correctness of a global fade detection. This is why results appear to be lower than what can be expected with such algorithms. Classifying a global effect as a local one is a confusion which is more frequent than the opposite one. This is why missed detections present higher rates. For example, among the 416 missed global cuts, 182 were actually detected but classified as local ones (Table 5.1).

Compared with the same shot-segmentation algorithm (without classification), better results are obtained when the effect classification is taken into account. This is due to the fact that most errors of segmentation are not retained as candidates for a global classification in the second case, and so are considered as potential local effects. Some of them can then be filtered when they do not respect the editing region conditions (compacity and repetition).

Therefore, precision and recall rates are improved for global transition detection when applying the classification into editing levels (Table 5.2).

The performance of the editing level classification has been studied in detail on a TV news programme (of 18 576 frames). The system detected 23 composition effects of type 'Create'. Among them, 16 were actual composition effects of this type, 3 were composition effects of type 'Delete' and 4 were global transitions. All of these results are summarized in Table 5.3.

Refinements on the knowledge we can now automatically extract from the document structure have still to be taken into account in macrosegmentation tools (see Section 5.5) or document classification into genre where composition effects can be used as a feature to characterize some shots (corresponding to interviews, trailers, splitscreens, etc.).

Table 5.2 Results on global effect detection and classification

Cuts		Fades	
False	Missed	False	Missed
38	416	76	137

Table 5.3 Confusion matrix of low editing levels classification

System	Reference			
	Global	Create	Delete	Internal
Create	4	16	3	0
Delete	3	3	15	0
Internal	5	0	1	20

5.3 Evaluation of Shot-Segmentation Algorithms

Four major problems can be found for the evaluation of temporal video segmentation systems:

- the selection and gathering of a corpus (test database);
- the definition of a reference segmentation for this corpus (ground truth);
- defining one or more 'quality-measures' criteria (measure);
- comparing the automated segmentation and the reference segmentation (evaluation).

Temporal video segmentation techniques can be used in a wide range of applications, each of which induces various requirements. Examples of such constraints can be described in terms of the type of the transition effects to be recognized, and the accuracy of their detection and location, and also in terms of induced computational time. All these constraints have an impact on the above mentioned problems, namely the selection of a corpus, the definition of the effects to be indexed and the selection of an appropriate quality criterion. There is therefore a strong need for the definition of a global evaluation protocol allowing one to test consistently various algorithms and systems for different targeted applications. Moreover, the experience of developing such evaluation protocols within the field of speech recognition [18] suggests that such a development will raise a set of questions on the problem of temporal video segmentation itself.

The corpus used to test the segmentation algorithms must be sufficiently heterogeneous in terms of effect types. It must contain some of the common cases of possible error of segmentation like sequences with fast motion, high illumination changes (explosions, flashes), different editing rhythms, and so on.

At the least, the corpus choice must take into account the documents' size, their heterogeneity in terms of effect types, the number of effects etc. Proposing a measure to characterize a transition effect, and consequently the difficulty of detecting it, is equivalent to finding a solution to improve the temporal segmentation. Therefore a measure can be proposed to evaluate the complexity of the corpus:

- Number of effective effects classified by type.
- Histograms of duration of gradual transition effects (like fades and wipes): the results of the segmentation algorithms are often affected by the transition duration.
- Editing rhythm: maximum and minimum (and maybe mean and variance) of shot duration.
- Histogram of distances between effects: distance in number of frames between two consecutive effects, and between two consecutive effects of the same type.

- Variation in illumination: it is important to take this parameter into account when evaluating the quality of the resulting segmentations since it may explain why errors were made at certain points and so decrease the importance of these errors. This parameter is measured using the maximum and minimum values of differences between histograms. If the difference between the frames located before and after an effect is above a certain threshold, the detection of the transition is easy and an error made at this location can be seriously considered. On the other hand, if this difference is important between two frames within a shot (e.g. flash), a transition effect insertion is likely to be made at this place.
- Motion quantities:
  - Histogram of the Spatial Derivative Difference: a difference exceeding a given threshold may denote the presence of motion or camera work, and a transition effect insertion at this location is likely to happen.
  - Histogram of the Difference between Boundary Phase.
  - Autocorrelation measure.

The selection of a corpus is (and should be) rather independent from the evaluation protocol since it mostly depends on a related application. However, the protocol must take into account all the corpus characteristics for the evaluation of the segmentation difficulty.

Precise rules have to be defined for the various effects and their possible subdivisions. Cuts or dissolves have an intuitive significance though their exact technical definition should be much more precise and sometimes complex, and may also depend on the target application. Experience (and the state-of-the-art) shows that giving such formal definitions is not always straightforward. For instance, temporal discontinuities within 'visual jingles' and stroboscopic effects are some rather complex cases for which it could be really difficult to define, even manually, where potential boundaries can be placed. This problem obviously applies when an effect occurs only on a sub-part of the screen (in the case of a composition). In that sense, the appearance and disappearance of superimposed text, small images and logos should not be counted as transition effects.

Different file formats have been used by the authors to store the results of the segmentation process. Such formats generally follow rdf or xml guidelines or some other proposals made by the MPEG-7 group (see Section 5.4).

In order to evaluate the reliability of a technique, one needs to compare the segmentation given by the system with the reference segmentation according to given rules. These rules must define the error and correctness in the specific context of the considered application. For example, a reference segmentation may describe cuts, fades, dissolves and wipes and the segmentation algorithm only be able to detect cuts and fades, without distinction between fades and cross-dissolves.

For each individual type of effect (cut, dissolve or other) and for any of their combinations, we need to count the insertions Ni (false detections) and the deletions Nd (missed effects) between the reference indexation and the indexation produced by the tested system. For this, additional rules (distinct from the ones used to determine whether an effect is actually present or not) must be defined in order to determine whether an effect is correctly matched between a segmentation and the other. For instance, it may be decided that a dissolve effect has been correctly detected if and only if it partly overlaps by at least 50% with the correct one, in order to allow for approximate boundary detection (and/or indexation).

Some other choices have been made in the literature, for example [19] considers a minimum overlap of one frame and measures the recall and precision of gradual transition.

The total number Nt of effects of each type as well as the total number Nf of frames in the document (or database) must also be counted in order to be able to compute the various quality criteria.

5.3.1 Selection of Quality-Measure Criteria

Different measures have been proposed to compute the error or success rate over the results of different segmentation methods. For a given formula that computes the error rate (e.g. the number of inserted transition effects over the number of real ones) the results vary significantly and depend on the definitions of error and success within the application domain. We recall below some examples of the formulas that have already been proposed to evaluate temporal segmentation methods.

Accuracy

A simple expression to compute the accuracy is proposed by Aigrain and Joly [17], which is also equivalent to a measure commonly used for the evaluation of speech recognition systems [18]:

$$\text{Accuracy} = \frac{\text{Nt} - (\text{Nd} + \text{Ni})}{\text{Nt}} = \frac{\text{Nc} - \text{Ni}}{\text{Nt}} \tag{5.1}$$

where Nt, Nd, Ni and Nc are, respectively, the number of actual transition effects present in the video database, and the number of transition effects deleted, inserted and correctly found by the tested system.

Counterintuitive results may be obtained using this measure, for example $\text{Accuracy} < 0$ when $\text{Ni} > \text{Nc}$ or $\text{Nd} + \text{Ni} > \text{Nt}$. This may happen when the number of errors is greater than or equal to the number of transitions. Moreover, it is important to include the size of the video sequence in the evaluation of a segmentation method since the number of errors may potentially be equal to the number of frames Nf.

Error rate

The previous measures do not take into account the complexity of the video sequence nor its size. In [20], a measure is proposed to evaluate the error rate (insertion and deletion of transition effects) over the whole results of the segmentation algorithm:

$$\text{ErrorRate} = \frac{\text{Nd} + \text{Ni}}{\text{Nt} + \text{Ni}} = \frac{\text{Nd} + \text{Ni}}{\text{Nc} + \text{Nd} + \text{Ni}} \tag{5.2}$$

Here again, this measure does not include the complexity and size of the test video sequence. Moreover, this measure is not adequate for the evaluation and the comparison of methods because it implicitly gives more importance to deleted transition effects than to inserted ones. This importance is not weighted with an explicit factor and is therefore difficult to assess. For example, for a video sequence containing 10 transition effects, we obtain an ErrorRate equal to 1/3 if the segmentation technique produces 5 insertions (i.e. Nt = 10; Nd = 0; Ni = 5). By contrast, the error rate increases (ErrorRate = 1/2) in the case of 5 deletions (i.e. Nt = 10; Nd = 5; Ni = 0).

Recall and precision

The measures used by Boreczky and Rowe in [21] can be applied in different contexts, and mainly in the context of information retrieval. They propose the Recall (which is the ratio between desired found items) and the Precision (which is the ratio of found items that are desired).

$$\text{Recall} = \frac{\text{Nc}}{\text{Nc} + \text{Nd}} \qquad \text{Precision} = \frac{\text{Nc}}{\text{Nc} + \text{Ni}} \tag{5.3}$$

The results produced by these formulas are not normalized and are therefore difficult to compare. However, an algorithm that makes many errors receives worse precision and recall scores than an algorithm that detects an effect for each frame. In this respect, graphs of the Recall are displayed as a function of the Precision for different threshold values. Different Recall values are given for a given Precision value since these measures are, in general, compensated: if the evaluated segmentation method is very strict, the number of deleted transition effects increases while the number of inserted effects decreases. The consequence is a decreasing Recall value against an increasing Precision value.

These two parameters are strongly correlated, so their global evaluation shows the same problems as in the previous measures.

A modification of these parameters is proposed in [19] to measure the detection of gradual transitions involving several frames: the recall and precision cover. The recall represents the ratio between the overlap (in frames) of the detected effect and the real one, over the real effect duration. The precision is the ratio between the overlap and the detected transition duration:

$$\text{Recall}_{\text{cover}} = \frac{b}{a} \qquad \text{Precision}_{\text{cover}} = \frac{b}{c} \tag{5.4}$$

where a is the duration of the real transition, c the detected one and b the overlap between both effects.

Time boundary and classification errors

Hampapur and colleagues have proposed a very interesting application-oriented measure [22]. They consider the following two types of errors in the detection of transition effects: the type of the recognized transition effects and the temporal precision of the segmentation. One can increase the weight corresponding to a given error type according to the segmentation application. To compute the error, their measure compares the results of the automated segmentation to those obtained with a manual segmentation (which is supposed to be the reference, containing only correct information):

$$\text{E}(\text{V}, \text{V}') = \text{E}_{\text{LS}} * \text{W}_{\text{LS}} + \text{E}_{\text{SC}} * \text{W}_{\text{SC}} \tag{5.5}$$

where V:{S1; S2; ...; SN} is the manual video segmentation in N segments, V′:{S′1; S′2; ...; S′K} is the automated video segmentation, E_{LS} is the error in terms of segment temporal limit defined by the transition effects, W_{LS} is the weight of the temporal limit error with regard to the application, E_{SC} is the error of misclassification of transition effects and W_{SC} is the weight of the classification error.

To compute the error, segments are matched one to another and the maximal overlap between corresponding segments in the two videos is computed.

5.3.2 *Performance Measure*

The performance measure proposed by Hampapur and colleagues in [22] is related to the application domain and is able to compute errors made by most of the segmentation methods. However, the boundary segment error is very difficult to evaluate in the case of dissolve effects. The manual segmentation is therefore not reliable in this case. The following subsections describe some measures which overcome these shortcomings.

Error probability

This measure computes the probability of making an error (deletion or insertion) when an error is possible. The temporal segmentation methods can make a detection error on each video frame:

$$\mathrm{P(e|ep)} = \frac{\mathrm{Nd + Ni}}{\mathrm{Nf}} \tag{5.6}$$

Insertion probability

The insertion probability is the probability that a transition effect is detected where no effect is present:

$$\mathrm{P(insertion)} = \mathrm{P(detection|noEffect)} = \frac{\mathrm{Ni}}{\mathrm{Nf - Nt}} \tag{5.7}$$

Deletion probability

This is the probability of failing to detect an effect when the effect exists:

$$\mathrm{P(deletion)} = \mathrm{P(noDetection|effect)} = \frac{\mathrm{Nd}}{\mathrm{Nt}} \tag{5.8}$$

Similar equations can be found in [23] to evaluate the story segmentation in news. It is based on a original formula from [24] used to evaluate the text segmentation.

$$\mathrm{Pmiss} = \frac{\sum_{i=1}^{N-k} \delta_{\mathrm{hyp}}(i, i+k)(1 - \delta_{\mathrm{ref}}(i, i+k))}{\sum_{i=1}^{N-k} (1 - \delta_{\mathrm{ref}}(i, i+k))} \tag{5.9}$$

$$\mathrm{PFalseAlarm} = \frac{\sum_{i=1}^{N-k} (1 - \delta_{\mathrm{hyp}}(i, i+k))\delta_{\mathrm{ref}}(i, i+k)}{\sum_{i=1}^{N-k} \delta_{\mathrm{ref}}(i, i+k)} \tag{5.10}$$

where $\delta(i, j) = 1$ if i and j are from the same story and is equal to 0 otherwise. The segmentation of news links every word to a story. The formula compares the results for the automatic (hyp) and reference (ref) segmentations.

In the case of the shot segmentation, words are frames and stories are shots. It should be noted that equations (5.9) and (5.10) are basically equivalent to equations (5.7) and (5.8), respectively.

In [23], an inter-human comparison rate is given to compare the manual segmentation results.

Correctness probability

This is the probability of detecting a transition effect when it exists and not detecting it when it does not exist. One can give more importance to either of these situations by using weights ($k1$; $k2$).

$$\begin{aligned} P(\text{correction}) &= k1 * P(\text{detection}|\text{effect}) + k2 * P(\text{noDetection}|\text{noEffect}) \\ &= k1 * (1 - P(\text{deletion})) + k2 * (1 - P(\text{insertion})) \\ &= (k1 + k2)(1 - (k1' * P(\text{deletion}) + k2' * P(\text{insertion}))) \end{aligned} \tag{5.11}$$

where $k1$; $k2$; $k1'$ and $k2'$ take values between 0.0 and 1.0 with ($k1 + k2 = 1$). These measures take into account both the total number of transition effects (Nt) and the total number of frames of the processed sequence (Nf). This makes these measures more robust to problems encountered with the previous definitions. For the results shown in the last section we have used $k1 = k2 = k1' = k2' = 0.5$.

5.3.3 Method Complexity Evaluation

In order to compare temporal segmentation methods, it may be important to use a measure of complexity in relation to the induced computational time, the need for learning etc.

Complexity of transition effect detection

The detection of a transition may not be consistently difficult. The complexity of this detection must therefore be included within the evaluation process in order to weight the possible types of errors. Cases of segmentation errors are different for the different segmentation algorithms. However, in general, an error is more important if the transition is clear: between very different frames, different characters, different places. On the other hand, illumination change and fast motion may produce insertions in most algorithms.

There are several measures to evaluate the algorithms in relation to three different difficult cases:

- Insertion, deletion and correction probabilities related to changes in image content. This results in defining the probability of the insertion of a transition effect over the histogram difference and the probability of correct detection over the histogram difference, respectively given as:

$$P(\text{insertion}|\,\Delta H) P(\text{deletion}|\Delta H) \text{ and } P(\text{correctDetection}|\Delta H)$$

- Insertion, deletion and correction probabilities related to motion quantity.
- Insertion, deletion and correction probabilities related to the rhythm. The rhythm can be given by the shot duration or the frequency ratio between two types of transition effect.

5.3.4 Temporal Video Segmentation Methods

Many automated tools for the temporal segmentation of video streams have already been proposed. It is possible to find some papers that are providing the state of the art of such methods (see e.g. [25, 26]).

In order to give an example of possible values produced by the previous measures, they have been applied on the results produced by a set of different classical segmentation tools.

Table 5.4 gives error rates for methods based on histogram difference (method 1), intensities difference (method 2), difference between the addition of intensities (method 3), histogram intersection (method 4), invariant moments difference (method 5), thresholded intensities variation (method 6), correlation rate (method 7), χ^2 formula (method 8) and χ^2 formula over blocks (method 9). All values are presented as percentages.

Objective comparison between different temporal video segmentation systems is possible once a common corpus, a reference segmentation of it, an automatic comparison tool of automatic segmentations, and an appropriate global quality criterion are commonly selected. But in order to be able to compare those tools in a deeper manner, more accurate definitions of transition effects must be given.

The main difficulty resides in reaching a consensus on a common definition of transition effects, a common reference file format and a common mean of evaluation. The experience acquired in the speech recognition domain should be used as guidance, and periodic comparative performance tests should similarly be set up for evaluating temporal video segmentation systems.

5.4 Formal Description of the Video Editing Work

The editing work is the architectural step of an audiovisual document production process. It defines a temporal structure of the document, establishing its rhythm and the organization of

Table 5.4 Several reliability measures of cut segmentation

Method	Accuracy	Precision	Recall	Error rate	Insertion probability	Deletion probability	Error probability	Correction probability
1	−157.2	24.88	78.0	76.75	2.11	22.03	2.29	87.92
2	37.0	65.48	78.1	44.67	0.37	21.90	0.56	88.86
3	77.8	90.85	86.5	20.39	0.07	13.45	0.19	93.23
4	−27.9	41.67	69.9	64.66	0.87	30.09	1.14	84.51
5	−38.5	40.88	86.5	61.55	1.12	13.45	1.23	92.70
6	−373.3	16.44	91.5	83.80	4.18	8.49	4.22	93.66
7	1.2	50.29	95.3	50.90	0.84	4.66	0.88	97.24
8	−137.2	21.92	53.6	81.56	1.71	46.38	2.11	75.94
9	−1462.5	3.50	55.2	96.59	13.65	44.81	13.93	70.76

its semantic content. Nowadays, different domains need a description of the editing work made on video. Applications related to cinema analysis need tools to correctly describe the editing work. Other applications, like information retrieval on video databases, need a precise and complete description of this work. Another important need is the description of the editing work itself—that is, the description of how the final document has been created from the captured audio and video sources. The sequence of operations performed during the editing is called the Edit Decision List (EDL). These lists can be generated and interpreted by operators and by different editing software.

Different languages and standards are used to describe both kinds of information. However, researchers are currently trying to find answers to the needs related to the exchange of editing work descriptions.

To specify how the editing is performed, an ad hoc Description Scheme (DS) that provides the structure of the description of the video editing work for analysis and synthesis applications is proposed by the MPEG-7 standard. Therefore, the chosen description language is the MPEG-7 DDL.

This DS is intended to be generic in the sense that it supports a large number of applications which need a description of editing work. Among other applications, video document analysis requires a definition of shots and transition effects as accurate and complete as possible. Providing these definitions is an ambitious task with regard to the evolution of editing software. In the literature, we find different analytic definitions of the items of video editing. Among others, we can find these shot definitions: 'Film portion recorded by the camera between the beginning and the end of a recording. In a final film, the pastes limit a shot' and 'An image sequence which presents continuous action, which appears to be from a single operation of the camera. In other words, it is the sequence of images that is generated by the camera from the time it begins recording images to the time it stops recording images'. The facilities brought by virtual editing make these definitions too ambiguous and sometimes false. An editing process consists of connecting and mixing different recordings at different times. We can mix, for example, sequences recorded at the same time by different cameras. An example is the well-known composite rush called splitscreen. The screen is split into several parts, each one displaying different rushes. This is made by compositing techniques. Recordings are connected by transition effects and mixed by compositing. Considering this, we propose the following definition:

> A shot is formed by one or several recordings if it contains composition transition effects between two global transition effects. A single recording is called a rush in the film industry.

This definition can be extended to a rush, which can be considered as a sequence of frames showing only text or a fixed image. It can be inserted in a composite rush and superimposed to another rush.

The initial shot definitions have to be modified to take into account elements like composite rushes, composition transition effects and internal transition effects. The main characteristic that allows distinguishing between a global transition effect (between shots) and an internal transition effect (between rushes on a composite rush) is the frame area that is affected by the effect. The definitions of these items are given below.

Definition of a rush

A rush is a frames sequence before being assembled with other rushes in order to build a shot (or another intermediate rush).

Definition of a composite rush

A composite rush is a particular case of a rush generated by a compositing work. It combines frames from several rushes in the same frame. One example is the result of the superimposition of the anchor's name over the image of the anchor in the TV news broadcast (picture-in-picture operation). Several compositions may appear in the same shot. They are linked by composition transition effects.

Definition of composition transition effect

Rushes used to create a composite rush are introduced by transition effects similar to the ones which link up the shots. Visually, the difference between those kinds of transition effects is the size, on the final frame, of the introduced segment. In the case of a composite rush, it takes only a sub-part of the frame, so rushes already present before the transition are still visible. The difference between global and composition transitions is also syntactic and semantic. Transitions creating compositions link video segments of a syntactic level lower than shots. There are three types of composition transitions:

- a composition transition effect that introduces a composite rush (in that case, the rush on the full frame before the effect becomes an element of the composite rush);
- a composition transition effect that ends a composite rush (the rush on the full frame at the end of the effect was an element of the composite rush);
- the last type is a transition between two composite rushes introducing a new composition element.

The regions taking part in a composite rush may move within the frame space. This motion is not considered as a composition transition effect introducing a new composite rush. A composition transition effect is considered only when a sub-region appears or disappears on the whole frame.

Definition of an internal transition effect

Internal transition effects occur in regions reserved to each rush in a composite rush. These effects have got all the characteristics of global transition effects excepted that they occur only in sub-regions of frames. They differ from global transition effects by the type of the involved video segments (only rushes for the internal one and shots for the second one).

Definition of a global transition effect

A global transition effect is a visual effect that covers the whole frame. This effect supposes that the rush (or all the rushes) on the screen before the effect disappears, and that new ones, which were not visible until that moment, appear after the effect.

Definition of a shot

A shot is a sequence of frames that forms a temporal unit in the video document. The temporal boundaries are defined by the presence of global transition effects which cover the whole frame. One or several rushes may compose a shot involving composition and internal transition effects.

Editing software is an application domain where a description of editing work is essential. A clear and precise description for the synthesis of transition and other editing effects is required. So, the definition of elements participating in editing work must be given in two ways: first, in relation to their semantic and syntactic level (analytic) and, second, in a synthetic way. Results of automatic segmentation tools will use the analytic format. But a synthetic format, carrying the exact information about editing operations which have actually been performed, may be used for the evaluation of these results. Several standards have been also proposed to describe EDL such as [27] or the Advanced Authoring Format [28].

In current editing tools, a large variety of effects is possible and the names of these effects are not always given. Even so, there is no standard way to describe these effects. Behind a given name, many different transition effects may actually be involved. A solution given by the Advanced Authoring Format, for example, is to define a dictionary giving the name of the effect and a textual definition to be comprehensible by the editor. These definitions will be linked to the synthetic description of transitions and other effects. No semantic or syntactic information is provided.

The structure of AV documents and the editing work must be given in a clear and unambiguous language. In the domain of information retrieval, description languages must be adapted to the content and also to the queries. For these purposes, different logical languages can be used (e.g. [29]).

The solution presented here consists in creating a generic DS to described transitions (Transition DS) that can be specialized with analytic descriptors which are standardized in MPEG-7 to describe every type of effect. To create this DS, the MPEG-7 Description Definition Language, which is based on XML Schema, has been chosen. This DS gives information on how the editing work has been produced (Figure 5.4).

The AnalyticEditedVideoSegment DS contains a set of sub-description schemes to describe the video editing work. It allows the description of shot position and their duration, the description of transition types and their classification in relation to their syntactic level. These

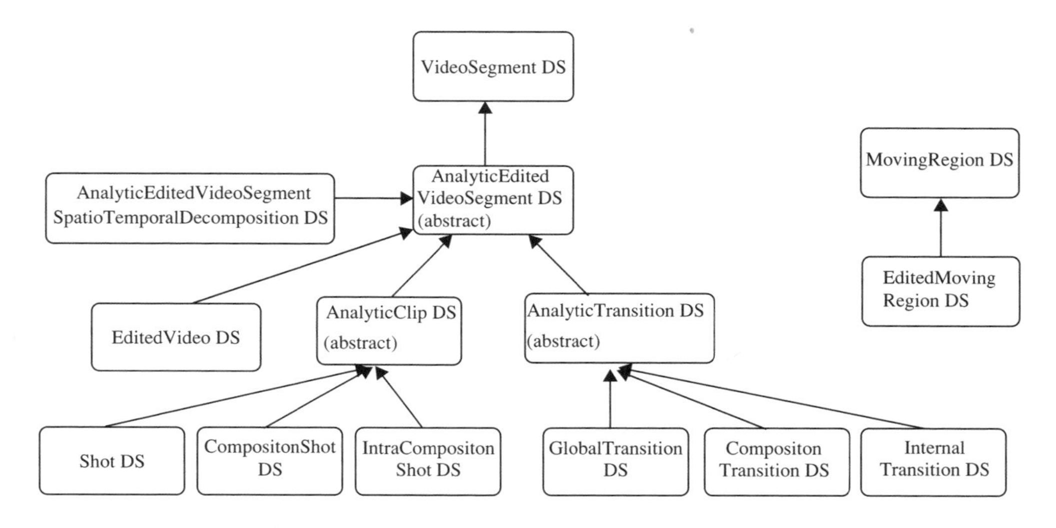

Figure 5.4 Analytic video editing work DS hierarchy in MPEG-7

elements of description are called analytic. They describe segments of a final document, which has already been edited. It consists of a hierarchy based on shots, transitions and their components, which are considered as video segments. The editing process draws up their positions in the final document. We define two kinds of video segments: clips and transitions, which can be sequentially linked. Clips and transitions can be of three types, with regard to the video editing work level in which they appear:

- Clips named shots are video segments at the top level. They can only be linked by global transitions.
- Composition shots can only appear inside an editing region. They are added or deleted by composition transitions.
- Intra-composition shots are linked by internal transitions inside an editing region.

All of these description schemes, intended to handle results of editing work analysis, contain a descriptor to inform about the reliability of the given elements. The automatically produced description can be used to build some indexes or to perform new analysis on the basis of the video segment localization and identification.

The basic DS to describe a transition effect is the Analytic Transition DS. The Analytic Transition DS represents the manner in which video segments are linked together. The type of the transition (cut, fade in, fade out, cross-dissolve, wipe) has to be chosen from a list of predefined terms. This mechanism allows the dictionary to be opened to other potential terms used to designate some complex digital effects, for example.

This Analytic Transition DS is an abstract one. It can not be instantiated. There are three different transition levels in relation to the syntactic meaning and consequently in relation to the type of video segments between which the transition is realized. The three levels are called global, composition and internal. The location of transitions is given by a time descriptor.

Global transitions link shots, so a description of Shot elements is required. For composition and internal transition, the linked segments which are required depend on the type of the composition: create, delete or change.

This DS can be used in a large number of applications from information retrieval to movie production. One of the first applications using this DS aims at representing results of video-to-shots segmentation algorithms.

The description of the editing work can be addressed from different points of view in different applications. But all applications, from video retrieval to movie post-production, need an editing work description. The DS presented in this chapter provides a unique, compact and complete solution to the need for description for the video editing work. It has been specifically designed in order to be extended to take into account synthetic pieces of descriptions of a video editing work (EDLs) and to be easily adapted to describe the audio editing in the same kind of structure.

5.5 Macrosegmentation

One hour of television provides typically about 700 shots per hour. In actual motion picture, there can often be 500 to 2000 shots per hour. Thus, temporal objects greater than shots must be extracted to enable quick browsing of the video contents, quick positioning in the document for interactive viewing, or to automatically construct abstracts of the document. Of course, for such kinds of applications, we want these macroscopic time objects to match sequences in the video that constitute a narrative unit or share the same setting. Only a few works have been presented in the field of temporal segmentation of videos to extract sequence boundaries in the general case. A sequence boundary can be set only according to a break in the semantic continuum. As a deep semantic analysis of audiovisual contents can still not be achieved, most works on sequence segmentation are based on the analysis of some subsets of audiovisual grammar rules. Automatic construction of a macrosegmentation is a relatively difficult problem—and the evaluation of such a tool is maybe a much more complex task.

As well as for shot segmentation, motivations for automatically extracting sequence (or semantic events) boundaries lay in the fact that human macrosegmentation can not be reliably performed by only accessing some representative images but requires access to the full video content, which is time consuming.

5.5.1 General Approaches

The first user interfaces developed to enable browsing of audiovisual content sets highlighted that different scales (or zoom levels) must be used to perform quick selection, for example. Therefore, some hierarchical browsers have been proposed where shots (or frames) are gathered to build homogeneous sets (in fact, a predetermined number of shots is gathered to build each set) [30, 31].

Yeung and colleagues [32] have proposed a technique for structuring the complete contents of a video in a directed graph, whose nodes are clusters of similar shots and whose edges represent temporal relations between shot clusters: there is an edge between cluster C1 and cluster C2 if a shot in C2 immediately follows a shot in C1. Shots are gathered in the same cluster when their visual content is quite similar (the similarity is based on a distance computed on key-frames histograms). This type of macro-structuration is very interesting, but it does not build a linear macrosegmentation suitable for hierarchical browsers.

In video, there is a wide range of possible temporal local clues for macroscopic changes, such as special transition effects, changes on the soundtrack, modification of the editing rhythm, changes in sets etc. Such clues are used by the viewer/listener in collaboration with his/her global semantic understanding of action or narration. These clues are more or less invariant depending on the type of programme or director. The method proposed by Yong Rui and colleagues [33] is based on the computation of an activity rate, a visual similarity and the 'time locality' (which aims at gathering distant shots that provide quite similar key frames). This last criterion is also used in the method proposed by [34]. This method also extracts information about the time of the action (night or day). Continuous shots that belong to the same time unity are gathered. Three other clues are used by the method proposed in [35], such as the editing rhythm, the camera work and elements on the soundtrack structure. This method is based on the application of rules expressing the local clues which are given by the medium contents to enable identification of more macroscopic changes. For example, a sequence boundary is set on the nearest shot boundary when music appear on the soundtrack and shots are gathered when their duration is directly correlated with the duration of previous ones (such rhythmic properties are produced during the editing step).

5.5.2 *TV News and Sport Events Segmentation*

Many works have been presented on TV news since HongJiang Zhang and colleagues proposed a method for the automatic macrosegmentation of such programmes [36]. This method is based on a specialized model for this type of programme, using recognition of particular types of shots (anchor person, anchor person with a logo, interview etc) and rules on the possible global temporal organization of a TV news programme.

More recently, many research works were done to classify, analyse and detect events in sport videos because of their broad distribution on networks, their large audience and mainly because they are highly structured. Considering the fact that sports event temporal structures do not follow the same strict rules used in TV news, the automatic tools which have been developed integrate a large set of low-level features and some strategies to exploit the stochastic nature of those low-level feature evolutions during some scenes of interest (called highlights).

Analysis of baseball and tennis videos

Di Zhong and Shih-Fu Chang [37] present a method to analyse the structure of sport videos such as tennis and baseball. Compared with other videos like television news and movies, sport videos have a well-defined temporal structure of content. A long game is often hierarchically divided into some segments and sub-segments (games, sets, quarters etc.). Moreover, sport videos are shot with a fixed number of cameras on the field. In a tennis game, when a service is made, the scene is usually commutated with an overview of the field. For broadcast events, there are some advertisements or some special information inserted between sections of play. The principal objective of the work presented in [37] is the automatic control of fundamental sights (for example, service) which indicate boundaries of high-level segments. On the base of the identification of those segment boundaries, useful applications, such as the automatic production of a synopsis, can be developed. The technique suggested to detect the basic units in a game, such as services in tennis and throws in baseball, exploits the idea that these units usually start with a specific view. An approach based on the dominant colour is mainly employed, and to improve the effectiveness, objects present in the picture are checked to remove false alarms. In particular, constraints of uniformity (related with colour, edges and movement) are considered to segment object moving in the foreground and objects in the background. Then, some specific characteristics are used to identify some scenes. For example, in a general view of a service in a set of tennis, there are horizontal and vertical lines on the court. The detection of these lines is taken into account to improve the identification.

An automatic analysis tool for tennis videos has been also proposed in [38]. The main goal consists in the identification of strokes in videos of tennis based on a hidden Markov model (HMM). The first step segments the player from the background, then, in the second step, HMMs are applied. The considered characteristics are the dominant colour and the shape description of the segmented player.

Extraction of highlights in baseball videos was also considered in [39], where a statistical model is built in order to explore the spatial and the temporal specific structure in videos of baseball. The suggested approach is based on two observations. First, the majority of highlights are composed of certain types of shots, like a run overview or the view of the audience. Although the visual content of the same type of shots may slightly differ, they present some common statistical properties. The second observation is that highlights are usually composed of a set of really specific successive shots (this work considers that only one point of view can be seen in a shot). For example, a typical home-run is usually composed of a view of the pitch followed by a view of the audience and then of a close-up. The authors propose different HMMs to take into account 'home-runs', 'nice hits', 'nice catches' and 'plays within the diamond'. The characteristics used in [39] are a descriptor of edges, the quantity of grass, the quantity of sand, the camera motion and the player heigth. Naturally, the context of all the home-runs may differ but it can be modelled by an HMM. In the suggested system, an HMM for each type of highlight is learned. A probabilistic classification is then made by combining a first classification of the view and the HMM. First, the system segments the digital video into shots. Each shot is then compared with the learned model, and its probability of belonging to one of the predefined classes is then computed. Then, considering the potential succession of views, the probability of each type of highlight can be computed with the qualified HMM.

Analysis of football (soccer) videos

In videos of football, most of the shots present similar characteristics of colour distribution. A simple clustering algorithm of shots would not be relevant to identify high-level elements in the game structure. Moreover, football videos contain less "canonical views" than baseball or tennis games, which could indicate some temporal boundaries of important events.

In [40], the method exploits positions of the player and the ball during the game as an input, and so relies on precise tracking of semantic information. In [41], the authors present a system which uses knowledge of the spatial structure of the field to analyse the contents in a football video. The document is split into semantic segments which are presented to an end user to select events of interest.

Phases of stop and play can be detected by types of views according to [42] and by characteristics of motion and colour according to [43]. In [42], a model is proposed to segment a football video into logical units of 'play' and 'break'. A proportion of grass colour is used to classify frames in three types of view: global, zoom-in and close-up. Global views are classified as 'play' segments and the close-ups as 'break' ones if they have a minimal length. The work described in [42] is refined in [43], where an algorithm to analyse the structure of football videos is proposed. First, the same two states of the play are defined: Play and Break. A set of features related to the field, a ratio of dominant colours and motion intensity is selected. Each state of the game has a stochastic nature which is modelled by a set of HMM. Then, techniques of standard dynamic programming are employed to obtain the maximum of probability to segment the whole game into the two states.

In [44], the authors proposed a framework to automatically analyse and summarize football videos by using object-based and cinematic features. The proposed framework includes algorithms of low-level video processing, such as detection of dominant colour areas, a robust shot segmentation algorithm, and the classification of types of views, as well as some algorithms of high level to detect a goal, the referee or the penalty box. The system can automatically generate three types of abstracts: (1) all slow-motion segments, (2) all the goals and (3) an extension of the previous abstracts, including results of penalty box and referee detection. The first two types of abstracts are based on the cinematographic features for fast computing reasons, whereas abstracts of the last type contain semantics of a higher level. The distinction of the colour of the referee's shirt is employed for his detection. The penalty zone is characterized by three parallel lines specific to this sector on a football field. The algorithm of goal detection is based on a cinematographic charter which should be fulfilled:

- Duration of the break: a break due to a goal does not last less than 30 and not more than 120 seconds.
- Occurrence of at least one close-up/out-of-field shot: this shot can be a close-up on a player or a view of the public.
- Occurrence of at least one slow-motion segment: the action is generally replayed once or several times.
- The relative position of the slow-motion segment: it follows the close-up/out-of-field shot.

Methods presented in [45, 46] are based on correlations between low-level descriptors and semantic events in a football game. In particular, in [45] the authors show that low-level descriptors are not sufficient, individually, to obtain satisfactory results. In [46, 47], they propose to exploit the temporal evolution of low-level descriptors in correspondence with semantic events with an algorithm based on a finite-state machine.

Analysis of basketball videos

A method for segmenting basketball videos into semantic events is presented in [48]. This method is also rules-based, involving some features of motion (vector directions and magnitudes), colour and edges. The system proposes a classification into nine most significant events. The experimental results give a precision rate of classification between 70% and 82%, and a recall rate between 76% and 89% for the nine classes.

In [49], motions in the video are analysed through their answer to a set of filters which are designed to react to some specific dominant movements. Then HMMs are involved to characterize models of motion evolution corresponding to16 different events. The rate of user satisfaction considering the results of this classification is about 75%.

Slow-motion segment detection

Considering work in sport events detection and macrosegmentation, it appears that detection of slow-motion segments is an important tool because those segments are always related to significant events, and so are useful to detect highlights and to generate abstracts.

Pan and colleagues [50] present a method for highlight detection for various types of sports mainly based on the idea that interesting events are usually replayed in slow-motion immediately after their occurrence in the video. They present an algorithm to compute the probability and to localize the occurrence of slow-motion segments which are modelled by an HMM. Five states are used to structure the HMM: slow-motion, still, normal replay, edit effect and normal play. Four features are used by this HMM. The fourth one, i.e. the RGB colour histograms, is used to capture gradual transitions. This feature is also used to filter commercials in sports programming.

In [51], an evolution of that work is proposed to detect slow-motion segments by locating logos in specific parts of the sports video near replay sections. The algorithm initially determines a template of the logo, which is then used to detect similar images in the video. Then, replay segments are identified by gathering detected sequences with the logo and slow-motion segments.

5.5.3 Audiovisual Segmentation Based on the Audio Signal

Whereas most current approaches for macrosegmentation of audiovisual data are focused on the video signal, the soundtrack can play a significant role for such a goal. This part describes some approaches which involve audio analysis. For a larger description of existing techniques for analysing audiovisual contents using the audio signal, refer to [52].

In the case of audiovisual sequences (of football, basketball or baseball games), the soundtrack is mainly composed of comments coexisting with some foreground and background noises. Background noises include the ambient noise of the crowd, segments of applause or cheering, and special events such as whistles or beats.

Rui and colleagues [53] detect highlights in baseball videos using only audio features. The characterization is carried out by considering a combination of some generic audio characteristics of sports events and specific characteristics of baseball, combined within a probabilistic framework. The audio track is a mix of the speech of the speaker, the noise of the crowd, and some noises of traffic and music, on which on automatic control of gain is applied to harmonize the audio level. They specifically detect baseball hits and excited speech to evaluate the probability of a segment of highlight. Thus, the same approach could also be applied in theory to some other sports such as golf. Audio features taken into account are related with some energy, phoneme, information complexity and prosodic features. The performance of the proposed algorithm has been evaluated by comparing its results with highlights chosen by a user in a collection of baseball videos.

Lefevre and colleagues [55] propose such a method, which segments the audio signal associated with a football video sequence into three classes (speaker, crowd and the whistle of the referee). The suggested method also employs cepstral analysis and HMM.

5.5.4 Audiovisual Segmentation Based on Cross-Media Analysis

Several approaches propose semantic event detection through the combination of audio and video features, mainly to improve performances.

In [56] the authors propose a method based on a cross-media analysis of audiovisual information. An analysis of the video structure of a tennis game is performed. Shots are classified with colour and motion features in four types of views (global and medium views, close-ups and views of the audience are characterized by different distributions of colour content and camera motion). From the audio stream, classes of sounds such as ball hits, applause or speech are identified. Audio and video features are simultaneously combined with HMMs. On the basis of these features, shots are classified into four types: first missed serve, rally, replay and break.

Han and colleagues [57] proposed an integrated system of baseball summarization. The system detects and classifies highlights in baseball videos. The authors observe that each sport game has a typical scene with a characteristic colour distribution, such as the pitch scene in baseball, the corner-kick scene in football and a serve in tennis. For these scenes, colour distribution is strongly correlated between similar events. Edge distribution is used to distinguish views of the public from views of the ground (with a lower edge density than for the previous class). Camera motion is estimated by a robust algorithm. Players are detected with colour, edge and texture features. The authors also consider some audio characteristics. In particular, the occurrence of silence, speech, music and noise and their combination in each shot is detected. To achieve this task, they use MFCCs in Gaussian mixture models. The authors also extract some words or some expressions from captions, regarding them as clues for highlight detection (such as field, base, score etc.). The multimedia features are then aggregated by using an algorithm based on maximization of entropy to detect and classify the highlights.

In order to reduce the number of false detections, a semantic indexing algorithm has been introduced, based on audio and video features, in [58]. The video signal is initially processed to extract visual low-level descriptors from MPEG compressed videos. A descriptor 'lack of motion' is evaluated with a threshold on the mean value of motion vector magnitudes. This descriptor is relevant to localize the beginning or the end of important phases in the game. Parameters of the camera motion, represented by horizontal factors of pan and zoom, are evaluated on vector fields of P-frames. Cuts are detected as an abrupt variation of motion information and a high number of Intra-Coded Macroblocks in P-frames. But the low-level features mentioned above are not sufficient, individually, to obtain satisfactory results. To detect some specific events, like goals or shoots towards the goal, the suggestion is to exploit the temporal evolution of motion features according to such events. A goal event is usually associated with a pan or a fast zoom followed by a lack of motion and a cut. In general, a fast pan occurs during a shoot towards the goalkeeper or during an ball exchange between distant players. Fast zooms are employed when interesting situations are likely to occur according to the perception of the cameraman. The evolution of visual low-level descriptors is modelled by a controlled Markov chain. A model to capture the presence of a goal and five other events (e.g. corner, penalty) is presented. Those events are then analysed in the second stage considering the transition of audio loudness between the pairs of consecutive selected shots. The method suggested is applied to the detection of a goal. The experimental results show the effectiveness of the suggested approach.

Petkovic and colleagues [59] proposed an algorithm to extract highlights in car race videos. The extraction is carried out by considering a cross-media approach which employs audio and video features combined with dynamic Bayesian networks (DBN). Four audio features are selected for speech detection; namely, the short time energy (STE), pitch, MFCC and pause rate. For the identification of specific keywords on speech segments, a keyword-spotting tool is used. In the video analysis, colour, shape and motion features are taken into account. The video is segmented into shots while computing colour histogram differences of colour among several successive frames. The presence of superimposed text on video is also involved in the analysis.

The motion magnitude is estimated and detectors of semaphore, dust, sand and replays are applied in order to characterize some specific events (such as start or fly-out events). Some other details of this algorithm are described in [60]. The given results show that taking into account audio and video improves highlight characterization in car race videos. For example, the audio DBN detect only 50% of highlights in the race. The integrated audiovisual DBN can detect 80% of all the relevant segments in the race.

5.6 Conclusion

Audiovisual macrosegmentation is a challenging field of research. Such work requires a large set of skills. While considering the state-of-the-art, contributions are usually gathering specialists of audio processing, video processing and, most of the time, specialists of data fusion and mining. The number and the diversity of conferences which are nowadays proposing some special session on that topic is increasing.

Today, works are mainly focused on really specific contents with generally rather small opportunities to address other kinds of documents. But this is only the first step in semantic audiovisual content analysis.

While considering efforts made in order to built a framework for the development of shot-segmentation algorithms, we can imagine that a lot of work has still to be achieved during the next years on audiovisual macrostructure identification. While trying to find some new ways of producing tools which could automatically summarize any kind of content, we should also consider some 'beside tools' which will soon be required by this technology. We could use our knowledge and past experiments on shot segmentation in order to prepare these tools and to quickly be able to propose some frameworks of evaluation. Hence, we will soon have to answer questions such as 'how can we distinguish a good summarizing tool from a bad one?', 'How can results which are generated during that kind of processing (metadata, detected events, even learnt events in trained HMMs) be shared in interoperable systems?'.

5.7 Acknowledgement

This chapter gathers research works developed with Rosa Ruiloba, Georges Quénot, Stéphane Marchand-Maillet and Zein Ibrahim. Parts of this work were developed during the AVIR (ESPRIT No. 28798) and the KLIMT (ITEA No. 000008) European projects.

References

[1] R. Ruiloba, P. Joly, Adding the concept of video editing levels in shot segmentation. Special Session on Intelligent Systems for Video Processing, In *Proceedings of IPMU 2002*, Annecy, France, July 2002.

[2] R. Ruiloba, P. Joly, Framework for evaluation of Video-To-Shots segmentation algorithms. *Networking and Information Systems Journal*, **3**(1), 77–103, 2000.

[3] R. Ruiloba, P. Joly, Description scheme for video editing work. In A.G. Teschar, B. Vasudev, V.M. Bore (eds) *Proceedings of SPIE—Multimedia Systems and Applications III*, Boston, MA, vol. 4209, pp. 374–385. SPIE, 2001.

[4] Z. Ibrahim, Etude des relations entre descripteurs multimedia dans les macrostructures audiovisuelles, Research report, IRIT.

[5] P. Aigrain, P. Joly, V. Longueville, Medium knowledge-based macro-segmentation of video into sequences, M.T. Maybury (ed.) *Intelligent Multimedia Information Retrieval*. MIT Press, Cambridge, MA, 1997.

[6] A. Nagasaka, Y. Tanaka, Automatic video indexing and full search for objects appearances. In *Proceedings of IFIP—Second Working Conference on Visual Database Systems*, Budapest, October 1991.

[7] H. Zhang, A. Kankanhalli, S.W. Smolliar, Automatic partitionning of full-motion video. *Multimedia systems*, **1**(1), pp. 10–28, 1993.

[8] J.M. Corridoni, A. Del Bimbo, Structured digital video indexing. In *Proceedings of the 13th International Conference on Pattern Recognition*, vol. 3, pp. 125–129. IEEE, 1996.

[9] R. Zabih, J. Miller, K. Mai, A feature-based algorithm for detecting and classifying scene breaks. In *Proceedings of ACM Multimedia '95*, pp. 189–200. ACM Press, New York, 1995.

[10] P. Bouthemy, M. Gelgon, F. Ganansia, A unified approach to shot change detection and camera motion characterization. *IEEE Transactions on Circuits and Systems for Video Technology*, **9**(7), 1030–1044, 1999.

[11] M. Wu, W. Wolf, B. Liu, An algorithm for wipe detection. In *Proceedings of the International Conference on Image Processing*, October 1998, pp. 893–897. IEEE, 1998.

[12] Y. Taniguchi, A. Akutsu, Y. Tonomura, PanoramaExcerpts: extraction and packing panoramas for video browsing. In *Proceedings of the Conference ACM Multimedia '97*, 8–14, November 1997, pp. 427–436. ACM Press, New York, 1997.

[13] J. Meng, Y. Juan, S. Chang, Scene change detection in a mpeg compressed video sequence. In *Proceedings of SPIE Digital Video Compression: Algorithms and Technologies*, vol. 2419, pp. 14–25. SPIE, 1995.
[14] R. Ruiloba, P. Joly, S. Marchand-Maillet, G. Quénot, Towards a standard protocol for the evaluation of video-to-shots segmentation algorithms. In *Proceedings of Content-Based Multimedia Indexing*, Toulouse, 1999.
[15] A. Dailianas, R. Allen, P. England, Comparison of automatic video segmentation algorithms. In *Proceedings of SPIE Photonics West*, Philadelphia, PA, October 1995.
[16] R. Lienhart, Comparison of automatic shot boundary detection algorithms. In *Proceedings of SPIE—Image and Video Processing VII*, vol. 3656, p. 29, 1999.
[17] P. Aigrain, P. Joly, The automatic real-time analysis of film editing and transition effects and its applications. *Computer and Graphics*, **18**(1), 93–103, 1994.
[18] *Proceedings of DARPA Speech and Natural Language Workshop*, January 1993.
[19] G. Lupatini, C. Saraceno, R. Leonardi, Scene break detection: a comparison. In *Proceedings of VIII International Workshop on Research Issues in Data Engineering: Continuous-Media Databases and Applications,* pp. 34–41, February 1998.
[20] J.M. Corridoni, A. Del Bimbo, Film semantic analysis. In V. Cantoni (ed.) *Computer Architectures for Machine Perception, CAMP '95*, Como, Italy, pp. 202–209. IEEE, Los Alamitos, CA, 1995.
[21] J.S. Boreczky, L.A. Rowe, Comparison of video shot boundary detection techniques. In *Proceedings of SPIE Electronic Imaging*, vol. 2670, pp. 170–179, San Jose, CA, 1996.
[22] A. Hampapur, R. Jain, E. Weymouth, Production model based digital video segmentation. *Multimedia Tools and Applications*, **1**, 9–46, 1995.
[23] A.G. Hauptmann, M.J. Witbrok, Story segmentation and detection of commercials in broadcast news video. In *Research and Technology Advances in Digital Libraries Conference*, Santa Barbara, CA, April 1998. IEEE, 1998.
[24] D. Beeferman, A. Berger, J. Lafferty, Text segmentation using exponential models. In *Proceedings of the Conference on Empirical Methods in Natural Language Processing 2 (AAAI '97)*, Providence, RI, 1997. ACL, 1997.
[25] F. Idris, S. Panchanathan, Review of image and video indexing techniques. *Journal of Visual Communication and Image Representation*, **8**, 146–166, 1997.
[26] I. Koprinska, S. Carrato, Temporal video segmentation: a survey. *Signal Processing, Image Communication*, **16**, 477–500, 2001.
[27] SMPTE, SMPTE Standard for Television. Transfer of Edit Decision Lists. ansi/smpte 258m-1993. American National Standard/Society of Motion Picture and Television Engineers (SMPTE) Standard. Approved 5, February 1993, pp. 1–36, 1993.
[28] AAF, Advanced Authoring Format (AAF). An Industry-Driven Open Standard for Multimedia Authoring. AAF Association, 2000.
[29] E. Ardizzone, M.-S. Hacid, A knowledge representation and reasoning support for modeling and querying video data. In Proceedings of the *11th IEEE International Conference on Tools with Artificial Intelligence,* 8–10 November 1999, Chicago, IL. IEEE, 1999.
[30] M. Cherfaoui, C. Bertin, Two-stage strategy for indexing and presenting video. In *Proceedings of SPIE '94, Storage and Retrieval for Video Databases*, San Jose, CA, February 1994, pp. 174–184. International Society for Optical Engineering, 1994.
[31] H. Zhang, S.W. Smoliar, J.H. Wu, Content-based video browsing tools. In *Proceedings of SPIE '95, Multimedia Computing and Networking*, San Jose, CA, February 1995, vol. 2417, pp. 389–398. International Society for Optical Engineering, 1995.
[32] M. Yeung, B.L. Yeo, W. Wolf, B. Liu, Video browsing using clustering and scene transitions on compressed sequences. In *Proceedings of SPIE '95, Multimedia Computing and Networking*, San Jose, CA, February 1995, vol. 2417, pp. 399–413. International Society for Optical Engineering, 1995.
[33] Y. Rui, S. Huang, S. Mehrota, Exploring video structures beyond the shots. In *Proceedings of the IEEE Conference on Multimedia Computing and Systems*, Austin, TX, 28 June–1 July 1998. IEEE, 1998.
[34] P. Faudemay, C. Montacié, M.J. Caraty, Video indexing based on image and sound. In *Proceedings of SPIE '97, Multimedia Storage and Archiving Systems II*, Dallas, TX, November 1997, pp. 57–69. International Society for Optical Engineering, 1997.
[35] P. Aigrain, P. Joly, V. Longueville, Medium knowledge-based macro-segmentation of video into sequences. In M.T. Maybury (ed.) *Intelligent Multimedia Information Retrieval*, pp. 159–173. MIT Press, Cambridge, MA, 1997.

[36] H. Zhang, S.W. Smoliar, J.H. Wu, Content-based video browsing tools. In *Proceedings of SPIE '95, Multimedia Computing and Networking*, San Jose, CA, February 1995, vol. 2417, pp. 389–398. International Society for Optical Engineering, 1995.

[37] D. Zhong, S.-F. Chang, Structure analysis of sports video using domain models. In *Proceedings of IEEE International Conference on Multimedia and Exposition (ICME 2001)*, Tokyo, Japan, 22–25 August 2001, pp. 920–923. IEEE, 2001.

[38] M. Petkovic, W. Jonker, Z. Zivkovic, Recognizing strokes in tennis videos using hidden Markov models. In *Proceedings of the IASTED International Conference on Visualization, Imaging and Image Processing*, Marbella, Spain, 2001.

[39] P. Chang, M. Han, Y. Gong, Extract highlights from baseball game video with hidden markov models. In *Proceedings of ICIP '2002*, Rochester, NY, September 2002, pp. 609–612.

[40] V. Tovinkere, R.J. Qian, Detecting semantic events in soccer games: toward a complete solution. In *Proceedings of IEEE International Conference on Multimedia and Exposition ICME 2001*, Tokyo, Japan, 22–25 August 2001, pp. 1040–1043. IEEE, 2001.

[41] Y. Gong, L.T. Sin, C.H. Chuan, H. Zhang, M. Sakauchi, Automatic parsing of TV soccer programs. In *Proceedings of ICMCS '95*, Washington, DC, May 1995.

[42] P. Xu, L. Xie, S-F. Chang, A. Divakaran, A. Vetro, H. Sun, Algorithms and system for segmentation and structure analysis in soccer video. In *Proceedings of IEEE International Conference on Multimedia and Exposition (ICME 2001)*, Tokyo, Japan, 22–25 August 2001. IEEE 2001.

[43] L. Xie, S-F. Chang, A. Divakaran, H. Sun, Structure analysis of soccer video with hidden Markov Models. In *Proceedings of IEEE International Conference on Acoustics, Speech, and Signal Processing (ICASSP 2002)*. IEEE, 2002.

[44] A. Ekin, M. Tekalp, Automatic soccer video analysis and summarization. In *Proceedings of SPIE 2003, Storage and Retrieval for Media Databases*, Santa Clara, CA, January 2003, pp. 339–350. International Society for Optical Engineering, 2003.

[45] A. Bonzanini, R. Leonardi, P. Migliorati, Semantic video indexing using MPEG motion vectors. In *Proceedings of EUSIPCO 2000*, Tampere, Finland, September 2000, pp. 147–150.

[46] A. Bonzanini, R. Leonardi, P. Migliorati, Event recognition in sport programs using low-level motion indices. In *Proceedings of IEEE International Conference on Multimedia and Exposition (ICME 2001)*, Tokyo, Japan, 22–25 August 2001. IEEE 2001.

[47] R. Leonardi, P. Migliorati, Semantic indexing of multimedia documents. *IEEE Multimedia*, **9**(2), 44–51, 2002.

[48] Wensheng Zhou, Asha Vellaikal, C.-C. Jay Kuo, Rule-based video classification system for basketball video indexing. In *Proceedings of ACM International Workshop on Multimedia Information Retrieval*, Los Angeles, CA, 4 November 2000, pp. 213–216.

[49] Gu Xu, Yu-Fei Ma, Hong-Jiang Zhang, Shiqiang Yang, Motion based event recognition using HMM. In *Proceedings of the International Conference on Pattern Recognition*, vol. 2, pp. 831–834. IEEE, 2002.

[50] H. Pan, P. van Beek, M.I. Sezan, Detection of slow-motion replay segments in sports video for highlights generation. In *Proceedings of IEEE International Conference on Acoustics, Speech, and Signal Processing (ICASSP 2001)*. IEEE, 2001.

[51] H. Pan, B. Li, M. Sezan, Automatic detection of replay segments in broadcast sports programs by detection of logos in scene transition. In *Proceedings of IEEE International Conference on Acoustics, Speech, and Signal Processing (ICASSP 2002)*, May 2002, Orlando, FL, May 2002. IEEE, 2002.

[52] T. Zhang, C.-C. J. Kuo, Audio content analysis for online audiovisual data segmentation and classification. *IEEE Transactions on Speech and Audio Processing*, **9**, 441–457, 2001.

[53] Y. Rui, A. Gupta, A. Acero, Automatically extracting highlights for TV baseball programs. In *Proceedings of ACM Multimedia 2000*, Los Angeles, CA, pp. 105–115.

[54] Dongqing Zhang, D. Ellis, Detecting sound events in basketball video archive. Columbia University, Electrical Engineering Department, New York City, NY.

[55] S. Lefevre, B. Maillard, N. Vincent, 3 classes segmentation for analysis of football audio sequences. In *Proceedings of ICDSP 2002*, Santorin, Greece, July 2002.

[56] E. Kijak, G. Gravier, P. Gros, L. Oisel, F. Bimbot, HMM based structuring of tennis videos using visual and audio cues. In *Proceedings of IEEE International Conference on Multimedia and Exposition (ICME 2003)*, Baltimore, MO, 6–9 July, 2003. IEEE, 2003.

[57] M. Han, W. Hua, W. Xu, Y. Gong, An integrated baseball digest system using maximum entropy method. In *Proceedings of ACM Multimedia 2002*, Juan Les Pins, France, December 2002, pp. 347–350.

[58] A. Bonzanini, R. Leonardi, P. Migliorati, Exploitation of temporal dependencies of descriptors to extract semantic information. In *Proceedings of VBLV 2001*, Athens, Greece, 11–12 October 2001.

[59] M. Petrovic, V. Mihajlovic, W. Jonker, S. Djordievic-Kajan, Multi-modal extraction of highlights from tv formula 1 programs. In *Proceedings of IEEE International Conference on Multimedia and Exposition (ICME 2002)*, Lausanne, Switzerland, August 2002. IEEE, 2002.

[60] V. Mihajlovic, M. Petrovic, Automatic annotation of formula 1 races for content-based video retrieval. Technical report, TR-CTIT-01-41, December 2001.

6

Object-Based Video Indexing

Jenny Benois-Pineau

6.1 Introduction

The Semantic Web can be understood as an enormous information pool containing very heterogeneous data coming from various domains of human activity. Since the beginning of the past decade, we have seen a progressive transformation of the scale and the nature of the data. From text and elementary graphics, the information content has evolved to multimedia collections of a large scale. Today users search for multimedia data for their leisure and professional activities. In the first case we can mention the well-established practice of searching for music in digital form and the newer practices of searching for movies and programmes in the context of Web TV and broadband Internet, and browsing through personal multimedia collections, such as those recorded by personal home devices from broadcast channels. In the second case, one can imagine information retrieval in distributed image databases, video archives in the domain of cultural heritage, surveillance video and so on. In order to be retrieved, all this rich and various multimedia data has to be indexed. From the early 1990s, indexing of video content for visual information retrieval [1] has been the focus of a large research field tightly related to the development of the Semantic Web concept: content-based multimedia indexing.

The problem of indexing multimedia data by content for efficient search and retrieval in the Semantic Web framework can be considered at different levels of knowledge. In the 'macro' view of content that is sequential in time, such as video, the Table of Content is usually understood as the description of the content in terms of chapters characterized by low-level (signal-based) homogeneity or high-level (semantic) uniformity. A vast amount of literature has already been devoted to this, and new methods continue to appear for segmenting video into homogeneous sequences, such as shots, or grouping them into semantic chapters (see for instance [1] for an overview of such methods).

The 'micro' view of the content can be understood as a set of objects evolving inside the chapters and is of much interest in particular for retrieval tasks; examples are searching for a particular player in a sports programme, recognition of a particular character in a movie and search for all scenes in which he appears, and extraction of scenes of high tension such as car

Multimedia Content and the Semantic Web Edited by Giorgos Stamou and Stefanos Kollias

chases. The usual query by object implies matching of descriptors of an example object or of a generic prototype imagined by the user with the description of objects in an indexed content.

Object-based indexing of multimedia content is a new and quickly developing trend which supposes analysis of various sources of multimedia data representing a content unit: the audio track, the video signal, the associated text information (e.g. the keywords extracted from an Electronic Programme Guide available on the Internet). In this chapter we will limit ourselves to video indexing and will try to give an overview of trends in this field.

The MPEG-7 [2] standard supplies a normalized framework for object description in multimedia content. An overview of MPEG-7 visual descriptors applicable for characterization of video objects is given in this chapter.

Nevertheless, the standard does not stipulate methods for automatic extraction of objects for the purpose of generating their normalized description. In the case of video content, these methods, named 'spatio-temporal segmentation,' have to combine both grey-level and motion information from the video stream. Some of these methods are presented in this chapter. Specifically, methods based on morphological grey-level and colour segmentation are the focus of attention. Here, colour homogeneity criteria and motion modelling and estimation methods are described.

In the framework of the knowledge-based web, the retrieval of video content concerns mainly indexed and compressed content. Furthermore, the situation when content is available for indexing in an already compressed form (MPEG-1, -2, -4) becomes more and more frequent, as in broadcast, archiving or video surveillance applications. This is why, in the remainder of the chapter, the object extraction from video in the framework of a new 'rough indexing' paradigm is presented.

6.2 MPEG-7 as a Normalized Framework for Object-Based Indexing of Video Content

MPEG-7, also called 'Multimedia Content Description Interface' [2,3], was designed to supply a standardized description of multimedia content, thus ensuring the interoperability of components and systems in the whole multimedia information search and retrieval framework. A number of different tools are comprised in the standard to achieve its objectives. These tools are Descriptors (D), Description schemes (DS), Description Definition Language (DDL) and a number of system tools.

- A Descriptor is a representation of a feature [3], where a feature is a distinctive characteristic of a data, such as, for instance, the shape of an object in a video. A Descriptor defines the syntax and semantics of the feature representation. A Descriptor allows the evaluation of the corresponding feature via the Descriptor value. It can be compared with the reference Descriptor by means of corresponding metrics in the retrieval scenario.
- A Description Scheme specifies the structure and semantics of the relationships between its components, which may be both Descriptors and Description Schemes. It provides a solution to model and describe multimedia content in terms of structure and semantics. A simple example is a feature film, temporally structured as scenes and shots, including textual Descriptors at the scene level, and colour, motion and audio amplitude Descriptors at the shot level.

- A Description Definition Language is a language that allows the creation of new Description Schemes and possibly Descriptors. It also allows the extension and modification of existing Description Schemes.
- Systems tools are related to the binarization, synchronization, transport and storage of descriptions, as well as to the management and protection of intellectual property.

The basic structure of an MPEG-7 content description follows the hierarchical structure provided by XML. The interpretation of descriptions is realized by parsing of XML documents, and as far as the Descriptors are concerned, it consists of extraction of their values.

A complete overview of MPEG-7 is largely beyond the objectives of this chapter. In the highlight of the object-based indexing of video, we will limit ourselves to an overview of descriptors which are suitable for describing objects in video content. They are normalized in the standard component MPEG-7 Visual [2]. These descriptors are qualified as 'low-level' as they are computed on the basis of video signal analysis and do not integrate any semantic information. On the other hand, learning of semantic categories of video objects is possible in an appropriate descriptor space. They can be used to compare, filter or browse video on the basis of non-textual visual descriptions of the content or in combination with common text-based queries. Relatively to the generic objects, these descriptors are subdivided into the following groups: (i) colour descriptors, (ii) texture descriptors, (iii) shape descriptors and (iv) motion descriptors. MPEG-7 also proposes a descriptor for a specific object common for a wide range of application of still images and video, that is a Face Descriptor.

As far as generic descriptors are concerned, most of them are designed for whole images or video frames and have to be specified for objects (such as Dominant Colour Descriptor (DCD) for instance). Nevertheless, some of them are explicitly designed to characterize objects (such as Contour-Shape Descriptor, Motion Trajectory Descriptor for moving objects).

6.2.1 Colour Descriptors

These descriptors reflect different aspects of the colour feature: richness and distribution of colour, colour layout in image plane, spatial structure of the colour, colour contrast. They are: [2]

- *Color Space Descriptor* defines the colour space used in the description among RGB, YCrCb, HSV and HMMD spaces admissible by MPEG-7. The associated Colour Quantization Descriptor specifies the partitioning of the chosen colour space into discrete bins.
- *Dominant Color Descriptor* defines a limited set of colours in a chosen space to characterize images or video. It gives their statistical properties such as variance and distribution. It can be applied to the whole frame and to a single object as well.
- *Scalable Color Descriptor* (SCD) uses Haar transform of image histogram computed in HSV space. Its main advantage consists in its scalability.
- *Group of Frames* or *Group of Pictures Descriptor* is the extension of SCD on a group of pictures or to a group of frames in a video. It is computed by aggregating histograms of individual frames. It can be, for instance, applied to describe a video object during its time-life in a video document.
- *Color Structure Descriptor* represents the image both by the colour distribution of the image, similar to the colour histogram, and by the local spatial structure of the colour. It is based on

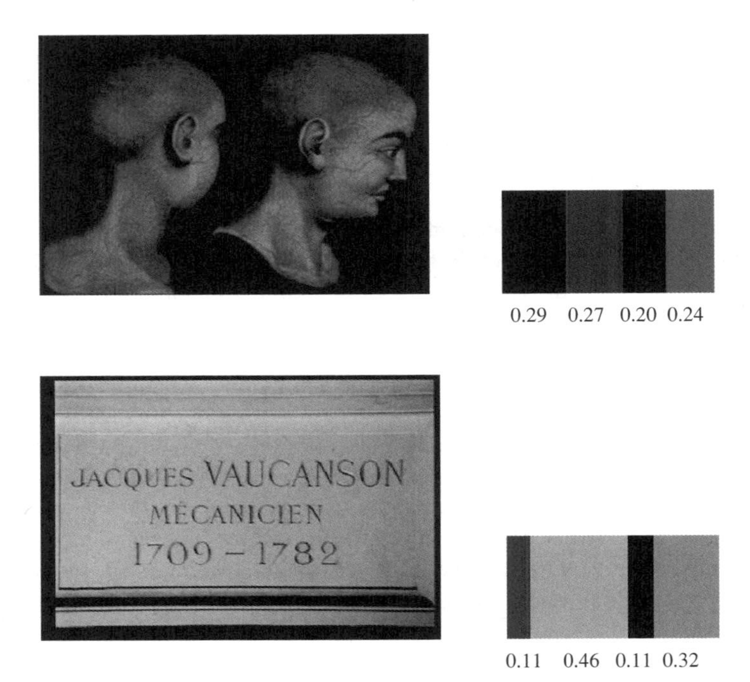

Figure 6.1 Dominant Colour Descriptor (La Joueuse de Tympanon, ©SFRS)

the colour structure histogram. The latter is computed by scanning the image by a structuring element. This descriptor captures the difference between images having exactly the same number of pixels of the same colour but distributed with different compactness.

- *Color Layout Descriptor* (CLD) reflects the spatial layout of representative colour in the image but also the contrasts of these colours as it is based on spectral representation of the colour by a discrete cosine transform (DCT).

The Dominant Colour Descriptor is computed as:

$$DCD = ((c_i, p_i, v_i)_{1 \leq i \leq n}, s),\ i = 1, \ldots, N$$

and is illustrated in Figure 6.1. Here c_i is the colour vector, p_i is the fraction of pixels corresponding to this colour, v_i is the variance and s is the parameter characterizing spatial homogeneity of the colour in the image. It is computed as a linear combination of spatial homogeneity parameters of each individual dominant colour. The number N of dominant colours in the image is limited. The maximal number of eight colours was found to be sufficient [3]. The DCD shown in Figure 2.1 contains four dominant colours; the p_i values are shown under the colour vectors at the right-hand side of the figure.

Colour Layout Descriptor computation is illustrated in Figure 2.2. It follows very much the philosophy of content encoding standards. In fact the MPEG-1, -2 and -4 standards, as we will see later in the chapter, use the DCT for blocks of 8 × 8 pixels in order to encode the original video frames in an intra-frame mode. The Colour Layout Descriptor first transforms the whole frame in a 8 × 8 thumbnail image by decomposing the original frame into 64

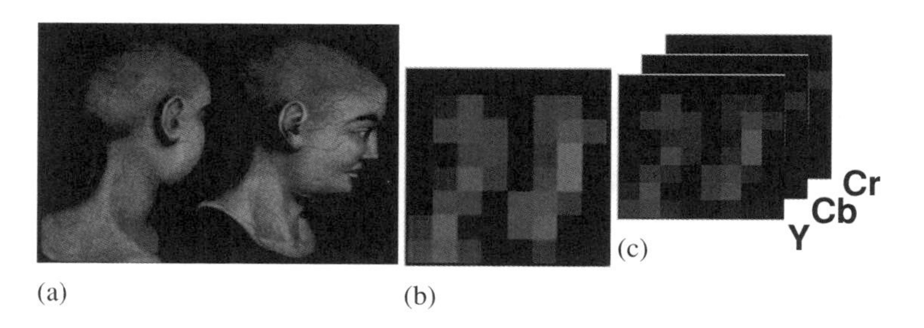

Figure 6.2 Colour Layout Descriptor computation: (a) original frame, (b) zoomed thumbnail 8 × 8 image and (c) YCrCb components of the thumbnail

blocks and replacing each block by its mean colour value. Such a thumbnail well reflects the structure of the colour in the original frame (see Figure 6.2a and b). Then the DCT is applied to the thumbnail and the first few DCT coefficients in YCrCb space are retained for the final descriptor. This allows preservation of the mean colour of the thumbnail but also captures the contrast between colours in image. The latter is ensured by means of DCT coefficients of the order higher then 0. To our opinion, it is quite reasonable to suppose that CLD applied to the bounding box of a video object will characterize well its colour layout. Furthermore, in order to retain only object colours in the bounding box and not use the background, techniques such as shape-adapted DCT, applied in the MPEG-4 standard, can be profitable. They proceed by filling of all pixels of the rectangular bounding box by only object pixels, thus excluding contrasts between object and background.

6.2.2 Shape Descriptors

MPEG-7 proposes shape descriptors for objects in video from various points of view. First of all, if an object is considered as a 2D entity with a known set of pixels, then the Region-Based Shape Descriptor can be retained. This descriptor is based on the complex 2D angular radial transformation (ART). It is an orthogonal unitary transform defined on a unit disk in polar coordinates. It is applied to the characteristic function of the object in polar coordinates. The *ART Descriptor* represents a set of magnitudes of complex ART spectral coefficients. It is invariant with regard to rotation and to scale transformations of the shape. A detailed description of it can be found in [3]. It is clear that its application to video objects has to be designed properly. In general, a video object changes its shape during its time-life. Therefore, the aggregation or selection rules have to be designed in order to compare two video objects each having their proper time-life and deformations in time.

Another possibility to characterize the shape of 2D video objects consists in using the *Contour-Based Shape Descriptor*. This descriptor is based on curvature scale-space (CSS) representation of the contour [2]. Extraction of the CSS representation involves calculation of the curvature of a closed contour [3]. For a parametric planar closed contour $C(u) = [x(u), y(u)]$, the curvature can be calculated as

$$\kappa(u) = \dot{x}(u)\ddot{y}(u) - \ddot{x}(u)\dot{y}(u) \tag{6.1}$$

The zero crossings of function (6.1) correspond to the characteristic points of the contour. By progressively smoothing a contour of an arbitrary object, a convex contour will be obtained.

The latter will not contain zero-crossings in its curvature function. The descriptor consists of the eccentricity and circularity of the initial and smoothed convex contours and also of the parameters characterizing the deformation of the contour during the smoothing process.

The use of this descriptor is rather challenging in video in the case of articulated objects (such as humans in sports programmes or in feature films). This is due to the strong shape variations during the object time-life.

In [6] comparison of the Region-Based and Contour-Based Shape Descriptors is reported in the case of similarity-based retrieval of objects. The Contour-Based Shape Descriptor is on average 15% better than the Region-Based Shape Descriptor. On the other hand, the Region-Based Shape Descriptor is more robust in the case of extreme scaling. It outperforms the Contour-Based Shape Descriptor by 7%.

MPEG-7 also proposes shape descriptors for the 3D shapes in video. Thus the *3D Shape Spectrum Descriptor* expresses characteristic features of objects represented as 3D polygonal meshes, and the *Multiple-View Descriptor* allows a combination of multiple 2D views of the same object in video. Today these 3D descriptors are mostly appropriate to the indexing and retrieval of specific video content, such as 3D objects in synthetic and augmented reality scenes or in stereo (or multiple view) conferencing applications.

6.2.3 Motion Descriptors

Taking into account the nature of the video as an image in a 2D+t space and its inherent property such as motion, MPEG-7 provides a large variety of descriptors able to capture motion of camera and objects in video. Starting from the *Motion Activity Descriptor* [4], which expresses the nature of the motion in image plan independently of its separation into objects and background, it gives a powerful tool to model the motion of the camera by the *Camera Motion Descriptor* and supplies *Motion Trajectory* and *Parametric Motion* for the characterization of moving regions in video; that is, for indexing of dynamic object behaviour.

The Motion Trajectory Descriptor represents the trajectory of regions/objects in video. In the framework of object-based indexing of video content it is specifically interesting. In the context of a known genre of video content semantics can be easily deduced from this descriptor (alone or in combination with other descriptors, e.g. Dominant Region Colour). Thus in a sports content it can allow a search for a specific event such as a goal. In video surveillance content indexing the changes in this descriptor could yield a generation of an alarm key-frame. In this descriptor, the successive spatio-temporal positions of the object are represented by the positions of one representative point of the object, such as its centre of gravity. The trajectory model is a first- or second-order piecewise approximation of the spatial positions of the representative point along time for each spatial dimension (x, y or z). The first-order interpolation is the default interpolation. Here the coordinate equations are:

$$x(t) = x_i + v_i(t - t_i) \quad \text{with} \quad v_i = \frac{x_{i+1} - x_i}{t_{i+1} - t_i} \tag{6.2}$$

The second-order approximation of the trajectory is:

$$x(t) = x_i + v_i(t - t_i) + \frac{1}{2}a_i(t - t_i)^2 \quad \text{with} \quad v_i = \frac{x_{i+1} - x_i}{t_{i+1} - t_i} - \frac{1}{2}a_i(t_{i+1} - t_i) \tag{6.3}$$

In equations (6.2) and (6.3) v_i and a_i have the sense of velocity and acceleration. A specific bit-flag indicates the order of interpolation in this descriptor. In the case of second-order interpolation the coefficient a_i is encoded. Another important component of this descriptor is a Boolean value specifying whether or not the camera follows the object. This flag is also very helpful for semantics derivation from this descriptor. Finally, information about the measurement units and the coordinate system used is included.

The *Parametric Motion Descriptor* approximates both objects' and global motions and deformations (camera motion). In the case of video objects, parametric motion can describe the motion of a whole object or of one of its regions, as the video objects are represented in MPEG-7 by a Moving Region Description Scheme. The latter allows for representation of connected and non-connected objects in video.

The Parametric Motion Descriptor represents the motion by one of the classical parametric motion models starting from pure translation—that is, a two-parameter motion model—up to a second-order deformation where the displacement of pixels constituting an object follows a 12-parameter polynomial model. Later in the chapter we will focus on motion modeling in video and present some models which can be selected for this descriptor.

6.2.4 Texture Descriptors

Texture descriptors proposed in MPEG-7 can be associated with regions in images and in video. From a mathematical point of view they are based on both spectral and spatial representation of the image signal. Thus the *Homogeneous Texture Descriptor* uses the decomposition of signal spectrum into channels using Gabor wavelets [2,3]. The *Texture Browsing Descriptor* specifies the perceptual characterization of a texture, which is similar to human characterization in terms of coarseness, regularity and directionality. It is also based on wavelet decomposition of the image signal. The third textural descriptor, the *Edge Histogram Descriptor*, is computed in the spatial domain by detecting edges of vertical, horizontal, 45° and 135° orientation, and computing their histograms in each of the 4×4 sub-images into which the original frame is subdivided. We think that this descriptor is not particularly suitable for an object in video and is rather useful to characterize a scene globally.

The richness of the descriptor set provided by MPEG-7, the description of the structure of objects by means of appropriate *Moving Region* and *Still Region* description schemes, and the possibility of associateing semantic textual information [5] make us believe that the future of semantic object-based indexing of video content is related to this standard.

6.3 Spatio-Temporal Segmentation of Video for Object Extraction

The MPEG-7 standard supplies a normalized, flexible and rich framework for object description in multimedia content. The richness of object descriptors, which has not been totally explored yet [6], will allow for an efficient search of similar objects in video documents in a 'query by example' scenario or for a semantic unification of chapters of a video content unit by the presence of characteristic (the same) objects. Thus in sports programmes the extraction of the dominant colour region [7] allows for the identification of the playing field and thus semantic indexing of game situations. In video production and post-production applications the detection of the clapperboard in cinematographic video sequences [8] allows separation of a semantically meaningful content from a 'technical' one. The problem of object-based indexing is therefore simplified in cases when an a priori model of the content can be built (e.g. sports programmes, news etc.). In the case of generic video content such as feature films or documentaries, when the scene model and object nature are not available, the extraction step, which is preliminary to the object description, still remains a challenging problem. The extraction of meaningful objects from video is also called in the literature [9] 'spatio-temporal segmentation' of video.

The large variety of methods of spatio-temporal segmentation which have been developed in the context of video coding and indexing can be roughly classified into two large groups: region-based methods and differential/variational methods. The latter include active contours, which represent flexible curves that, pushed by spatio-temporal forces, will evolve to the borders of moving objects [10] or minimize domain-based criteria in the framework of the variational approach [11].

Region-based methods can be characterized as purely motion based and both motion and colour based. The approach, probably the most popular in purely motion-based methods, is based on Markov random field modelling of the label field of the segmentation [12–14].

Finally, the region-based methods which were developed first from video coding applications [6] use two types of homogeneities: the homogeneity of regions constituting articulated objects with regard to motion-based criteria and the homogeneity of the colour.

These methods of spatio-temporal segmentation allow extraction of a semantic object from video frames as a set of connected regions, each having a homogeneous colour. The whole set can be differentiated from the background due to motion [9] or depth [15]. In most cases these methods consist of two steps. The first one represents a grey-level or colour-based segmentation which allows to splitting of the image plane of video frames into a set of colour homogeneous regions. Their borders constitute external object borders, internal borders, and the borders of homogeneous areas in the background. Then the problem of object extraction will consist in selecting regions belonging to the object. This selection can be done by motion-based merging [9], aggregation by relative depth [15] or by user interaction. Despite various methods used for grey-level and colour segmentation for the first phase of the approach, such as those proposed in [9,17], morphological methods have become the most popular now. Since the fundamental work [18], a large variety of these methods have appeared and been applied to object extraction [19,20]. This is why we think it interesting to present an overview of these methods to the reader, without, nevertheless, pretending to be exhaustive. We can refer the reader to [16,18] for the theoretical aspects of mathematical morphology.

6.3.1 Morphological Grey-Level and Colour-Based Segmentation of Video Frames

Mathematical morphology is a theory allowing an analysis of structures and shape processing. The origin of this theory is in the works by G. Matheron and J. Serra [16]. First applied to the processing of black-and-white images, it was then used for segmentation of grey-scale images [18], colour images [21] and even motion vector fields [22]. In the framework of object-based indexing of video content, colour-based morphological segmentation has become a necessary stage on the way to object extraction.

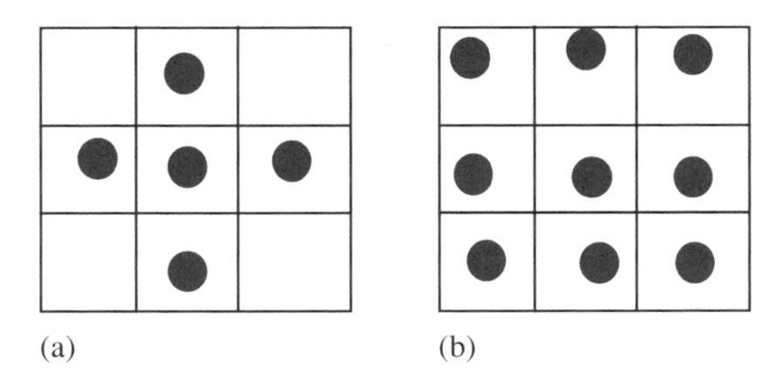

Figure 6.3 Structuring elements: (a) 4-disk and (b) 8-disk

Fundamentals of morphological segmentation

The fundamental tools for image transformation in the morphological approach represent morphological operators. In the context of image processing, morphological operators allow a geometrical transformation of a grey-scale or binary image by means of a 'structuring element', using set theory concepts.

A structuring element S is a mask of some predefined shape. The elements of the mask form a pattern. In the case of grey-level or binary morphology flat structuring elements are the most frequently used. This means that all pixels of the pattern of S have the same value. In practice, for images or video frames defined on a square grid, the most widely used structuring elements are 4-disk (Figure 6.3a), and 8-disk (see Figure 6.3b) of radius 1. In Figure 6.3 the filled circles denote pixels which constitute the pattern of the structuring element.

The fundamental morphological operators, which allow a synthesis of morphological filters for images, are the dilation and the erosion.

Morphological dilation

Morphological dilation $\delta_n(f)$ of an application f by a structuring element S_n (of radius n) is defined as:

$$\delta_n(f)(x) = \text{Max}(f(x - k), k \in S_n)$$

Here $f(x - k)$ denotes the translation of the application f by the vector $-k$.

Applied to the luminance of an image or to one of the chrominance components f this operator can be understood as follows. The value of a pixel $f(x)$ is replaced by the maximal value in its neighbourhood on a pixel grid defined by the structural element centred on the pixel x.

Morphological erosion

Morphological erosion $\varepsilon_n(f)$ of an application f by a structuring element S_n (of radius n) is defined as:

$$\varepsilon_n(f)(x) = \text{Min}(f(x + k), k \in S_n)$$

Here $f(x + k)$ denotes the translation of the application f by the vector k.

Applied to the luminance of an image or to one of the chrominance components f this operator can be understood as follows. The value of a pixel $f(x)$ is replaced by the minimal value in its neighbourhood on a pixel grid defined by the structural element centred on the pixel x.

The operators of dilation and erosion are dual; that is:

$$\delta_n(f)(x) = -\varepsilon_n(-f)$$

Here the sign '$-$' denotes the complement [18]. In the case of image luminance or chrominance components the complement of $f(x)$ means $f_{\max} - f(x)$, with $f_{\max}$ the maximal possible value of the component.

The two fundamental operators of dilation and erosion can be applied directly to the images in order to simplify the image signal. Thus the dilation will eliminate dark spots and enlarge bright areas and small details if their radius does not exceed the radius of the structuring element. On the other hand, the erosion will eliminate bright details and enlarge dark areas. They can also be used to synthesize more sophisticated morphological filters such as morphological opening and morphological closing.

Morphological closing

Morphological closing ϕ is the composition of the dilation followed by the erosion with the same structuring element:

$$\phi = \varepsilon_n \circ \delta_n$$

Morphological opening

Morphological opening γ is defined as:

$$\gamma = \delta_n \circ \varepsilon_n$$

These filters allow for removal of dark (respectively light) details with surfaces smaller than those of the structuring elements used. Nevertheless these operations deform the borders of objects in images, this deformation being dependent on the shape of the structuring element.

This is why the geodesic operators, which can preserve the shape of significant details in images, are more advantageous.

Geodesic dilation

The geodesic dilation of radius 1 of the application f conditionally to a reference r called marker denoted $\delta_1^g(f, r)$ is defined as:

$$\delta_1^g(f, r) = \mathrm{Min}(\delta_1(f), r)$$

Geodesic erosion is defined via the complements as:

$$\varepsilon_1^g(f, r) = -\delta_1^g(-f, -r)$$

These operators are very seldom used as such. They serve to define the so-called reconstruction filters. Thus the reconstruction by dilation is defined as:

$$\gamma^{rec}(f, r) = \delta_\infty^g(f, r) = \ldots \delta_1^g(\ldots \delta_1^g(f, r) \ldots, r)$$

Symmetrically the reconstruction by erosion is defined as:

$$\phi^{rec}(f, r) = \varepsilon_{\infty}^{g}(f, r) = \dots \varepsilon_{1}^{g}(\dots \varepsilon_{1}^{g}(f, r) \dots, r)$$

In the latter equations the sign ∞ means reiterating the operation up to idempotence.

In the case that the original application f is chosen as a marker image and the application to transform is a result of erosion (respectively dilation) of f, then the morphological filters of reconstruction by opening (respectively closing) are obtained:

$$\gamma^{rec}(\varepsilon_n \circ f, f) = \dots \delta_1^g(\dots \delta_1^g(\varepsilon_n \circ f, f) \dots, f) \tag{6.4}$$

and

$$\phi^{rec}(\delta_n \circ f, f) = \dots \varepsilon_1^g(\dots \varepsilon_1^g(\delta_n \circ f, f) \dots, f) \tag{6.5}$$

Here $(\delta_n \circ f)$ (respectively$(\varepsilon_n \circ f)$) denotes the dilation (respectively erosion) by a structuring element of radius n.

In classical segmentation approaches the segmentation process can be divided into the following phases: filtering of image signal in order to avoid an over-segmentation, extraction of the border elements and finally extraction of internal areas—closed regions in image plan. In a region-based version of a segmentation system the closed regions would be obtained by a region growing process allowing extraction of closed borders as the locus of pixels or contour elements in a dual space. The morphological approach gives its proper solution for these three segmentation steps. The filtering is fulfilled by morphological operators described above. The extraction of contour elements is realized by gradient computation. The morphological gradient is computed as:

$$Grad_n(f) = \delta_n(f) - \varepsilon_n(f)$$

Taking into account the definition of the erosion and the dilation operators, it is clear that inside the flat areas in the image (with a constant signal) the gradient is null. The gradient will have its maxima in the areas of contours. The thickness of them will depend on the radius of the structuring element. The null areas in the gradient correspond to the interior of homogeneous regions. The maxima of gradient constitute the areas of uncertainty, where the final contour of regions should pass. In order to obtain the unambiguous borders, the morphological approach proposes the watershed method. The watershed method represents an image as a geodesic surface in which the crest lines correspond to the contours of regions (Figure 6.4). The valleys between the crest lines are called the watersheds. If the gradient image is considered, the watersheds represent the minima of the gradient that is the interior of regions. Then the region growing can be performed in the same way as water fills in the geodesic surface if a tiny hole has been pierced in each minimum of the surface. The place where different watersheds touch each other will represent contours of regions.

The reference algorithm for region growing by watershed on the gradient surface was proposed in [23]. These methods allow a uniform region growing in the sense that the 'water' fills all the pixels uniformly depending on the pixel 'elevation'; that is, the gradient magnitude. Nevertheless, many authors (e.g. [18]) stated a strong over-segmentation when the gradient surface is used for region growing. In the literature, the 'modified watershed' methods have appeared using the original (filtered) grey-level or colour image [22]. In this case the gradient

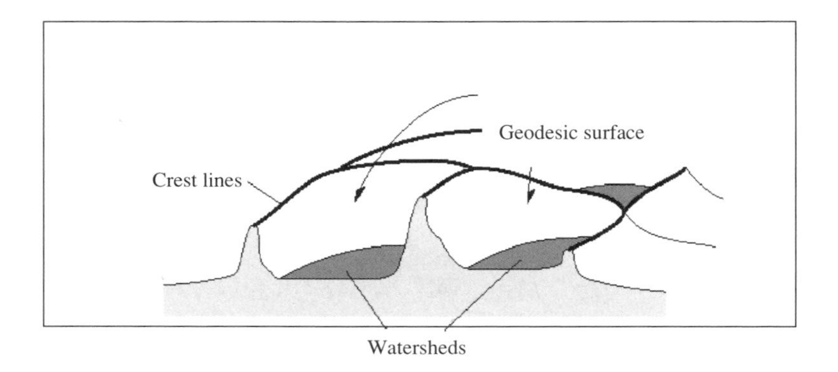

Figure 6.4 Illustration of the watershed principle

surface allows for marker extraction. The marker is a binary image in which the zeros of gradient correspond to 1 and the uncertainty is marked by 0. Then the region growing can be performed joining pixels to each connected region of the marker if their grey-level (colour) satisfies the homogeneity criterion with the region which is growing. In the following we will present a morphological colour-based segmentation which lies in the mainstream of the morphological approach and can serve as a powerful tool for object extraction in video.

Morphological colour-based segmentation with adaptive thresholding

When a colour image has to be segmented in the framework of the morphological approach, the first step, morphological simplification of the original signal, has to be done in order to avoid a strong over-segmentation. Here the three components of the image in the original or adapted colour system have to be filtered. In the segmentation method developed in the framework of [21], a scalar morphological filtering by partial reconstruction has been applied to each component of the image in the RGB system. The partial reconstruction means that in reconstruction filters (6.4) and (6.5), the outer operator (geodesic dilation or erosion, respectively) is not repeated up to idempotence of the operation, but is reiterated m times, with $m \gg n$. Here n is the radius of the structuring element of the inner operation (erosion or dilation, respectively). A typical result of such a filtering is given in Figure 6.5.

After the computation of the morphological gradient, a proposed marker extraction follows a thresholding scheme. Thus all pixels in the marker image will be marked as 'inner' pixels of future regions if the gradient magnitude is less than the predefined threshold. In thresholding of the gradient in order to extract contours, classically relative thresholds are chosen with regard to the theoretical dynamic range of the gradient magnitude. Marker computation with different thresholds is illustrated in Figure 6.6.

It can be seen that increasing the threshold will reduce the number of uncertainty pixels and thus will accelerate the modified watershed region growing.

In order to ensure the modified watershed, the white connected components in the marker image are then labelled. Each component will represent the seed for the future region. The final number of regions thus will represent exactly the number of connected components in the marker image.

Figure 6.5 Results of filtering by partial reconstruction $n = 2$, $m = 8$, sequence 'Children', MPEG-4. (a) Original colour frame and (b) filtered frame

Figure 6.6 Marker extraction: (a) with a threshold of 1.5% and (b) with a threshold of 3%. Sequences: Akyio, Children, Stefan. Courtesy to A. Mahboubi

The modified watershed method developed in the framework of [21] uses the differential sensibility of the human visual system (HVS) to the contrast. The watershed region growing is fulfilled in the colour space according to the following iterative algorithm.

A pixel with spatial coordinates (x, y) and colour coordinates $(I^{C_1}(x, y), I^{C_2}(x, y), I^{C_3}(x, y))^T$ is associated to its adjacent region R if the following condition is satisfied:

$$\left|I^{C_1}(x, y) - m_R^{C_1}\right| + \left|I^{C_2}(x, y) - m_R^{C_2}\right| + \left|I^{C_3}(x, y) - m_R^{C_3}\right| < S_R \tag{6.6}$$

Here $\left(m_R^{C_1}, m_R^{C_2}, m_R^{C_3}\right)^T$ is the mean colour vector of the growing region and s_R is the region adapted threshold. This threshold is based on a simplified model of the contrast sensitivity of the HVS. In [24] and numerous subsequent works it has been shown that there exists a nonlinear dependence of the sensitivity of the HVS from the contrast. The HVS is the most reactive in the middle of grey-level dynamics (in the case of 8-bit grey-level quantizing this means the values close to 128) and the least reactive in 'dark' (close to zero) and bright (close to 255) ranges. To use it in determining the threshold, a piecewise linear function of the mean grey-level $Z(m)$ of a region is introduced:

$$Z(m) = |m - 127| + 128$$

If segmenting a grey-level image (only one absolute difference in equation (6.6)) the threshold s_R can then be computed as:

$$s_R^g = Z(m_R)\Delta$$

Here $0 < \Delta < 1$ and has the meaning of the percentage of the mean value of a region.

In the case of colour images, the mean grey-level of the region is replaced by the mean of mean colour coordinates of a region:

$$m_R^{col} = \left(m_R^{C_1} + m_R^{C_2} + m_R^{C_3}\right)/3$$

In this case the colour-based threshold will be computed as:

$$s_R = 3^* Z\left(m_R^{Col}\right)\Delta$$

This thresholding scheme means that all pixels adjacent to already labelled regions, for which the colour difference is lower than the threshold, will be associated to the regions. Other pixels, which either do not have labelled pixels in their neighbourhood or are too different from the adjacent regions, remain uncertain. In order to label all pixels by the watershed growing process, a 'hitting' of the glacier is simulated. After assignment of all possible pixels to their adjacent regions according to equation (6.6) for the given set of region-dependent threshold values s_R, the temperature increases and the water level rises in all watersheds according to their mean level. This is simulated by increasing the value of Δ. A nonlinear and linear laws can be proposed here. Good results were observed with a simple linear process; that is, at each iteration:

$$\Delta^i = \Delta^{i-1} + \delta\Delta$$

An example result of the process of watershed with 'hitting' is presented in Figure 6.7 for the video sequence 'Children'.

It can be seen by comparing Figure 6.7a and 6.7b that the proposed colour-based method gives neater contours (see figure 6.7a) of resulting regions compared to the classical watershed on the gradient surface.

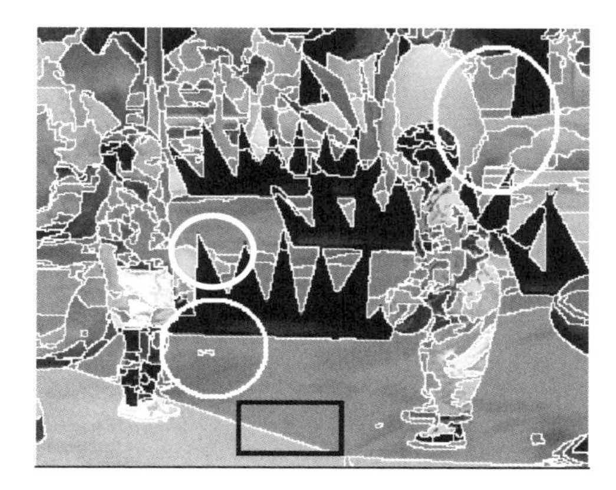
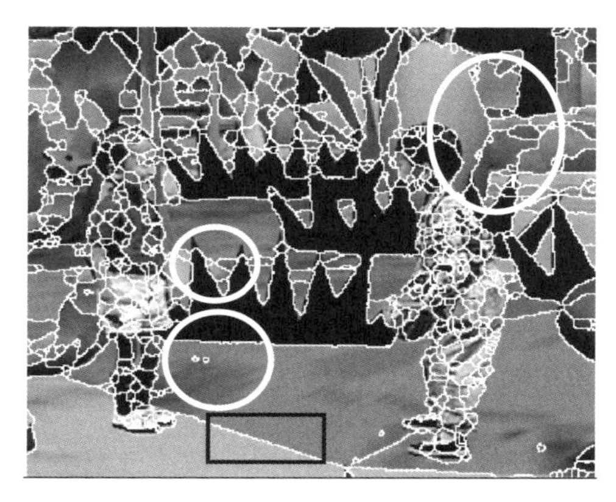

Figure 6.7 Results of watershed with hitting on the sequence 'Children': (a) modified watershed with hitting, and (b) watershend on gradient surface

Nevertheless, when the objective of the still-frame segmentation is to serve as a basis for extraction and indexing of objects of interest, then morphological segmentation methods have the disadvantage of giving too fine a segmentation. In this case in the same manner as pixels were merged with the adjacent regions, the resulting adjacent regions can be merged if their colour descriptors are similar.

Classically, such region-merging is fulfilled on the region adjacency graph $RAG = \{\{R\}, \{E\}\}$. The RAG is a (usually) non-oriented graph in which the set of nodes $\{R\}$ corresponds to the set of regions in the frame partition. The set of edges $\{E\}$ expresses the connectivity of the partition. Hence if a pair of nodes is connected by an edge then the regions in the frame partition corresponding to the nodes are neighbours in the image plan according to the chosen discrete topology. The reciprocal is also satisfied. The RAG is a valuated graph, in which edges are weighted by a similarity measure of the colour descriptors of incident region nodes. The region-merging process can be designed as an optimal graph search, as the construction of binary partition trees as in [25]. In the context of [21] the region merging is realized as an arbitrary graph search. Here a pair of regions is merged if the criterion of similarity of their mean colour vectors is satisfied:

$$\frac{\|\vec{m}_i - \vec{m}_j\|}{(\|m_i\|^2 + \|m_j\|^2)^{1/2}} < s_{RAG}^k$$

The process is also iterative and the relative threshold s_{RAG}^k is incremented linearly as the watershed threshold up to the desired level of detail in the segmentation still keeping the integrity of the object of interest. The results of the process for the sequence 'Stefan' are given in Figure 6.8. The upper left image represents the result of the watershed region growing. The lowest right image represents the result of merging when the object is no more connected. Therefore, the choice of the final partition is possible from the first to the seventh image in the set of nested segmentations obtained by merging of adjacent regions according to the criterion.

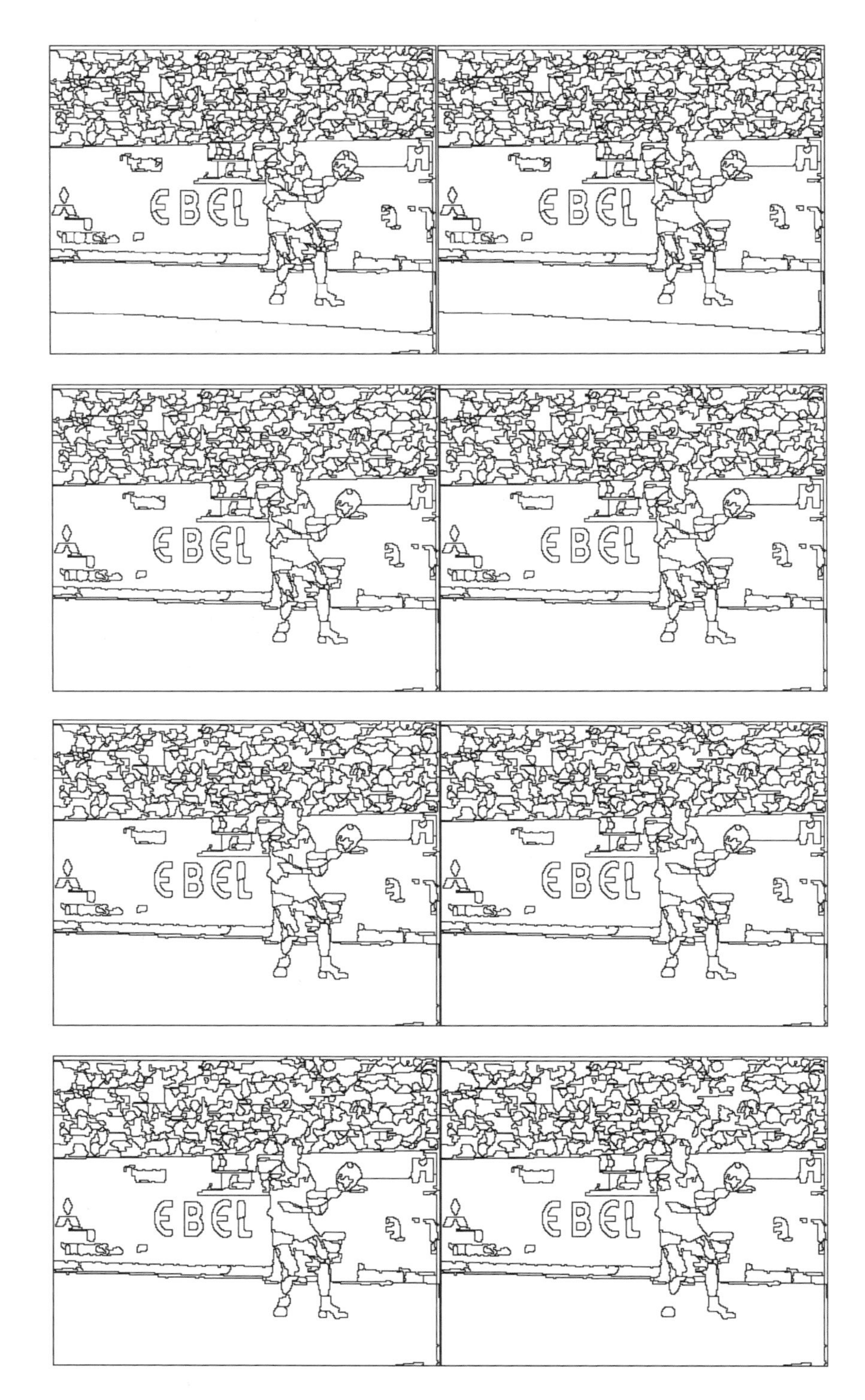

Figure 6.8 Results of merging regions in the initial morphological partition by search on the RAG. Courtesy to A. Mahboubi

6.3.2 Motion Modelling for Object Extraction and Indexing

Motion of regions in the partition of video frames is the descriptor which will allow automatic distinguishing between the object of interest and the background. The motion measured in the background reflects the camera work, while the motion measured in each region belonging to the object of interest reflects the deformation or motion of parts of complex articulated objects in the 3D space.

In order to use object motion as a discriminating descriptor for object extraction and indexing we will consider first different levels of motion modelling in video.

The motion that can be observed in the plane of video frames is a so-called 'apparent motion' [27] which in the ideal case represents a projection of a real 3D motion into a 2D image plane by means of an optical system of the camera. Generally speaking, the apparent motion can be observed due to luminance changes between two successive images in the video.

These changes can be induced not only by motion of objects but also by changes in lighting. On the other hand, if a physical 3D motion is present inside objects of a homogeneous luminance, then the changes in luminance cannot be observed in video and hence the apparent motion can not be measured.

Several levels of motion modelling of objects can be proposed. The local characterization of motion will consist in an elementary displacement vector for each pixel (x, y) of a video object $\vec{d}(x, y) = (dx, dy)^T$. This vector characterizes the displacement of the pixel between two consecutive frames in video. In the case of compressed video documents (in the case of access to Internet-based archives of advertising content, broadcast audiovisual content or video surveillance content encoded by MPEG-1, -2, -4 or H.26X coding standards) one can suppose that an object in a video represents a combination of squared blocks of constant (MPEG) or variable (H.264) size. In this case the elementary displacement vector is considered the same for all pixels of the block. When indexing at the level of objects, it is too redundant to represent object motion by a pixel-wise or block-wise displacement vector. If the object, is articulated, e.g. a human or a complex compound object, it is interesting to index the motion of each part of the object which when projected into the image plane represents a region. Furthermore, region-based motion descriptors would allow a distinction between regions belonging to the background and those belonging to objects as in [9] and [15]. Therefore it is interesting to propose a global motion model for a given region in the image plane.

Developing the elementary displacement vector $\vec{d}(x, y)$ in a Taylor series around a region gravity centre $G(x_g, y_g)$ and neglecting terms of order higher then 1, we have [27]:

$$\vec{d} = \begin{pmatrix} dx \\ dy \end{pmatrix} = \begin{pmatrix} a_0 \\ b_0 \end{pmatrix} + \begin{pmatrix} a_1 & a_2 \\ b_1 & b_2 \end{pmatrix} \begin{pmatrix} x - x_g \\ y - y_g \end{pmatrix} \tag{6.7}$$

The coefficients a_1, a_2, b_1, b_2 can be interpreted as the partial derivatives of the displacement vector $a_1 = \frac{\partial dx}{\partial x}$, $a_2 = \frac{\partial dx}{\partial y}$, $b_1 = \frac{\partial dy}{\partial x}$, $b_2 = \frac{\partial dy}{\partial y}$, the coefficients a_0, b_0 being the coordinates of the displacement vectors of the gravity centre G. Thus a six-parameter affine motion model for a region is introduced $\Theta aff = (a_0, b_0, a_1, a_2, b_1, b_2)^T$. Several authors [11,12] considered this model a good compromise between the complexity of motion modelling and also the adequacy to observed motion in the 2D plane of video frames. The authors of [12] give another formulation to this model:

$$\vec{d} = \begin{pmatrix} a_0 \\ b_0 \end{pmatrix} + \frac{1}{2}\begin{pmatrix} div\cdot(x-x_g) - rot\cdot(y-y_g) + hyp1\cdot(x-x_g) + hyp2\cdot(y-y_g) \\ div\cdot(y-y_g) + rot\cdot(x-x_g) - hyp1\cdot(y-y_g) + hyp2(x-x_g) \end{pmatrix}$$

here *div* is the divergence of the displacement vector field: $div(\vec{d}) = \frac{\partial \mathrm{d}x}{\partial x} + \frac{\partial \mathrm{d}y}{\partial y}$, *rot* is its rotor $rot(\vec{d}) = \frac{\partial \mathrm{d}y}{\partial x} - \frac{\partial \mathrm{d}x}{\partial y}$, and *hyp*1 and *hyp*2 are the hyperbolic terms expressing deformations of motion field, $hyp1 = \frac{\partial \mathrm{d}x}{\partial x} - \frac{\partial \mathrm{d}y}{\partial y}$, $hyp2 = \frac{\partial \mathrm{d}y}{\partial x} + \frac{\partial \mathrm{d}x}{\partial y}$. If the latter are weak, the six-parameter motion model can be reduced to a four-parameter model, expressing physically interpretable motions in the image plane, such as translation, rotation and zoom:

$$\begin{cases} \mathrm{d}x = t_x + k(x-x_g) - \theta(y-y_g) \\ \mathrm{d}y == t_y + \theta(x-x_g) + k(y-y_g) \end{cases} \tag{6.8}$$

Here t_x, t_y are the translation parameters along the Ox and Oy axes in the image plane, k is a zoom factor and θ is the rotation angle around an axis orthogonal to the image plane.

Several authors, such as [28] consider even more simple models in order to characterize motion in the background of a scene. Thus the three- parameter model (translation and zoom factors) will be expressed as:

$$\begin{cases} \mathrm{d}x = t_x + k(x-x_g) \\ \mathrm{d}y = t_y + k(y-y_g) \end{cases}$$

Finally, the poorest model—a simple translational model can be mentioned in relation to a local characterization of motion inside an object—is a block-based characterization. Here the model expression is:

$$\begin{cases} \mathrm{d}x = t_x \\ \mathrm{d}y = t_y \end{cases} \tag{6.9}$$

Note, that the models (6.7), (6.8) and (6.9) form part of the Parametric Motion Descriptor of MPEG-7.

If the object is neither articulated nor deformable, only one model can be considered for the whole object in a video. Nevertheless, this situation is rather rare and in most cases the unique model is supposed only for camera motion such as in [29]. In the case of object motion we will remain in a region-based framework and will briefly describe the motion estimation on a region basis.

6.3.3 Object- and Region-Based Motion Estimation

Considering an object as a single region in an image plane or, in a more realistic way, a set of connected regions, the problem consist now of proposing methods to estimate the chosen motion model.

Generally speaking, motion estimation is based on the invariance of luminance hypothesis, which says that in the absence of noise and lighting changes, a pixel conserves its intensity along its trajectory. This hypothesis is expressed by the following. Denoting by $I(x, y, t)$ the intensity of an image point (pixel) (x, y) at time moment t, and considering the displacement vector at this pixel $\vec{d}(x, y) = (\mathrm{d}x, \mathrm{d}y)^T$ between moments of time t and $t + \Delta t$ we will introduce a measure called 'displaced frame difference' as:

$$DFD(x, y, \vec{d}) = I(x + \mathrm{d}x, y + \mathrm{d}y, t + \Delta t) - I(x, y, t)$$

The hypothesis of the conservation of pixel intensity will then be expressed as:

$$DFD(x, y, \vec{d}) = 0 \tag{6.10}$$

Developing $I(x + \mathrm{d}x, y + \mathrm{d}y, t + \Delta t)$ in a Taylor series in the vicinity of point (x, y, t), neglecting terms of order higher than 1 we will get:

$$I(x + \mathrm{d}x, y + \mathrm{d}y, t + \mathrm{d}t) = I(x, y, t) + \mathrm{d}x\frac{\partial I}{\partial x} + \mathrm{d}y\frac{\partial I}{\partial y} + \mathrm{d}t\frac{\partial I}{\partial t}$$

Finally, using the hypothesis of conservation of pixel intensity (6.10) the following equation will be deduced:

$$I_x u + I_y v + I_t = 0 \tag{6.11}$$

with

$$\frac{\mathrm{d}x}{\mathrm{d}t} = u, \frac{\mathrm{d}y}{\mathrm{d}t} = v.$$

This equation is called the optical flow equation (OFE) as it links the optical flow, i.e. the velocity field, with image spatial gradient and temporal derivative.

The problem of motion estimation consists in estimation of displacement field or optical flow, which satisfies (6.10) or (6.11), respectively. In reality (6.10) and (6.11) are never satisfied because of the noise present in video. Therefore, the estimation methods are based on the minimization of the left-hand sides of either (6.10) or (6.11) or some criteria based on them.

The motion estimation problem is ill-posed. In fact considering u and v as independent variables, only one equation is formulated for them. Another interpretation consists in rewriting the equation (6.11) in a vector form as:

$$\nabla\vec{I} \cdot \vec{w} = -I_t \tag{6.12}$$

with $\vec{w} = (u, v)^T$ the velocity vector. Let us decompose $\vec{w} = (u, v)^T$ according to two orthogonal directions $\vec{w} = \vec{w}_{||} + \vec{w}_{\perp}$, the first $\vec{w}_{||}$ being parallel to the contours in the image and thus orthogonal to the image gradient $\nabla\vec{I}$, the second $\vec{w}_{\perp}$ being parallel to the image gradient. Then from (6.12) it can be seen that only $\vec{w}_{\perp}$ is observable. This component of optical flow is called the 'normal optical flow'. Furthermore, if the image gradient is null—that is, a flat area in the video frame is observed—then any optical flow satisfies equation (6.12).

A vast literature has been devoted to the problems of a direct or parametric motion estimation. An overview can be found in [27]. In the first case a motion vector is estimated directly to each pixel in a video frame. In the second, the motion vector field is supposed to follow a predefined model $\vec{d} = \vec{d}(\Theta)$ and the model parameters Θ have to be estimated.

Without being exhaustive in the description of direct methods for motion estimation, we have to outline here a so-called 'block-based' motion estimation or 'block-matching', as it is the most widespread now in the case of compressed video content. This estimation is a part of the MPEG and H.26X compression algorithms. Here the whole image plane of a video frame is split into square blocks. The displacement vector is supposed to be constant for all pixels in the same block. Then the optimal displacement vector per block is estimated between the current frame in the video stream and the reference frame as that optimizing the criterion based on DFD, such as MAD:

$$\mathrm{MAD} = \min_{\vec{d}\in W} \sum_{\vec{p}\in B} \left| I(\vec{p}, t) - I(\vec{p} + \vec{d}, t - \Delta t) \right|$$

Here W is a square 'window', that is the possible variation range for both coordinates of the displacement vector, B is a block, $\vec{p} = (x, y)$, $I(\vec{p}, t)$ is the current frame and $I(\vec{p} + \vec{d}, t - \Delta t)$ is the reference frame, usually from the past. When the vector $\vec{d}$ gets all values from W, the estimation method is called 'full search'.

When considering region-based or object-based estimation, the parametric motion estimation is usual. According to a chosen motion model, the vector of its parameters has to be estimated. Here the methods can follow a 'two-step' or 'one-step' schemes. In a two-step scheme low-level motion descriptors, such as displacement vectors per pixel or per block, are computed first. Then the global motion model can be deduced based on these descriptors as initial measures. Thus in [30] indexing of particular objects—human faces—is addressed. After face detection, the elementary displacement vectors $\vec{d}(x_i, y_i)$are estimated for n characteristic points situated on a face border by block-matching. Then these vectors are considered as initial measures for estimation of a six-parameter linear model of the whole face by least square scheme. Here the model equation will be:

$$Z = H\Theta + V \tag{6.13}$$

with $Z = (\mathrm{d}x_1, \ldots, \mathrm{d}x_n, \mathrm{d}y_1, \ldots, \mathrm{d}y_n)^T$, V the uncorrelated noise and H the observation matrix, corresponding to the motion model (6.7):

$$H = \begin{pmatrix} 1 & x_1 - x_g & y_1 - y_g & 0 & 0 & 0 \\ \cdots & \cdots & \cdots & \cdots & \cdots & \cdots \\ 1 & x_n - x_g & y_n - y_g & 0 & 0 & 0 \\ 0 & 0 & 0 & 1 & x_1 - x_g & y_1 - y_g \\ \cdots & \cdots & \cdots & \cdots & \cdots & \cdots \\ 0 & 0 & 0 & 1 & x_n - x_g & y_n - y_g \end{pmatrix}$$

The closed-form solution for the motion model of the 'face' object is then found as:

$$\Theta = ((H^T H^{-1}) \cdot (H^T Z))$$

This model allows for tracking of the frontal face object along the time and indexing the trajectory of the human.

One-step schemes for region-based and object-based motion estimation use numerical methods, such as differential methods of first and higher order. The gradient descent methods have proved to be a good compromise between the computational cost and the quality of estimation. The criterion to minimize in these schemes has to be derivable. This is why quadratic criteria are usually employed. Thus in [17], the estimation of four-parameter simplified affine model

(6.9) per region is based on the optimization of the criterion of mean square error (MSE) of motion compensation:

$$\mathrm{MSE} = \frac{1}{N} \sum_{(x,y)\in R} DFD^2(x, y, \mathrm{d}\,(\Theta, x, y)) \tag{6.14}$$

Here N is the number of pixels of the region.

The estimation method is based on the gradient descent with an adaptive gain, and in case of the MSE (6.14) criterion can be expressed as:

$$\Theta^{i+1} = \Theta^i - \frac{\varepsilon_\Theta}{2 \cdot N} G^i \tag{6.15}$$

Here i is the iteration number, $G^i = (\frac{\partial MSE}{\partial t_x}, \frac{\partial MSE}{\partial t_y}, \frac{\partial MSE}{\partial k}, \frac{\partial MSE}{\partial \theta})^T$ is the gradient vector of the error functional MSE.

The gain matrix ε_Θ is developed to take into account the spatial gradient in video frames as:

$$\varepsilon_\Theta = \frac{1}{\left|\nabla I(\vec{p}^i + \vec{d}(\Theta^i), t)\right|^2 + \alpha^2} \begin{bmatrix} \varepsilon_{tx} & 0 & 0 & 0 \\ 0 & \varepsilon_{ty} & 0 & 0 \\ 0 & 0 & \varepsilon_k & 0 \\ 0 & 0 & 0 & \varepsilon_\theta \end{bmatrix} \tag{6.16}$$

Here the gains $\varepsilon_{tx}, \varepsilon_{ty}$ have different order than the gains $\varepsilon_{div}, \varepsilon_{rot}$ for the following reason. In the absence of rotational and zoom components in the motion of region, the translational terms have to be comparable to the sampling step of the frame grid. Therefore, in order to get a fast convergence, the same value of order 0.1 is proposed for $\varepsilon_{tx}, \varepsilon_{ty}$ terms. The gains $\varepsilon_k, \varepsilon_\theta$ correspond to the parameters of lower order, and typical values for them are chosen as 0.001–0.01.

Gradient-based optimization schemes suppose a convex error function and need an appropriate initialization. A satisfactory solution for this represent a multi-resolution initialization. Thus several versions of the initial pair of frames are computed by low-pass filtering and subsampling as:

$$I^l(t) = h{*}I^{l-1}(t), I^l(t-1) = h{*}I^{l-1}(t-1) \tag{6.17}$$

Here $l = 0, \ldots L$, with $l = 0$ corresponding to the full-resolution, h is the low-pass filter and $*$ is the convolution operation. In a subsampling operation, a new image is computed from that resulting from a low-pass filtering by conserving each sth pixel in a row and in a column. The parameter s is called a subsampling factor. Thus a sort of pyramid for the pair of consecutive frames is built: at the top the lowest resolution and smallest size images are situated; at the bottom are full resolution frames.

The multi-resolution initialization consists in optimizing the criterion at a higher level of the pyramid l. Then the optimal parameter vector Θ_l is propagated to the lower level and is used for the optimization process as the initialization: Θ^0_{l-1}. The propagation of motion parameters in the case of a four-parameter affine model (6.8) is expressed as:

$$t_x^{l-1} = s \cdot t_x^l, \qquad t_y^{l-1} = s \cdot t_y^l, \qquad k^{l-1} = k^l, \qquad \theta^{l-1} = \theta^l \tag{6.18}$$

The same multi-resolution strategy turns out to be very efficient in cases when more complete motion models are to be estimated for regions inside an object or for the whole object mask (e.g. the six-parameter affine model (6.7)).

Hence, in this section we discussed methods for extraction of objects from video for object-based video indexing of raw video content.

An object here is considered as a set of connected regions and is isolated from the background of the visual scene. The borders of regions adjacent to the background form the object border. Thus the shape of the objects can easily be indexed using shape descriptors (e.g. those of MPEG-7). The inner area of regions constituting the object is also known. This will allow computing colour and texture descriptors inside the object. Finally, a motion model has been estimated for each region of the object, or for the whole object if it is simple enough and can be represented by only one region. This will allow tracking of the object along the time and computing its dynamic descriptors, such as trajectory.

In the whole of this section we supposed that the content was available in a base-band; that is, it has not been compressed. If the video content is available in compressed form, then in order to apply the methods described above the video has to be decompressed. This will yield a complementary computational cost. Therefore, it is interesting to design methods for object-based indexing directly in the compressed domain.

6.4 Rough Indexing Paradigm for Object-Based Indexing of Compressed Content

The multimedia content which a user can search for on the web is rarely available today in a raw, base-band, form. When accessing the content on the server the client has to decode one of the standardized coding formats for content representation and coding. In the case of broadcast content this is mainly MPEG-2 compressed stream. In the case of streaming of video, browsing the access copies in video archives and consulting video surveillance content, these are MPEG-1, H.263, H.264, Realvideo and other international and industrial standards. They are all characterized by a lossy encoding of video data. Furthermore, when ingesting the content into web-based media archives, an automated compression of it is fulfilled. Thus the indexing of the content either during its ingestion or consumption should be done on already degraded data.

Recently, a new trend in analysis methods for indexing multimedia content has appeared, which we can qualify as the 'rough indexing' paradigm. Many authors (such as [31]) are interested in fast and approximate analysis of multimedia content at a poor or intentionally degraded resolution. Coded multimedia streams give a rich background for development of these methods, as low resolution data can be easily extracted from MPEG and H.26X compressed streams without complete decoding. Thus many authors have dealt with extraction of moving foreground objects from MPEG-2 compressed video with still background [28]. Independently, numerous works [26,29] have been devoted to the estimation of the global camera model from compressed video. Thus the 'rough data'—noisy motion information and partially decoded colour information—are available from partially parsed MPEG streams, to fulfil the fundamental operation of an object-based indexing system, such as object extraction from video. In [32] a new rough indexing paradigm is introduced. It can be expressed as 'the most complete model' on rough data and at a rough resolution (both spatial and temporal). The paradigm is designed to index compressed video streams without complete decoding. This allows quick access to the essence of the content—the dynamics of the scene, camera motion and meaningful objects—by means of the methods of compressed content analysis developed in this paradigm.

In order to describe these methods in detail we will need to briefly introduce the principle of hybrid motion-compensated and transform-based coding. Without the loss of generality, this can be done using the example of the MPEG-1 and MPEG-2 standards.

6.4.1 Description of MPEG-1 and -2 Standards

The MPEG-1 and MPEG-2 standards, as well as the recommendations for video coding H.261, H.262, H.263 and H.264, implement the scheme of a so-called hybrid video coding. This means that video frames are encoded due both to motion compensation and spatial colour coding in the transform domain (DCT) [33]. The architecture of MPEG-1 and -2 bit streams comprises frames encoded in three modes: Intra-coded (I-Frames), Predicted (P-frames) and Bidirectional (B-Frames).

Intra-coded (I-frames)

The frame is encoded independently from others. The coding method is the one used in the JPEG standard for still images [34]. Here the video frame is split into blocks of size 8×8 pixels. The same DCT transform is applied to the luminance and chrominance components of colour frames. Then the transform coefficients are quantized and coded in a lossless manner by an entropic encoder.

The DCT transform is computed as [35]:

$$F(u, v) = \frac{2}{N} C(u)C(v) \sum_{x=0}^{N-1} \sum_{y=0}^{N-1} f(x, y) \cos \frac{(2x+1)u\pi}{2N} \cos \frac{(2y+1)v\pi}{2N} \tag{6.19}$$

with

$$C(u), C(v) = \begin{cases} \frac{1}{\sqrt{2}} & \text{for } u, v = 0 \\ 1 & \text{otherwise} \end{cases} \tag{6.20}$$

Here $f(x, y)$ is the original centred signal of luminance or chrominance components of the frames, N is the linear size of the block, $N = 8$, and u and v are the spatial frequencies.

The DCT coefficients are organized in blocks of the same size $N \times N$ according to the variation of u and v. As can be seen from (6.19) and (6.20), the value of the coefficient $F(u, v)$ when $u = 0$ and $v = 0$ is:

$$F(0, 0) = N \cdot \bar{f}_{N \times N}$$

Where $\bar{f}_{N \times N}$ is the mean value of the signal in the current block. This coefficient is called a 'direct coefficient' or DC and will play an important role in the video indexing methods in the framework of the rough indexing paradigm.

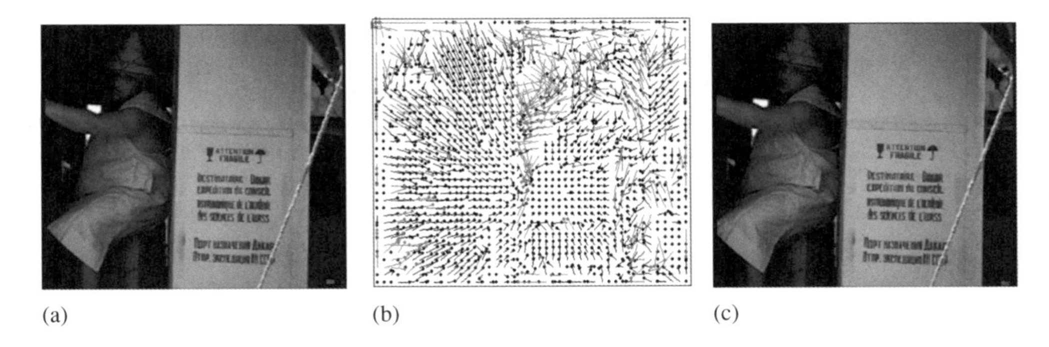

(a) (b) (c)

Figure 6.9 Macroblock displacement field for a P-frame in MPEG-2-encoded video: (a) the reference frame, (b) motion vectors and (c) the current P-frame

Predictive-coded (P-frames)

The frame is predicted from a reference frame by motion compensation. Here the P-frame is split into 16×16 macroblocks and a motion vector is encoded per macroblock. An example of the macroblock displacement field for an MPEG-2-encoded feature documentary is given in Figure 6.9.

The difference between the original video frame at time t (current P-frame) and the same frame compensated by motion from the reference frame—that is, from the previous P-frame or I-frame—is called the error signal. It is also DCT-transformed as the I-frame and encoded. Sometimes the macroblocks in the P-frame are badly compensated by motion because of the poorness of the translational model. In this case the encoder will take the decision on intra-frame coding of the current macroblock. The complete scheme of I-frames is applied: the macroblock is split into four 8×8 blocks and the centred luminance and chrominance signals are encoded.

Bidirectional-coded (B-frames)

These frames are also encoded per macroblocks. Generally speaking, two main modes are observed here: (i) predictive with regard to the past or to the future reference P- or I-frame, and (ii) bidirectional—the macroblock is interpolated by motion compensation from the previous and from the future reference frames.

An example of the general architecture of the MPEG-1 and MPEG-2 bitstream with IBP frames with I–P distance of 3, which means that two B-frames are situated between two P-frames or I- and P-frames, is shown in Figure 6.10.

Usually the sequence of frames between two I-frames consecutive in time corresponds to a standard entity called a group of pictures (GOP).

Despite this common architecture, the MPEG-1 and MPEG-2 standards have a lot of different functionalities, such as the possibility of encoding of the interlaced video by MPEG-2, which is absent in MPEG-1, the improved error resilience in MPEG-2, scalability and so on. Nevertheless; they are not in the focus of our interest with regard to the rough indexing paradigm. The most important features are those we have already mentioned, a specific architecture of the bitstream, the availability of macroblock motion vectors in the coded bitstream and the direct coefficients available for I-frames.

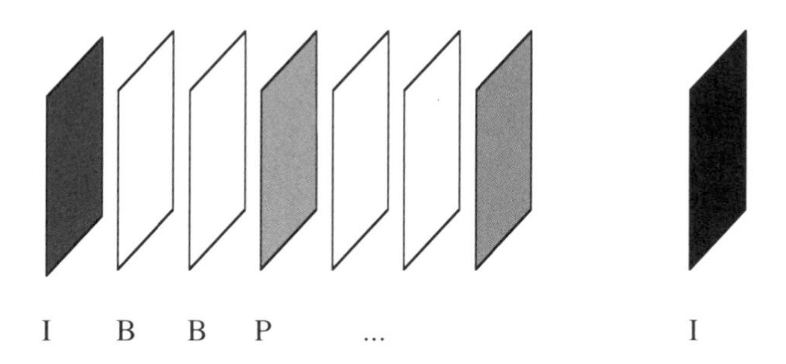

Figure 6.10 Example of the architecture of the bitstream of the MPEG-1 and -2 standards

As we mentioned above, the rough indexing paradigm supposes the use of rough data. These data represent first of all motion vectors of macroblocks in P-frames. They are noisy, as standard motion estimators in MPEG encoders cannot correctly estimate motion vectors on the occluding borders of moving objects, on the flat areas in the video frames. This is due to the fact that the motion estimation problem is ill-posed (see Section 6.3.3). Furthermore, on the borders of video frames the estimation is also noisy as the parts of the scene which 'enter' the frame due to the camera motion have no antecedents in the reference frame. Finally, the motion vectors are not available for all macroblocks in P-frames as there are some macroblocks which are 'intra-coded'.

The second kind of rough data we use represents the colour data, that is DC coefficients decoded from I-frames.

All of these data are available at a rough resolution. Hence motion vectors are obtained at a temporal resolution of P-frames. This means that for a PAL/SECAM frame rate (25 fps) with a typical architecture of two B-frames between P-frames (I–P distance of 3) and two GOPs per second, this temporal resolution will be of 8 fps in average. The colour information is available at 2 fps as 2 I-frames are encoded per second for this video format and MPEG architecture.

As far as spatial resolution is concerned, the motion vectors are available for macroblocks of 16×16 pixels. This can be interpreted as a dense optical flow, that is one motion vector per pixel for a frame of 16 times lower spatial resolution compared to the original frame. The colour information is available for frames of 8 times lower resolution than the original frames.

These rough data at the rough resolution can be used to accomplish all functionalities of a standard video indexing system; that is, segmenting the raw video into shots, into homogenous sequences of camera motion (micro-shots), clustering shots into scenes and so on. In this book we are particularly interested in object-based indexing of video content. Thus we will present the object extraction method in the framework of the rough indexing paradigm we started to develop in [32].

6.4.2 Foreground Object Extraction Scheme

In order to extract foreground objects from compressed video at rough spatial and temporal resolution we combine both motion information—the complete first-order camera motion descriptor of the MPEG-7 standard—and region-based colour segmentation. Unlike the situation of a fixed camera largely presented in the literature (see for example [28]), the method handles arbitrary camera motion, as well as variable object shape and motion.

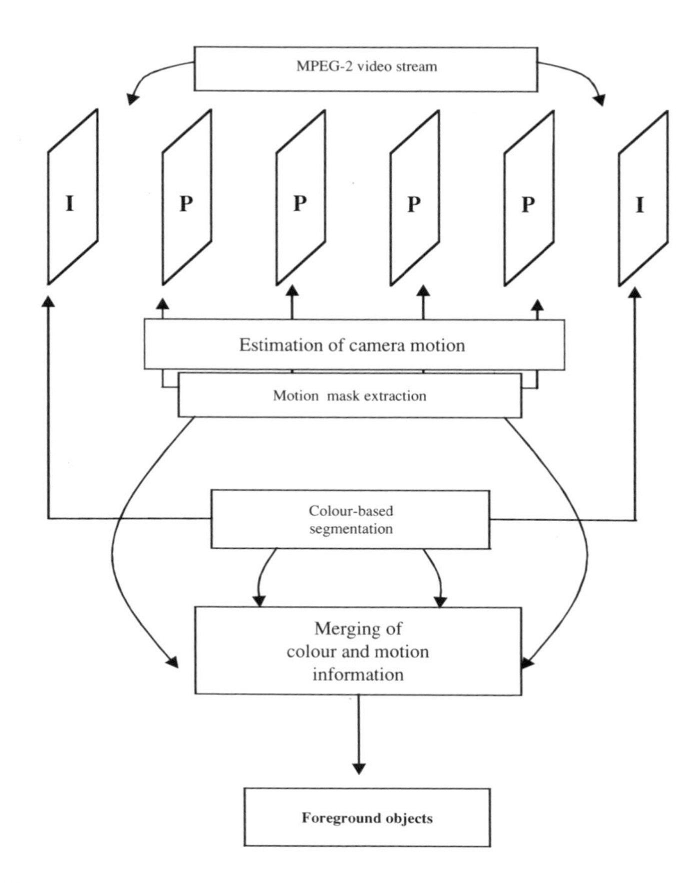

Figure 6.11 Block diagram of a foreground object extraction in the framework of the rough indexing paradigm of compressed video

Figure 6.11 displays the global scheme of the approach. Considering the MPEG-2 stream inside each GOP limited by intra-coded I-frames we utilize noisy macroblock motion vectors in P-frames to estimate camera motion according to the affine model (6.7) and separate 'foreground blocks' which do not follow it. These blocks then serve for motion mask extraction from P-frames. From the I-frame, instead, we extract all colour information; that is, we apply a colour segmentation algorithm to the DCT coefficients of the I-frame to subdivide the image into colour homogeneous regions.

Once obtained, colour and motion information are merged together at I-frame moment of time to extract the foreground objects.

Hence the exact shape of objects is available at DC spatial and I-frame temporal resolution. Their trajectory can be computed at I-frame resolution or interpolated using motion information at a P-frame resolution in video.

As can be seen from Figure 6.11, the estimation of camera motion is a fundamental step in the whole method, therefore we will discuss this in detail in the next section.

6.4.3 Camera-Motion Estimation from MPEG-2 Optical Flow

When estimating camera motion in the context of object-based indexing of video content, the objective is two-fold. First of all, it is important to get the camera motion descriptors, as together with other indexes they will help a semantic interpretation of video. Furthermore, the knowledge of global motion, as we have already said, will help for object separation. In these tasks, robust motion estimators developed by the authors of [29] are specifically interesting. Robust motion estimation limits the contribution of measures which are too far from the estimated model; that is, outliers. Simultaneously, it allows for labelling outliers in the image plane. It is reasonable to suppose that the areas occupied by objects with proper motion will contain a concentration of outliers.

Robust motion estimation usually follows a parametric scheme. Instead of optimizing a mean square error (MSE) criterion (6.14) they use robust criteria. Thus, instead of optimizing MSE—that is, mathematical expectance of the squared errors—one can optimize their median. Such an estimator is called M-estimator, and is robust when up to 50% of outliers are present in the whole set of measures. Nevertheless, its computational cost prevents its use in real-world problems.

An estimator expresses the likelihood of the model with regard to the observations. The estimation of model parameters Θ is then interpreted as maximization of likelihood $f(\Theta, r)$, where r are the observations, for instance the *DFD* values used in (6.14). Maximizing the likelihood turns out the same as minimizing the opposite of logarithmic likelihood $-\log(f(\Theta, r))$. Let us suppose that residuals r follow Gaussian law with zero mean and variance σ^2. In this case, the estimator will be $\rho(r, \sigma) = \frac{r^2}{2\sigma^2}$; that is, the least square estimator. The authors of [29] show that the influence of outliers depends on the derivative of the estimator. In the case of the least square estimator, this influence is unlimited as the derivative is not limited. Estimators widely used for motion estimation are that of Geman and McClure used in [36]:

$$\rho(r, \sigma) = \frac{r^2}{\sigma + r^2} \tag{6.21}$$

and the bi-weight estimator of Tuckey used by the authors of [29]:

$$\rho(r, C) = \begin{cases} \frac{r^6}{6} - 2 \cdot \frac{C^2 r^4}{4} + \frac{C^4 r^2}{2}, \text{if } |r| < C \\ \frac{C^6}{6}, \text{otherwise} \end{cases} \tag{6.22}$$

The derivatives $\psi = \rho'$ of both estimators are limited. For the estimator (6.21) its derivative is $\psi(r, \sigma) = \frac{2r\sigma}{\sigma + r^2}$ and $\lim_{r \to \infty} \psi(r, \sigma) = 0$. The derivative of the estimator (6.22) is:

$$\psi(r, C) = \begin{cases} r(r^2 - C^2)^2_{;} \text{if } |r| < C \\ 0, \text{otherwise} \end{cases} \tag{6.23}$$

As can be seen, it is limited and depends on the constant C.

The problem of estimation with the estimator (6.22) is formulated by the authors of [29] as the weighted least square estimation. In fact, here the problem is to minimize the sum of

$\rho(r)$ in the whole video frame. This can be formulated as the minimization of a weighted least square sum:

$$\sum_i \rho(r_i) = \sum_i \frac{1}{2} w_i r_i^2 \tag{6.24}$$

Following the usual development, by differentiating (6.24) we will obtain:

$$w_i = \frac{\psi(r_i)}{r_i} \tag{6.25}$$

That is:

$$\psi(r, C) = \begin{cases} (r^2 - C^2)^2, \text{if } |r| < C \\ 0, \text{otherwise} \end{cases} \tag{6.26}$$

The authors of [29] estimate motion model in base-band using as residuals r the DFD values in pixels of the current video frame $DFD\,(x, y, \mathrm{d}(\Theta, x, y))$.

In the case of MPEG-2-encoded video, there is no need to decode frames in order to re-estimate motion. The motion vectors of macroblocks in P-frames will be used as initial measures. In this case r is the residual between the motion vector extracted from MPEG-2 and that obtained by the model Θ [32]:

$$r = \|\vec{v}_\Theta - \vec{v}_{MB}\|$$

Here the affine motion model with six parameters $\theta = (a_1, a_2, a_3, a_4, a_5, a_6)^T$ (6.7) is considered as admissible in the MPEG-7 Parametric Motion Descriptor. In the case of macroblock optical flow it is expressed as follows:

$$\begin{aligned} \mathrm{d}x_i &= a_1 + a_2 x_i + a_3 y_i \\ \mathrm{d}y_i &= a_4 + a_5 x_i + a_6 y_i \end{aligned} \tag{6.27}$$

Here $(x_i, y_i)^T$ is the position of the ith macroblock centre in the current image and $(\mathrm{d}x_i, \mathrm{d}y_i)^T$ is the motion vector pointing from the current position to the macroblock centre of the reference frame.

The weighted least square estimation of the optimal parameter vector gives the closed-form solution as:

$$\hat{\Theta} = \left(H^T W H\right)^{-1} H^T W Z \tag{6.28}$$

Here the observation matrix H is the same as in (6.13); it contains the coordinates of the macroblock centres. The matrix W is diagonal and contains the weights for the measured motion vectors computed according to (6.25):

$$W = \begin{pmatrix} w_1 & 0 & \dots & 0 & 0 \\ 0 & w_2 & \dots & 0 & 0 \\ \dots & \dots & \dots & \dots & \dots \\ 0 & 0 & \dots & w_{2N-1} & 0 \\ 0 & 0 & \dots & 0 & w_{2N} \end{pmatrix} \tag{6.29}$$

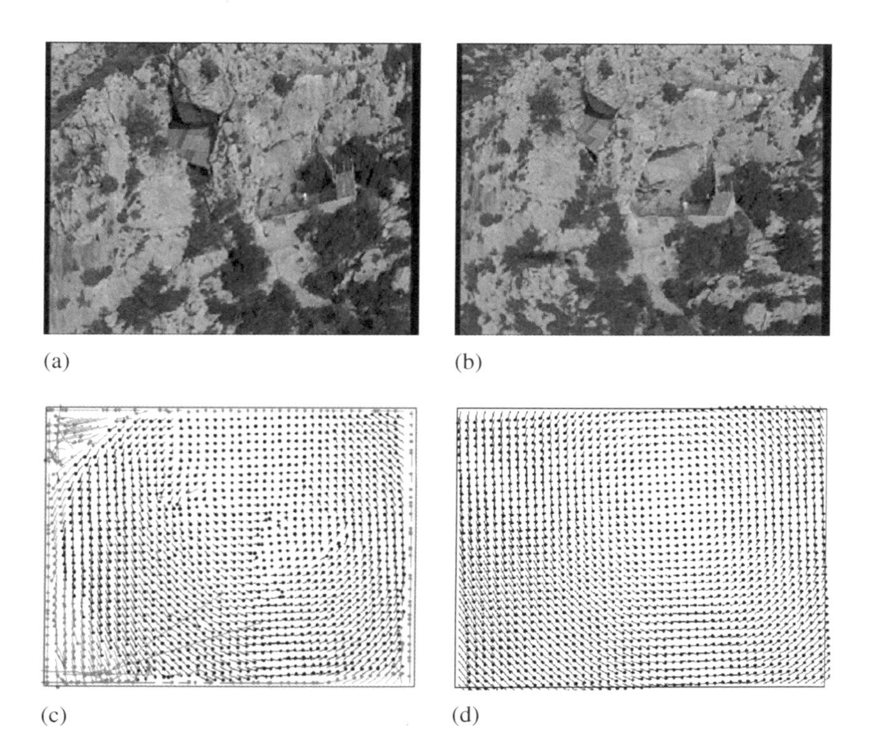

Figure 6.12 Results of robust motion estimation with rotational, zoom and translational components: (a) reference frame, (b) current frame, (c) macroblock motion vectors and (d) motion vectors according to estimated model

Here N is the number of macroblocks in a P-frame. All weights satisfy $w_i \geq 0$. When w_i is close to 0, this means that the macroblock has a different motion from the global one. For the macroblocks coded in mode Intra, the weights are settled to be zero.

As can be seen from (6.26), a constant C tunes the weights. In order to choose it adequately, a multi-resolution scheme is proposed. First of all, a multi-resolution pyramid of motion vectors is computed by low-pass filtering of macroblock motion vectors. At the highest level of the pyramid all weights are considered equal to 1; this means that the value of the constant at the lowest resolution level is taken as $C = \infty$. Then the propagation of the constant is realized based on the statistics of the residuals, namely, $C_{l-1} = k\sigma_l$, where σ_l is the standard deviation of residuals at the higher level of the pyramid.

This estimation of motion model ensures an adequate expression of camera motion with regard to that observed in the video. An example of optical flow conformant to the estimated six-parameter model for a pair of P-frames from the documentary 'Homme de Tautavel' (SFRS) is given in Figure 6.12. Here the motion field adequately reflects a combination of rotation, zoom and translation.

As seen from the motion estimation method with weighted least squares, each measure has its proper weight. These weights are used for separation of macroblocks belonging to moving objects from the background.

6.4.4 Object Extraction at Rough Spatial and Temporal Resolution

The availability of the motion model and of the map of macroblock weights at a P-frame resolution allows for extraction of objects area at this temporal resolution. Then DC images extracted from I-frames will be used to get more precise shape and colours of objects.

Let us first focus on moving object extraction from P-frames.

Motion mask extraction

The weights obtained for macroblocks in the motion estimation process can be normalized to fit the interval $[0, \ldots, 1]$ and then scaled to the interval $[0, \ldots, 255]$. Thus a grey-level image $I_w(x, y)$ of weights is obtained at P-frame temporal resolution. Its spatial resolution is *V/MacroBlockSize*H/MacroBlockSize*, where V and H are the vertical and horizontal resolution of the original video frames, respectively. Low pixel values in this image correspond to the macroblocks which do not follow dominant (camera) motion and thus most likely belong to the foreground objects. A simple thresholding of this image results in a binary image in which macroblock outliers are presented as pixels in white (see Figure 6.13). In Figure 6.13 the initial macroblock motion field decoded from the MPEG-2 P-frame is shown. The noisy motion vectors can be observed corresponding to the foreground objects and areas entering in the camera view field. The result of thresholding of weight image $I_w(x, y)$ is shown in

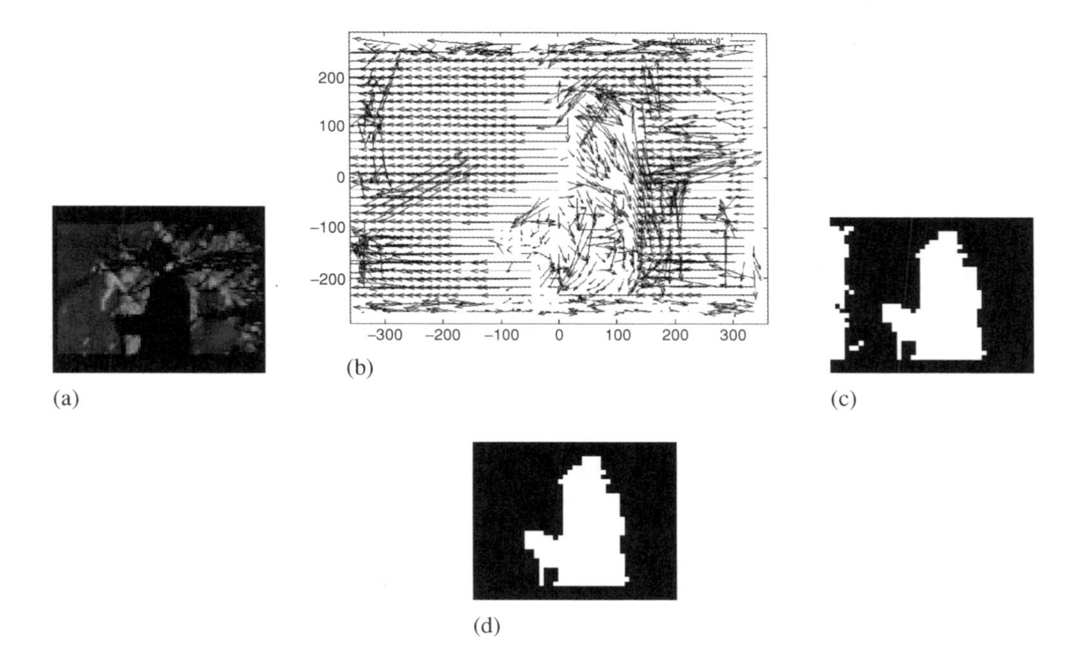

Figure 6.13 Object mask extraction from P-frames. 'De l'Arbre à l'Ouvrage', ©SFRS: (a) original frame at DC resolution, (b) raw macroblock vectors, (c) binary image of weights and (d) binary mask after motion-based filtering

Figure 6.13c. Here the white pixels are situated in the object area but also on the border of the frame due to the camera motion. Therefore, a supplementary filtering is needed to eliminate these 'false' object macroblocks.

To filter the outliers on the frame border, the estimated camera motion model can be used. In fact with forward prediction motion coding such as MPEG encoding, the displacement vector of a macroblock in the current frame is related to the coordinates of pixel $(x_c, y_c)^T$ in the current frame and its reference pixel $(x_p, y_p)^T$ in the reference frame as:

$$\begin{aligned} dx &= x_p - x_c \\ dy &= y_p - y_c \end{aligned} \tag{6.30}$$

Now using the model (6.27) these equations can be solved for x_c and y_c taking as reference pixels the corners of the reference frame. Thus the reference frame will be warped to the current frame. Hence the geometry of the zone entered in the frame is obtained. If some outliers are present in that zone they are supposed to be caused by camera motion and are no longer considered as an object mask. The result of this filtering is given in Figure 6.13d. Compared to the binary mask from Figure 6.13c, all white pixels on the border of the frame are eliminated.

Repeating the method described above for all P-frames inside a single video shot motion masks are obtained for foreground objects in the shot. This method requires preliminary segmentation of video content into shots, as at the border of shots, MPEG motion vectors are massively erroneous and the estimated camera motion does not correspond to the reality. Hence the first guess of objects is obtained at reduced temporal resolution accordingly to the rough indexing paradigm.

Nevertheless, masks in each pair of P-frames were obtained independently from each other. This is why they remain noisy in time. To improve the detection one can profit from the temporal coherence of a video and smooth the detection along the time. To do this the object area can be modelled as a 3D volume in (x, y, t) space. Here, the characteristic function of objects $f(x, y, t)$ is known at time moment corresponding to P-frames, as is depicted in Figure 6.14a.

Let us consider two consecutive GOPs in the MPEG-2 stream inside a shot. To smooth $f(x, y, t)$ along the time a 3D segmentation algorithm will be applied to such pairs of GOPs. The result of this segmentation is a 3D volumetric mask that highlights the region inside which a foreground object is probably located and moves. In the same way as morphological segmentation turns out to be a powerful tool for segmenting the colour frame, as we have already seen, it efficiently resolves the problem of such a 3D segmentation of binary image $f(x, y, t)$. In this work we used a 3D morphological segmentation algorithm developed in [37]. The algorithm follows a usual morphological scheme: filtering, gradient computation and region growing by watershed in a 3D space. In morphological operations, a 3D six-connected structuring element was used (see Figure 6.14b).

The 3D segmentation allows for smoothing of initial noisy characteristic function of objects. Considering the volume slices $I_b(x, y)$ of the filtered characteristic function $\tilde{f}(x, y, t)$, we obtain a 2D mask for each P-frame analysed.

The extracted motion masks represent a good guess of the object shape in P-frames.

Nevertheless, the problem of I-frames remains. As the MPEG-2 stream does not contain motion vectors for the I-frame, the mask cannot be obtained through motion analysis.

From the structure of the MPEG-2 compression standard, it can be seen that, considering two consecutive GOPs, the I-frame is usually situated between two B-frames. If only P-frames

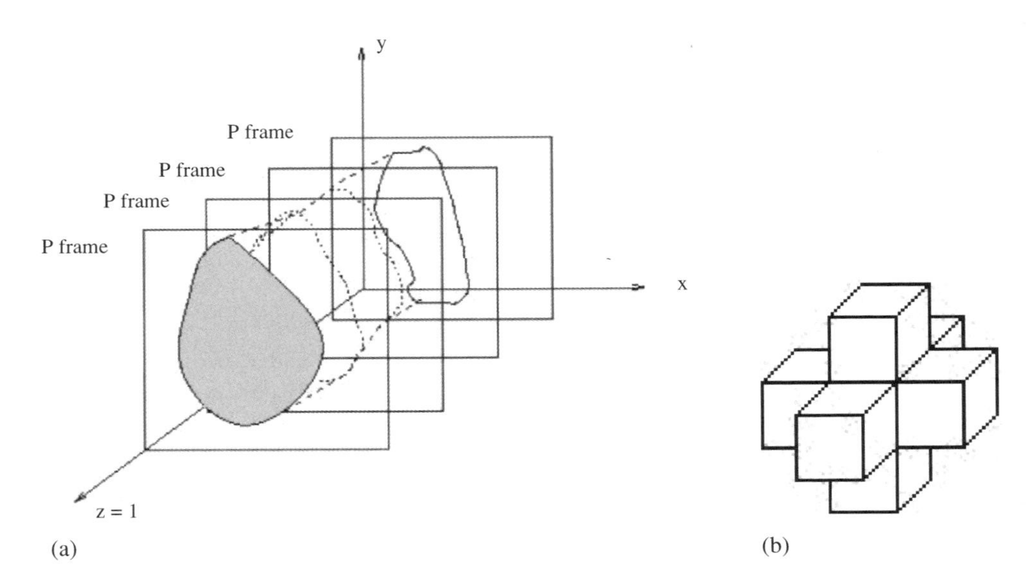

Figure 6.14 3D segmentation of object masks: (a) 3D segmentation along the volume composed by P-frames and (b) the six-connected element used for morphological segmentation

are considered, as in the case of this work, it is enclosed by two P-frames. Therefore, in order to calculate the mask for the I-frame, we can consider the P-frame that comes before the I-frame and the following one and then interpolate the two images (see Figure 6.10 for the temporal alignment of the frames in the MPEG stream). The interpolation can be fulfilled by two approaches: (i) a motion-based one, where the region masks are projected into the frame to be interpolated, and (ii) a spatio-temporal segmentation without use of motion. For the sake of low computational cost, a spatio-temporal interpolation is used. We interpolate masks by morphological filtering. In fact, the resulting binary mask in I-frame $I^b{}_t(x, y)$ will be computed as:

$$I_t^b(x, y) = \min(\delta I_{t-1}^b(x, y), \delta I_{t+1}^b(x, y))$$

Here δ denotes the morphological dilation with four-connected structural element of radius 1 as shown in Figure 6.3a.

In this way the mask for the I-frame is obtained that exhibits the approximate position of the objects.

Hence, the location of foreground objects is defined in a video content inside each video shot at a temporal resolution corresponding to the distance between two consecutive P-frames. The method as presented allows for object mask extraction without any constraints on the topology and object interaction. Nevertheless it does not manage the conflicts in the case of overlapping objects. An exact object identification is possible if each object can be presented as a separate volume in 3D (x, y, t) space during its time-life inside a video shot. Nevertheless, the method can be easily extended to more complex situations with overlapping objects by a supplementary topological spatio-temporal analysis of 3D segmentation maps.

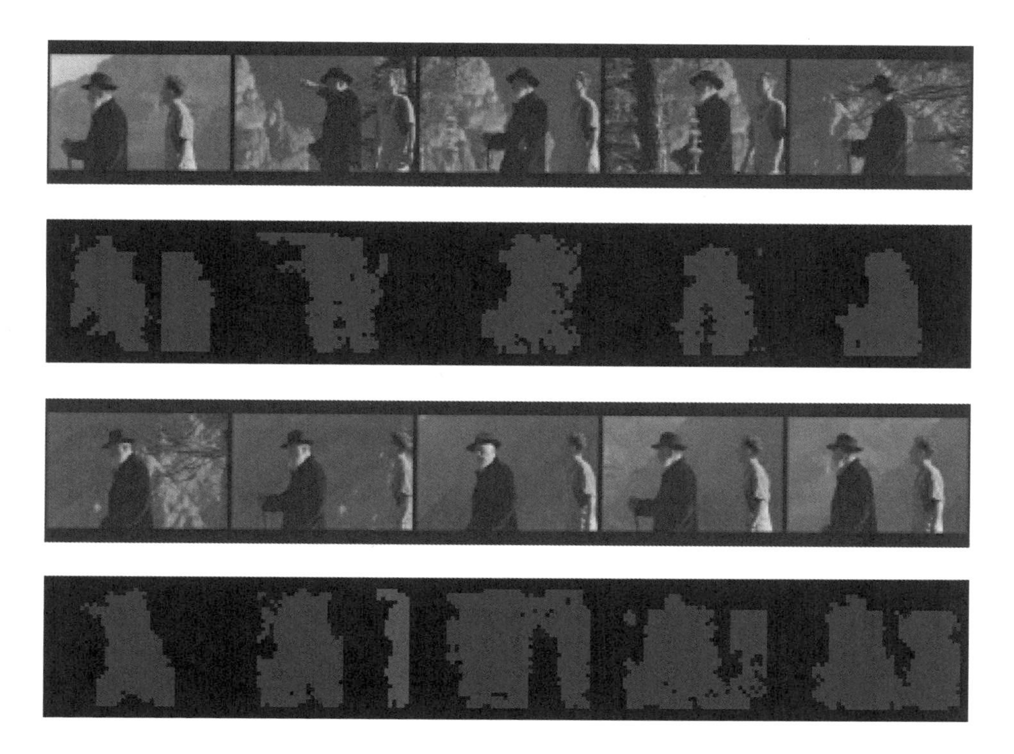

Figure 6.15 Results of object mask extraction at P-frame temporal resolution in video. 'De l'Arbre à l'Ouvrage', ©SFRS

The original frames and motion masks of objects are presented in Figure 6.15. Here the frames and the corresponding masks are given at DC spatial resolution.

The knowledge of object location, even if an approximate one, directly allows for indexing objects by their trajectory as defined by MPEG-7, or by a rough shape. In order to be able to extract the objects more precisely and access other features, such as colour and texture, it is necessary to access the objects located inside masks. This is realized at I-frame temporal resolution by means of colour segmentation.

Colour-based segmentation of objects at a rough resolution

The objective of this segmentation is two-fold: first of all the knowledge of spatial borders of regions will help refining of the object mask, and secondly, the colour and texture descriptors of the object can be computed more precisely.

Hence we implement a colour segmentation process for the I-frame to subdivide it into regions. Then regions are selected overlapping with the motion mask we have calculated before. The set of overlapped regions forms the objects of interest. Remaining in the rough indexing paradigm framework, we will use only DC coefficient of I-frames shaped into DC images. The DC coefficients are easy to extract from the MPEG-2 stream without complete decoding of it. When working with 4:2:0 and 4:2:2 chroma formats [35] an interpolation of

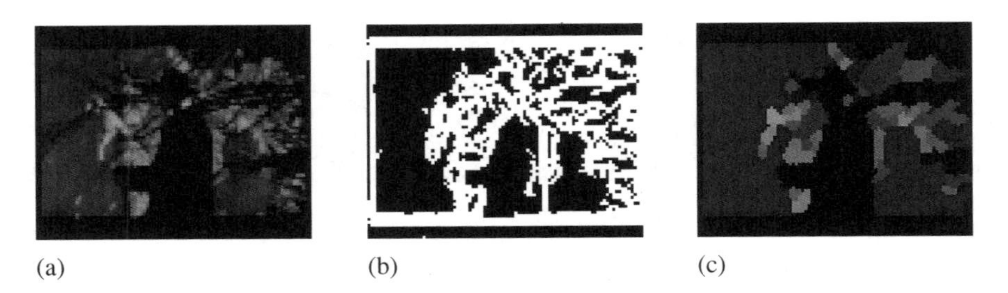

(a) (b) (c)

Figure 6.16 Results of morphological segmentation in YUV space at DC images of an I-frame. 'De l'Arbre à l'Ouvrage', ©SFRS: (a) original DC image, (b) thresholded gradient and (c) segmentation map

chromatic components should be done. The chromatic components U and V or Cr and Cb are subsampled with regard to the resolution of the luminance component Y. In the case of 4:2:0 format, the subsampling factor is of 4; in the case of 4:2:2 it is of 2. We use a 0-order interpolation; that is, the repetition of pixels of chromatic components.

For the colour segmentation we remain in the morphological framework and apply it to the still colour DC frame.

The morphological segmentation with morphological filtering, gradient computation and modified watershed developed for full-resolution video frames (see Section 6.3.1) turns to be efficient for low-resolution frames such as DC images of I-frames. Instead of RGB colour space, in the case of rough indexing, the YUV space is used as the 'direct' space of video encoding.

The results of this segmentation are shown in Figure 6.16. Here the original DC image from the I-frame is displayed in Figure 6.16a, the result of morphological gradient is shown in Figure 6.16b, and the result of the segmentation with each region displayed with its mean colour is presented in Figure 6.16c

Merging of the results of segmentation with motion masks of objects consists in superimposing resulting maps. The regions which are mostly inside the motion mask are supposed to belong to objects of interest.

Another benefit of merging of both results, motion masks and colour segmentation, consists in facilitating filtering of false objects. We have already mentioned that the problem of motion estimation is ill-posed. Any displacement vectors can be obtained in the flat areas of images, the spatial gradient being null in these areas. Such areas often correspond to sky or water in video landscapes, walls in interior scenes. Commercial MPEG encoders produce very.erroneous macroblock vectors inside flat areas. They are considered as outliers by the robust motion estimator and thus are included in the motion mask. An example of such a flat area is presented in Figure 6.17. Here the motion vectors are chaotic in the right upper corner of the frame (Figure 6.17b), which corresponds to the sky region (Figure 6.17a). Consequently, the motion mask contains a component in the same location. The result of object extraction is presented in Figure 6.17c: the false object that is the sky region is extracted.

The filtering of flat areas is based on computing the mean energy of gradient inside the connected components R of the mask:

$$\bar{E}(R) = \frac{1}{card(\mathrm{R})} \sum_{x,y} \left\| \bar{G}(R) \right\|^2$$

If the energy is weak, then the area is flat and has to be excluded from the mask.

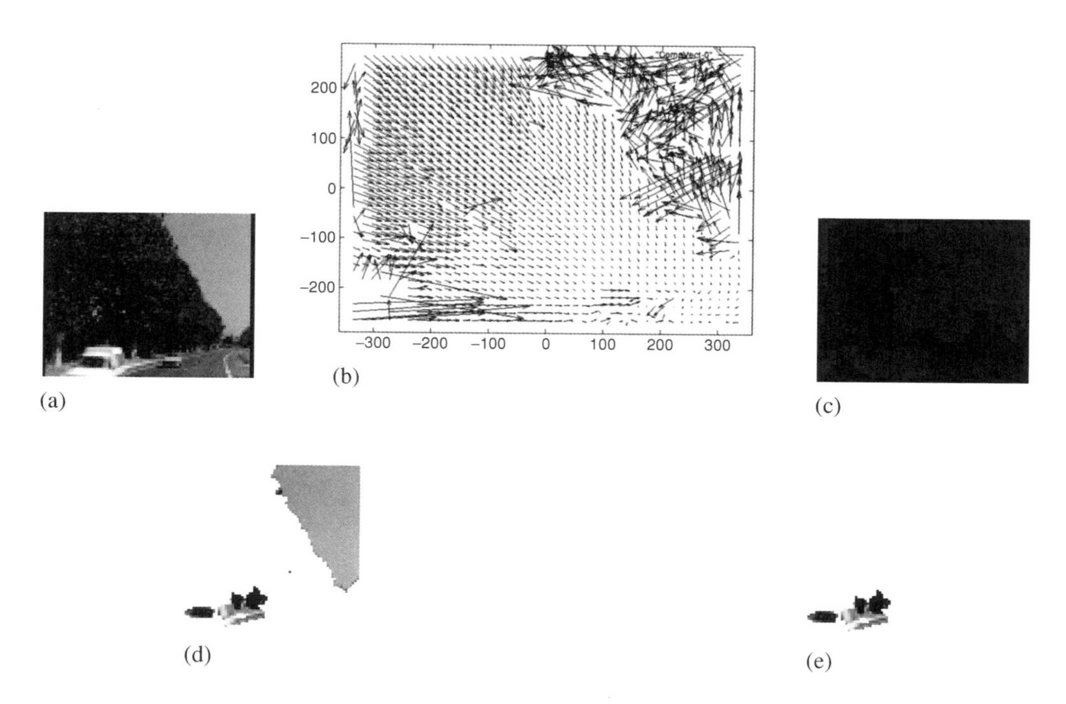

Figure 6.17 Filtering of flat areas. 'Chancre coloré de platanes', ©SFRS: (a) original DC image, (b) MPEG-2 macroblock motion vectors, (c) motion mask, (d) object extraction and (e) flat area filtering

Other knowledge on the content can be applied as well. Thus, if it is known that the video scene was shot outside, then the location of the flat area can also be taken into consideration.

Thus, combining the motion analysis and colour-based segmentation, a refined object can be extracted from MPEG-compressed video at I-frame temporal resolution. This resolution is quite sufficient to compute 'spatial' indexes of objects in order to enable the video search system to process object-based queries on video databases.

The results of object extraction at I-frame resolution in 'De l'Arbre à l'Ouvrage' documentary is given in Figure 6.18. It can be seen that the object extraction remains rather rough, as parts of the objects can be lost due to the filtering of flat areas. Furthermore, if the objects of interest do not move in P-frames enclosing the I-frames, then they cannot be detected. Generally speaking the method shows performance around 80% of recall and misses the detection in the situations mentioned above, and also if the objects are too small, that is they cover less than a few (5–8 approximately) macroblocks in full resolution frame. The over-detection of the method is not expressed as the presence of false objects (they are well filtered by warping frames and flat area removal) but in the over-estimation of objects masks. Nevertheless it is fast—it performs at approximately 1 GOP per second on a Pentium 4 PC—and fulfils well the objective of a rough indexing: the presence of objects, their approximate trajectory and their general appearance.

Figure 6.18 Results of the extraction of objects of interest at I-frame resolution. 'De l'Arbre à l'Ouvrage'

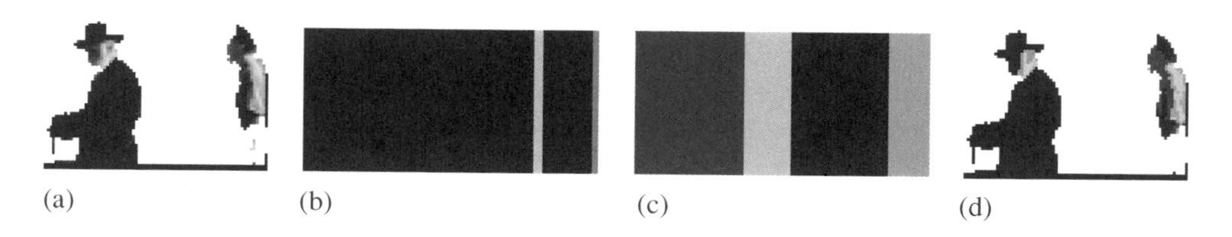

Figure 6.19 Dominant colour descriptor of objects of interest at DC resolution: (a) extracted objects, (b) DCD of the left object, (c) DCD of the right object and (d) objects reconstructed with their DCDs

Once extracted, the objects at a rough resolution can be indexed, for instance by colour. Figure 6.19 displays the dominant colour descriptor (DCD) computed for objects at a rough resolution. Taking into account the small size of objects and rather strong filtering of the original signal in DC images, only four colours are computed by split LBG method as suggested in [2].

Here, the image in Figure 6.19d is obtained by reconstructing the pixels of the extracted object by the nearest colour vector of DCD. It shows a good fit to the originally extracted objects.

6.5 Conclusion

In this chapter we gave a short overview of a problem which, in the context of the Semantic Web and, in particular, in search and retrieval of multimedia information, remains a challenging one. Object-based indexing of multimedia content appeals for a large variety of media analysis methods. Restricted to video, as in this chapter, it requires a combination of image and signal processing methods for extraction of objects from generic video content. In this chapter we presented main trends, from our point of view, in colour segmentation and motion analysis which will serve as the basis for the technology in the future.

Today, it is clear that the semantics-based search for multimedia information requires interoperability of systems and normalization of content description. This is why we devoted the first section of this chapter to a short overview of the MPEG-7 standard and specifically its descriptors, which should constitute a part of normalized description in order to allow access and retrieval of the semantic objects in various query scenarios.

We also outlined how such descriptors can be used to deduce the semantics of the content from its low-level indexes.

Taking into account the application range and the sources of multimedia data in the framework of the Semantic Web, we believe that indexing of video in compressed form will be a must. This is why we proposed and described in this chapter a new rough indexing paradigm for object-based video indexing in the compressed domain. We think that in order to realize semantic queries in a large application domain, we first of all need a general overview and the most relevant and common characteristics of video objects. This is why this paradigm has a direct implication in the field of the Semantic Web. Despite the maturity of video analysis tools developed in numerous research works, the gap between low-level and high-level semantic indexing of video content is far from being filled. This is the perspective of research in video indexing.

References

[1] A. Del Bimbo, *Visual Information Retrieval*. Morgan Kaufmann, San Francisco, CA, 1999.

[2] B.S. Manjunath, P. Salembier, T. Sikora (eds) *Introduction to MPEG-7 Multimedia Content Description Interface*. John Wiley & Sons, Chichester 2002.

[3] ISO/IEC 15938-3: 2001, Multimedia Content Description Interface—Part 3: Visual, Version 1.

[4] S. Jeannin, A. Divakaran, MPEG7 visual motion descriptors. *IEEE Transactions on Circuits and Systems for Video Technology*, **11**(6), 720–725, 2001.

[5] A.B. Benitez, H. Rising, C. Jorgensen, R. Leonardi *et al.*, Semantics of multimedia in MPEG-7. In *Proceedings of IEEE International Conference on Image Processing, ICIP 2002*, Rochester, NY, September 2002. IEEE, 2002.

[6] J.-H. Kuo, J.-L. Wu, An MPEG-7 content-based analysis/retrieval system and its applications. In *Proceedings of Visual Communications and Image Processing, VCIP 2003*, SPIE vol. 5150, Lugano, July 2003, pp. 1800–1810. SPIE, 2003.

[7] A. Ekin, A.M. Tekalp, Robust dominant colour region detection and colour-based application for sports video. In *Proceedings of IEEE International Conference on Image Processing, ICIP 2003*, Barcelona, Spain, September 2003, pp. 11–19. IEEE, 2003.

[8] J. Stauder, B. Chupeau, L. Oisel, Object detection in cinematographic video sequences for automatic indexing. In *Proceedings of Visual Communications and Image Processing, VCIP 2003*, SPIE vol. 5150, Lugano, July 2003, pp. 449–457. SPIE, 2003.

[9] J. Benois-Pineau, F. Morier, D. Barba, H. Sanson, Hierarchical segmentation of video sequences for content manipulation and adaptive coding. *Signal Processing*, **66**, 181–201, 1998.

[10] N. Paragios, R. Deriche, A PDE-based level-set approach for detection and tracking of moving objects. In *Proceedings of 6th International Conference on Computing Vision*, Bombay, India, January 1998, pp. 1139–1145. IEEE, 1998.

[11] S. Jeahn-Besson, M. Barlaud, G. Aubert, Region-based active contours for video object segmentation with camera compensation. In *Proceedings of IEEE International Conference on Image Processing, ICIP 2001*, Thessaloniki, Greece, September 2001. IEEE, 2001.

[12] E. François, P. Bouthemy, Motion segmentation and qualitative dynamic scene analysis from an image sequence. *International Journal of Computer Vision*, **10**(2), 1993.

[13] C. Kervrann, F. Heitz, A hierarchical Markov modeling approach for the segmentation and tracking of deformable shapes. *Graphical Models and Image Processing*, **60**, 173 – 195, 1998.

[14] P. Csillag, L. Boroczky, Segmentation for object-based video coding using dense motion fields. In *Proceedings of VLBV '98*, Urbana-Champaign, IL, October 1998, pp. 145–148. University of Illinois, 1998.

[15] M. Pardas, Video object segmentation introducing depth and motion information. In *Proceedings of IEEE International Conference on Image Processing, ICIP '98*, Chicago, IL, October 1998, vol 2, pp. 637–641. IEEE, 1998.

[16] J. Serra, *Image Analysis and Mathematical Morphology*. Academic Press, London, 1982.

[17] L. Wu, J. Benois-Pineau, D. Barba, Spatio-temporal segmentation of image sequences for object-oriented low bit-rate coding. *Signal Processing: Image Communication*, **8**(6), 513–544, 1996.

[18] P. Salembier, Morphological multiscale segmentation for image coding. *Signal Processing*, **38**, 359–386, 1994.

[19] Y. Tsaig, A. Averbuch, Automatic segmentation of moving objects in video sequences: a region labeling approach. *IEEE Transactions on Circuits and Systems for Video Technology*, **12**(7), 597–612, 2002.

[20] M.A. El Saban, B.S. Manjunath, Video region segmentation by spatio-temporal watersheds. In *Proceedings of IEEE International Conference on Image Processing, ICIP 2003*, Barcelona, Spain, September 2003. IEEE, 2003.

[21] A. Mahboubi, J. Benois-Pineau, D. Barba, Tracking of objects in video scenes with time varying content. *EURASIP Journal on Applied Signal Processing*, **2002**(6), 582–584, 2002.

[22] C. Gu, Multivalued morphology and segmentation-based coding, PhD Thesis, Ecole Polytechnique Fédérale de Lausanne, 1995.

[23] L. Vincent, P. Soille, Watersheds in digital spaces: an efficient algorithm based on immersion simulations. *IEEE Transactions on Pattern Analysis and Machine Intelligence*, **13**(6), 583, 1991.

[24] K. Chehdi, Q.M. Liao, A new approach to the improvement of texture modelisation. In *ISEIS*, Kobe, Japan, 10 June 1991, pp. 1–24.

[25] P. Salembier, L. Garrido, Binary partition tree as an efficient representation for image processing, segmentatation and information retrieval. *IEEE Transactions on Image Processing*, **9**(6.4), 561–576, 2000.
[26] Y.P. Tan, D.D. Saur, S.R. Kulkarni, P.J. Ramadge, Rapid estimation of camera motion from compressed video with application to video annotation. *IEEE Transactions on Circuits and Systems for Video Technology*, **10**(1), 133–146, 2000.
[27] M. Barlaud, C. Labit, *Compression et codage des images et des videos*, pp. 133–176. Hermes, Lavoisier, 2002.
[28] N.H. AbouGhazaleh, Y. El Gamal, Compressed video indexing based on object motion. In *Proceedings of Visual Communications and Image Processing, VCIP 2000*, Perth, Australia, June 2000, pp. 986–993. SPIE, 2002.
[29] P. Bouthemy, M. Gelgon, F. Ganansia, A unified approach to shot change detection and camera motion characterisation. *IEEE Transactions on Circuits and Systems for Video Technology*, **9**(7), 1030–1044, 1999.
[30] L. Carminati, J. Benois-Pineau, M. Gelgon, Human detection and tracking for video surveillance applications in low density environment. In *Proceedings of Visual Communications and Image Processing, VCIP 2003*, SPIE vol. 5150, Lugano, July 2003. SPIE, 2003.
[31] J.Fauquier N. Boujemaa, Region-based retrieval: coarse Segmentation with fine signature. In *Proceedings of IEEE International Conference on Image Processing, ICIP 2002*, Rochester, NY, September 2002. IEEE, 2002.
[32] F. Manerba, J. Benois-Pineau, R. Leonardi, Extraction of foreground objects from a MPEG2 video stream in 'rough-indexing' framework, In *Multimedia Processing, SPIE EI 2004*, San-José, CA, January 2004. SPIE, 2004.
[33] D.J. Legall, The MPEG video compression algorithm. *Signal Processing: Image Communication*, **4**(2), 129–140, 1992.
[34] ISO/IEC 10918—Information technology—Digital Compression and Coding of Continuous Tone Still Images (JPEG).
[35] ISO/IEC 13818-2: 2000. Information Technology—Coding of Moving Pictures and Associated Audio Information: Video.
[36] J.M. Black, P. Anandan, Robust dynamic motion estimation over time. In *Proceedings of IEEE CCVPR*, Maui, Hawaii, June 1991, pp. 296–302. IEEE, 1991.
[37] S. Benini, E. Boniotti, R. Leonardi, A. Signoroni, Interactive segmentation of biomedical images and volumes using connected operators. In *Proceedings of IEEE International Conference on Image Processing, ICIP 2000*, Vancouver, Canada, September 2000. IEEE, 2000.

7

Automatic Extraction and Analysis of Visual Objects Information

Xavier Giró, Verónica Vilaplana, Ferran Marqués and Philippe Salembier

7.1 Introduction

This chapter is structured as follows. Section 7.2 introduces the main concepts of the semantic class model to be used in this work. In Section 7.3, the usefulness of a region-based image representation is further discussed and the binary partition tree is presented. Section 7.4 develops the bases of the perceptual model whereas Section 7.5 details the structural model. In both cases, the extraction of human frontal faces is used as an example to illustrate the usefulness of the models. Note that we apply both models for the same class to clearly explain the ideas behind the two approaches but, in practice, the use of a particular model is related to the perceptual variability of the class instances for a given application. In the sequel, the semantic class describing a human frontal face will be referred to as the semantic class 'face'. Finally, Section 7.6 presents the conclusions of this work.

7.2 Overview of the Proposed Model

In our work, a semantic class (SC) represents the abstraction of a semantic object. We use two types of models to describe a semantic class: a *perceptual model* and a *structural model*. A semantic class can be described by both types of models. This concept is illustrated in Figure 7.1 where a semantic class (represented by a grey ellipse) is associated with both descriptions. Typically, semantic classes with a limited amount of perceptual variability can be handled with a perceptual model whereas more complex semantic classes require a structural model.

Multimedia Content and the Semantic Web Edited by Giorgos Stamou and Stefanos Kollias

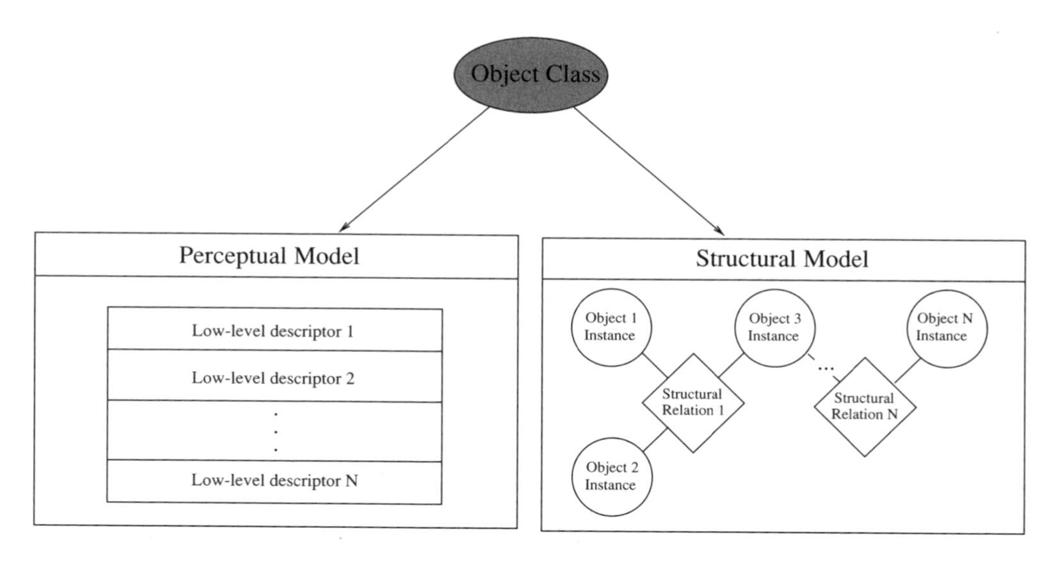

Figure 7.1 Perceptual and structural models of an object class

Perceptual model

A semantic class can be characterized by a set of low-level visual descriptors defining the perceptual characteristics of all class instances. In this work, a low-level descriptor is a descriptor that can be directly evaluated on the signal (e.g. a histogram or the shape of a region). The perceptual model is a list of low-level descriptors whose combination defines the semantic class. Therefore, perceptual models actually bridge the gap between the perceptual and semantic descriptions of the entities in the image.

Structural model

In order to deal with the perceptual variability, a semantic class can be decomposed into its simpler parts (parts that form the object) and the relations among these parts. In turn, these simpler parts are instances of simpler semantic classes (e.g. every wheel in the description of the semantic class 'car' is an instance of the simpler semantic class 'wheel'). Moreover, in this work, the relations among simpler parts are assumed to be only structural (e.g. two wheels can be associated by a structural relation such as 'near to', but semantic relations such as 'similar to' are not allowed). The structural model that represents the instances of these simpler semantic entities and their structural relations (SRs) is described by means of a graph, the so-called description graph. This concept is illustrated in Figure 7.1, where an example description graph is presented. In it, instances of simpler semantic class are represented by white circles while rhombi correspond to structural relations.

The description of a semantic class, however complex, ultimately relies on a set of perceptual models. Every instance in a description graph is associated to a simpler semantic class that may be described by a perceptual or/and a structural model. If the simpler semantic class is

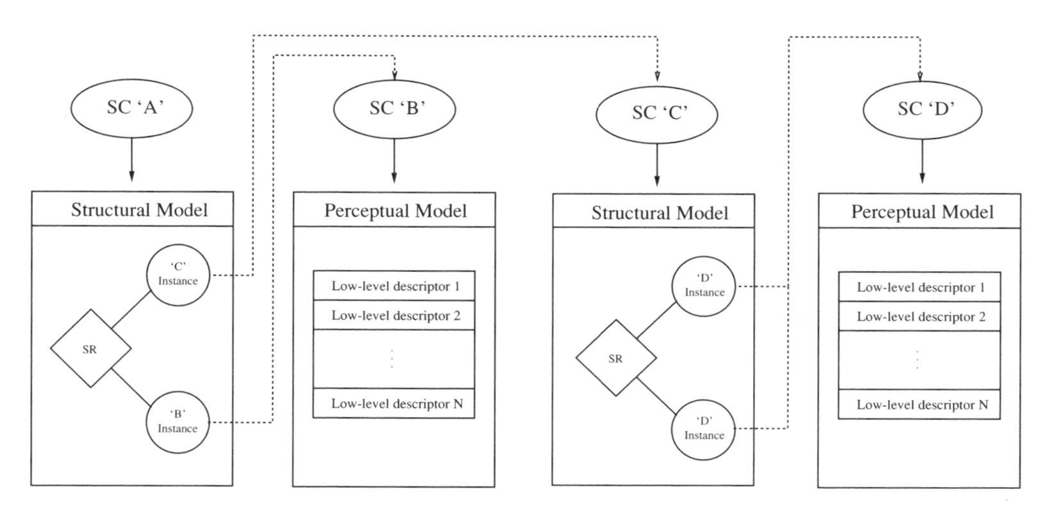

Figure 7.2 Example of decomposition of a structural model into simpler structural and perceptual models

described only by a structural model, the previous decomposition can be iterated until reaching the simplest possible level of semantic classes, which can only be described by perceptual models. This concept is illustrated in Figure 7.2: the SC 'A' is described by a structural model. This semantic class is described by two instances of simpler semantic classes ('B' and 'C') and their structural relations. The SC 'B' is represented by a perceptual model. However, SC 'C' is decomposed, in turn, into two instances of a single SC 'D' and their structural relations. Finally, the SC 'D' is represented by a perceptual model.

7.3 Region-Based Representation of Images: The Binary Partition Tree

In most object analysis tasks, one of the first difficulties to be faced is related to the raw representation of the original data built around a rectangular array of pixels. Detecting objects directly on this representation is difficult in particular because one has to detect not only the presence of the object but also its position and its scale. In this section, we discuss the interest in binary partition trees [1] as a region-based representation that can be used for a large number of object analysis and recognition applications. The idea is to perform a first step of abstraction from the signal by defining a reduced set of regions at various scales of resolution that are moreover representative of the perceptual features of the image. Instead of looking at all possible pixel locations and all possible scales, the object recognition algorithm will base its analysis strategy on this reduced set of candidate regions.

An example binary partition tree is shown in Figure 7.3. The lower part of the figure presents an original image (left) and an initial partition corresponding to this image (centre). Note that this partition is composed of 100 regions and would be considered as over-segmented for a large number of applications. Here, it only defines the lowest resolution scale in term of regions.

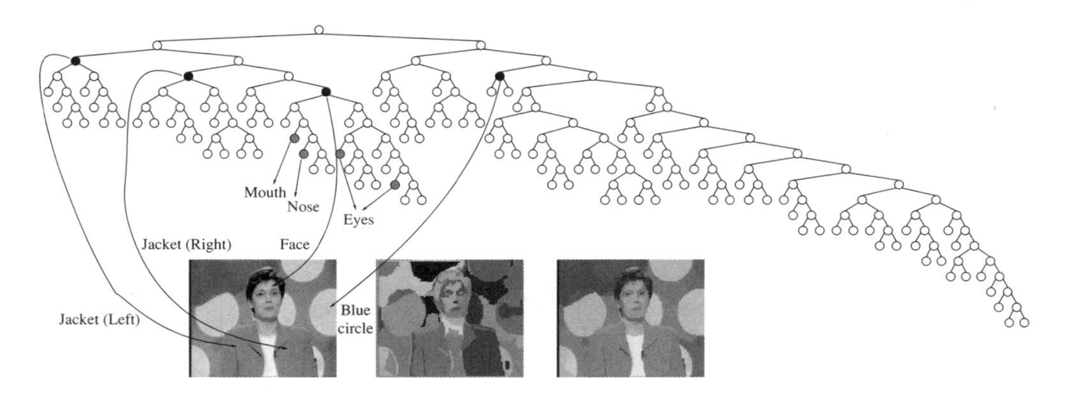

Figure 7.3 Example binary partition tree (top) together with the original image (left), the initial partition with 100 regions (centre) and the regions of the initial partition represented by their mean value (right)

Ideally, it should represent all elements that are perceptually relevant. The image obtained by filling the regions of the initial partition with their mean grey value is shown in the right part of Figure 7.3. As can be seen, almost all details of the original image are visible. Information about the similarity is encoded in the tree shown in the upper part of the figure. The tree leaves represent the regions of the initial partition. The remaining tree nodes are parent nodes. They represent regions that can be obtained by merging the regions represented by the child nodes. The information about the similarity is encoded by the tree structure itself: the set of regions that are the most similar to a given region R_0 is given by the set of siblings nodes of R_0. As a result, the similarity between regions represented in the lower (upper) part of the tree is very high (rather low).

Several approaches can be followed to create the tree. An attractive solution relies on a region-based merging algorithm that follows a bottom-up approach. Starting from an initial partition, the algorithm recursively merges neighbouring regions based on a homogeneity criterion until one region is obtained. As described by Figure 7.3, the tree leaves represent the regions of the initial partition. The remaining nodes represent the regions that are obtained by merging the regions associated to the two children nodes. In this representation, the root node represents the entire image support. Note that the resulting tree is binary since, at each step, only two neighbouring regions are merged.

Using region-based merging algorithms such as [2–5] the binary partition tree is created by keeping track of the regions that are merged at each iteration. The homogeneity criterion used in the example of Figure 7.3 is based on colour similarity. It should be noticed, however, that the homogeneity criterion has not to be restricted to colour. For example, if the image for which we create the binary partition tree belongs to a sequence of images, motion information should also be used to generate the tree. In a first stage, regions are merged using a colour homogeneity criterion. At a given stage, the difference in colour between a region and all its neighbours is so high that its merging would be meaningless. This is when a different feature such as motion can be used to assess the similarity between regions and continue the merging process (see [1, 6] for more details). In practice, one can create a hierarchy of generic (that is, application independent) similarity criteria: colour, motion, depth etc., that are successively used to create the binary partition tree.

Note that in many cases, if the criteria used to create the tree are generic, it is unlikely for complex objects to be represented as individual nodes. Only simple objects that are homogeneous in terms of the criteria used to compute the tree can be expected to be represented in the tree as single nodes. The notion of extended nodes discussed in Section 7.4.3 can be used to tackle this issue. As an example, the tree represented in Figure 7.3 has been created with a colour homogeneity criterion. As a result, only objects that are homogeneous in colour can be expected to be represented as single nodes. Some examples are indicated in Figure 7.3.

Once the tree has been created, the remaining analysis steps will directly work on its nodes. The number of nodes is dramatically smaller than the number of pixels. Moreover, the set of nodes spans all possible scales in terms of regions. Let us explain now how perceptual models can be used to extract semantic information from the binary partition tree.

7.4 Perceptual Modelling of a Semantic Class

The purpose of the perceptual modelling is to describe a semantic class, here an object, using features that can be directly measured on the signal. The detection of an instance of a particular object is therefore done by extracting and analysing the low-level features that indicate the presence of the object of interest. These low-level features may be more or less complex and represent signal attributes that do not need any interpretation such as pixel distributions or geometrical features.

In many cases, the object detection cannot be reliably performed using only a single feature. While several features may suggest the presence of the object with different reliabilities, none of them may be sufficient for describing the class. However, the likelihoods provided by individual features can be combined to build a more robust detector.

The definition of a semantic class by a perceptual model then involves the selection of a set of useful features followed by a learning stage where these attributes are characterized or described by statistical or other kind of models using sample data. Finally, a combination rule has to be defined to merge the information provided by the individual features.

The object detector based on these models analyses different candidate regions: for each region it extracts the features modelling the semantic class, computes the likelihood value for each descriptor and then combines these values into a final class likelihood. The output of the detector is a list of regions that more likely belong to the class of interest.

7.4.1 Perceptual Model Definition

Selection of low-level features

The first step in the definition of a perceptual model for a semantic class is the selection of a set of low-level visual features. The goal is to characterize the semantic class by measurements whose values are very similar for instances of the same class and very different for instances of other semantic classes. This leads to the idea of seeking distinguishing features that are invariant to certain transformations of the input signal (translation, rotation, scale, occlusions, projective distortion, non-rigid deformations, illumination etc.). The choice of distinguishing features is a critical design step that requires knowledge about the problem domain.

Many different visual features may be employed to describe a semantic class. They can be grouped into attributes dealing with the distribution of pixels and attributes addressing geometrical features.

- *Features related to the pixel distribution:* they take into account the value of the pixel components and their relative positions in the area of analysis. Within this class of features, the most commonly used ones are colour (for example, characterizing the pixel distribution in an area by the colour histogram) and texture (for example, using a wavelet or a principal components analysis). The evaluation of such features on a segmented image improves the robustness of the estimation since the boundaries of the areas of similar pixel distribution have already been computed.
- *Geometrical features:* they account for the structural characteristics of the area of support of the object being analysed. Within this class of features, the most commonly used ones analyse the shape of the various connected components (for example, using a curvature scale space representation) and the pattern that these components may form (for example, using a graph to describe their relations). As previously, the evaluation of such features on a segmented image improves the robustness of the estimation since the boundaries of the connected components are already available.

Feature models and decision functions

The next steps in the definition of a perceptual model are the modelling of the feature values for a set of class representatives and the design of a function or decision rule to evaluate if a candidate region is an instance of the class.

The feature models may be built using prior knowledge about the class or by a learning process using training data. When learning is employed, models are usually expressed in terms of statistics or probability distributions associated to the values of the descriptors for the sample data. The training set has to be carefully selected in order to allow good generalization but to avoid over-training [7]. Following the classification proposed by Tax [8], the learning of the descriptor models may be approached with different techniques:

- *Density methods:* they are the most straightforward methods. Density methods estimate the probability density function for the selected feature using the training data. The most frequently used distribution models are Gaussian, mixtures of Gaussians and Parzen density estimation [9, 10].
- *Boundary methods:* these methods avoid the estimation of the complete density of the data and only focus on the boundary of the data. The K-centres technique, the nearest neighbour method and the support vector data description belong to this category of approaches [8, 11].
- *Reconstruction methods:* in some cases, the object generating process itself can be modelled. It is then possible to encode a candidate region with the model, to reconstruct the measurements for the candidate and to use the reconstruction error to measure the fit of the candidate to the model. It is assumed that the smaller the reconstruction error, the better the region fits to the model and the more likely that it is not an outlier. Representative reconstruction methods are principal component analysis, k-means clustering and learning vector quantization [12].

For each visual descriptor, a function f has to be inferred so that if x is the value of the descriptor for a given region, $f(x)$ is an estimate of the likelihood or probability that the region is an instance of the class. The function f may also estimate the distance or resemblance of the region to the target class.

Density methods estimate the probability density of the class $f(x) = p(x/\omega)$, where ω is the target class. On the other side, boundary and reconstruction methods fit a model to the data and define a distance between a test instance x and the model, $f(x) = d_\omega(x)$. In some applications, votes or binary outputs are preferred and the function f is an indicator function: $I(p(x/\omega) \geq \theta)$ or $I(d_\omega(x)) \leq \theta)$ where θ is a decision threshold.

Combining rules

As mentioned above, in many cases, it is unlikely that a single feature can be used to characterize a class optimally. Using the best feature (the feature that leads to the maximum likelihood) and overlooking the other descriptors might give poor results. To improve the algorithm performance, different descriptors can be combined [8].

Descriptors can be combined using class posterior probabilities. The posterior probability for the class ω, $p(\omega/x)$, can be computed from $p(x/\omega)$ using Bayes rule:

$$p(\omega/x) = \frac{p(x/\omega)p(\omega)}{p(x)} = \frac{p(x/\omega)p(\omega)}{p(x/\omega)p(\omega) + p(x/\bar{\omega})p(\bar{\omega})} \qquad (7.1)$$

where $\bar{\omega}$ is the set of all possible objects that are different from ω. Information about this set is generally not available, and thus the distribution $p(x/\bar{\omega})$ is unknown, and so are the priors $p(\omega)$ and $p(\bar{\omega})$. This means that the class posterior probability cannot be computed.

However, if a uniform distribution is assumed for $p(x/\bar{\omega})$, i.e. considering that it is independent of x, it can be proved that $p(\omega/x_1) < p(\omega/x_2)$ if and only if $p(x_1/\omega) < p(x_2/\omega)$ [7]. In this case, $p(x/\omega)$ can be used instead of $p(\omega/x)$.

The descriptors will then be combined using likelihoods. For descriptors that provide distances $d_\omega(x)$ instead of probabilities, the distances must first be transformed into likelihoods. This transformation may be done by fitting sample descriptor values to some distribution or by applying a mapping like $\tilde{p}(x/\omega) = \frac{1}{c_1} exp(-d_\omega(x)/c_2)$, which models a Gaussian distribution around the model if $d_\omega(x)$ is a squared Euclidean distance.

In the sequel, it is assumed that a class model is defined by R descriptors, with feature spaces $\aleph^k$ and decision functions f_k, for $k = 1, ..., R$. It is also assumed that the output values have been normalized to likelihoods, and $f_k : \aleph^k \rightarrow [0, 1]$ for all k. When a candidate region is evaluated, R measurement vectors x_k, $k = 1, ..., R$ are obtained. Let $\mathbf{x} = (x_1, ..., x_R)$ be the set of measurements for the candidate. With these notations, typical combination rules are:

- Weighted sum of estimated likelihoods: $f_{wsl}(\mathbf{x}) = \sum_{k=1}^{R} w_k f_k(x_k)$, where $\sum_{k=1}^{R} w_k = 1$
- Product combination of estimated likelihoods: $f_{pl}(\mathbf{x}) = \frac{\prod_{k=1}^{R} f_k(x_k)}{\prod_{k=1}^{R} f_k(x_k) \prod_{k=1}^{R} \theta_k}$, where θ_k is an approximation of the conditional distribution (uniform) of $\bar{\omega}$. Using $p(x/\omega)$ instead of $p(\omega/x)$, the rule becomes $f_{pl}(\mathbf{x}) = \prod_{k=1}^{R} f_k(x_k)$

More complex combination rules may be used. For example, it could be interesting to define a combination rule where a descriptor is used only if its associated likelihood is significantly high.

7.4.2 Object Detection

For the detection, our strategy relies on a region-based approach. Images are segmented into homogeneous regions (for some convenient homogeneity criterion) and a binary partition tree is constructed from the initial partition. The candidates for the object detection are the regions represented by the nodes of the tree. For every node in the binary partition tree, descriptor values and likelihoods are computed and likelihoods are combined into a global class probability. The most likely regions are considered as object instances. This region-based approach reduces the computational burden of an exhaustive search and increases the robustness of the feature extraction.

7.4.3 Example: Face Detection

The proposed approach is illustrated with a human face detector based on a perceptual model of the semantic class face (frontal faces). A face can be associated with a set of homogeneous regions. Consequently, it should be possible to find a face by properly selecting a set of regions from a segmented image [13].

Selection of candidates with a binary partition tree

The initial image partition is created using a region growing technique, where regions are merged until a given peak signal-to-noise ratio (PSNR) is reached. Then a binary partition tree is built. The merging order is based on a colour similarity measure between regions. Although the use of colour as a similarity measure helps to construct meaningful regions in the binary partition tree, the presence of the desired regions (faces) as nodes is not ensured as they are not homogeneous in colour.

To overcome this problem and provide the binary partition tree with more flexibility, the tree analysis uses information from regions associated to tree nodes as well as from neighbouring regions. The strategy relies on the notion of *extended nodes*. In an extended node, the area of support of a node is determined by the shape of the object to detect, frontal faces in this case. The region corresponding to a node is extended by enlarging its area of support. The new area is formed by the regions of the initial partition contained in a face shape model placed on the node regions. Figure 7.6 illustrates this concept and demonstrates the convenience of the method: an extended node may represent objects that are not completely represented as individual nodes in the tree. Figure 7.6b and 7.6c present the nodes that are selected in the binary partition tree, whereas Figure 7.6e and 7.6f show the extended nodes.

Face class modelling

In this work, the face class is defined with the following set of low-level visual descriptors. In all cases f_{feat} denotes the normalized decision function derived for the model of feature *feat*.

- *Colour* (f_c): the descriptor is the mean value of the region for U and V components in the YUV colour space. To characterize face colour, the skin colour distribution is modelled using a Gaussian distribution in the (u, v) space.
- *Aspect ratio* (f_{ar}): this descriptor measures the aspect ratio of the bounding box of the region. Its distribution is also modelled with a Gaussian.

- *Shape* (f_{sh}): a face shape model (A) is compared to the shape of the extended node (B). This comparison relies on the modified Hausdorff distance between the contour points of both shapes:

$$H(A, B) = max(h(A, B), h(B, A)) \tag{7.2}$$

where $h(A, B)$ is the modified directed Hausdorff distance proposed in [14]:

$$h(A, B) = \frac{1}{|A|} \sum_{a \in A} min_{b \in B} \|a - b\| \tag{7.3}$$

Before computing the distance, both shapes are normalized to ensure scale invariance.

The last two descriptors are texture features that use principal component analysis (PCA) to describe the global appearance of regions. Given a collection of n by m pixel training images represented as vectors of size $N = nm$ in a N-dimensional space, the PCA finds a set of orthonormal vectors that capture the variance of the data in an optimal way (in the squared error sense). The eigenvectors of the data covariance matrix are computed and the eigenvectors with largest eigenvalues, which point in the direction of the largest variances, are preserved. These vectors are used as basis vectors to map the data.

- *Distance in the feature space* (f_{difs}): a Gaussian model is assumed for the face class in the subspace spanned by the first M eigenvectors of a PCA computed on the training dataset. The similarity measure between a candidate x and the face class is the Mahalanobis distance in the subspace (the distance between x and the sample mean $\bar{x}$).

$$d_{difs}(x) = \sum_{i=1}^{M} \frac{y_i^2}{\lambda_i} \tag{7.4}$$

where y_i is the projection of the mean normalized vector $x - \bar{x}$ on the i-eigenvector and λ_i is the i-eigenvalue [15].
- *Distance from the feature space* (f_{dffs}): the face class is modelled as the subspace spanned by the first M eigenvectors of a PCA. Another similarity measure between a candidate and the face class is the reconstruction error, the Euclidean distance between the candidate and its projection on the subspace [15].

$$d_{dffs}(x) = \|x - \bar{x}\|^2 - \sum_{i=1}^{M} y_i^2 \tag{7.5}$$

The distances defined by these two descriptors are transformed into likelihoods by fitting them to the face class distributions obtained through the training data (a χ^2 distribution for the *difs* (f_{difs}) and a gamma distribution for the *dffs* (f_{dffs})).

Density methods are used to learn the feature models for colour, shape, aspect ratio and *difs* descriptors, whereas *dffs* is based on a reconstruction method. The training of the feature models is performed with a subset of 400 images from the XM2VTS database [16]. $M = 5$ eigenvectors are used in *difs* and *dffs*. The descriptors are combined by a product combination of estimated likelihoods:

$$f_{pl} = f_c f_{ar} f_{sh} f_{difs} f_{dffs} \tag{7.6}$$

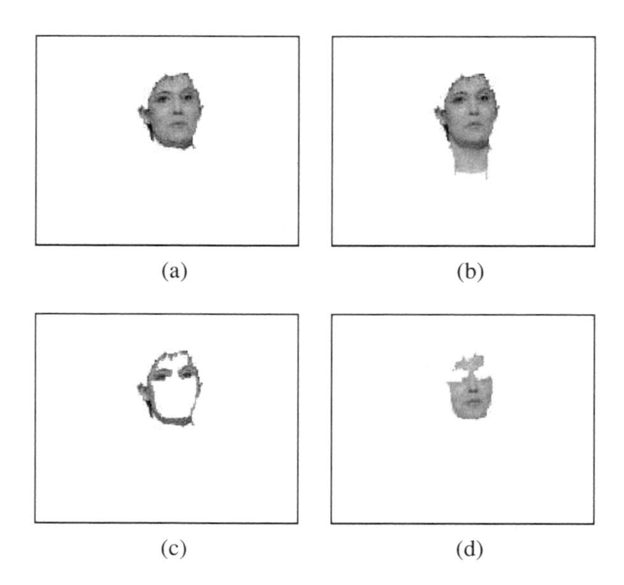

Figure 7.4 Results of the face detection on the example of Figure 7.3: (a) detected face, (b) region corresponding to the father node of the face region, (c) and (d) regions corresponding to the children nodes of the face region.

Face detection

The face detector computes the descriptors for the candidates defined by the binary partition tree. As too small regions do not contain enough information to be analysed, nodes that are smaller than a given threshold are not taken into account.

The likelihoods are combined and the candidates with highest confidence values are proposed as face instances. The result of this procedure is presented in Figure 7.4 where the selected node in the example of Figure 7.3 is shown. To show the accuracy of the selected node, Figure 7.4b, c and d present the father and children nodes of the selected one. As can be seen, the selected node is the best representation of the face in the scene that can be obtained.

Figures 7.5 and 7.6 show the binary partition tree and the related images of another example, respectively. Two faces are present in the original image. In the binary partition tree of Figure 7.5, the subtrees associated to the selected nodes are marked (the background face in grey and the foreground face in black). In this case, the detection of the faces illustrates the usefulness of evaluating the extended nodes. Figure 7.6b and 7.6c present the nodes that are selected in the tree, whereas Figure 7.6e and 7.6f show the extended nodes. The complete shape of the faces can be extracted in this example thanks to the extension of the nodes since complete faces do not appear as single nodes in the tree.

7.5 Structural Modelling of a Semantic Class

7.5.1 Definition of Structural Models with Description Graphs

A second approach for the modelling of semantic classes is to treat them as a structure of simpler semantic classes instead of as a whole. As shown in Figure 7.1, a structural model of

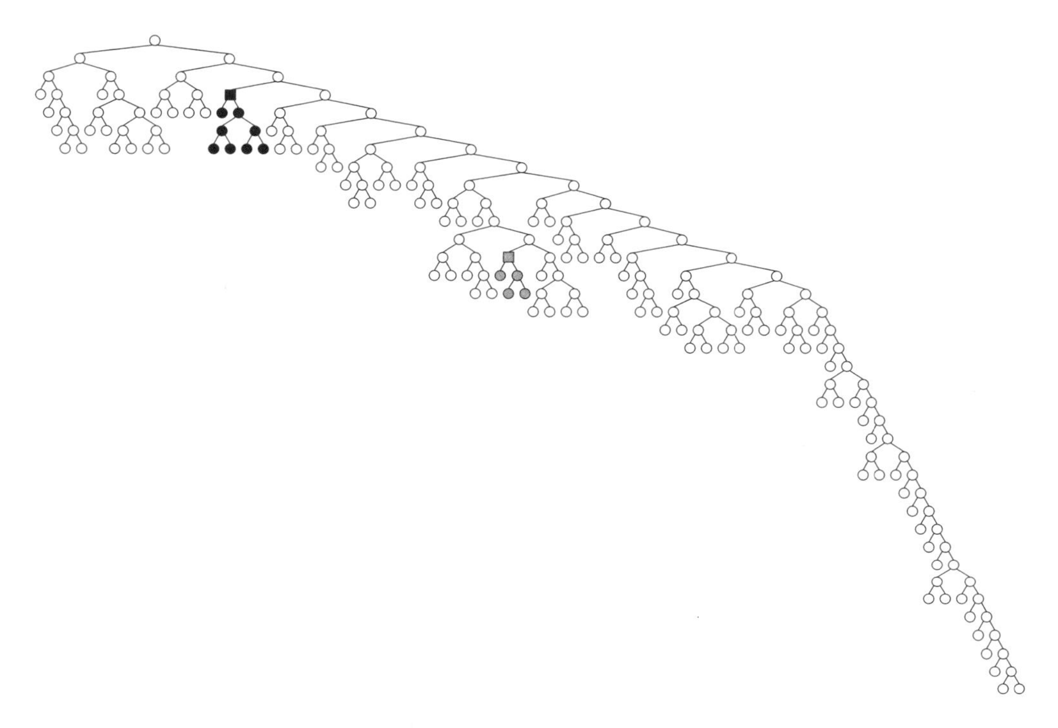

Figure 7.5 Binary partition tree of the original image of Figure 7.6a. The shaded subtrees correspond to the two face regions (see Figure 7.6).

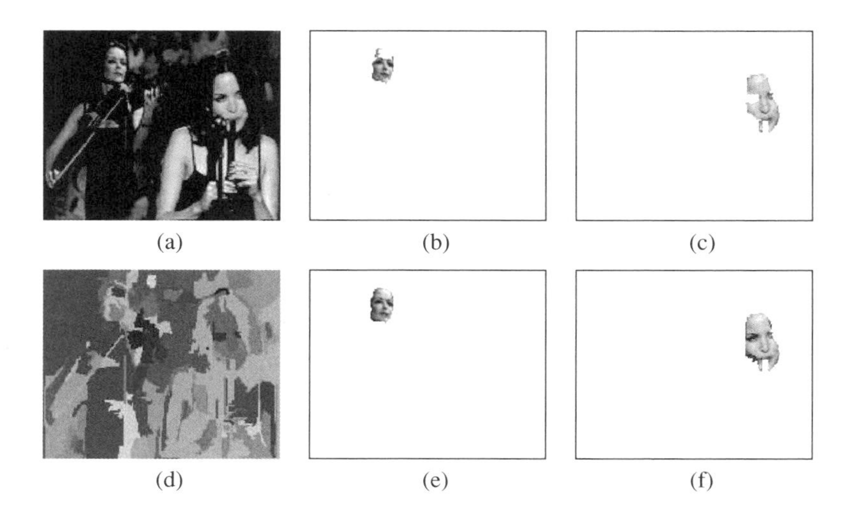

Figure 7.6 Example of face detection: (a) original image, (b) and (c) detected face regions corresponding to the shaded subtrees of Figure 7.5, (d) original partition, (e) and (f) extended nodes of the two face regions.

a semantic class is formed by instances of other semantic classes that satisfy certain *structural relations* (SR) among them. The proposed approach organizes these instances in terms of a description graph (DG) [17]. A generic graph is a group of V objects called *vertices* and a group of E non-ordered pairs of vertices called *edges*. A description graph assigns structural relations and instances of semantic classes to its vertices. An SC vertex is represented by a circle whereas an SR vertex is represented by a rhombus. Description graph edges create connections between the two types of vertices, creating a complex model that describes the common structure of the semantic class. The use of graphs as a tool to express meaning in a form that is logically precise, humanly readable and computationally tractable has been widely treated in the work of Sowa on conceptual graphs [18]. Previous experiences on object and event modelization [19, 20] have shown their applications on the indexing of visual data.

Although it could seem natural to assign structural relations to edges in the graph, such an approach would limit the range of possible structural relations since graph edges connect only two vertices. Therefore, only binary relations could be defined. Structural relations as associated to vertices allow the definition of richer relations while avoiding the use of hyper graphs. Structural relations are computed over descriptors that belong to their associated instances. Note that these descriptors may not appear in the perceptual model of the semantic class itself. For example, a structural relation such as 'above' might be evaluated based on the centre of mass of the two related instance vertices, understanding the centre of mass as a descriptor of the region associated to the semantic class instance. However, this descriptor should not belong to the perceptual models of the instanced semantic class.

Description graph vertices are classified into necessary and optional. Necessary vertices correspond to those parts of the object that must be represented in all instances. On the other hand, the presence of the optional vertices is not mandatory. Following the 'face' example, Figure 7.7 shows a description graph for the structural model of a face. In it, two necessary instances of the SC 'eye' and a necessary instance of the SC 'mouth' are structured with an SR 'triangle'. The model is completed with an optional instance of the SC 'moustache'

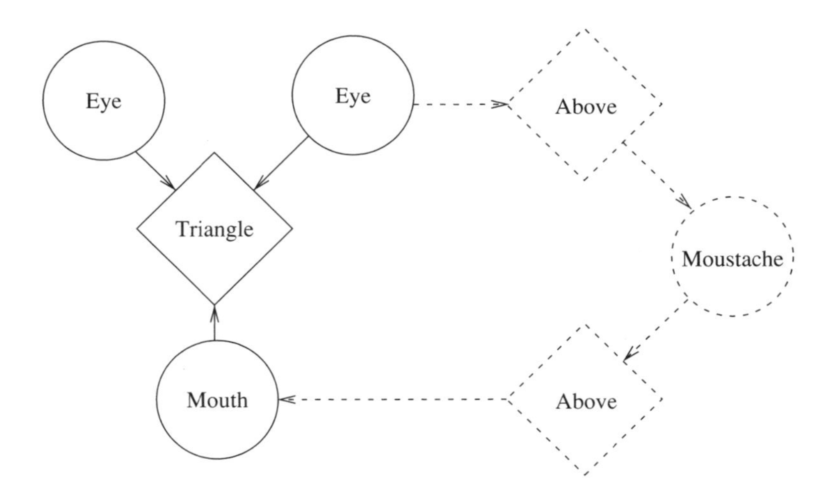

Figure 7.7 Description graph for the structural model of the object class 'face'.

related with the eyes and mouth by two necessary vertices of an SR 'above'. This example illustrates the fact that although any complete instance of a face must contain two eyes and a mouth, it may also include a moustache. The presence of a moustache in a face is considered optional, because its absence does not decrease the belonging to the SC 'face'. Nevertheless, the detection of a moustache in an image reinforces the probability of a correct face detection.

7.5.2 Likelihood Function

Analogously to the low-level descriptors of the perceptual model, description graphs also require a likelihood function to measure the *probability* (f) of a set of regions of being an instance of a semantic class. It combines the individual probabilities of each vertex with their weights. The *weight* (w) of a vertex expresses its relevance in the description graph, taking values between 0 and 1, where 0 denotes irrelevant and 1 very relevant. Weights can be set manually or as a result of a learning algorithm [21], and they should be considered as part of the model. Following the previous face example, if the weight of the vertex 'mouth' is higher than the sum of the weights of the vertices 'eye', a query for a face to an image database will retrieve images depicting only a mouth before those showing only two eyes.

The proposed likelihood function for description graphs is shown in equation (7.7). The global probability f is computed by combining the probabilities of the instance vertices weighted according to the model. The expression is normalized by the sum of the weights. Sums for the N necessary vertices and the O optional vertices are expressed separately for clarity.

$$f = \frac{\sum_{k=0}^{N} w_k f_k + \sum_{l=0}^{O} w_l f_l}{\sum_{k=0}^{N} w_k + \sum_{l=0}^{O} w_l} \tag{7.7}$$

While all necessary vertices must be included when computing f, there is no previous information to decide which optional vertices must be considered. A possible selection criterion is to consider only those optional vertices whose inclusion in the expression increases f. Equation 7.9 shows that this condition is accomplished when the likelihood of an optional vertex is higher than the current f. With this criterion, the design of the decision algorithm is simple. First, all optional vertices are sorted in decreasing order according to their associated f. Then, the ordered list is scanned adding optional vertices to the expression until reaching a vertex k with a f_k smaller than the updated f.

$$\frac{\sum_{k=0}^{N} w_k f_k + \sum_{l=0}^{n-1} w_l f_l + w_n f_n}{\sum_{k=0}^{N} w_k + \sum_{l=0}^{n-1} w_l + w_n} > \frac{\sum_{k=0}^{N} w_k f_k + \sum_{l=0}^{n-1} w_l f_l}{\sum_{k=0}^{N} w_k + \sum_{l=0}^{n-1} w_l} \Leftrightarrow \tag{7.8}$$

$$\Leftrightarrow f_n > \frac{\sum_{k=0}^{N} w_k f_k + \sum_{l=0}^{n-1} w_l f_l}{\sum_{k=0}^{N} w_k + \sum_{l=0}^{n-1} w_l} \tag{7.9}$$

7.5.3 Extraction Algorithm

The process of detecting an object in an image can be understood as a graph matching between a subset of binary partition tree nodes and the description graph vertices describing a semantic class. For each semantic class vertex in the description graph, the likelihood of every node in the binary partition tree to be its instance should be computed. Therefore, the number of possible combinations may be huge, making it inadvisable to check every possibility one by one. For this reason, a heuristic algorithm is proposed.

The approach is based on searching first instances of the necessary vertices in the model. Only when this process is finished, the extraction algorithm starts looking for instances of the optional vertices. In both cases, the searching order is from higher to lower weights.

Therefore, the algorithm starts analysing the necessary SC vertex with highest weight. If the associated semantic class is defined by a perceptual model, the instance likelihood can be computed directly on the low-level descriptors of the binary partition tree node under analysis. Those binary partition tree nodes whose low-level descriptors are similar to the perceptual model ones will be associated to the instance vertex.

On the other hand, if the semantic class associated to the necessary SC vertex is defined by a structural model, the algorithm is iterated looking for instances of the vertices of this new structural model in the subtree below the node under analysis. Those binary partition tree nodes whose subtrees contain nodes that match the structural model will be associated to the instance vertex.

Once the necessary SC vertex with highest weight has been analysed, the analysis of the following SC vertex can be performed relying on the SR vertex information. Structural relations provide prior knowledge that reduces the number of binary partition tree nodes to consider, a restriction that decreases significantly the computation effort. Therefore, only those nodes in the binary partition tree representing regions that fulfill the structural relations are analysed. For example, a structural relation like ‘include’ may dramatically minimize the binary partition tree nodes to consider.

As shown in the face example of Figure 7.8, structural and perceptual models complement each other to allow the description of a complex semantic class.

7.5.4 Example: Face Detection

This study case shows a structural approach for the automatic detection of faces in images. While in the previous section a combination of low-level descriptors is used as a basic tool, the following example applies a different strategy to reach the same goal. In this case, the detection of a face is based on the previous extraction of the individual facial features, a process driven according to the algorithm described in the previous section.

Figure 7.9 shows the description graph of the considered structural model of face. ‘mouth’, ‘eyes’ and ‘skin’ are chosen as necessary vertices, while ‘eyebrows’ and ‘nostrils’ are considered optional because they are not always visible in a human face image. All semantic classes in the description graph have their own perceptual model based on colour and shape descriptors.

The search for the face starts from the most relevant SC vertex of the description graph, in our case, the mouth. Those binary partition tree nodes satisfying the mouth perceptual model are marked as candidates. The following step consists of looking for the next most relevant

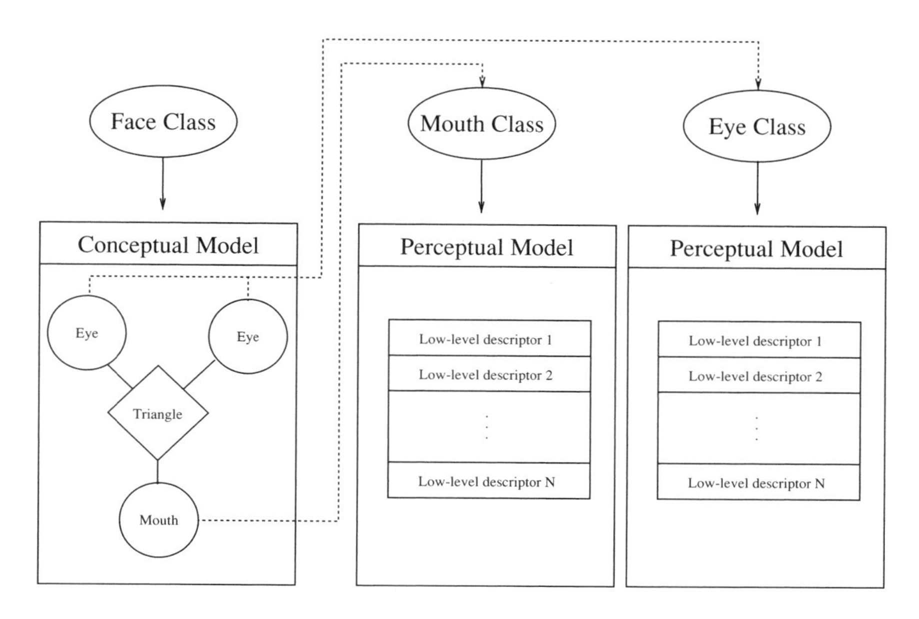

Figure 7.8 Object instances in description graph refer to models of an object class

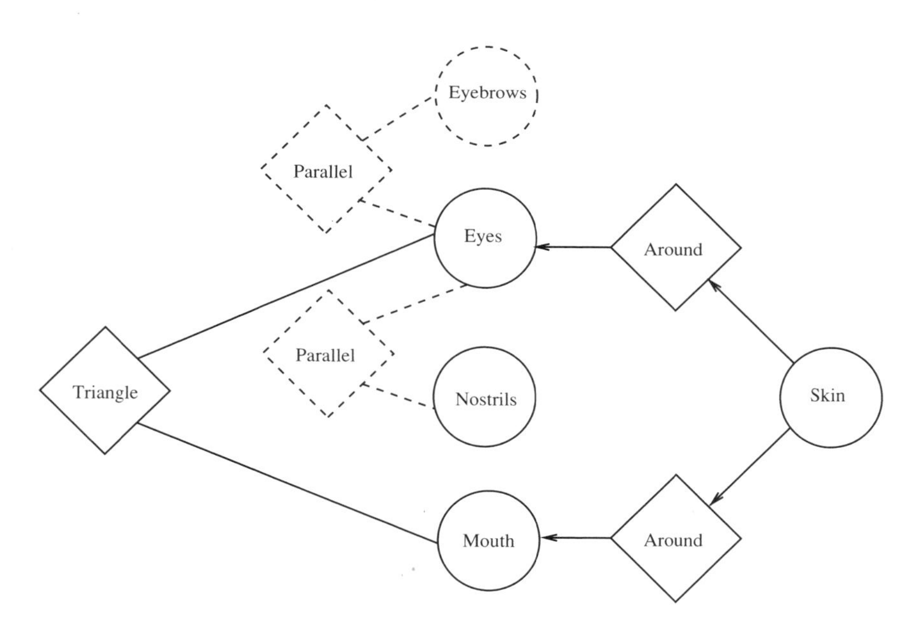

Figure 7.9 Description graph for SC 'face' used in the example

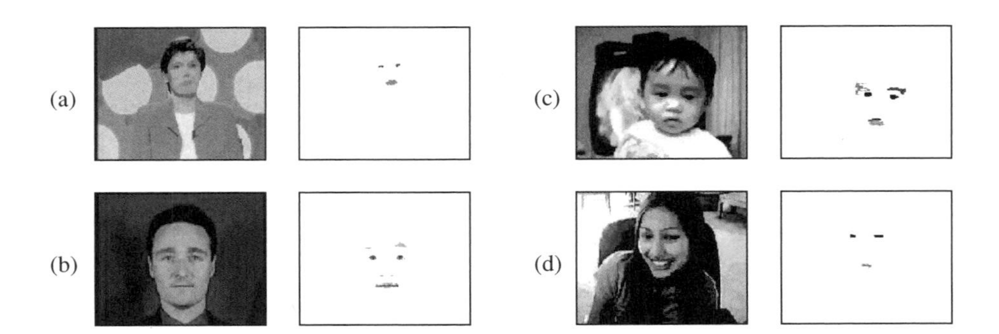

Figure 7.10 Initial partitions and extracted facial features of four study cases

semantic class, which is the eyes. As the eyes must satisfy a triangular structural relation with respect to the mouth, those regions marked as mouth candidates are used as anchor points to find the eye candidates. This approach reduces drastically the total amount of node candidates to be considered. Several trios of candidates are found, which are ordered according to their global probability. At this point, it is important to notice that when three binary partition tree nodes form a candidate instance of a face, none of them can be part of another instance of face. In general, the algorithm considers that a single binary partition tree node cannot be part of two different instances (unless one of them is part of the other). This restriction is basic to discarding many of the candidates. After the eyes, the algorithm looks for the skin around the mouth and eyes. When the algorithm has been applied on the necessary vertices, the search for the optional ones starts. In this example, optional vertices are those related to eyebrows and nostrils parallel to the eyes. The result of the search is a list of instance candidates ordered according to their final probability. Finally, a threshold is applied to select those candidates accepted as valid instances of the SC 'face'.

Figure 7.10 shows some results of the algorithm on four examples. Their associated numerical results are shown in Table 7.1. The training of the weights was performed with a subset of

Table 7.1 Numerical results of the examples of Figure 7.10

Semantic class	Necessary	w	Case (a)	Case (b)	Case (c)	Case (d)
mouth	N	0.20	1.00	1.00	1.00	1.00
nostrils	O	1.00	0.00	1.00	0.00	0.00
eyes	N	0.20	1.00	1.00	1.00	1.00
skin	N	0.20	0.92	0.79	0.58	0.60
eyebrows	O	1.00	0.00	1.00	0.00	0.00
triangle	N	0.20	1.00	1.00	1.00	1.00
around(mouth)	N	0.20	1.00	1.00	1.00	1.00
around(eyes)	N	0.20	1.00	1.00	1.00	1.00
parallel(nostrils)	O	1.00	—	1.00	—	—
parallel(eyebrows)	O	1.00	—	1.00	—	—
f			0.99	0.99	0.97	0.93

400 images from the XM2VTS database [16], initially fixing low values to necessary vertices and high values to the optional ones. Notice that structural relations linked to non-present optional instances cannot be computed.

7.6 Conclusions

In this chapter, we have presented an algorithm for object detection in images. It relies on a combination of two models for characterizing the semantic classes: a perceptual model based on the low-level features, and a structural model that exploits the decomposition of a semantic class into its simpler parts and their structural relations. The application of both models for object detection relies on a region-based description of the image.

The use of a region-based representation of the image allows reducing the set of candidate positions and scales to analyse within the image while improving the robustness of the analysis since all features are estimated on homogeneous regions. In this work, a binary partition tree representation is chosen which not only proposes an initial partition of the image but a set of binary mergings based on a specific criterion (for instance, colour homogeneity).

Complex objects are very unlikely to appear in a binary partition tree as nodes since, commonly, generic criteria are used for its creation. This difficulty is circumvented by separating the selection of the positions and scales that are to be analysed (node selection) from the definition of the exact area of analysis (node extension). The extension of a node allows including object geometrical information in the low-level descriptor evaluation. As illustrated in Section 7.4, this strategy improves the object detection performance when applying perceptual models.

The binary partition tree representation is useful as well when using structural models. The characterization of a binary partition tree node as an instance of a semantic class represented by a structural model requires the analysis of the regions that form this node. The binary partition tree allows the analysis to be restricted to the subtree associated to this node.

Perceptual models are used in this work to bridge the semantic gap; that is, to characterize objects by means of low-level descriptors. Given the perceptual model of a semantic class, the way to combine its low-level descriptors may vary depending on the application. In some applications, the resolution at which the object is represented in the image can make unfeasible the evaluation of a specific descriptor. Furthermore, if the detection algorithm is to be applied online, very time-consuming low-level descriptor extraction algorithms might have to be avoided.

In the examples presented in Section 7.4, it can be seen that, even for a structured object like a human face, perceptual models can be very useful. Results in that section have been selected to demonstrate the robustness of the proposed approach even when the scene has illumination problems or faces are partially occluded, tilted or not correctly focused.

Structural models are used in this work to cope with the variability of the instances of complex objects. Due to this variability, the low-level characterization of the object may become unfeasible and higher-level information may be necessary. In this work, this information is related to the decomposition of the objects into simpler parts and their structural relations. Note that, since we are detecting objects, we do not require other type of relations (e.g. semantic relations) to be included in the model.

As in the perceptual case, the way to combine the information within the model is very important in the structural case. The distinction between necessary and optional vertices in the model and their nonlinear combination lead to a very robust object characterization. The results presented in Section 7.5 have been chosen to demonstrate this robustness. This way, we have selected images with illumination problems, with necessary features represented by very small regions or with optional features that are partially or not represented.

Finally, it should be stressed that the powerfulness of the proposed approach is based on the combination of both models. Structural models eventually rely on perceptual models since the latter provide the initial semantic description of the image. In turn, perceptual models require the use of a higher level of description to characterize variable objects, which is provided by the structural models.

Acknowledgements

This work has been partially supported by the FAETHON project of the European Commission (IST-1999-20502) and the Revise project of the Spanish Governement (TIC2001-0996).

References

[1] P. Salembier, L. Garrido, Binary partition tree as an efficient representation for image processing, segmentation and information retrieval. *IEEE Transactions on Image Processing*, **9**(4), 561–576, 2000.
[2] O. Morris, M. Lee, A. Constantinidies, Graph theory for image analysis: an approach based on the shortest spanning tree. *IEE Proceedings, F*, **133**(2), 146–152, 1986.
[3] T. Vlachos, A. Constantinidies, Graph-theoretical approach to colour picture segmentation and contour classification. *IEE Proceedings, I*, **140**(1), 36–45, 1993.
[4] J. Crespo, R.W. Shafer, J. Serra, C. Gratin, F. Meyer, A flat zone approach: a general low-level region merging segmentation method. *Signal Processing*, **62**(1), 37–60, 1997.
[5] L. Garrido, P. Salembier, D. Garcia. Extensive operators in partition lattices for image sequence analysis. *EURASIP Signal Processing*, **66**(2), 157–180, 1998.
[6] P. Salembier, F. Marqués, Region-based representations of image and video: segmentation tools for multimedia services. *IEEE Transactions on Circuits and Systems for Video Technology*, **9**(8), 1147–1169, 1999.
[7] R. Duda, P. Hart, D. Stork, *Pattern Classification*, 2nd edn. Wiley Interscience, New York 2001.
[8] D. Tax. One-Class Classification: Concept Learning in the Absence of Counter-Examples, PhD thesis, TU Delft, 2001.
[9] C. Bishop, Novelty detection and neural network validation. In *IEE Proceedings on Vision, Image and Signal Processing. Special Issue on Applications of Neural Networks*, **141**, 271–222, 1994.
[10] L. Parra, G. Deco, S. Miesbach, Statistical independence and novelty detection with information preserving nonlinear maps. *Neural Computation*, **8**, 260–269, 1996.
[11] M. Moya, D. Hush, Network constraints and multi-objective optimization for one-class classification. *Neural Networks*, **9**(3), 463–474, 1996.
[12] C. Bishop, *Neural Networks for Pattern Recognition*. Oxford University Press, Oxford, 1995.
[13] F. Marqués, V. Vilaplana, Face segmentation and tracking based on connected operators and partition projection. *Pattern Recognition*, **35**(3), 601–614, 2002.
[14] M.P. Dubuisson, A.K. Jain, A modified haussdorff distance for object matching. In *ICPR-94*, vol. A, pp. 566–568, Jerusalem, Israel, 1994.
[15] B. Moghaddam, A. Pentland, Probabilistic visual learning for object representation. *IEEE Transactions on Pattern Analysis and Machine Intelligence*, **19**(7), 696–710, 1997.
[16] K. Messer, J. Matas, J. Kittler, J. Luettin, G. Mairse, XM2VTSbd: the extended M2VTS database. In *Proceedings of the 2nd Conference on Audio and Video Based Biometric Person Authantication*. Springer Verlag, New York, 1999.
[17] X. Giró, F. Marqués, Semantic entity detection using description graphs. In *Workshop on Image Analysis for Multimedia Application Services (WIAMIS'03)*, pp. 39–42, London, April 2003.
[18] J.F. Sowa, *Knowledge Representation: Logical, Philosphical, and Computational Foundations*. Brooks Solidous Cole, Pacific Grove, CA, 2000.

[19] A. Ekin, A. Murat Tekalp, R. Mehrotra, Integrated semantic-syntactic video event modeling for search and retrieval. In *IEEE International Conference on Image Processing (ICIP)*, vol. 1, pp. 22–25, Rochester, New York, September 2002. IEEE, 2002.
[20] M.R. Naphade, I.V. Kozintsev, T.S. Huang. Factor graph framework for semantic video indexing. *IEEE Transactions on Circuits and Systems for Video Technology*, **12**(1), 40–52, 2002.
[21] M.R. Naphade, J.R. Smith, Learning visual models of semantic concepts. In *IEEE International Conference on Image Processing (ICIP'03)*, vol. 2, pp. 445–448, Barcelona, Spain, September 2003. IEEE, 2003.

8

Mining the Semantics of Visual Concepts and Context

Milind R. Naphade and John R. Smith

8.1 Introduction

There is increasing focus on semantic access to media content. The emergence of the MPEG-7 standard makes it possible to describe content with rich semantic descriptors and use these descriptors for semantic access. However, the automatic population of these descriptors is a challenging problem. The difficulty lies in bridging the gap between low-level media features and high-level semantics. Towards this end, automatic techniques for learning models of the desired semantics and their application for populating MPEG-7 are gaining ground [2]. Early attempts at learning models for detecting audiovisual events include the work of Naphade and colleagues (explosion, waterfall [3], speech, music [4], helicopter [5], laughter, applause [6]), semantic visual templates [7] and rule-based systems [8]. We proposed a probabilistic framework of concept and context modelling to enable semantic concept detection [3]. In particular, we proposed generic trainable systems for three kinds of concepts: objects (car, man, helicopter), sites (outdoor, beach) and events (explosion, man walking). We termed these probabilistic multimedia representations of semantic concept as multijects. We modelled the presence or absence of a particular semantic concept in a frame or a shot as a binary random variable and learnt probabilistic models for the two hypotheses using training content annotated in this binary fashion. Thus each concept can be modelled directly from the positive and negative training samples. However, such direct modelling overlooks the presence of contextual constraints within concepts that occur together.

Intuitively, it is straightforward that the presence of certain semantic concepts suggests a high possibility of detecting certain other multijects. Similarly, some concepts are less likely to occur in the presence of others. The detection of sky and greenery boosts the chances of detecting a Landscape, and reduces the chances of detecting Indoors. It might also be possible to detect some concepts and infer more complex concepts based on their relation with the detected ones. Detection of human speech in the audio stream and a face in the video stream

Multimedia Content and the Semantic Web Edited by Giorgos Stamou and Stefanos Kollias

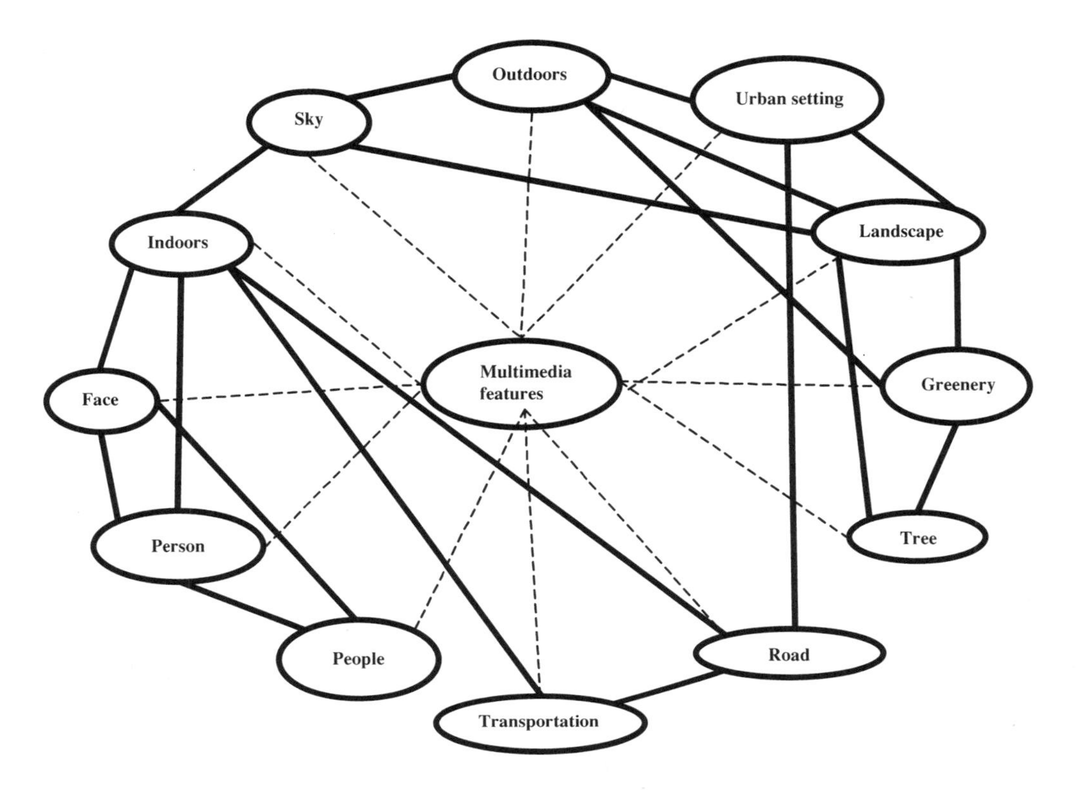

Figure 8.1 A conceptual multinet showing relations between 12 visual concepts. Nodes in the network represent semantic concepts. Edges represent relationship. Symbols on the edges can be interpreted as indicative of the nature of interaction between concepts linked by the edge

may lead to the inference of human talking. To integrate all the concepts and model their interaction, we proposed the network of multijects [3], which we termed as a multinet [9]. A conceptual figure of a multinet is shown in Figure 8.1, with positive (negative) signs indicating positive (negative) interaction.

In this chapter we present the implementation of a hybrid framework for modelling concepts and context. Discriminant training has been shown to improve detection performance of the multijects [10]. On the other hand, the graphical multinet that we proposed [9] for a small number of concepts demonstrated performance improvement over generatively trained multijects. In this chapter we combine multiject models built using discriminant training along with a graphical multinet that models joint probability mass functions of semantic concepts. We use the multijects to map low-level audiovisual features to the high-level concepts. We also use a probabilistic factor graph framework, which models the interaction between concepts within each video shot as well as across the video shots within each video clip. Factor graphs provide an elegant framework to represent the stochastic relationship between concepts, while the sum–product algorithm provides an efficient tool to perform learning and inference in factor graphs. Using exact as well as approximate inference (through loopy probability propagation) we show that there is significant improvement in the detection performance.

Using the TREC Video 2002 Benchmark Corpus and some of the benchmark concepts [11], we show that explicit modelling of concept relationships and the use of this model for enforcing inter-conceptual and temporal relationships leads to an improvement in detection performance. We also provide efficient approximations to the context models that are scalable in terms of number of concepts being modelled. We show that these approximations retain the significant improvement in detection performance.

8.2 Modelling Concepts: Support Vector Machines for Multiject Models

The generic framework for modelling semantic concepts from multimedia features [2, 3] includes an annotation interface, a learning framework for building models and a detection module for ranking unseen content based on detection confidence for the models (which can be interpreted as keywords) (Figure 8.2). Suitable learning models include generative models [12] as well as discriminant techniques [10]. Positive examples for interesting semantic concepts are usually rare. In this situation it turns out that discriminant classification using support vector machines (SVM) [13] performs better [10]. The SVMs project the original feature dimension using nonlinear kernel functions and attempt to find that linear separating hyperplane in the higher dimensional space which maximizes generalization capability. For details see [13].

8.2.1 Design Choices

Assuming that we extract features for colour, texture, shape, structure etc., it is important to fuse information from across these feature types. One way is to build models for each feature type

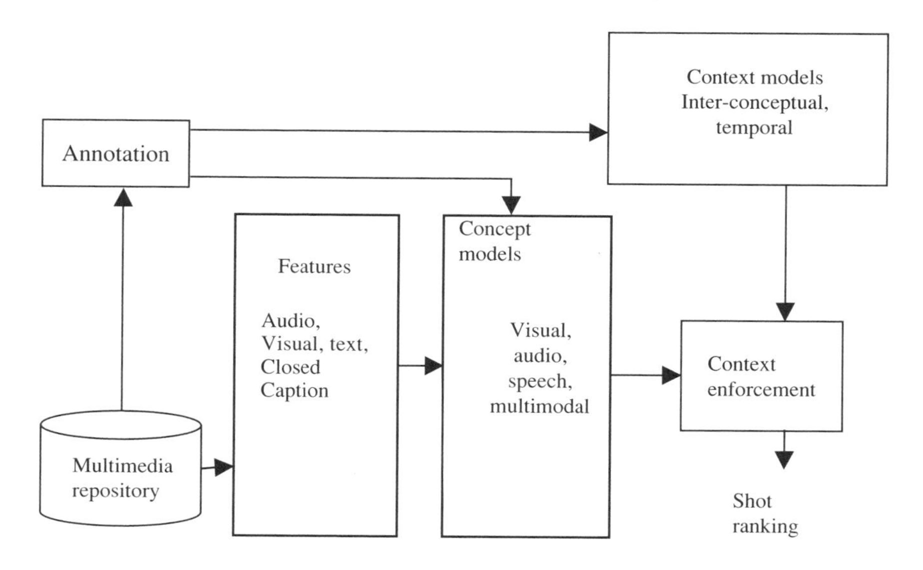

Figure 8.2 The concept and context modelling framework

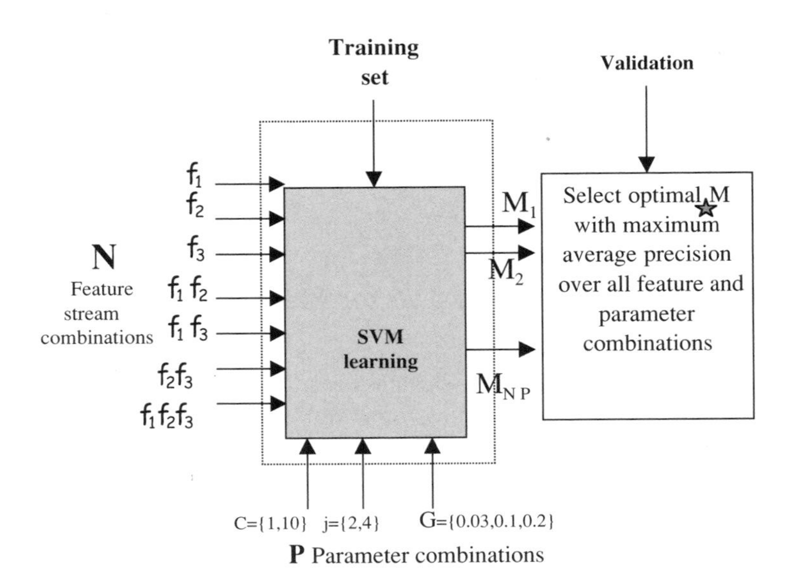

Figure 8.3 SVM learning: optimizing over multiple possible feature combinations and model parameters

including colour, structure, texture and shape. We also experiment with early feature fusion by combining multiple feature types at an early stage to construct a single model across different features. This approach is suitable for concepts that have a sufficiently large number of training set exemplars and feature types that are correlated and dependent. We can simply concatenate one or more of these feature types (appropriately normalized). Different combinations can then be used to construct models and the validation set is used to choose the optimal combination (Figure 8.3). This is feature selection at the coarse level of feature types.

To minimize sensitivity to the design choices of model parameters, we experiment with different kernels, and for each kernel we build models for several combinations of the parameters. Radial basis function kernels usually perform better than other kernels. In our experiments we built models for different values of the RBF parameter gamma (variance), relative significance of positive versus negative examples j (necessitated also by the imbalance in the number of positive versus negative training samples) and trade-off between training error and margin c. While a coarse to fine search is ideal, we tried three values of gamma, two values of j and two of c. Using the validation set, we then performed a grid search for the parameter configuration that resulted in highest average precision.

8.3 Modelling Context: A Graphical Multinet Model for Learning and Enforcing Context

To model the interaction between multijects in a multinet, we proposed a *factor graph* [1,9] framework. Factor graphs subsume graphical models like Bayesian nets and Markov random fields and have been successfully applied in the area of channel error correction coding [1] and,

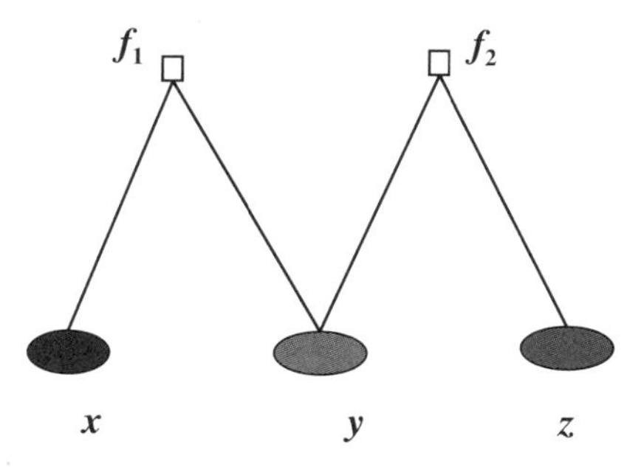

Figure 8.4 An example of function factorization $f(x, y, z) = f_1(x, y) * f_2(y, z)$

specifically, iterative decoding. Let $\boldsymbol{x} = \{x_1, x_2, \ldots, x_N,\}$ be a vector of variables. A *factor graph* visualizes the factorization of a global function $f(\boldsymbol{x})$. Let $f(\boldsymbol{x})$ factor as:

$$f(x) = \prod_i f_i(x_i) \tag{8.1}$$

where $\boldsymbol{x}_i$ is the set of variables of the function f_i. A factor graph for f is defined as the bipartite graph with two vertex classes V_f and V_v of sizes m and n, respectively, such that the ith node in V_f is connected to the jth node in V_f if and only if f_i is a function of $\boldsymbol{x}_j$ (Figure 8.4). For details please see [1]. (We will use rectangular blocks to represent function nodes and circular nodes to represent variable nodes.)

Many signal processing and learning problems are formulated as optimizing a global function $f(\boldsymbol{x})$ marginalized for a subset of its arguments. The *sum–product algorithm* allows us to perform this efficiently, though in most cases only approximately. It works by computing messages at the nodes using a simple rule and then passing the messages between nodes according to a reasonable schedule. A message from a function node to a variable node is the product of all messages incoming to the function node with the function itself, marginalized for the variable associated with the variable node. A message from a variable node to a function node is simply the product of all messages incoming to the variable node from other functions connected to it. If the factor graph is a tree, exact inference is possible using a single set of forward and backward passage of messages. For all other cases inference is approximate and the message passing is iterative [1], leading to loopy probability propagation. Because relations between semantic concepts are complicated and in general contain numerous cycles (e.g. see Figure 8.1), this provides the ideal framework for modelling context.

8.3.1 Modelling Inter-Conceptual Context in a Factor Graph

We now describe a shot-level factor graph to model the probabilistic relations between various frame-level semantic features F_i obtained by using the distance of the test set examples from the separating hyperplane as a measure of confidence. To capture the co-occurrence relationship between the twelve semantic concepts at the frame level, we define a function node which is connected to the 12 variable nodes representing the concepts as shown in Figure 8.5. Since we will present results using 12 concepts, the multinets in this chapter depict 12 variable nodes.

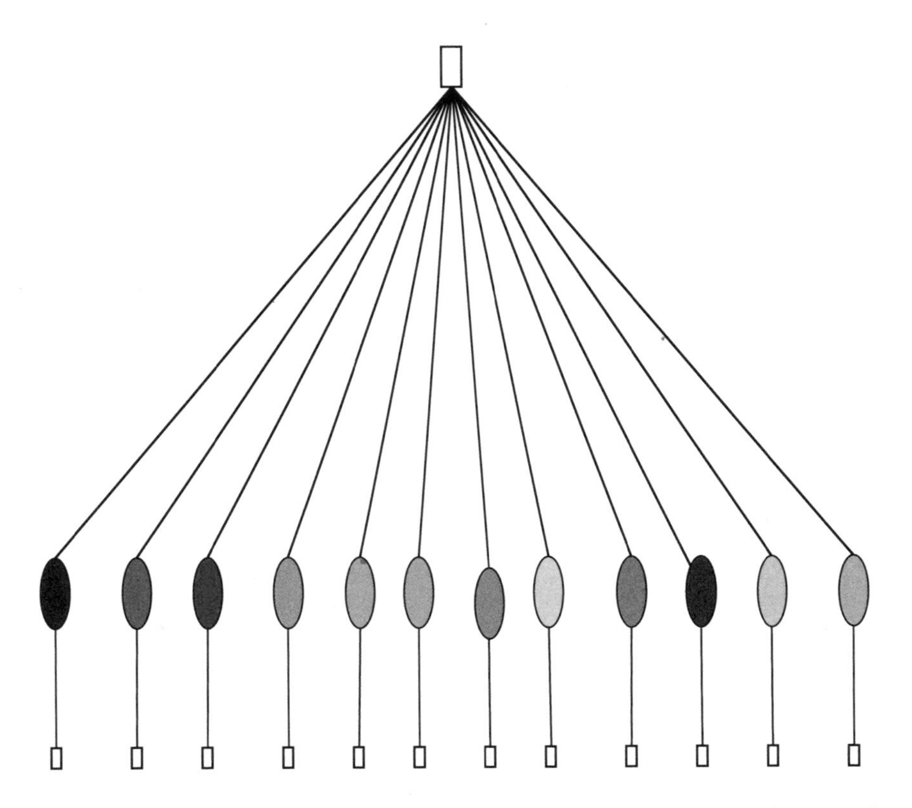

Figure 8.5 A factor graph multinet with 12 concepts represented by variable nodes, 12 concept detectors denoted by the function nodes at the bottom and one global context function denoted by the function node at the top. In this chapter the global context function used is the joint mass function $P(F_1, \ldots, F_N)$

The function node depicted by the rectangular box at the top represents the joint mass function $P(F_1, .., F_N)$. The function nodes below the variable nodes provide the individual SVM-based multijects' confidences, i.e. $(P(F_{\_i} = 1|X), P(F_{\_i} = 0|X))$, where F_i is the ith concept and X is the observation (features). These are then propagated to the function node. At the function node the messages are multiplied by the joint mass function estimated from the training set. The function node then sends back messages summarized for each variable. This modifies the soft decisions at the variable nodes according to the high-level relationship between the 12 concepts. In general, the distribution at the function node in Figure 8.5 is exponential in the number of concepts (N) and the computational cost may increase quickly. To alleviate this we can enforce a factorization of the function in Figure 8.5 as a product of several local functions where each local function accounts for co-occurrence of two variables only. This modification is shown in Figure 8.6.

Each function at the top in Figure 8.6 represents the joint probability mass of those two variables that are its arguments (and there are C_2^N such functions) thus reducing the complexity from exponential to polynomial in the number of concepts. For 12 concepts, there are 66 such local functions and the global function is approximated as a product of these 66 local functions. The

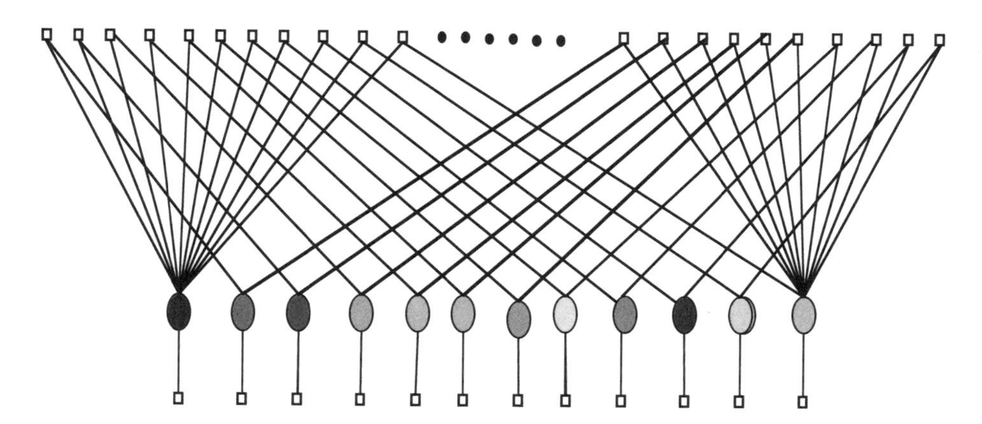

Figure 8.6 An efficient implementation by a loopy approximation. The 12 concepts are represented by variable nodes, 12 concept detectors denoted by the function nodes at the bottom and the global function at the top in Figure 8.5 is replaced by a product of C_2^N local functions, where N is the number of variable nodes for concepts. Each local function has two variables (concepts) as its argument. For $N = 12$, there are 66 local functions covering all possible pairs of concepts

factor graph is no longer a tree and exact inference becomes hard as the number of loops grows. We then apply iterative techniques based on the sum–product algorithm to overcome this.

8.3.2 Modelling Inter-Conceptual and Temporal Context in a Factor Graph

In addition to inter-conceptual relationships, we can also incorporate temporal dependencies. This can be done by replicating the slice of factor graph in Figure 8.5 or Figure 8.6 as many times as the number of frames within a single video shot and by introducing a first-order Markov chain for each concept.

Figures 8.7 and 8.8 show two consecutive time slices and extend the models in Figures 8.5 and 8.6, respectively. The horizontal links in Figures 8.5 and 8.6 connect the multinet instances in consecutive shots.

In general, if each concept represented by the variable node within each shot is assumed to be binary then a state vector of N such binary concepts can take 2^N possible configurations and thus the transition matrix $\boldsymbol{A}$ is $2^N \times 2^N$ dimensional.

However, we make an independence assumption across time which results in a simplified transition matrix. We assume that each concept's temporal context can be modelled independently of the other concepts in the past and the future. Since we are already accounting for within-concept dependencies using the multinet at each time instant or shot, this assumption is not harmful. This results in the simplification of the temporal context estimation. Instead of estimating the entire $\boldsymbol{A}$ matrix, we now estimate N 2×2 transition matrices. Each variable node for each concept in a time slice can thus be assumed to be connected to the corresponding variable node in the next time slice through a function modelling the transition probability $\boldsymbol{A}_c$, $c = \{1, \ldots, N\}$. This framework now becomes a dynamic probabilistic network. For inference, messages are iteratively passed locally within each slice. This is followed by message passing

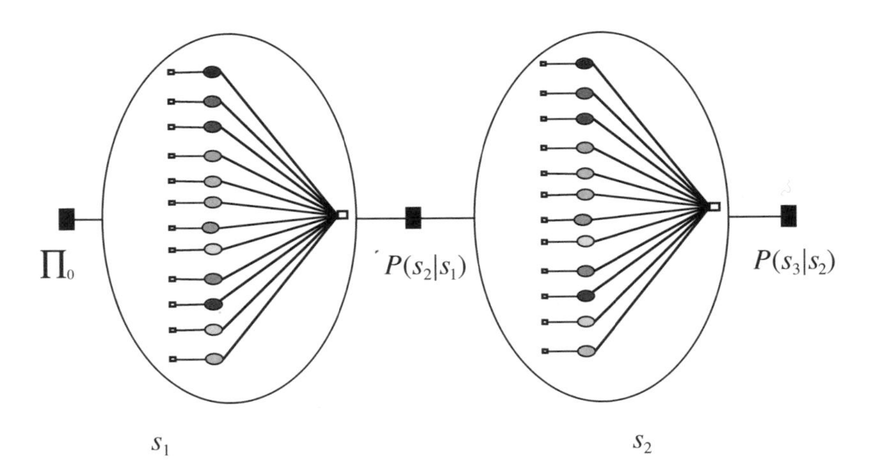

Figure 8.7 Replicating the multinet in Figure 8.5 for each shot (keyframe) and introducing temporal dependencies between the value of each concept in consecutive shots. Within each state the unfactored global multinet of Figure 8.5 is used. The temporal transitions are modelled by a stochastic transition matrix $\boldsymbol{A}$. The function nodes shown in between the shots use this matrix $\boldsymbol{A}$ to determine $P(s_i|s_{i-1})$. $\prod_i$ is the prior on the state i

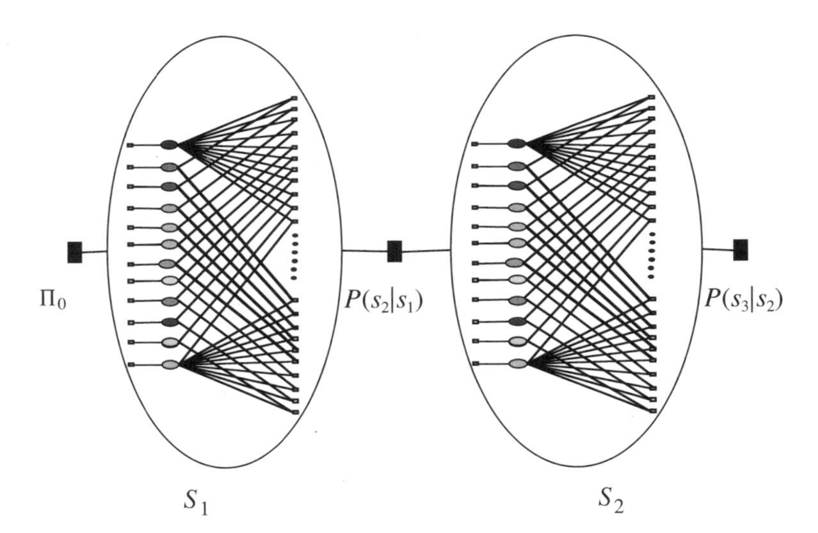

Figure 8.8 Replicating the multinet in Figure 8.6 for each shot (keyframe) and introducing temporal dependencies between the value of each concept in consecutive shots. Within each state the factored multinet of Figure 8.6 is used. The temporal transitions are modelled by a stochastic transition matrix $\boldsymbol{A}$. The function nodes shown in between the shots use this matrix $\boldsymbol{A}$ to determine $P(s_i|s_{i-1})$. $\prod_i$ is the prior on the state i

across the time slices in the forward direction and then in the backward direction. Accounting for temporal dependencies thus leads to temporal smoothing of the soft decisions within each shot.

8.4 Experimental Set-up and Results

8.4.1 TREC Video 2002 Corpus: Training and Validation

The National Institute for Standards and Technology (NIST) has constituted a benchmark for evaluating the state of the art in semantic video detection and retrieval. For this Video TREC 2002 Concept Detection Benchmark NIST provided a data set of 24 hours of MPEG video for the concept development and later tested the detectors using a 5-hour test set (see [11,14] for details of the corpus). We partitioned the NIST development set into a 19-hour training set and a 5-hour validation set. NIST defined non-interpolated average precision over a fixed number of retrieved shots as a measure of retrieval effectiveness. Let R be the number of true relevant documents in a set of size S; L the ranked list of documents returned. At any given index j let R_j be the number of relevant documents in the top j documents. Let $I_j = 1$ if the jth document is relevant and 0 otherwise. Assuming $R < S$, the non-interpolated average precision (AP) is then defined as:

$$\frac{1}{R}\sum_{j=1}^{S}\frac{Rj * Ij}{j} \tag{8.2}$$

8.4.2 Lexicon

We created a lexicon with more than a hundred semantic concepts for describing events, sites and objects [11]. An annotation tool that allows the user to associate the object-labels with an individual region in a key-frame image or with the entire image was used [15] to create a labelled training set using this lexicon. For experiments reported here the models were built using features extracted from key-frames. The experiments reported here were confined to 12 concepts that had the support of 300 or more examples in the training set. These 12 concepts are:

- *Scenes*: Outdoors, Indoors, Landscape, Cityscape, Sky, Greenery
- *Objects*: Face, Person, People, Road, Transportation vehicle, Tree

8.4.3 Feature Extraction

After performing shot boundary detection and key-frame extraction [16], each key-frame was analysed to detect the five largest regions described by their bounding boxes. The system then extracts the following low-level visual features at the frame level or global level, as well as at the region level for the entire frame and each of the regions in the key-frames:

- *Colour histogram (72)*: 72-bin YCbCr color space ($8 \times 3 \times 3$).
- *Colour correlogram (72)*: single-banded auto-correlogram coefficients extracted for 8 radii depths in a 72-bin YCbCr colour space correlogram.
- *Edge orientation histogram (32)*: using a Sobel filtered image and quantized to 8 angles and 4 magnitudes.

- *Wavelet texture (12)*: based on wavelet spatial-frequency energy of 12 bands using quadrature mirror filters.
- *Co-occurrence texture (48)*: based on entropy, energy, contrast, and homogeneity features extracted from grey-level co-occurrence matrices at 24 orientations (see [17]).
- *Moment invariants (6)*: based on Dudani's moment invariants [18] for shape description modified to take into account the grey-level intensities instead of binary intensities.
- *Normalized bounding box shape (2)*: the width and the height of the bounding box normalized by that of the image.

8.4.4 Results

Having trained the 12 multiject models using SVM classifiers, we then evaluate their performance on the validation set. This is used as the baseline for comparison with the context-enforced detection. Details of the baseline detection are presented elsewhere (see [2]). The detection confidence is then modified using the multinet to enforce context. We evaluate performance of the globally connected multinet of Figure 8.5, the factored model of Figure 8.6 as well as the conceptual-temporal models of Figures 8.7 and 8.8. The comparison using the average precision measure (equation (8.2)) is shown in Table 8.1.

The improvement in mean average precision using the global unfactored multinet is 17%. The approximation of the global multinet using the factored version also improves the average precision by 14%. Also, it is worth noticing that across all concepts the maximum improvement is as much as 117% in the case of Road, with almost no concept suffering any deterioration in performance. Using the temporal context in conjunction with the inter-conceptual context improves mean average precision by 20.44% for the unfactored multinet and 20.35% for the factored approximation. Thus, the factored approximation results in performance almost identical to the global function. Importantly, this improvement is achieved without using any additional training data or annotation.

Table 8.1 Concept detection performance measure listed in decreasing order of number of positive examples in a training set of 9603 keyframes

Semantic concept	SVM baseline	Global multinet	Factored multinet	Global temporal multinet	Factored temporal multinet
Outdoors	0.653	0.692	0.681	0.686	0.699
Person	0.235	0.295	0.285	0.237	0.251
People	0.417	0.407	0.415	0.403	0.424
Sky	0.506	0.59	0.562	0.545	0.545
Indoors	0.476	0.483	0.477	0.61	0.604
Face	0.282	0.442	0.445	0.385	0.446
Cityscape	0.174	0.241	0.204	0.202	0.166
Greenery	0.208	0.229	0.207	0.245	0.2
Transportation	0.294	0.38	0.38	0.391	0.411
Tree	0.146	0.16	0.166	0.199	0.169
Road	0.209	0.334	0.33	0.446	0.453
Landscape	0.497	0.537	0.499	0.585	0.562
Mean average precision	0.3414	0.3991	0.3877	0.4112	0.4109

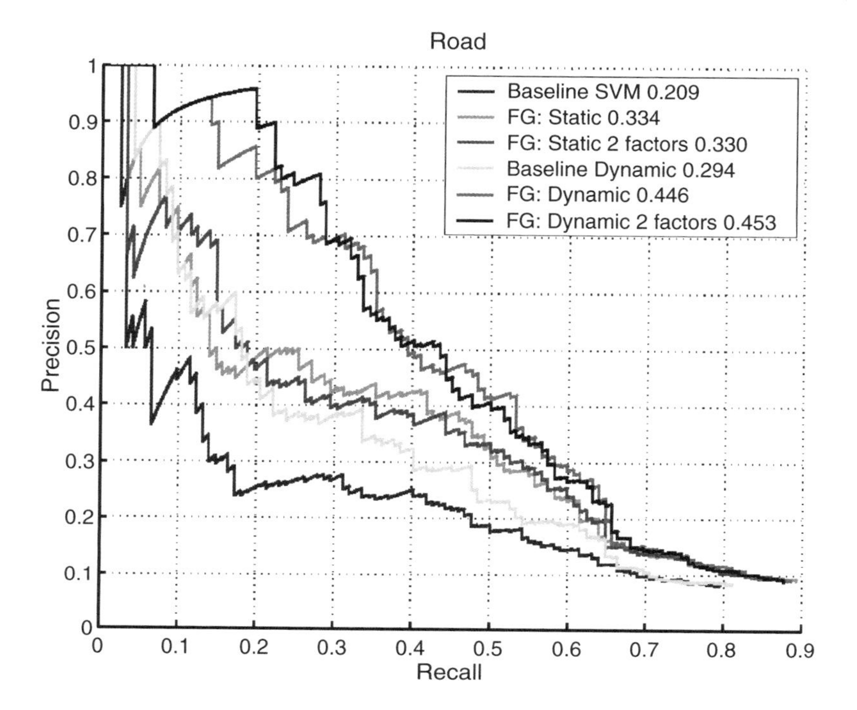

Figure 8.9 Precision recall curves and corresponding average precisions with and without the various proposed context models for the concept Road. The average precision as well as the curves show that detection is significantly improved by accounting for context. Inter-conceptual context improves detection, and accounting for temporal context in addition improves detection further. Also, the factored context models result in performance very similar to the unfactored context models in the static as well as dynamic case.

Figure 8.9 shows a set of precision recall curves with and without the various proposed context models, for the concept Road.

In Figure 8.9 the average precision as well as the precision recall curves show that detection is significantly improved by accounting for context. Inter-conceptual context improves detection, and accounting for temporal context in addition improves detection further. Also, the factored context models result in performance very similar to the unfactored context models in the static as well as dynamic case. Inter-conceptual and temporal context modelling and enforcement thus improves detection performance for Road by 117%.

8.5 Concluding Remarks

We present a hybrid framework that can combine discriminant or generative models for concepts with generative models for structure and context. Using the TREC Video 2002 benchmark corpus and 12 semantic concepts from this corpus, we show that robust models can be built for several diverse visual semantic concepts. We use a novel factor graphical framework to model inter-conceptual context. Using the sum–product algorithm [1] for approximate or exact inference in these factor graph multinets, we attempt to correct errors made during isolated concept detection by forcing high-level constraints. This results in a significant improvement in the overall detection performance. Enforcement of this probabilistic context model enhances the detection performance further by 17% over the baseline concept detection. Adding temporal context to the inter-conceptual context model and enforcing this results in an improvement of 20% over baseline. The approximate factored context model proposed also results in a similar performance improvement. This improvement is achieved without using any additional training data or separate annotations. Also, it is worth noticing that across all concepts the maximum improvement is as much as 117%. The maximum deterioration, on the other hand, is insignificant, with almost no concept suffering any deterioration in performance. Future research aims at improving the efficiency of loopy propagation by pruning local functions corresponding to uncorrelated variable pairs. Future research also aims at modelling relationships other than correlations such as layout.

Acknowledgement

We would like to thank the IBM TREC team for annotation, shot detection, classifier averaging and feature extraction.

References

[1] F. Kschischang, B. Frey, H. Loeliger, Factor graphs and the sum-product algorithm. *IEEE Transactions on Information Theory*, **47**(2), 498–519, 2001.

[2] M. Naphade, J. Smith, Learning visual models of semantic concepts. In *Proceedings of the IEEE International Conference on Image Processing, ICIP 2003*, Barcelona, Spain, September 2003. IEEE, 2003.

[3] M. Naphade, T. Kristjansson, B. Frey, T.S. Huang, Probabilistic multimedia objects (multijects): a novel approach to indexing and retrieval in multimedia systems. In *Proceedings of the IEEE International Conference on Image Processing, ICIP 1998*, Chicago, IL, October 1998, vol. 3, pp. 536–540. IEEE, 1998.

[4] M.R. Naphade, T.S. Huang, Stochastic modeling of soundtrack for efficient segmentation and indexing of video. In *Proceedings of SPIE Storage and Retrieval for Multimedia Databases*, January 2000, vol. 3972, pp. 168–176. SPIE, 2000.

[5] M.R. Naphade, T.S. Huang, Recognizing high-level audio-visual concepts using context. In *Proceedings of the IEEE International Conference on Image Processing, ICIP 2001*, Thessaloniki, Greece, October 2001, vol. 3, pp. 46–49. IEEE, 2001.

[6] M. Naphade, T. Huang, Discovering recurring events using unsupervised methods. In *Proceedings of the IEEE International Conference on Image Processing, ICIP 2002*, Rochester, NY, September 2002. IEEE, 2002.

[7] S.F. Chang, W. Chen, H. Sundaram, Semantic visual templates—linking features to semantics. In *Proceedings of the IEEE International Conference on Image Processing, ICIP 1998*, Chicago, IL, October 1998, vol. 3, pp. 531–535. IEEE, 1998.

[8] T. Zhang, C. Kuo, An integrated approach to multimodal media content analysis. In *Proceedings of SPIE, IS&T Storage and Retrieval for Media Databases*, San Jose, CA, January 2000, vol. 3972, pp. 506–517. SPIE, 2000.

[9] M.R. Naphade, I. Kozintsev, T.S. Huang, A factor graph framework for semantic video indexing. *IEEE Transactions on Circuits and Systems for Video Technology*, **12**(1), 40–52, 2002.

[10] M. Naphade, J. Smith, The role of classifiers in multimedia content management. In *Proceedings of SPIE, IS&T Storage and Retrieval for Media Databases*, San Jose, CA, January 2003, vol. 5021. SPIE, 2003.

[11] W.H. Adams, A. Amir, C. Dorai, S. Ghoshal, G. Iyengar, A. Jaimes, C. Lang, C.Y. Lin, M.R. Naphade, A. Natsev, C. Neti, H.J. Nock, H. Permutter, R. Singh, S. Srinivasan, J.R. Smith, B.L. Tseng, A.T. Varadaraju, D. Zhang, IBM research TREC-2002 video retrieval system. In *Proceedings of the Text Retrieval Conference TREC*, Gaithersburg, MD, November 2002.

[12] M. Naphade, S. Basu, J. Smith, C. Lin, B. Tseng, Modeling semantic concepts to support query by keywords in video. In *Proceedings of the* IEEE International Conference on Image Processing, Rochester, NY, September 2002. IEEE, 2002.
[13] V. Vapnik, *The Nature of Statistical Learning Theory*. Springer-Verlag, New York, 1995.
[14] TREC Video Retrieval, 2002, National Institute of Standards and Technology, http://www-nlpir.nist.gov/projects/trecvid.
[15] C. Lin, B. Tseng, J. Smith, VideoAnnEx: IBM MPEG-7 annotation tool for multimedia indexing and concept learning. In *Proceedings of the IEEE International Conference on Multimedia and Expo*, 2003.
[16] S. Srinivasan, D. Ponceleon, A. Amir, D. Petkovic, What is that video anyway? In search of better browsing. In *Proceedings of the IEEE International Conference on Multimedia and Expo*, New York, July 2000, pp. 388–392.
[17] R. Jain, R. Kasturi, B. Schunck, *Machine Vision*. MIT Press/McGraw-Hill, New York, 1995.
[18] S. Dudani, K. Breeding, R. McGhee, Aircraft identification by moment invariants. *IEEE Transactions on Computers*, **C-26(1)**, 39–45, 1997.

9

Machine Learning in Multimedia

Nemanja Petrovic, Ira Cohen and Thomas S. Huang

9.1 Introduction

Today, when the amount of video data is already overwhelming, automatic video understanding remains the ultimate research task. Video understanding is needed for managing video databases, content-based video retrieval, attaching semantic labels to video shots (indexing), human–computer interaction and understanding video events, to name a few. Solving this problem would give rise to a number of tremendously exciting and useful applications. For example, video understanding would lead to video compression with unprecedent compression rates. Often, video understanding is based on extracting low-level features and trying to infer their relation to high-level semantic concepts (e.g. moving car on a busy street). Many current semantic video classifiers are based on automatically extracted features, like colour histograms, motion, colour regions or texture, and subsequent mapping from features to high-level concepts. However, this approach does not provide semantics that describe high-level video concepts, and this problem is referred to as the 'semantic gap' between features and semantics.

Modelling difficult statistical problems in multimedia and computer vision often requires building complex probabilistic models with millions of (hidden or observable) random variables. Observable variables in most cases are pixel values or extracted features. Hidden variables often correspond to some mid-level or high-level concepts (class membership; position, pose or shape of an object; an activity or event in video, to name a few). Two common tasks associated with probabilistic models are: (a) answering of probabilistic queries of the form *P(hidden variables|observable variables)* often referred to as probabilistic inference; in terms of a probabilistic system with measurable output and unknown input this task corresponds to an *inverse problem*; and (b) learning the parameters of the model. In system theory this correspond to *system identification*. Certain properties of models (factorization) make them suitable for graphical representation, which proved to be useful for manipulation of underlying distributions, hence the name graphical probabilistic models. Recently, graphical models [1–3] have attracted a lot of attention.

Multimedia Content and the Semantic Web Edited by Giorgos Stamou and Stefanos Kollias

Machine learning is currently characterized by two prevailing learning paradigms: discriminative learning and generative learning. The discriminative models, like SVMs [4], neural networks [5] and logistic regression, are trying to directly map input variables (data) to output variables (labels) in order to perform classification, and thus the aim is to maximize classification accuracy. Learning in discriminative models is based on minimizing the empirical risk. Generative models are trying to model the distribution of features, and learning is based on maximizing the likelihood of the data given the parameterized model. Generative models are generally suitable for unsupervised learning, incorporation of priors and domain-specific knowledge. For recent comparison of generative (sometimes called *informative*) and discriminative learning, see e.g. [6].

As pointed out in [4], discriminative methods are methods of choice for classification since they *directly* construct mapping from the data to labels upon which the decisions are based, as opposed to generative methods that model a different problem (joint distribution) and then *indirectly* try to manipulate it into finding posterior distribution $P(label|data)$. However, the power of generative models comes when there is a need to explicitly model causes of uncertainty in the data and to include prior knowledge in the model. Bayesian networks, are a valuable tool for learning and reasoning about probabilistic relationships. Bayesian networks are also a great framework to incorporate prior knowledge in the model in the form of proper structure. For example, it is very hard to see how discriminative methods, deficient of modelling power and priors, might be used to separate background from moving foreground objects in a video sequence. Generative models elegantly model both the underlying physical phenomenon and the causes of variability in the data.

The properties of probabilistic graphical models make them particulary suited for problems in multimedia and vision. Graphical models may include many explanations of the sources of variability in the video data that discriminative models simply ignore. That helps model the middle tier between features and semantics. Graphical models are today widely applied at all three levels of multimedia understanding: core-level algorithms such as segmentation and tracking, fusion-level algorithms extending across multiple sensors, and top-most event-level analysis.

9.2 Graphical Models and Multimedia Understanding

Recently, several discriminative machine learning methods in multimedia and vision have shown superb performance for training and classification. Examples are SVMs for face and handwritten digit detection [6], linear regression, SVMs for gender [7] and age face classification, SVMs for detecting anchors in news videos etc. However, generative models are very appealing due to their ability to include priors and handle missing data, and learning in the unsupervised manner. Generative models are defining joint probability distribution over *all* variables in the model, and using several manipulation tools (Bayes rule, normalization, marginalization) compute the quantities of interest.

A graphical model is a graph $G = (V, E)$ consisting of a set of nodes V that correspond to (hidden or observable) random variables in a probabilistic model, and a set of edges E representing the dependencies between the variables (e.g. conditional probability distributions). Theoretically, *all graphical models are generative*, since they all specify joint distribution over all variables. The usefulness of graphical models stems from the factorization property: joint distribution in a complex probabilistic model can often be factored out in a number of terms, each involving only several variables. That is due to the fact that a variable in a model usually depends only on a small number of other variables. In this view, a graphical model is a useful visualization of the complex probabilistic model. But the real power of graphical models stems from distribution manipulation on the graph, which can give rise to a new class of powerful algorithms.

The research community is slowly coming to a common ground in understanding and treating different graphical models and developing inference tools. Graphical models, like Bayes nets (BNs) [1, 3, 8], Markov random fields (MRFs) [8–10] and factor graphs (FGs) [11] represent different families of distributions, but they can all be mutually converted [12]. This allows for extra flexibility when designing complicated probabilistic models. A direct consequence of that is that inference algorithms for one type of graphical model are readily applicable to others. Among several tools for learning the parameters of models or inferring the values of hidden variables, the most commonly used are junction tree inference algorithm [13, 14], exact EM algorithm [15], approximate (variational) EM algorithm [16, 17], loopy probability propagation [12] and Gibbs sampling [18].

The graphical representation has several advantages. Among them are the existence of algorithms for inferring the class label (and in general to complete missing data), the ability to intuitively represent fusion of different modalities with the graph structure [19, 20], the ability to perform classification and learning without complete data, i.e. missing feature values or unlabelled data. One of the advantages of graphical models is that they are suitable for unsupervised learning, minimizing the labour necessary for data normalization and labelling. For example SVMs for face detection usually work well if the faces are previously normalized with respect to shift, scale and light change. While discriminative methods typically require (manual) data normalization as a pre-processing step, graphical models can directly encode the knowledge of the type of transformation that affects data. Similarly, training a discriminative model and labelling the data can be laborious tasks. Based on the amount of data labelling required, machine learning methods can be divided into three broad classes:

1. Supervised: data samples are manually labelled and used for training. As a representative example we will review (Section 9.4.2) training of audio, visual and audiovisual multijects using SVMs, Gaussian mixture models and hidden Markov models, followed by semantic context modelling using factor graphs. The proposed solution was used to detect several high-level concepts (outdoor scenes, sky, rocks, snow, greenery). Other successful applications of unsupervised learning include detection of explosions, gunshots, rocket launches and human activities. [21–23].
2. Semi-supervised: in this setting, data samples are only partially labelled, and along with unlabelled data used for training. In order to reduce the burden of data labelling other modalities were proposed including multiple instance learning (using information at a different granularity) and active learning (actively selecting the samples that need to be annotated for training). In Section 9.3 we review in more detail learning with both labelled and unlabelled data. Representative work on facial expression recognition (Section 9.4.5) uses semi-supervised learning.
3. Unsupervised: the work in this area was focused on discovering the most dominant and frequent recurring patterns. Often, a recurring pattern was the frame itself, leading to different techniques for shot cut detection and hierarchical clustering of shots into scenes, episodes and stories. Successful applications include detection of explosions, laughter and soccer play [24, 25].

9.3 Learning Classifiers with Labelled and Unlabelled Data

Using unlabelled data to enhance the performance of classifiers trained with few labelled data has many applications in pattern recognition such as computer vision, HCI, data mining, text recognition and more. To fully utilize the potential of unlabelled data, the abilities and limitations of existing methods must be understood. In the domain of semi-supervised learning with both labelled and unlabelled data there were some attempts to build discriminative classifiers [26] together with a few ad hoc methods. But, it was shown that unlabelled data are helpful only with generative models—for discriminative models they do not add any information.

Many pattern recognition and HCI applications require the design of classifiers. Classifiers are either designed from expert knowledge or using training data. Training data can be either labelled to the different classes or unlabelled. In many applications, obtaining fully labelled training sets is a difficult task; labelling is usually done using human expertise, which is expensive, time consuming and error prone. Obtaining unlabelled data is usually easier since it involves collecting data that is known to belong to one of the classes without having to label it, e.g., in facial expression recognition, it is easy to collect videos of people displaying expressions, but it is very tedious and difficult to label the video to the corresponding expressions. Learning with both labelled and unlabelled data is known as semi-supervised learning.

Cozman and colleagues [27] provide a general analysis of semi-supervised learning for probabilistic classifiers. The goal of the analysis is to show under what conditions unlabelled data can be used to improve the classification accuracy. Using the asymptotic properties of maximum likelihood estimators, they prove that unlabelled data helps in reducing the estimator's variance. They further show that when the assumed probabilistic model matches the true data generating distribution, the reduction in variance leads to an improved classification accuracy, which is not surprising, and has been analysed before [28, 29]. However, the surprising result is that when the assumed probabilistic model does not match the true data generating distribution, using unlabelled data can be detrimental to the classification accuracy.

This new result emphasizes the importance of using correct modelling assumptions when learning with unlabelled data. Cohen *et al.* [30] show that for Bayesian network classifiers, this new analysis of semi-supervised learning implies a need to find a structure of the graph that matches the true distribution generating the data. While in many classification problems simple structures learned with just labelled data have been used successfully (e.g. the Naive-Bayes classifier [31, 32]), such structures fail when trained with both labelled and unlabelled data [33]. Cohen *et al.* [30] propose a classification-driven stochastic structure search algorithm (SSS), which combines both labelled and unlabelled data to train the classifier and search for a better performing Bayesian network structure.

9.4 Examples of Graphical Models for Multimedia Understanding and Computer Vision

In the following sections we will go through several representative studies that utilize graphical models for different multimedia understanding and computer vision tasks. We begin with an important task in video understanding: classification based on the predefined topic (news, movies, commercials). It proved to be useful tool for browsing and searching video databases by topic. In Section 9.4.1 we review content-based video classification work that uses Bayes net to automatically classify videos into meaningful categories. Further, we will go through two graphical model frameworks in Sections 9.4.2 and 9.4.3 that utilize content-based video retrieval techniques, such as query by example (QBE), visual feature query and concept query (query by keyword). In Section 9.4.4 we will review an elegant solution as a post-processing step of one of the fundamental problems in computer vision (recovering the shape of an object) using graphical models. Section 9.4.5 demonstrates the use of Bayesian network classifiers in two applications: facial expression recognition from video sequences and frontal face detection from images under various illumination. For both applications, general issues regarding learning Bayesian networks arise. The final subsection discusses in more details the issues that arise when learning with partially labelled data sets (known as semi-supervised learning).

9.4.1 Bayesian Network for Video Content Characterization

This work [34] is a neat showcase of bridging the gap between low-level features and high-level semantics using the Bayes net to incorporate prior knowledge about the structure of the video content (Figure 9.1a). It describes a Bayesian architecture for content characterization of movies. Even though this is a daunting task in general, certain rules in video production lend themselves to be exploited in the content characterization. Namely, there are very well known rules between the stylistic elements of the scene and the semantic story. Thus, incorporating priors into Bayes nets in a natural manner helps drive inference results to the part of probabilistic space that is consistent with data *and* our previous belief as to what the result should be.

The idea is to characterize content in terms of four semantic attributes: *action, close-up, crowd* and *setting*. The bottom layer of the Bayesian network consist of three visual sensors: (1) activity-level sensor based on motion energys, (2) skin colour detector that helps to discriminate close-up shots and (3) a heuristic natural scene detector that performs wavelet decomposition

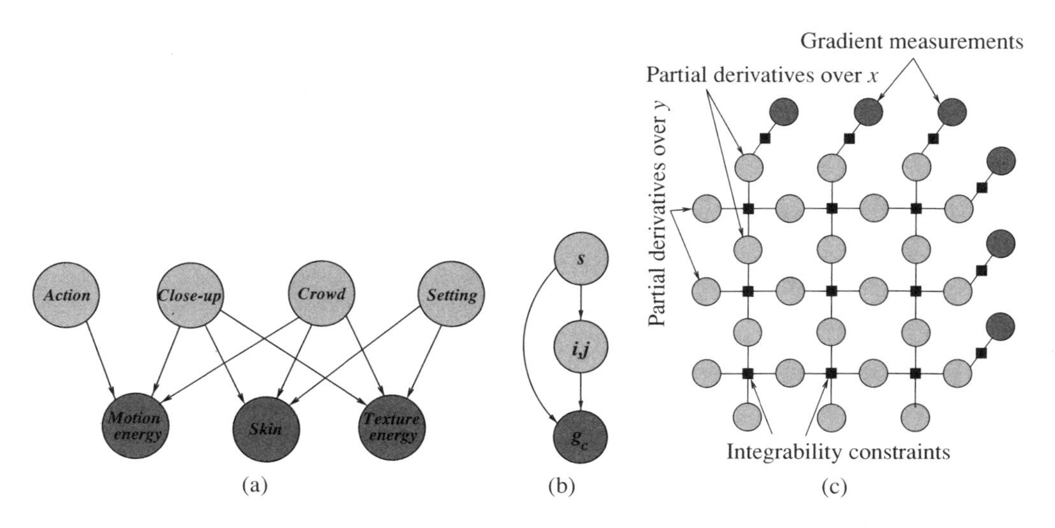

Figure 9.1 (a) Adaptive video fast forward (AFF) graphical model, (b) video content characterization graphical model and (c) graphical model for enforcement of integrability constraint

and computes the ratio of energies in horizontal and vertical bands (natural scenes tend to have this ratio close to one). Although this network models the structure well, the network is not trained from samples and conditional densities are fixed heuristically (due to the small size of the net). Once the state of the node is observed, the marginal probabilities of hidden states is computed using message passing in the junction tree.

This work introduces a very flexible retrieval system with support for direct specification of the semantic attributes. The details and results are given in [34].

9.4.2 Graphical Models for Semantic Indexing and Retrieval in Video

This work [35] is a pioneering approach for semantic video analysis using the probabilistic framework and graphical models. It introduces the probabilistic framework that comprises multimedia objects within the Bayesian multinet. This work introduces the concepts of multijects and multinets as probabilistic multimedia objects and networks which can be trained using both supervised and unsupervised learning to detect the presence or absence of high-level semantic concepts in video. Although it is a very versatile concept, in this work it is used for detecting sites (locations) in video.

Multijects are probabilistic multimedia objects (binary random variable) representing semantic features or concepts. Multijects can belong to any of three categories: (1) objects (face, car, animal, building), (2) sites (sky, mountain, outdoor, cityscape) and (3) events (explosion, waterfall, gunshot, dancing). In this work the following site multijects were developed: sky, water, forest, rocks and snow. Audio multijects include music, human speech and aircraft flying. A multinet is a probabilistic network of multijects which accounts for the interaction between high-level semantic concepts.

Learning multijects starts with video segmentation into shots using shot-boundary detection. Second, segmentation is performed in the space–time domain to obtain video regions. Each region is then manually labelled for the presence/absence of the high-level semantic concept (one or more semantic concepts are assigned to each region). Also, several dozen features are extracted for each region. Figure 9.2a illustrates a typical multiject. For all so labelled regions, conditional distribution of a concept given the feature vector is fitted by mixture of

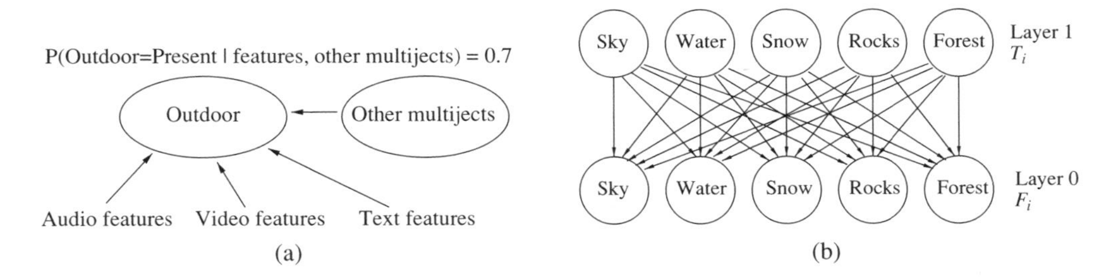

Figure 9.2 (a) Multiject for a semantic concept ('Outdoor') and (b) Bayesian multinet. All nodes are binary random variables indicating presence/absence of a multiject (semantic concept). Layer 0: frame-level multiject-based semantic features. Relationships, co-occurrence and exclusion among them are ignored at this level. Layer 1: inference from layer 0, takes into account relationships among individual multijects

Gaussians (for site multijects). Once these distributions are learned, it is straightforward to compute the probability of the presence of each of the semantic concepts in a region. Testing each region for a semantic concept and combining the results for all regions in the frame, the information is composed at the frame level. Binary variables F_i in Figure 9.2b correspond to the presence/absence of concepts (multijects) in the frame. But they are computed independently for each frame and do not take into account the tendency that concepts coexist or exclude each other. Another set of binary variables T_i is introduced that takes into account interactions of concepts (multijects) at the frame level. In Figure 9.2b each T_i is therefore connected to all lower level variables. In the training phase, the Bayesian multinet is trained with F_i (obtained from labelled regions of the frame) in conjunction with the ground truth. In the testing phase the multinet is used for inference, for example to find $P(T_i = 1|F_1, \ldots, F_N)$ for N concepts. Thus it automatically detects particular high-level concepts in the frame.

More details and performance measures of multinets for concept detection are given in [35].

9.4.3 Graphical Model for Adaptive Video Fast Forward (AFF)

We review a pixel-based generative model [36] that can be used as a measure of similarity for query-by-example video retrieval [37, 38]. Query by example is a popular video retrieval technique in which the user is asked to provide the query clip and the system retrieves all clips similar to the query given some similarity measure. In this work, the generative model is used to correctly explain the source of variability in video (movement of the foreground objects) and to construct a similarity measure that proved useful for video retrieval tasks.

This model builds on the well-known pFinder [39] system successfully used to track people and body parts in virtual environments, in a controlled setting: known background, fixed lighting, extensive model training etc. However, in the case of the general class of videos, most of the conditions can not be controlled: the background is unknown, the number of moving objects in the scene is a priori unknown, and training is not feasible. The AFF model (Figure 9.1b) takes a query clip that is several seconds long and assumes the decomposition of a scene in a detailed background model (represented as a Gaussian appearance model with independent pixels) and a coarse foreground model represented with several Gaussian blobs, each having its spatial mean and covariance, and colour mean and covariance.

The generation of feature vectors $\mathbf{g}_c(i, j)$, where c is one of the C features modelled for each pixel at the location (i, j) is given by the model (Figure 9.1b). These features, measured in the neighbourhood around (i, j) could include colour, texture, edge strengths or orientations, but in this work they are limited to R, G and B colour channels. The image features can be generated from several models indexed by s. One of these models ($s = 0$) generates each pixel with a separate mean and variance in each colour channel, while remaining models ($s > 0$) are blobs similar to [39]. The blobs are modelled as mixture of Gaussians with their spatial and colour means and variances. The variable s is referred to as the *segmentation* variable.

The pixel is modelled to either belong to the background (that is defined by its mean image and variance), or the pixel is sampled from the foreground blob mixture model (that is defined by spatial means and variances of blobs and their colour means and variances). The pixel generation is assumed to start with the selection of the object model s (by drawing from the prior $p(s)$), followed by sampling from appropriate distributions over the space and colour according to $p([i, j]|s)$ and $p(\mathbf{g}_c(i, j)|s, i, j)$.

Pixel position is chosen according to:

$$p([i, j]|s = 0) = \mathcal{U}(\text{uniform distribution})$$
$$p([i, j]|s > 0) = \mathcal{N}([ij]; \gamma_s, \Gamma_s), \tag{9.1}$$

where $\mathcal{N}$ denotes a Gaussian (normal) distribution. For the background model pixel position is arbitrarily chosen within the frame; for the foreground model, pixel position is more likely to be chosen around the spatial centre of the Gaussian blob with mean position γ_s, and position variance Γ_s.

Pixel feature vector (colour) is sampled according to:

$$p(\mathbf{g}_c|s = 0, i, j) = \mathcal{N}(\mathbf{g}_c; \mu_{0,c}(i, j), \Phi_{0,c}(i, j))$$
$$p(\mathbf{g}_c|s > 0) = \mathcal{N}(\mathbf{g}_c; \mu_{s,c}, \Phi_{s,c}). \tag{9.2}$$

Here, c indexes a particular feature (colour) in the colour vector $\mathbf{g}_c$; $\mu_{0,c}(i, j)$ is the mean of the cth feature of the background pixel (i, j); $\Phi_{0,c}(i, j)$ is the variance of the cth feature of the background pixel (i, j). The cth feature of sth blob is denoted as $\mu_{s,c}$, and its variance as $\Phi_{s,c}$.

There is a number of possible extensions to this model. For example, blobs can be allowed to move keeping their spatial covariance matrix fixed; camera motion can be directly incorporated by treating the frame as a part of the scene much larger in resolution than the field of view etc. In its basic form, the model can be trained using the exact EM algorithm in several seconds for a 10-second query clip. Testing target frames against the model is done many times faster than real-time playback speed, thus allowing rapid skipping of frames dissimilar to query clip.

The details of adaptive video fast forward and results are given in [36].

9.4.4 Graphical Model for Enforcing Integrability for Surface Reconstruction

Recovering the 3D shape of objects, referred to as a shape-from-X technique, is a classic computer vision problem. Shape-from-X refers to the reconstruction of the shape from stereo, motion, texture, shading, defocus etc. In this chapter we focus our attention on the shape recovery from images of a static object made under different lighting conditions, also known as photometric stereo (PMS).

Photometric stereo is a shape from shading method using several images to invert the reflectance map. It was first introduced by Woodham [40]. The basic algorithm for PMS estimates the surface gradients and then integrates the gradients along arbitrary paths to recover the 3D surface. One problem in many PMS algorithms is that the so-called integrability constraint is not strictly enforced. The integrability constraint should be satisfied for a reconstructed surface to be a valid 3D surface of an object.

The photometric stereo algorithm estimates surface normals from multiple images of the same object made under different lighting conditions (e.g. position of the light source is changing). Assuming Lambertian surface, reflected light intensity is proportional to the cosine of the angle between surface normal and the direction of incident light [41]. If the albedo of the surface is assumed uniform, theoretically two images are sufficient for the least square estimate of the surface normals. The computation of surface gradients in the x and y directions ($\{p, q\}$) is straightforward once surface normals are known [41].

A valid set of gradients in an $N \times M$ image must satisfy the 'integrability' constraint:

$$p(x, y) + q(x + 1, y) - p(x, y + 1) - q(x, y) = 0 \tag{9.3}$$

for all $x \in \{1, \ldots, N - 1\}$, $y \in \{1, \ldots, M - 1\}$.

However, due to the noise and discretization the set of estimated gradients often violates the integrability constraint. The given vector field is not a gradient field and needs to be projected into the subspace of gradient (integrable) fields. Enforcing the integrability constraint is then reformulated as the denoising problem: given noisy measurements of surface gradients, we try to estimate 'clean' gradients that are integrable and also close to noisy measurements in some sense.

We assume the local observation model to be Gaussian with some variance σ^2:

$$P(p|\tilde{p}) = \mathcal{N}(p; \tilde{p}, \sigma^2) \tag{9.4}$$

The probabilistic model is defined as the joint distribution over all variables that can be factorized, as follows:

$$P(\{p, q\}, \{\tilde{p}, \tilde{q}\}) \propto \prod_{x=1}^{N-1} \prod_{y=1}^{M-1} \delta(p(x, y) + q(x + 1, y) - p(x, y + 1) - q(x, y))$$
$$\prod_{x=1}^{N-1} \prod_{y=1}^{M} e^{-(p-\tilde{p})^2/2\sigma^2} \prod_{x=1}^{N} \prod_{y=1}^{M-1} e^{-(q-\tilde{q})^2/2\sigma^2} \tag{9.5}$$

Two terms in the joint probability distributions penalize gradients that are not integrable in any 2×2 elementary loop, and also penalize gradients that are far away from its noisy measurements. A seemingly complicated joint distribution can be factored out in a fashion that is convenient for graphical representation.

The conditional distribution of an unknown gradient (say $p(x, y)$) given noisy measurements is simply the marginal of the above joint distribution:

$$P(p(x, y) = k|\tilde{p}, \tilde{q}) \propto \int_{p(x,y)=k} P(\{p, q\}, \{\tilde{p}, \tilde{q}\}) \tag{9.6}$$

As we can see, a complicated joint distribution can be factored out in a number of terms and efficiently visualized using the graphical model (factor graph in this case) shown in Figure 9.1c. Observable nodes correspond to noisy measurements $\tilde{p}, \tilde{q}$ whereas hidden nodes correspond to unknown, integrable gradients p, q. An efficient algorithm to calculate the marginal (equation (9.6)) of complicated joint distribution (equation (9.5)) is the belief propagation algorithm in a richly connected graph. The details of the algorithm and results are given in [41].

9.4.5 *Bayesian Network Classifiers for Facial Expression Recognition*

It is argued that to truly achieve effective human–computer intelligent interaction (HCII), there is a need for the computer to be able to interact naturally with the user, similar to the way human–human interaction takes place. Humans interact with each other mainly through speech, but also through body gestures, to emphasize a certain part of the speech, and display of emotions. Emotions are displayed by visual, vocal and other physiological means.

Facial expressions are probably the most visual method by which humans display emotions. This section focuses on learning how to classify facial expressions with video as the input, using Bayesian networks. We have developed a real-time facial expression recognition system [42, 43]. The system uses a model-based non-rigid face tracking algorithm to extract motion features that serve as input to a Bayesian network classifier used for recognizing the different facial expressions. There are two main motivations for using Bayesian network classifiers in this problem. The first is the ability to learn with unlabelled data and infer the class label even when some of the features are missing (e.g. due to failure in tracking because of occlusion). Being able to learn with unlabelled data is important for facial expression recognition because of the relatively small amount of available labelled data. Construction and labelling of a good database of images or videos of facial expressions requires expertise, time and training of subjects and only a few such databases are available. However, collecting, without labelling, data of humans displaying expressions is not as difficult. The second motivation for using Bayesian networks is that it is possible to extend the system to fuse other modalities, such as audio, in a principled way by simply adding subnetworks representing the audio features.

Overview of facial expression recognition research

We begin by describing the problem. Since the early 1970s, Paul Ekman and colleagues have performed extensive studies of human facial expressions [44]. They found evidence to support universality in facial expressions. These 'universal facial expressions' are those representing happiness, sadness, anger, fear, surprise and disgust. They studied facial expressions in different cultures, including preliterate cultures, and found much commonality in the expression and recognition of emotions on the face. Matsumoto [45] reported the discovery of a seventh universal facial expression: contempt. Babies seem to exhibit a wide range of facial expressions without being taught, thus suggesting that these expressions are innate [46]. Ekman and Friesen [47] developed the Facial Action Coding System (FACS) to code facial expressions where movements on the face are described by a set of action units (AUs). Each AU has some related muscular basis. This system of coding facial expressions is done manually by following a set of prescribed rules. The inputs are images of facial expressions, often at the peak of the expression. This process is very time consuming. Ekman's work inspired many researchers to analyse facial expressions by means of image and video processing. By tracking facial features and measuring the amount of facial movement, they attempt to categorize different facial expressions. Recent work on facial expression analysis and recognition [48–52] has used these 'basic expressions' or a subset of them. We have previously used hierarchical HMMs [53] to classify the six basic expressions while automatically segmenting the video. Pantic and Rothkrantz [54] provide an in-depth review of much of the research done in automatic facial expression recognition in recent years.

Facial expression recognition system

In the remainder of the section, we describe an example of a real-time facial expression recognition system [30]. The system is composed of a face tracking algorithm which outputs

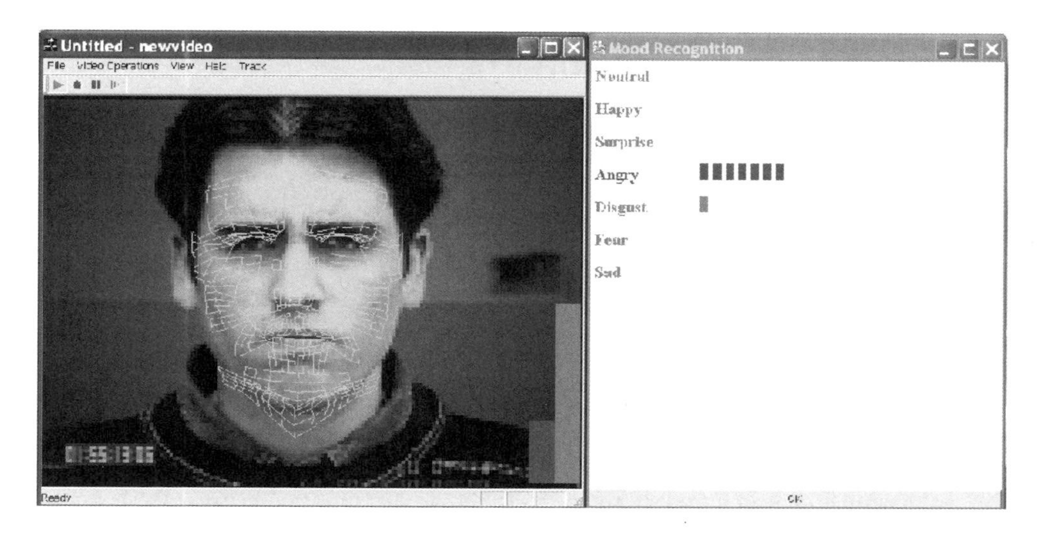

Figure 9.3 A snapshot of our real-time facial expression recognition system. On the right side is a wireframe model overlayed on a face being tracked. On the left side the correct expression, Angry, is detected (the bars show the relative probability of Angry compared to the other expressions). The subject shown is from the Cohn-Kanade database

a vector of motion features of certain regions of the face; the features are used as inputs to a Bayesian network classifier. A snapshot of the system, with the face tracking and recognition results, is shown in Figure 9.3.

The face tracking shown is based on a system developed by Tao and Huang [55] called the piecewise Bézier volume deformation (PBVD) tracker. The face tracker uses a model-based approach where an explicit 3D wireframe model of the face is constructed. In the first frame of the image sequence, landmark facial features such as the eye corners and mouth corners are selected interactively. The generic face model is then warped to fit the selected facial features. The face model consists of 16 surface patches embedded in Bézier volumes. The surface patches defined this way are guaranteed to be continuous and smooth. The shape of the mesh can be changed by changing the locations of the control points in the Bézier volume.

The recovered motions are represented in terms of magnitudes of some predefined motion of various facial features. Each feature motion corresponds to a simple deformation on the face, defined in terms of the Bézier volume control parameters. We refer to these motion vectors as motion-units (MUs). Note that they are similar but not equivalent to Ekman's AUs, and are numeric in nature, representing not only the activation of a facial region, but also the direction and intensity of the motion. The MUs are used as the features for the Bayesian network classifiers.

Experimental analysis

We investigate here the effect of using both labelled and unlabelled data for Bayesian network classifiers for facial expression recognition. In [30] we consider experiments where all the data is labelled and show limited experiments for person-dependent tests. We also show experiments for the more general problem of person-independent expression recognition.

We use two different databases, a database collected by Chen [48] denoted as the Chen-Huang database, and the Cohn-Kanade AU coded facial expression database [56].

The first is a database of subjects that were instructed to display facial expressions corresponding to the six types of emotions. All the tests of the algorithms are performed on a set of five people, each one displaying six sequences of each one of the six emotions, starting and ending at the neutral expression. The video sampling rate was 30 Hz, and a typical emotion sequence is about 70 samples long (~2 s).

The Cohn-Kanade database [56] consists of expression sequences of subjects, starting from a neutral expression and ending in the peak of the facial expression. There are 104 subjects in the database. Because not all of the six facial expressions sequences were available to us for some of the subjects, we used a subset of 53 subjects, for which at least four of the sequences were available. For each subject there is at most one sequence per expression with an average of 8 frames for each expression.

Accuracy is measured with respect to the classification result of each frame, where each frame in the video sequence was manually labelled to one of the expressions (including neutral). This manual labelling can introduce some 'noise' in our classification because the boundary between neutral and the expression of a sequence is not necessarily optimal, and frames near this boundary might cause confusion between the expression and the neutral. A different labelling scheme is to label only some of the frames that are around the peak of the expression, leaving many frames in between unlabelled. We did not take this approach because a real-time classification system would not have the information of the peak of the expression available to it.

Experiments with labelled and unlabelled data

We perform person independent experiments, training the classifiers with labelled and unlabelled data. To generate the mixed sets of labelled and unlabelled data, we first partition the data to a training set and test set (2/3 training, 1/3 for testing), and choose at random a portion of the training set, removing the labels. This procedure ensures that the distributions of the labelled and the unlabelled sets are the same.

We compared several algorithms, including: Naive Bayes with labelled data only (NB-L); Naive Bayes with labelled and unlabelled data using the Expectation Maximization algorithm (EM-NB); Tree-Augmented Naive Bayes [57] with labelled data only (TAN-L); TAN with labelled and unlabelled learned using EM [30, 58] (EM-TAN); Cheng-Bell-Liu structure learning algorithm trained with labelled and unlabelled [59] (EM-CBL1); and our stochastic structure search algorithm [30] (SSS).

We then train Naive Bayes and TAN classifiers, using just the labelled part of the training data and the combination of labelled and unlabelled data. We use the SSS and EM-CBL1 algorithms to train a classifier using both labelled and unlabelled data (we do not search for the structure with just the labelled part because it is too small for performing a full structure search).

Table 9.1 shows the results of the experiments. We see that with NB and TAN, when using 200 and 300 labelled samples, adding the unlabelled data degrades the performance of the classifiers, and we would have been better off not using the unlabelled data. We also see that EM-CBL1 performs poorly in both cases. Using the SSS algorithm, we are able to improve the results and utilize the unlabelled data to achieve performance which is higher than using

Table 9.1 Classification results for facial expression recognition with labelled and unlabelled data

Dataset	Train		Test	NB-L	EM-NB	TAN-L	EM-TAN	EM-CBL1	SSS
	Labelled	Unlabelled							
Cohn-Kanade	200	2980	1000	72.5 ± 1.4	69.1 ± 1.4	72.9 ± 1.4	69.3 ± 1.4	66.2 ± 1.5	**74.8 ± 1.4**
Chen-Huang	300	11982	3555	71.3 ± 0.8	58.5 ± 0.8	72.5 ± 0.7	62.9 ± 0.8	65.9 ± 0.8	**75.0 ± 0.7**

just the labelled data with NB and TAN. The fact that the performance is lower than in the case when all the training set was labelled (about 75% compared to over 80%) implies that the relative value of labelled data is higher than that of unlabelled data, as was shown by Castelli [60]. However, had there been more unlabelled data, the performance would be expected to improve.

9.5 Conclusions

With the continuing proliferation of cameras and sensors, multimedia understanding, human–computer interaction and computer vision will become indispensable parts of our everyday lives. In the near future we will have ubiquitous vision with cameras that see everything, everywhere, all the time. Networks with thousands of sensors that cover large geographic areas and track hundreds of objects will be reality. New, intelligent solutions are needed to bring HCI to a more natural level, and improve our tools for managing video databases.

Undoubtedly, advances in these areas will be inspired by advances in artificial intelligence. But machine learning, and especially graphical models, with their modeling power, have the potential to bring exciting solutions in the much shorter term.

References

[1] R.G. Cowell, A.P. Dawid, S.L. Lauritzen, D.J. Spiegelhalter, *Probabilistic Networks and Expert Systems*. Springer-Verlag, New York, 1999.

[2] S.L. Lauritzen, D.J. Spiegelhalter, Local computations with probabilities on graphical structures and their application to expert systems. *Journal of the Royal Statistical Society B*, **50**, 1988.

[3] J. Pearl, *Probabilistic Reasoning in Intelligent Systems*. Morgan Kaufmann, San Mateo, CA, 1988.

[4] V.N. Vapnik, *Statistical Learning Theory*. John Wiley & Sons, Chichester, 1998.

[5] C. Bishop, *Neural Networks for Pattern Recognition*. Oxford University Press, Oxford, 1996.

[6] A.Y. Ng, M.I. Jordan, On discriminative vs. generative classifiers: a comparison of logistic regression and naive bayes. In T.G. Dietterich, S. Becker, Z. Ghahramani (eds) *Advances in Neural Information Processing Systems 14: Proceedings of the 2001 Conference*. MIT Press, Cambridge, MA, 2002.

[7] B. Moghaddam M.-H. Yang, Gender classification with support vector machines. In *Proceedings of IEEE International Conference on Automatic Face and Gesture Recognition*, March 2000, pp. 306–311. IEEE, 2000.

[8] S.L. Lauritzen. *Graphical Models*. Oxford University Press, Oxford, 1996.

[9] J.M. Hammersley, P. Clifford, Markov fields on finite graphs and lattices, unpublished manuscript, 1971.

[10] R. Kindermann, J.L. Snell, *Markov Random Fields and their Applications. Contemporary Mathematics 1*. American Mathematical Society, Providence, RI, 1980.

[11] F.R. Kschischang, B.J. Frey, H.-A. Loeliger, Factor graphs and the sumproduct algorithm. *IEEE Transactions on Information Theory, Special Issue on Codes on Graphs and Iterative Algorithms*, **47**(2), 498–519, 2001.

[12] J. Yedidia, W.T. Freeman, Y. Weiss, Understanding belief propagation and its generalizations. In G. Lakemeyer, B. Nebel (eds) *Exploring Artificial Intelligence in the New Millennium*, pp. 239–236. Morgan Kaufmann, San Francisco, CA, 2003.

[13] F. Jensen, *An Introduction to Bayesian Networks*. Springer-Verlag, Berlin, 1996.

[14] F. Jensen, S. Lauritzen, K. Olesen, Bayesian updating in recursive graphical models by local computations. *Computational Statisticals Quarterly*, **4**, 269–282, 1990.

[15] A.P. Dempster, N.M. Laird, D.B. Rubin, Maximum likelihood from incomplete data via the EM algorithm. *Journal of the Royal Statistical Society, Series B*, **39**(1), 1–38, 1977.

[16] Z. Ghahramani, M.J. Beal, Graphical models and variational methods. In M. Opper and D. Saad (eds) *Advanced Mean Field Methods*. MIT Press, Cambridge, MA, 2001.

[17] M.I. Jordan, Z.Ghahramani, T. Jaakkola, K. Saul, An introduction to variational methods for graphical models. *Machine Learning*, **37**(2), 183–233, 1999.

[18] D. Spiegelhalter, A. Thomas, W. Gilks, BUGS, bayesian inference using gibbs sampling, Technical Report, MRC Biostatistics Unit, Cambridge, UK, 1993.
[19] A. Garg, V. Pavlovic, J. Rehg, T.S. Huang, Multimodal speaker detection using error feedback dynamic Bayesian networks. In *Proceedings of IEEE Computer Vision and Pattern Recognition Conference, CVPR 2000*, Hilton Head Island, SC, 13–15 June 2000. IEEE, 2000.
[20] N. Oliver, E. Horvitz, A. Garg, Layered representations for learning and inferring office activity from multimodal information. *Computer Vision and Image Understanding*, **96**(2), 163–180, 2004.
[21] M. Brand, N. Oliver, A. Pentland, Coupled hidden markov models for complex action recognition. In *Proceedings of IEEE Computer Vision and Pattern Recognition Conference, CVPR '97*, San Juan, June 1997, pp. 994–999. IEEE, 1997.
[22] I. Kozintsev, M. Naphade, T.S. Huang, Factor graph framework for semantic video indexing. *IEEE Transactions on Circuits and Systems for Video Technology*, **12**(1), 40–52, 2002.
[23] V. Pavlovic, J.M. Rehg, T.-J. Cham, K. Murphy, A dynamic bayesian network approach to figure tracking using learned dynamic models. In *International Conference on Computer Vision*, vol. 1, pp. 94–101. IEEE, 1999.
[24] M. Naphade, T.S. Huang, Discovering recurring events in video using unsupervised methods. In *Proceedings of IEEE International Conference on Image Processing*. IEEE, 2002.
[25] P. Xu, L. Xie, S.-F. Chang, A. Divakaran, A. Vetro, H. Sun, Algorithms and systems for segmentation and structure analysis in soccer video. In *Proceedings of the IEEE International Conference on Multimedia and Expo, (ICME)*. IEEE, 2001.
[26] K.P. Bennett, A. Demiriz, Semi-supervised support vector machines. In M.S. Kearns, S.A. Solla, D.A. Cohn (eds) *Advances in Neural Information Processing Systems 11: Proceedings of the 1998 Conference*, pp. 368–374. MIT Press, Cambridge, MA, 1999.
[27] F.G. Cozman, I. Cohen, M.C. Cirelo, Semi-supervised learning of mixture models. In S. Haller, G. Simmons (eds) *Proceedings of the Fifteenth International Florida Artificial Intelligence Research Society Conference*, pp. 327–331. AAAI Press, Menlo Park, CA, 2002.
[28] B. Shahshahani, D. Landgrebe, Effect of unlabeled samples in reducing the small sample size problem and mitigating the Hughes phenomenon. *IEEE Transactions on Geoscience and Remote Sensing*, **32**(5), 1087–1095, 1994.
[29] T. Zhang, F. Oles, A probability analysis on the value of unlabeled data for classification problems. In P. Langley (ed.) *Proceedings of the International Conference on Machine Learning (ICML)*, Stanford, CA, July 2000, pp. 1191–1198. Morgan Kaufmann, San Francisco, CA, 2000.
[30] I. Cohen, N. Sebe, F.G. Cozman, M.C. Cirelo, T.S. Huang, Learning Bayesian network classifiers for facial expression recognition using both labeled and unlabeled data. In *Proceedings of IEEE Computer Vision and Pattern Recognition Conference, CVPR '98*, Santa Barbara, CA, 23–25 June 1998. IEEE, 1998.
[31] S. Baluja, Probabilistic modelling for face orientation discrimination: learning from labeled and unlabeled data. In M.S. Kearns, S.A. Solla, D.A. Cohn (eds) *Advances in Neural Information Processing Systems 11: Proceedings of the 1998 Conference*, pp. 854–860. MIT Press, Cambridge, MA, 1999.
[32] R. Kohavi, Scaling up the accuracy of naive-Bayes classifiers: a decision-tree hybrid. In E. Simoudis, J. Han, U. Fayyad (eds) *Proceedings of the Second International Conference on Knowledge Discovery and Data Mining*, Portland, OR, August 1996, pp. 202–207. AAAI Press, Menio Park, CA, 1996.
[33] I. Cohen, F.G. Cozman, A. Bronstein, On the value of unlabeled data in semi-supervised learning based on maximum-likelihood estimation, Technical Report HPL-2002-140, HP Labs, 2002.
[34] N. Vasconcelos, A. Lippman, Bayesian modeling of video editing and structure: semantic features for video summarization and browsing. In *Proceedings of IEEE International Conference on Image Processing, ICIP '98*, Chicago, IL, 4–7 October 1998. IEEE, 1998.
[35] M.R. Naphade, T.S. Huang, A probabilistic framework for semantic indexing and retrieval in video. In *Proceedings of the IEEE International Conference on Multimedia and Expo (ICME)*, pp. 475–478. IEEE, 2000.
[36] N. Petrovic, N. Jojic, T.S. Huang, Scene generative models for adaptive video fast forward. In *Proceedings of International Conference on Image Processing, ICIP 2003*, Barcelona, Spain, 14–17 September 2003. IEEE, 2003.
[37] A. Yoshitaka, T. Ichikawa, A survey on content-based retrieval for multimedia databases. *IEEE Transactions on Knowledge and Data Engineering*, **11**(1), 81–93, 1999.
[38] M.M. Zloof, QBE/OBE: a language for office and business automation. *Computer*, **14**(5), 13–22, 1981.
[39] C. Wren, A. Azarbayejani, T. Darrell, A. Pentland, pFinder: real-time tracking of the human body. *IEEE Transaction on Pattern Analysis and Machine Intelligence*, **19**(7), 780–785, 1997.

[40] R.J. Woodham, Photometric method for determining surface orientation from multiple images. *Optical Engineering*, **19**(1), 139–144, 1980.

[41] N. Petrovic, I. Cohen, B.J. Frey, R. Koetter, T.S. Huang, Enforcing integrability for surface reconstruction algorithms using belief propagation in graphical models. In *Proceedings of IEEE Computer Vision and Pattern Recognition Conference, CVPR 2001*, Kauai, Hawaii, 8–14 December 2001. IEEE, 2001.

[42] I. Cohen, N. Sebe, A. Garg, M. Lew, T.S. Huang, Facial expression recognition from video sequences. In *Proceedings of the IEEE International Conference on Multimedia and Expo (ICME)*, Lausanne, Switzerland, August 2002, pp. 121–124. IEEE, 2002.

[43] N. Sebe, I. Cohen, A. Garg, M.S. Lew, T.S. Huang, Emotion recognition using a Cauchy naive Bayes classifier. In R. Kasturi (ed.), *16th International Conference on Pattern Recognition: Proceedings*, 11–15 August 2002, Québec City, Canada, pp. 17–20. IEEE Computer Society Press, Los Alamitos, CA, 2002.

[44] P. Ekman, Strong evidence for universals in facial expressions: a reply to Russell's mistaken critique. *Psychological Bulletin*, **115**(2), 268–287, 1994.

[45] D. Matsumoto, Cultural influences on judgments of facial expressions of emotion. In *Proceedings of the 5th ATR Symposium on Face and Object Recognition*, Kyoto, Japan, pp. 13–15, 1998.

[46] C.E. Izard, Innate and universal facial expressions: evidence from developmental and cross-cultural research. *Psychological Bulletin*, **115**(2), 288–299, 1994.

[47] P. Ekman, W.V. Friesen, *Facial Action Coding System: Investigator's Guide*. Consulting Psychologists Press, Palo Alto, CA, 1978.

[48] L.S. Chen, Joint processing of audio-visual information for the recognition of emotional expressions in human–computer interaction, PhD thesis, University of Illinois at Urbana-Champaign, Urbana, IL, 2000.

[49] G. Donato, M.S. Bartlett, J.C. Hager, P. Ekman, T.J. Sejnowski, Classifying facial actions. *IEEE Transactions on Pattern Analysis and Machine Intelligence*, **21**(10), 974–989, 1999.

[50] I.A. Essa, A.P. Pentland, Coding, analysis, interpretation, and recognition of facial expressions. *IEEE Transactions on Pattern Analysis and Machine Intelligence*, **19**(7), 757–763, 1997.

[51] J. Lien, Automatic recognition of facial expressions using hidden Markov models and estimation of expression intensity, PhD thesis, Carnegie Mellon University, Pittsburg, PA, 1998.

[52] N. Oliver, A. Pentland, F. Bérard, LAFTER: A real-time face and lips tracker with facial expression recognition. *Pattern Recognition*, **33**(8), 1369–1382, 2000.

[53] I. Cohen, N. Sebe, A. Garg, L. Chen, T.S. Huang, Facial expression recognition from video sequences: temporal and static modeling. *Computer Vision and Image Understanding*, **91**(112), 160–187, 2003.

[54] M. Pantic, L.J.M. Rothkrantz, Automatic analysis of facial expressions: the state of the art. *IEEE Transactions on Pattern Analysis and Machine Intelligence*, **22**(12), 1424–1445, 2000.

[55] H. Tao, T.S. Huang, Connected vibrations: a modal analysis approach to non-rigid motion tracking. In *Proceedings of IEEE Computer Vision and Pattern Recognition Conference, CVPR '98*, Santa Barbara, CA, 23–25 June 1998, pp. 735–740. IEEE, 1998.

[56] T. Kanade, J. Cohn, Y. Tian, Comprehensive database for facial expression analysis. In *Proceedings of the 4th IEEE International Conference on Automatic Face and Gesture Recognition (FG'00)*, pp. 46–53, 2000. IEEE, 2000.

[57] N. Friedman, D. Geiger, M. Goldszmidt, Bayesian network classifiers. *Machine Learning*, **29**(2), 131–163, 1997.

[58] M. Meila, Learning with mixture of trees, PhD thesis, Massachusetts Institute of Technology, Boston, MA, 1999.

[59] I. Cohen, Automatic facial expression recognition from video sequences using temporal information, MS Thesis, University of Illinois at Urbana-Champaign, Urbana, IL, 2000.

[60] V. Castelli, The relative value of labeled and unlabeled samples in pattern recognition, PhD thesis, Stanford University, Palo Alto, CA, 1994.

Part Three

Multimedia Content Management Systems and the Semantic Web

10

Semantic Web Applications

Alain Léger, Pramila Mullan, Shishir Garg and Jean Charlet

10.1 Introduction

Through the conquering, pervasive and user-friendly digital technology within the information society, fully open web content emerges as multiform, inconsistent and very dynamic. This situation leads to abstracting (ontology) this apparent complexity and the offer of new and enriched services able to reason on those abstractions (semantic reasoning) via automata, e.g. web services. This abstraction layer is the subject of very dynamic worldwide research, industry and standardization activity in what is called the Semantic Web (e.g. DARPA, IST-Ontoweb, IST-NoE KnowledgeWeb, W3C). The first applications focused on information retrieval with access by semantic content instead of classical (even sophisticated) statistical pattern matching, and integration of enterprise legacy databases for leveraging company information silos (semantic information integration). The next large field of applications is now focusing on the seamless integration of companies for more effective e-work and e-business.

This new technology has its roots in the cognitive sciences, machine learning, natural language processing, multi-agent systems, knowledge acquisition, mechanical reasoning, logics and decision theory. It adopts a formal and algorithmic approach for common-sense reasoning in relation to its strong roots in human language conceptualization via the linking of computational machinery and human communications.

We give here some key applications areas either fielded or deployed where clear positive feedback is recognized.

10.2 Knowledge Management and E-Commerce

10.2.1 Knowledge Management

Knowledge is one of the key success factors for the enterprises of today and tomorrow. Therefore, company knowledge management (KM) has been identified as a strategic tool for enterprises. However if, information technology is one foundation of KM, KM is interdisciplinary by nature, and includes human resource management, enterprise organization and culture.

Multimedia Content and the Semantic Web Edited by Giorgos Stamou and Stefanos Kollias

Some definitions:

> Knowledge management is the systematic, explicit, and deliberate building, renewal and application of knowledge to maximize an enterprise's knowledge related effectiveness and returns from its knowledge assets [1].
>
> Knowledge management is the process of capturing a company's collective expertise wherever it resides in databases, on paper, or in people's heads and distributing it to wherever it can help produce the biggest payoff [2].
>
> KM is getting the right knowledge to the right people at the right time so they can make the best decision [3].

Moreover, in a recent paper [4], KM is summarized as follows:

> One organization is composed of people interacting for common objectives, in a given structure—either a 'formal structure' as in a company or an administration, or 'informal' as in community of interest or community of practices, within a closed environment and with an open environment.

KM is the management of the activities and the process aiming at leveraging the use and the creation of knowledge in organizations for two main objectives: capitalization of the corporate knowledge and durable innovation, as:

- access, sharing, reuse of knowledge (explicit or implicit, private or collective);
- creation of new knowledge.

KM must be guided by a strategic vision fully aligned with the strategic objectives of the organization as illustrated in Figure 10.1.

Time	Environment: Internal	Environment: External
Past	Preserve past knowledge for reuse. Avoid knowledge loss.	Keep trace of the lessons learned from interactions with the external world.
Present	Improve knowledge sharing and cooperation between members of the organization. Disseminate the best practices in the company.	Enhance relationships with the external environment (e.g. customers, competitors etc.).
Future	Enhance the learning phase for newcomers. Increase the quality of projects. Increase creativity.	Anticipate the evolution of the external environments (customers, competitors etc.). Improve reactivity to unexpected events and to crisis situations.

Figure 10.1 The role of KM in relation to time and environment

So a KM policy must rely on a deep understanding of the organization, its corporate culture, what kind of knowledge exists (either individually, or collectively in an internal group or in the whole organization), how the organization's intellectual capital can be assessed, how the past can explain the present and help to prepare for the future, and the strategic objectives of the organization and how they can be achieved according to the corporate culture and the environment of the end users [5].

In an organization, knowledge can be individual or collective; it can be explicit, implicit or tacit. In Nonaka's model, organizational learning relies on transformation between these different types of knowledge [6]. Collective knowledge can also emerge in a community of practice. Tacit knowledge can be transmitted without any language or external support (e.g. through observations), but in order to be transmitted to other persons, explicit knowledge generally needs a medium (i.e. document, database etc.) so that people can create their own knowledge either by interacting with each other or by retrieving information from explicit traces and productions of other colleagues' knowledge. Knowledge can also be distributed among several knowledge sources in the organization, with possibly heterogeneous viewpoints.

There are three significant aspects to be tackled:

- People (i.e. their knowledge, their organizational functions, their interest centers, their relational networks, their work environment etc.): any KM solution must be compatible with the end users' cognitive models and work environment.
- Organization (i.e. its objectives, its business processes, the corporate culture, its corporate strategy etc.): any KM solution must be compatible with the organizational strategy and culture.
- Information technologies to support the intended KM: the chosen technologies will depend on the KM objectives and on the intended end users' environment.

A CEN/ISSS project was initiated in 2002 with the goal of reaching a consensus on the state of the art on good practices to ease the deployment of KM in Europe. The project began in October 2002 through a call for inputs on the European Commission public KM portal KnowledgeBoard (http://www.knowledgeboard.com), and closed with a final set of CEN recommendations in autumn 2003 entitled 'European Guide to Good Practice in Knowledge Management' (http://www.cenorm.be/cenorm/index.htm).

The European KM Framework is designed to support a common European understanding of KM, to show the value of this emerging approach and to help organizations towards its successful implementation. The Framework is based on empirical research and practical experience in this field from all over Europe and the rest of the world. The European KM Framework addresses all relevant elements of a KM solution and serves as a reference basis for all types of organizations which aim to improve their performance by handling knowledge in a better way. The KM Framework relates the various components of KM (people, process, technology) to each other. It provides a schematic picture of how these various aspects depend on each other and it helps to position KM projects or activities.

The benefits of KM

In the past, information technology for knowledge management has focused on the management of knowledge containers using text documents as the main repository and source of knowledge. In the future, Semantic Web technology, especially ontologies and machine-processable metadata, will pave the way to KM solutions that are based on semantically related pieces of knowledge. The knowledge backbone is made up of ontologies that define a shared conceptualization of the application domain at hand and provide the basis for defining metadata, that have a precisely defined semantics, and that are therefore machine processable. Although the first KM approaches and solutions have shown the benefits of ontologies and related methods, a large number of open research issues still exist that have to be addressed in order to make Semantic Web technologies a complete success for KM solutions:

- Industrial KM applications have to avoid any kind of overhead as far as possible. Therefore, a *seamless integration* of knowledge creation, e.g. content and metadata specification, and knowledge access, e.g. querying or browsing, into the working environment is required. Strategies and methods are needed to support the creation of knowledge, as side effects of activities that are carried out anyway. These requirements mean *emergent semantics*, e.g. through ontology learning, are needed, which reduces the current time-consuming task of building up and maintaining ontologies.
- Access as well as presentation of knowledge has to be *context-dependent*. Since the context is set up by the current business task, and thus by the business process being handled, a tight integration of business process management and knowledge management is required. KM approaches can manage knowledge and provide a promising starting point for smart push services that will proactively deliver relevant knowledge for carrying out the task at hand more effectively.
- *Conceptualization* has to be supplemented by *personalization*. On one hand, taking into account the experience of the user and his/her personal needs is a prerequisite in order to avoid information overload, and on the other hand to deliver knowledge on the right level of granularity.

The development of knowledge portals serving the needs of companies or communities is still more or less a manual process. Ontologies and related metadata provide a promising conceptual basis for generating parts of such knowledge portals. Obviously, among others, conceptual models of the domain, of the users and the tasks are needed. The *generation of knowledge portals* has to be supplemented with the (semi-)automated evolution of portals. As business environments and strategies change rather rapidly, KM portals have to be kept up-to-date in this fast changing environment. Evolution of portals should also include some mechanism to 'forget' outdated knowledge.

KM solutions will be based on a combination of intranet-based functionalities and mobile functionalities in the very near future. Semantic Web technologies are a promising approach to meet the needs of mobile environments, such as, for example, location-aware personalization and adaptation of the presentation to the specific needs of mobile devices, i.e. the presentation of the required information at an appropriate level of granularity. In essence, employees should have access to the KM application *anywhere* and *anytime*.

Peer-to-peer computing (P2P), combined with Semantic Web technology, will be an interesting way of getting rid of the more centralized KM solutions that are currently used in ontology-based solutions. P2P scenarios open up the way to derive consensual conceptualizations among employees within an enterprise in a bottom-up manner.

Virtual organizations are becoming more and more important in business scenarios, mainly due to decentralization and globalization. Obviously, semantic interoperability between different knowledge sources, as well as trust, is necessary in inter-organizational KM applications.

The integration of KM applications (e.g. skill management) with *e-learning* is an important field that enables a lot of synergy between these two areas. KM solutions and e-learning must be integrated from both an organizational and an IT point of view. Clearly, interoperability and integration of (metadata) standards are needed to realize such integration.

KM is obviously a very promising area for exploiting Semantic Web technology. Document-based KM solutions have already reached their limits, whereas semantic technologies open the way to meet the KM requirements in the future.

Knowledge-based KM applications [7–10]

In the context of geographical team dispersion, multilingualism and business unit autonomy, usually the company wants a solution allowing the identification of strategic information, the secured distribution of this information and the creation of transverse working groups. Some applicative solutions allowed the deployment of an intranet intended for all the marketing departments of the company worldwide, allowing a better division and a greater accessibility to information, but also capitalization on the total knowledge of the company group. There are three crucial points that aim to ease the work of the various marketing teams of the company group: automatic competitive intelligence of the web, skill management and document management.

Thus, the system connects the 'strategic ontologies' of the company group (brands, competitors, geographical areas etc.) with the users, via the automation of related processes (research, classification, distribution, representation of knowledge). The result is a dynamic, 'Semantic Web' system of navigation (research, classification) and collaborative features.

The KM server is designed for organizations that need to manage large, rich, voluminous professional document repositories (commercial, economic, technical, financial, legal, medical, cultural etc.) and documentation of business procedures. So KM increases efficiency for both professionals and lay people by providing customized knowledge access for employees, R&D groups, subcontractors and business partners. Users access relevant and up-to-date knowledge resources. Information assets can be organized either automatically or by users using a common organization model and business terminology.

The implemented technology brings a real added value, as it allows the representation of the organizational dimension of information via 'active ontologies' (i.e. the conceptual knowledge bases which are used to support the automatic process).

From a functional point of view, a KM server organizes skill and knowledge management within the company, in order to improve interaction, collaboration and information sharing. This constitutes a virtual workspace which facilitates work between employees who speak different languages, automates the creation of work groups, organizes and capitalizes on structured and unstructured, explicit and tacit data of the company organization, and offers advanced features of capitalization. Each work group (project, community etc.) is a centralized and organized data space, allowing its members to research, exchange, share and then publish information (according to the roles and rights allotted to them) while eliminating geographical and temporal constraints (distributed KM). Furthermore, the semantic backbone also makes possible crossing of qualitative gap by providing cross-lingual data. Indeed, the semantic approach allows ontologies to overcome language barriers (culture and language differences).

The incoming information flow (from the web, internal documents, newswire etc.), representing typically thousands of elements each day, is sent directly to the correct projects and people who receive and consult only documents related to their projects. A reduction factor on research time of the order of 1000 to 1 is claimed to be possible by ontologies.

Some lessons learnt [11,12]:

- The main strong benefits for the enterprise are high productivity gains and operational valorization of knowledge legacy
- Productivity: automation of knowledge base maintenance, automation of content indexing, augmented productivity in publication cycle (commercial proposals, reports etc.), search efficiency.
- Quality and operational valorization of knowledge legacy: unified management of heterogeneous resources, information relevancy, capacity to represent complex knowledge, gains in development and maintenance of knowledge and content management solution, generic and evolvable solution.
- Human factors are key difficulties in full groupware functionalities of the KM solution towards the employees of the company, so adopt a step-by-step approach.
- Access to the information portal must be well designed and must be supported by a group of people dedicated to information filtering and qualifying (P2P is possible).

10.2.2 E-Commerce [13]

E-commerce is mainly based on the exchange of information between involved stakeholders using a telecommunication infrastructure. There are two main scenarios: business-to-customer (B2C) and business-to-business (B2B).

B2C applications enable service providers to promote their offers, and allow customers to find offers which match their demands. By providing a single access to a large collection of frequently updated offers and customers, an electronic marketplace can match the demand and supply processes within a commercial mediation environment.

B2B applications have a long history of using electronic messaging to exchange information related to services previously agreed among two or more businesses. Early plain-text telex communication systems were followed by electronic data interchange (EDI) systems based on terse, highly codified, well-structured, messages. Recent developments have been based on the use of less highly codified messages that have been structured using the eXtensible Markup Language (XML).

A new generation of B2B systems is being developed under the ebXML (electronic business in XML) label. These will use classification schemes to identify the context in which messages have been, or should be, exchanged. They will also introduce new techniques for the formal recording of business processes, and for the linking of business processes through the exchange of well-structured business messages. ebXML will also develop techniques that will allow businesses to identify new suppliers through the use of registries that allow users to identify which services a supplier can offer.

The coding systems used in EDI systems are often examples of limited scope, language-independent, mini-ontologies that were developed in the days when decimalized hierarchical classification systems were the most sophisticated form of ontology. There is a strong case for the redesign of many of these classification schemes based on current best practice for ontology development. ebXML needs to include well-managed multilingual ontologies that can be used to help users to match needs expressed in their own language with those expressed in the service provider's language(s).

Within Europe many of the needs of B2C applications match those of B2B applications. Customers need to use their own language to specify their requirements. These need to be matched with services provided by businesses, which may be defined in languages other than those of the customer. Businesses may or may not provide multilingual catalogues. Even where multilingual catalogues are supplied, they may not cover all European languages. For a single market to truly exist within Europe, it must be possible for customers to be able to request product and sales terms information in their own language, possibly through the use of online translation services. It is anticipated that services providing multilingual searching of sets of catalogues will act as an intermediary between businesses and their potential customers.

What does Knowledge-based E-Commerce benefit?

At present, ontologies, and more generally ontology-based systems, appear as a central issue for the development of efficient and profitable Internet commerce solutions. They represent a way to access a large range of Internet information (professional, business, leisure etc.) spaces with efficiency and optimization, which will be more and more prominent features of most business, governmental and personal information activity in the near future. However, because of a lack of standardization for business models, processes and knowledge architectures, it is currently difficult for companies to achieve the promised ROI from e-commerce.

Moreover, a technical barrier exists that delays the emergence of e-commerce, lying in the need for applications to meaningfully share information, taking into account the lack of reliability and security of the Internet. This fact may be explained by the variety of enterprise and e-commerce systems employed by businesses and the various ways in which these systems are configured and used. As an important remark, such interoperability problems become particularly acute when a large number of trading partners attempt to agree and define the standards for interoperation, which is precisely a main condition for maximizing the ROI.

Although it is useful to strive for the adoption of a single common domain-specific standard for content and transactions, such a task is still difficult to achieve, particularly in cross-industry initiatives, where companies cooperate and compete with one another.

In addition to this:

- Commercial practices may vary in a wide range and, consequently, cannot always be aligned for a variety of technical, practical, organizational and political reasons.
- The complexity of the global description of the organizations themselves, their products and services (independently or in combination), and the interactions between them remains a formidable obstacle.
- It is usually very difficult to establish, a priori, rules (technical or procedural) governing participation in an electronic marketplace.

- Adoption of a single common standard may limit business models which could be adopted by trading partners, and then, potentially, reduce their ability to fully participate in Internet commerce.

Ontologies appear as really promising for e-commerce for all the aforementioned reasons. Indeed, alternative strategies may consist of sharing foundational ontologies, which could be used as the basis for interoperation among trading partners in electronic markets. An ontology-based approach has the potential to significantly accelerate the penetration of e-commerce within vertical industry sectors, by enabling interoperability at the business level, reducing the need for standardization at the technical level. This will enable services to adapt to the rapidly changing online environment.

The following uses for ontologies, and classification schemes that could be defined using ontologies, have been noted within e-commerce applications:

- categorization of products within catalogues;
- categorization of services (including web services);
- production of yellow page classifications of companies providing services;
- identification of countries, regions and currencies;
- identification of organizations, persons and legal entities;
- identification of unique products and saleable packages of products;
- identification of transport containers, their type, location, routes and contents;
- classification of industrial output statistics.

Many existing B2B applications rely on the use of coded references to classification schemes to reduce the amount of data that needs to be transmitted between business partners. Such references have overcome the problems introduced by the natural ambiguity of words that have more than one meaning (polysemy) or can apply to more than one object (e.g. personal names such as John Smith). By providing a separate code for each different use of the term it is possible to disambiguate messages to a level where they can be handled without human intervention.

Very few of the existing classification schemes used within e-commerce applications have been defined as formal ontologies, or have been formally modelled to ensure that the relationships between terms are fully described. To date, most of the techniques introduced by ontologies have been applied to general linguistic situations, such as those involved in specific academic disciplines, rather than to the language adopted by specific industries.

Knowledge-based E-Commerce applications [14–17]

According to Zyl and Corbett [18], applications of this kind use one or more shared ontologies to integrate heterogeneous information systems and allow common access for humans or computers. This enforces the shared ontology as the standard ontology for all participating systems, which removes the heterogeneity from the information system. The heterogeneity is a problem because the systems to be integrated are already operational and it is too costly to redevelop them. A linguistic ontology is sometimes used to assist in the generation of the shared ontology, or is used as a top-level ontology, describing very general concepts like space, time, matter, object, event, action etc. for the shared ontologies to inherit from it. The benefits are the integration of heterogeneous information sources, which can improve interoperability, and more effective use and reuse of knowledge resources.

Yellow Pages and product catalogue are direct benefactors of a well-structured representation, which, coupled to multilingual ontology, clearly enhances the precision/recall of products or services search engines. The ONTOSEEK system (1996–1998) is the first system being prototyped associating domain ontology (in KR conceptual graph (CG) with very limited expressiveness) to a large multilingual linguistic ontology (SENSUS–WORDNET) for natural language search of products [19].

ONTOSEEK searchs products by mapping natural language human requests to domain ontology. Unlike traditional e-commerce portal search functions, the user is not supposed to know the vocabulary used for describing the products and, thanks to the SENSUS ontology, is able to express himself in his own vocabulary.

The main functional architectural choice of ONTOSEEK involves:

- use of a general linguistic ontology to describe products;
- great flexibility in expressing the request thanks to the semantic mapping offered between the request and the offers;
- Interactive guided request formulation through generalization and specialization links.

The CG KR is used internally to represent request and products. The semantic matching algorithm is based on a simple subsumption on the ontology graph and does not make use of a complex graph endomorphism.

ONTOSEEK has not been deployed commercially but in its trial period has fully demonstrated the potential benefits of making use of preliminary Semantic Web tools.

The MKBEEM prototype and technology (Multilingual Knowledge Based European Electronic Marketplace, IST-1999-10589, 2000–2003) concentrates on written language technologies and its use in the key sector of worldwide commerce. Within the global and multilingual Internet trading environment, there is an increasing pressure on e-content publishers of all types to adapt content for international markets. Localization—translation and cultural adaptation for local markets—is proving to be a key driver of the expansion of business on the web. In particular, MKBEEM is focusing on adding multilingualism to all stages of the information cycle, including multilingual content generation and maintenance, automated translation and interpretation, and enhancing the natural interactivity and usability of the service with unconstrained language input. On the knowledge technology side, the MKBEEM ontologies provide a consensual representation of the e-commerce field in two typical domains, B2C tourism and B2C mail order, allowing the commercial exchanges to be transparent in the language of the end user, the service provider or the product provider. Ontologies are used for classifying and indexing catalogues, filtering a user's query, selecting relevant products and providers, facilitating multilingual human–machine dialogues between user and software 'agent', and inferring information relevant to the user's request and trading needs.

In particular, MKBEEM has fully demonstrated how adding multilingual services to the following stages of the information cycle for multilingual B2C E-Commerce services:

- mapping the human language query onto an ontological representation;
- producing composite services from content provider catalogues;

- mapping ontological queries onto the ontology of the catalogues;
- describing the domain knowledge: used for classifying and indexing products, facilitating multilingual human–machine dialogues and inferring information relevant to the user's request.
- ease of integration of new content provider through components-based e-services for rapid updating of catalogue products in multilingual context offerings.

The effectiveness of the developed generic solutions has been tested in Finnish, French, Spanish and English in the domains of travel booking (SNCF French Rail services) and mail order sales (Redoute-Ellos).

10.3 Medical Applications

The medical domain is a favourite target for Semantic Web applications, just as the expert system was for artificial intelligence applications 20 years ago. The medical domain is effectively very complex: medical knowledge being difficult to represent in a computer, which makes the sharing of information difficult. Semantic Web solutions become very interesting in this context.

Thus, one of the main mechanisms of the Semantic Web, resource description using annotation principles, is of major importance in the medical informatics (or bioinformatics) domain, especially as regards the sharing of these resources (e.g. medical knowledge in the web or genomic database). Through the years, the information retrieval domain has been developed by medicine: the medical thesauri are enormous (1 000 000 terms for UMLS) and are principally used for bibliographic indexation. Nevertheless, the MeSH thesaurus (Medical Subject Headings) or the Unified Medical Language System (UMLS-http://www.nlm.nih.gov/research/umls/umlsmain.html) is used in the Semantic Web paradigm with varying degrees of difficulty. Finally, web services technology allows us to imagine some solutions to the interoperability problem, which is substantial in medical informatics.

We will describe current research, results and expected perspectives in theses three biomedical informatics topics in the context of the Semantic Web.

10.3.1 Resource Share

In the functional genomics domain, it is necessary to have access to several databases and knowledge bases that are accessible via the web but are heterogeneous in their structure as well as in their terminology. Among such resources, we can cite SWISSPROT, where gene products are annotated by GENEONTOLOGY, GENBANK, etc. In comparing the resources, it is easy to see that they propose the same information in different formats. The XML language, described as the unique common language of these databases, proposes as many Document Type Definitions (DTD) as resources and does not resolve the interoperability problem.

The solution comes from the Semantic Web with the mediator approach [20], which allows accessing of different resources with an ontology used as interlingua pivot. For example, in another domain than genomics, the mediator mechanisms of the NEUROBASE project [21] allows to federate different neuroimagery information bases situated in different clinical or research areas. The proposal consists of defining an IT architecture that allows access to and sharing of experimental results or data treatment methodologies. It would be possible to search in the various databases for similar results, or for images with certain features or to perform data mining analysis between several databases. The mediator of NEUROBASE is tested on decision support systems in epilepsy surgery.

10.3.2 Indexation and cataloguing

The PubMed site (http://www.ncbi.nlm.nih.gov/PubMed/) from the National Library of Medicine (NLM) provides access to the biggest database of scientific articles in the bioinformatics domain. These articles are indexed with MeSH (http://www.nlm.nih.gov/mesh/meshhome.html), a medical thesaurus composed of approximately 22,000 keywords (e.g. abdomen, hepatitis) and 84 qualifiers or subheadings (e.g. diagnosis, complications) [22]. The maintenance of PubMed highlights the main problem, which consists of choosing the pertinent index in order to represent an article. This is the same problem as in the CISMeF project (See below): the indexing task is all the more difficult since it is performed a posteriori. The NLM is developing an automatic indexing project based on the analysis of title and abstract of a given article and the analysis of the index of the articles listed in its bibliography [23].

The objective of CISMeF (French acronym for catalogue and index of French-language health resources) is to describe and index the main French-language health resources to assist health professionals and consumers in their search for electronic information available on the Internet [24]. CISMeF is a quality-controlled subject gateway initiated by the Rouen University Hospital (RUH-http://www.chu-rouen.fr/cismef). CISMeF began in February 1995. In March 2004, the number of indexed resources totalled over 13 500, with an average of 50 new resources indexed each week. The indexing thesaurus of CISMeF is MeSH. From the point of view of metadata, the choice of CISMeF is to use several metadata element sets: (1) the Dublin Core metadata format [25] to describe and index all the health resources included in CISMeF, (2) some elements from IEEE 1484 Learning Object Metadata for teaching resources [26], (3) specific metadata for evidence-based medicine resources which also qualify the health content, and (4) the HIDDEL metadata set [27], which will be used to enhance transparency, trust and quality of health information on the Internet in the EU-funded MedCIRCLE project.

CISMeF implies research in the information retrieval domain, in particular for request expansion. Recently, the KnowQuE (Knowledge-based Query Expansion) prototype system has been proposed [28], which includes:

- a morphological knowledge base in cooperation with Zweigenbaum and Grabar, which will benefit from the UMLF consortium in charge of developing the French Specialist Lexicon [29], e.g. the query asthmatic children will be derived into asthma AND child;
- a knowledge base of association rules extracted using the data mining knowledge discovery process, e.g. breast cancer/diagnostic $\Rightarrow$ mammography or hepatitis/prevention and control $\Rightarrow$ hepatitis vaccines [30];
- a formalized CISMeF terminology using the OWL language to benefit from its powerful reasoning mechanisms [31].

These applications, particularly CISMeF, bring to light questions about the comparative advantages of the utilization of thesauri versus ontologies for indexation. Even though thesauri reach their limits with the organization of medical concepts that are ambiguous or incoherent, the development of ontologies is time consuming. Moreover, an ontology mobilizes concepts with a description granularity that is not easily accessible to a practitioner in the context of his daily work. Solutions like the GALEN project [32] exist, which propose the concept of terminology server: in such a proposition, the terms of a thesaurus are described using the concepts and relations of the ontology. In this way, it is possible to maintain the thesaurus, to augment it with new terms (in describing it with the concepts/relations of the ontology), to verify the validity of a defined concept, to translate a conceptual expression in its canonical form etc. The GALEN project has been used in France in order to build and structure the new thesaurus of medical care (Classification commune des actes médicaux, CCAM). The terminology servers are close to the semantic thesauri developed in other domains [33].

10.3.3 Some Web Services for Interoperability

Web services technology can propose some solutions to the interoperability problem. We now describe a new approach based on 'patient envelope' and draw conclusions on the implementation of this envelope with web services technologies.

The patient envelope is a proposition of the Electronic Data Interchange for Healthcare group (EDI-Santé) (http://www.edisante.org/) with an active contribution from the ETIAM society (http://www.etiam.com/).

The objective of the work has been to focus on filling the gap between 'free' communication, using standard and generic Internet tools, and 'totally structured' communication as promoted by CEN (http://www.centc251.org/) or HL7 (http://www.hl7.org/). After a worldwide analysis of existing standards, the proposal consists of an 'intermediate' structure of information, related to one patient, and storing the minimum amount of data (i.e. exclusively useful data) to facilitate interoperability between communicating peers. The 'free' or the 'structured' information is grouped into a folder and transmitted in a secure way over existing communication networks [34]. This proposal has reached widespread celebrity with the distribution by Cegetel.rss of a new medical messaging service, called 'Sentinelle', fully supporting the patient envelope protocol and adapted tools.

After this milestone, EDI-Santé is promoting further developments based on ebXML and SOAP (Simple Object Access Protocol) in specifying exchange and medical properties:

1. Separate what is mandatory to the transport and the good management of the message (patient identification etc.) from what constitutes the 'job' part of the message.
2. Provide a 'container', collecting the different elements—text, pictures, videos etc.
3. The patient as a unique object of the transaction. Such an exchange cannot be anonymous. It concerns a sender and an addressee, who are involved in the exchange and responsible. The only way to perform this exchange between practitioners about a patient who can demand to know the content of the exchange implies the retention of a structure which is unique, a triplet {sender, addressee, patient}.
4. Conservation of the exchange semantics. The information about a patient is multiple. It comes from multiple sources and has multiple forms and supports (database, free textual document, semi-structured textual document, pictures etc.). It can be fundamental to maintain the existing links between elements, to transmit them together, e.g. a scanner and the associated report, and to prove it.

The interest in such an approach is that it prepares the evolution of the transmitted document, from free document (from proprietary ones to normalized ones with XML) to elements respecting HL7v3 or EHRCOM data types.

10.3.4 For the Future?

These different projects and applications highlight the main consequence of the Semantic Web, expected by the medical communities—the sharing and integration of heterogeneous information or knowledge. The answers to the different issues are the mediators, the knowledge-based system, and the ontologies, all based on normalized languages such as RDF, OWL or others. The work of the Semantic Web community must take into account these expectations. Finally, it is interesting to note that the Semantic Web is an integrated vision of the medical community's problems (thesaurus, ontology, indexation, inference) and provides a real opportunity to synthesize and reactivate some research [35].

10.4 Natural Language Processing

'The meaning is the weak point of the natural language studies and will remain till our knowledge on semantic has advanced far from the actual state' anticipated by Bloomfield [36].

10.4.1 Linguistic Ontology used in NLP Applications

Human language is built from individual words (lexical level), which can have many senses, and also belong to different lexical categories or part of speech. The texts in natural language are very structured objects presenting strong inter-phrase and intra-phrase cohesion [37].

Computational semantics addresses the question of modelling the important semantic phenomena occurring in human language data (quantifiers and scope, plurals, anaphora, ellipsis, adjectives, comparatives, temporal reference, verbs, attitudes, questions, events etc.). Traditionally, formal approaches to human language semantics have focused on the sentential level, which needs to be enlarged, as demonstrated in discourse representation theory (see for example [38]).

When a hearer receives some utterance from a speaker, he usually tries to understand what it is and why that speaker has produced it. From his own language competence, from knowledge about the world and the situation, his own previous beliefs etc. the hearer has to build an internal representation of what he thinks the speaker's proposition is, in order to choose the proper reaction.

To perform that building process, he must share his beliefs and use knowledge with the speaker, such as:

- speech recognition to split the sound into a sequence of word units (if the message is oral);
- lexical knowledge (attributing features to the word items);
- grammar rules (mainly dealing with the structure of the utterance and the word groupings, syntactic and semantic relations between the word units);
- semantic knowledge;
- conversational rules (mainly about the 'why' aspect and the discourse coherence);
- contextual knowledge (about the state of affairs, and the dialogue history).

All of these knowledge sources are not independently solicited (the correct recognition of words implies at least lexical and structural knowledge of the language). During the understanding process, the auditor builds up local hypotheses and checks their validity until he/she has a global view and a reliable interpretation. An utterance meaning involves three kinds of elements: situational, communicative and rhetoric.

- Situational meaning refers to the utterance context. It includes as extra-linguistic knowledge, the mental states of speaker (beliefs, attitudes, knowledge).
- Communicative intentions: this part takes into account the evolution of the situation that the speaker wants to point out:
  1. from the initial situation to the one he wants to reach;
  2. from what he asserts or presupposes to what he argues about;
  3. what he wants to focus on, and what remains in the background.
- Rhetoric sense (utterance style).

The meaning distribution among words, phrases, propositions and sentences can be represented by a net (oriented connex graph) with labels on nodes (semantemes) which is one of the senses labelled in a dictionary. Semantemes can be atomic or composite (equivalent to a semantic graph). Thus result from the message organisation (intention, focus, new/given etc.).

During a dialogue, an utterance allows a speaker to create new relations between discourse objects, to introduce his beliefs and requirements and, is doing so, to modify the universe of discourse. The hearer not only computes an interpretation of the intended meaning, but also updates his beliefs and knowledge.

According to Zyl and Corbett [18], over the past few years, there have been a number of reports of applications using a linguistic ontology. In addition to the more traditional use of these ontologies for natural language generation and for machine translation, these applications are being used for specifying the meaning of text in a specific domain, for intelligent information retrieval, and for integration of heterogeneous information systems. Ontology serves as a neutral format that other applications can access, or that one or more persons can understand in different natural languages. Some applications rely on one natural language and others are multilingual and allow translation between more than one natural language. Linguistic ontologies usually have the purpose of solving problems such as how knowledge of the world is to be represented, and how such organizations of knowledge are to be related to natural language generation (NLG) system levels of organization such as grammar and lexicons. Applications using a linguistic ontology often allow the users to express their queries in a natural language, and the information provided to the user might also be in a natural language. Other benefits of this approach include interoperability of software tools, and more effective use and reuse of knowledge resources.

First of all, we will take a look at a couple of ontology-based machine translation applications. An NLG application supports machine translation by providing language-neutral terms to which lexical terms of different languages can be attached. These applications are often knowledge-based machine translation (KBMT) systems, which translate the meanings (semantics) of text in one natural language to another natural language. The representations of meanings are captured by the language-independent ontology, which acts as an 'interlingua'. The principal reasons for using an ontology are to provide a grounding for representing text meaning in an interlingua and to enable lexicons for different languages to share models. The resulting ontology enables source language analysers and target language generators to share knowledge [18].

10.4.2 Automated Translation: PANGLOSS® and MIKROKOSMOS®

Analysis and generation in an automated multilingual translation may make use of a neutral representation (pivotal, interlingua) on which we map—as well as possible—the concept of the neutral representation to words in the multilingual lexical base. We refer this technique to KBMT, translating one source language to target language(s) via conceptual meaning (pivotal representation). The meaning representation is modelled in a language-independent ontology which plays the role of an interlingua.

WORDNET and EUROWORDNET [39] are prototypical examples. Unlike WORDNET, which is dedicated to English, EUROWORDNET is a multilingual base (German, Dutch, French, Italian, Spanish, Czech and Estonian). Each lexical base is organized in 'synsets' (set of synonymous words). Those synsets are then linked to an interlingua (Inter-Lingual-Index) base. So, a semantic bridge between similar words is possible through this mechanism.

The PANGLOSS® system [40] translates Spanish texts to English. The linguistic ontology used is SENSUS (as used in ONTOSEEK).

The MIKROKOSMOS® system [41,42] translates Spanish and Chineese texts into English. It makes use of an Interlingua (Text Meaning Representation, TMR), which provides a semantic representation for the three languages. It also provides an editing tool environment and an API to access the MIKROKOSMOS ontology.

The Consortium for Speech Translation Advanced Research (C-Star) has conducted research on an interlingua approach for speech-to-speech translation systems. They have developed, for instance, an interlingua designed for travel planning. The C-Star interlingua is based on domain actions.

Those systems are prototypical examples in the NLP domain for automated translation; however, ontology-based NLP could be or is used in many other areas, such as text mining, ontology building and information retrieval, to name a few.

10.5 Web Services

10.5.1 Introduction

The business community has evolved from a centralized to a network-based distributed community of collaborating partners. The role of the internet in fostering tighter partnership between businesses and their interactions with consumers has created an opportunity for innovations in both technology and service offerings.

The technology thrust has been in the evolution from distributed middleware architectures to a decentralized web services infrastructure. Certainly, web services infrastructures fulfil the promise of easier development and integration across heterogeneous service platforms and from that perspective contribute to improvements in operational expenses and bundled service offerings to improve a customer's experience with existing businesses. However, along with this technology comes a set of unique new revenue opportunities for service providers to enter into the centre of this B2B and B2C collaboration by facilitating services between businesses and consumers. The service providers in this space can potentially be traditional service providers (e.g. ISPs, mobile carriers, wireline phone service providers) as well as existing technology providers attempting to enter new markets with a services offering.

10.5.2 Web Services Defined

The W3C Web Services Architecture document [43] defines web services as follows:

> A Web service is a software system designed to support interoperable machine-to-machine interaction over a network. It has an interface described in a machine-processable format (specifically WSDL). Other systems interact with the Web service in a manner prescribed by its description using SOAP messages, typically conveyed using HTTP with an XML serialization in conjunction with other Web-related standards.

A more generic definition from looselycoupled.com is as follows:

> Automated resources accessed via the Internet. Web services are software-powered resources or functional components whose capabilities can be accessed at an internet URI. Standards-based web services use XML to interact with each other, which allows them to link up on demand using loose coupling.

Web services enable a service oriented architecture (SOA). An SOA is a system for linking resources on demand. In an SOA, resources are made available to other participants in the network as independent services that are accessed in a standardized way. This provides for more flexible loose coupling of resources than in traditional systems architectures.

Nonetheless, the automation that occurs today with web services is primarily through the APIs that are exposed. Therefore, the users of the web services are still required to have a lot of knowledge about the parameters exchanged through the APIs. With this limitation, the vision of true loose coupling of resources and machine interpretable use of web services has not yet been delivered on. The integration of Semantic Web techniques with web services aims at resolving these current problems.

10.5.3 Web Services Standards

Figure 10.2 shows the web services stack in terms of the standards that are defined.

In general, protocol stacks have multiple layers, each sitting on top of another and assuming availability of information packets from the underlying layer and transferring the information to the upper layer. This stack is no different. Additionally, there are open issues related to web services that are transversal across layers. The web services stack is therefore divided into two parts above the TCP/IP layer: (1) web services stack showing availability of matured protocols at each layer of the stack and (2) business issues, which are applicable to the web services stack as a whole.

The three key pillars with matured protocols above the TCP/IP layer are SOAP, WSDL and UDDI.

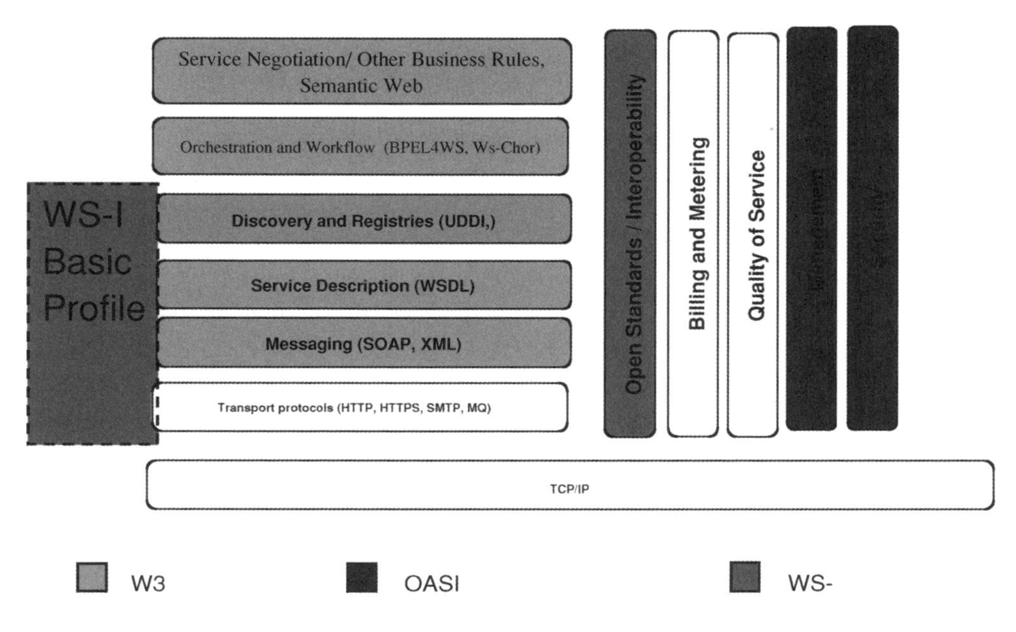

Figure 10.2 The web services stack

Standards-based web services communicate by sending messages via SOAP, a protocol standard defined by the W3C community. Based on XML, SOAP defines an envelope format and various rules for describing its contents. Seen (with WSDL and UDDI) as one of the three foundation standards of web services, it is the preferred protocol for exchanging web services, but by no means the only one; proponents of REST say that it adds unnecessary complexity.

WSDL (Web Services Description Language) is the standard format for describing a web service. Expressed in XML, a WSDL definition describes how to access a web service and what operations it will perform.

UDDI (Universal Description, Discovery and Integration protocol) is a directory model for web services. UDDI is a specification for maintaining standardized directories of information about web services, recording their capabilities, location and requirements in a universally recognized format.

Web Service Orchestration (WSO) promises the ability to integrate and assemble individual web services into standards-based business processes. WSO has become a key enabler for an enterprise's SOA and a critical layer in the web service technology stack. With WSO, these loosely coupled business processes will be designed, integrated, executed and managed similar to the way proprietary enterprise application integration (EAI) and business process management (BPM) tools operate today.

WSO is the path over which application server vendors tackle the BPM market. Today, major application server vendors Microsoft, IBM, BEA, Oracle and Sun are integrating WSO into their integrated application server offerings. Application developers will be using WSO-based tools to build applications which tie human activities, enterprise applications and business partners into an end-to-end business process based on SOA.

Web services management standardization is another area of activity at OASIS, focused on the problem of being able to manage the risks of uncontrolled deployment of web services. The WS-Distributed Management (DM) group is working to define management using web services as well as management of web services. Management includes areas such as transport, QoS and SLAs, integration, error handling, message routing etc.

10.5.4 Web Services Industry Adoption

Adoption of middleware architectures such as CORBA by enterprise IT organizations was aimed at application integration independent of platform or programming language. The premise was that the adoption of this middleware technology would reduce costs and deployment time. The premise was never achieved simply with the availability of these middleware technologies. The adoption of XML and web services as an integration layer above middleware technologies by IT organizations is aimed at significantly reducing the integration and deployment costs while making these applications Internet ready out of the box. To date, the largest successful deployment of this technology has been in this area. Hence, traditional platform vendors such as IBM, Sun, Microsoft, Bea Systems and others provide web services platforms for use by enterprise IT organizations. These web services platforms are being used with enterprises in a number of vertical market segments to integrate their IT systems. Examples of vertical markets include telecommunications, healthcare, financial systems, automotive industry, manufacturing, and government. Typically it is anticipated that reductions in IT costs can be anywhere from 50% to 60% of the cost of development and reductions in time to market can be expected to be in the 30–40% range. In addition, many of these enterprises are using web services across organizational boundaries as a way of tightly integrating their partners and their IT systems into their business processes.

Simultaneously, the late 1990s dotcom boom resulted in the creation of a large number of portals aimed at meeting the needs of the consumer market. The initial assumptions behind this were that the web sites and portals would be the new revenue opportunities. Today we find that the number of hits to a portal is no longer a meaningful representation of potential revenue opportunities. Instead the revenue opportunities lie with the actual services being offered through the portal. The emphasis has shifted towards XML-based content management support with portal server products and service interoperability. This enables ease of introduction of new services into existing portals and sharing of existing services between different portal environments. As a result, we find that the portal providers are looking at ways to move up the value chain by providing differentiated services by integrating web services. In addition, pure play portal vendors are looking to expand their offerings into customized BPM solutions and other areas. Corporate portals are also leveraging web services to enable them to deliver a suite of unique customized product offerings and to push relevant content directly at their customers and partners.

The increased use of web services in enterprise applications and the proliferation of content as web services lend themselves to the use of a mediator [44] to facilitate interactions between consumers of the web services and web service providers on the Internet. Such a mediator would provide services to enable the discovery, access to and associated billing for such services. Mediation represents a key opportunity for traditional telecommunications service providers to leverage their customer base in providing a new set of services and associated revenues and transforming their business from a data transport pipe to a value-added enabler of collaboration. A number of telecommunications and wireless providers are experimenting with service offerings aimed at providing a mediation service and associated platform that brings together consumers and third-party providers. Similar approaches are also being attempted in the electronic marketplaces where e-commerce providers and electronic auctioneers are attempting to bring together consumers and retailers through the use of web services based platforms. In this application, web services enable easy integration of third-party products for resale to consumers.

10.5.5 The Benefits of Semantics in Web Services?

The previous sections offer a high-level explanation of what web services are and their salient features. This section introduces the value of having semantics associated with web services.

Web services offer a level of interoperability that is strictly at the syntactic layer. By using the standardization offered by the WS-I Basic Profile that leverages SOAP 1.1, WSDL 1.1, UDDI 2.0 [45] etc. the messages sent between two nodes can be parsed, but prior knowledge of the interpretation of the message needs to exist. Semantic web services stand to improve on this capability to provide for a more seamless and simple computing environment.

However, at the end of the day, web services by themselves only offer a low-level language with which to communicate various aspects of messages such as requests, responses and message exchange patterns, as well as the descriptions of such information. Standards such as SOAP, WSDL and UDDI are a low-level set of standards that enable the exchange of these types of artefacts.

Bringing this type of message exchange to a level where it becomes meaningful to industrial applications requires some semantic understanding and interoperability as well. Standards such as DAML-S [46] and, more recently, OWL-S [47] are coming around to help develop such a layer, in conjunction with existing web services standards.

Some of the benefits of OWL-S towards enabling Semantic Web services include:

- A profile that describes what the service does (ServiceProfile), which is a high-level description that provides a publication and a discovery framework for clients and service providers. This facilitates the automated discovery of specific web services based on constraints defined in the requirements. The ServiceProfile has been designed to take into account the design of registries, and examples of services being deployed on a UDDI registry already exist today.
- A process model that describes how the service works in terms of IOPEs (input, output, preconditions and effects). This is an abstract description of a service. Processes are of three types: Atomic, Composite and Simple.
- A grounding which describes how to access and interact with actual services, and requires compliance with both OWL-S and WSDL. It is easy to extend WSDL bindings with OWL-S. This covers protocols, message formats, URIs and ports etc.

There is a definite advantage in being able to define orchestrated web services using process-related constructs in OWL-S. These include:

- Inputs and outputs: Parameters and its subclasses Input and ConditionalOutput are used to describe these artefacts

- Preconditions and effects: preconditions define the constraints on a service that need to be confirmed before an invocation to the service can be made. Effects are the resulting changes that will take place around the service to entities in virtual proximity to the service
- ConditionalOutput and ConditionalEffect: allow a number of conditions to be associated with the output and effect, respectively
- Conditions: OWL-S does not mandate a language to express conditioning statements, but it is understood that conditions are deeply ingrained in the effective use of Semantic Web services.

The types of processes include AtomicProcess, SimpleProcess and CompositeProcess. The process upper ontology defines the following set of controls: Sequence, Split, Split & Join, Choice, Unordered, Condition, If-Then-Else, Iterate, Repeat-While and Repeat-Until. Using these controls, OWL-S allows complex long running processes to be composed from other web service calls to be described. This is an obvious advantage over WSDL, which has a more highly grained focus and describes only individual web methods.

The unknown from a standards perspective is how OWL-S, for example, will interact with some of the upcoming specifications in the business process or in the orchestration space, such as the OASIS BPEL deliverables.

The above-mentioned artefacts are important in allowing for web services to be used in an industrial strength environment. The irony is that while Semantic Web technologies have been inadvertently labelled as a highly academic field, and web services on the other hand have evolved from a very low-level, pragmatic requirement for interoperability in a cross-platform, cross-language world, it is now the emergence of Semantic Web services technologies such as OWL-S that has the potential to significantly improve the business case for using web services.

OWL-S is also an effective means to define the model for WSDL and UDDI to work together, over and above the existing material around WSDL and UDDI that has evolved from the UDDI TC within the web services community [48].

Semantic Web services are moving forward at the standardization level with discussions around adding semantic data and inferencing capabilities available from within UDDI in an optional manner that is not imposing on users of UDDI who want to stay with just the core UDDI functionality as currently defined in the specifications. On a UDDI standardization perspective, the current versions are UDDI V2 [49], which is available in several versions from UDDI vendors, and UDDI V3 [50], which is more recent and is starting to see adoption at the UDDI vendor level.

10.5.6 Specific Application Field Adoption

We have seen some interesting examples of the adoption of web services, and have also seen several areas where the Semantic Web world is going in terms of adoption and application areas. In this section, we are going to take a closer look at the potential for a converged environment where web services and the Semantic Web technologies, coming together under the ‘Semantic Web Services’ banner, will be used.

Some of the potential for Semantic Web Services is in the enhancement of search services. Several search applications today are being offered over the web as web services, and these systems rely heavily on either proprietary or third-party metadata management tools. By utilizing standards-based Semantic Web Services, these services will be able to be offered to a wider audience in a format that is standards based. A great example of Semantic Web based search is available on the W3C website [51], which shows the value of context-based search. Combined with the ability to interact with the syntactically standardized interfaces that web services standards provide, this will allow for enhanced search functionality to be delivered to consuming Internet applications.

In the area of multimedia applications, the world is moving rapidly towards the use of semantic metadata, as against individual companies in the industry using their own proprietary approaches to metadata management. The ability to quickly establish B2B relationships and leverage partner metadata due to the existence of Semantic Web Services standards is going to improve time to market and cost constraints. Another example in the multimedia space can be found in the W3C Ontology Working Group's Use Cases and Requirements [52] document.

Other areas where Semantic Web Services will make an impact include BPM, voice services, e-commerce etc. For example, being able to leverage an existing KM system as part of a business process that involves multiple partners, in being able to represent adequate knowledge related to an organization and in making decisions about the right partners and relationships to leverage in a particular process, are all going to be invaluable contributions that Semantic Web Services can be anticipated to provide in the near future.

10.6 Conclusions

As shown in the preceding sections, the Semantic Web has the potential to significantly change our daily life due to the hidden intelligence provided for accessing services and large volumes of information. It will have a much higher impact on e-work and e-commerce than the current web. However, there is a long way to go before we can transform the vision from an academic adventure into a technology mastered and deployed industry. We have given some of the most prominent examples in intranet community portals and KM, e-commerce and information retrieval, natural language processing and machine translation, and lastly healthcare.

From a business viewpoint, such technology may really provide benefit to society as a whole, first to the economy, allowing companies to interoperate better and quickly find new or better opportunities, and second, to citizens or customers because it will help them to support their daily work, leisure and interactions with organizations. As a consequence of this, interest in showing the value of knowledge-based technology is growing due to the increased maturity of the field and also to business pressure for the evaluation of the measurable costs and benefits of that technology. This is the major objective of the recently launched IST-NoE Knowledgeweb project [53].

However, the technology is still immature in some key areas:

- scale-up to the web;
- heterogeneous and distributed contexts on the web;
- dynamicity of all the application fields.

The real take-up of Semantic Web technology is only visible in a few areas such as semantic data integration in large companies and knowledge management. Other areas will follow as soon as proof of the concept is done and business models show clear benefits (ROI) and new market (even in niches) creation.

References

[1] K. Wiig, Knowledge management: where did it come from and where will it go? Journal of Expert Systems with Applications, 13(1), 1–14, 1997.
[2] J. Hibbard, Knowledge management—knowing what we know. Information Week, 20 October 1997.
[3] G. Petrash, Managing knowledge assets for value. In Proceedings of the Knowledge-Based Leadership Conference, Boston, MA, October 1996. Boston, MA: Linkage.
[4] R. Dieng-Kuntz, Capitalisation des connaissances via un web sémantique d'entreprise. In Management des connaissances en entreprise, ch. 12. Hermes Science Publications, March 2004.
[5] R. Dieng-Kuntz, N. Matta (eds), *Proceedings of ECAI 2004 Workshop on Knowledge Management and Organizational Memories*, Valencia, Spain, 23–24 August 2004. Kluwer, Dordrecht, 2004.
[6] I. Nonaka, N. Konno, The concept of "Ba": building a foundation for knowledge creation. California Management Review, 40(3), 40–54, 1998.
[7] Arisem, http://www.arisem.com.
[8] Mondeca, http://www.mondeca.com.
[9] On-To-Knowledge, http://www.ontoknowledge.com.
[10] Atos Origin, http://www.si.fr.atosorigin.com.
[11] *Le Monde Informatique*, 11 July 2003.
[12] Mondeca, http://www.mondeca.com.
[13] M. Bryan, A. Léger, Public Deliverables D21–D22, 21, OntoWeb, web site of the EC project IST-OntoWeb, http://www.ontoweb.org and SIG4, http://sig4.ago.fr.
[14] Chemdex, http://www.chemdex.com.
[15] The Alice Project, http://kmi.open.ac.uk/projects/alice/.
[16] SMART-EC, http://www.telecom.ntua.gr/smartec/.
[17] MKBEEM, Multilingual Knowledge Based European Electronic Marketplace, http://www.mkbeem.com/.
[18] J. Zyl, D. Corbett, A framework for comparing the use of a linguistic ontology in an application. In *Proceedings of ECAI 2000 Workshop on Applications of Ontologies and Problem-solving Methods*, Berlin, August 2000.
[19] N. Guarino, C. Masolo, G. Vetere, OntoSeek: using large linguistic ontologies for accessing on-line yellow pages and product catalogs. In *Proceedings of Workshop on Artificial Intelligence for Electronic Commerce*, Orlando, FL, July 1999, pp. 118–119. American Association for Artificial Intelligence, Menlo Park, CA, 1999.
[20] G. Wiederhold, Mediators in the architecture of future information systems, *Computer*, **25**(3), 38–49, 1992.
[21] C. Barillot, L. Amsaleg, F. Aubry, J.-P. Bazin, H. Benali, Y. Cointepas, I. Corouge, O. Dameron, M. Dojat, C. Garbay, B. Gibaud, P. Gros, S. Kinkingnéhun, G. Malandain, J. Matsumoto, D. Papadopoulos, M. Pélégrini, N. Richard, E. Simon, Neurobase: management of distributed knowledge and data bases in neuroimaging. In *Human Brain Mapping*, vol. 19, p. 726, Academic Press, New York, 2003.
[22] Medical Subject Headings, http://www.nlm.nih.gov/mesh/meshhome.html.
[23] A.R. Aronson, O. Bodenreider, H.F. Chang, S.M. Humphry, J.G. Mork, S.J. Nelson, T.C. Rindflesch, W.J. Wilbur, The NLM indexing initiative. In J.M. Overhage (ed.) *Proceedings of the 2000 Annual Symposium of the American Medical Informatics Association*, Los Angeles, CA, November 2000, pp. 17–21. Hanley & Belfus, Philadelphia, PA, 2000.
[24] S.J. Darmoni, J.P. Leroy, F. Baudic, M. Douyere, J. Piot, B. Thirion, CISMeF: a structured health resource guide. *Methods of Information in Medicine*, **39**(1), 30–35, 2000.
[25] S.L. Weibel, T. Koch, The Dublin Core Metadata Initiative. *D-Lib Magazine*, 2000. Available at http://www.dlib.org/dlib/december00/weibel/12weibel.html.
[26] IEEE 1484, IEEE Learning Technology Standards Committee (LTSC). Available at: http://ltsc.ieee.org/.
[27] G. Eysenbach, A metadata vocabulary for self- and third-party labeling of health web sites: Health Information Disclosure, Description, and Evaluation Language (HIDDEL). In S. Bakken (ed.) *Proceedings of the 2001 Annual Symposium of the American Medical Informatics Association*, Washington, DC, November 2001, pp. 169–173. Hanley & Belfus, Philadelphia, PA, 2000.
[28] L.F. Soualmia, C. Barry, S.J. Darmoni, Knowledge-based query expansion over a medical terminology oriented ontology. In M. Dojat, E. Keravnou, P. Barahona (eds) *Artificial Intelligence in Medicine. Proceedings of the 9th*

Conference on Artificial Intelligence in Medicine in Europe, AIME 2003, pp. 209–213. Springer-Verlag, Berlin, 2003.
[29] P. Zweigenbaum, R. Baud, A. Burgun, F. Namer, E. Jarrousse, N. Grabar, P. Ruch, F. Le Duff, E. Thirion, S. Darmoni, Towards a unified medical lexicon for French. In R. Baud, M. Fieschi, P. Le Beux, P. Ruch (eds), *Proceedings of the 18th International Congress of the European Federation of Medical Informatics, Medical Informatics Europe (MIE) 2003: The New Navigators: From Professionals to Patients*, Saint Malo, pp. 415–420. IOS Press, 2003.
[30] L.F. Soualmia, C. Barry, S.J. Darmoni, Knowledge-based query expansion over a medical terminology oriented ontology on the web. In P. Barahona (ed.) *Proceedings of the 9th Conference on Artificial Intelligence in Medicine in Europe*, Protaras, Cyprus, October, pp. 209–213. Springer-Verlag, Berlin, 2003.
[31] L.F. Soualmia, S.J. Darmoni, Combining knowledge-based methods to refine and expand queries in medicine. In H. Christiansen (ed.) *Proceedings of the 6th International Conference on Flexible Query Answering Systems*, Lyons, June 2004, pp. 243–255. Springer-Verlag, Berlin, 2004.
[32] A.L. Rector, Terminology and concept representation languages: where are we? Artificial Intelligence in Medicine, **15**(1), 1–4, 1999.
[33] C. Roussey, S. Calabretto, J.-M. Pinon, SyDoM: a multilingual information retrieval system for digital libraries. In A. Hubler, P. Linde, J.W.T. Smith (eds) *Proceedings of the 5th International ICCC/IFIP Conference on Electronic Publishing (ELPUB 2001)*, Canterbury, July 2001, pp. 150–164. IOS, 2001.
[34] E. Cordonnier, S. Croci, J.-F. Laurent, B. Gibaud, Interoperability and medical communication using 'patient envelope'-based secure messaging. In R. Baud, M. Fieschi, P. Le Beux, P. Ruch (eds) *Proceedings of the 18th International Congress of the European Federation of Medical Informatics, Medical Informatics Europe (MIE) 2003: The New Navigators: From Professionals to Patients*, Saint Malo, pp. 230–236. IOS Press, 2003.
[35] J. Charlet, E. Cordonnier, B. Gibaud (2002) Interopérabilité en médecine: quand le contenu interroge le contenant et l'organisation. *Revue Information, interaction, intelligence*, 2(2).
[36] L. Bloomfield, *Language*. Holt, Rinehart & Winston, New York, 1933.
[37] *IJCAI-97 Workshop WP24 on Ontologies and Multilingual NLP*, Nagoya, Japan, 23 August 1997.
[38] FraCaS Project (1998) Language analysis and understanding. In *Survey of the State of the Art in Human Language Technology*, ch 3.
[39] P. Vossen (ed.) *EuroWordNet: A Multilingual Database with Lexical Semantic Networks*. Kluwer, Dordrecht, 1998.
[40] K. Knight, I. Chancer, M. Haines, V. Hatzivassiloglou, E.H. Hovy, M. Iida, S.K. Luk, R.A. Whitney, K. Yamada, Filling knowledge gaps in a broad-coverage MT system. In *Proceedings of the 14th IJCAI Conference*, Montreal, 1995.
[41] E. Viegas, An overt semantics with a machine-guided approach for robust LKBs. In *Proceedings of SIGLEX99 Standardizing Lexical Resources* (part of ACL99), University of Maryland, 1999. Available at acl.ldc.upenn.edu/W/W99/W99-0509.pdf.
[42] K. Mahesh, S. Nirenburg, A situated ontology for practical NLP. In *Proceedings of the Workshop on Basic Ontological Issues in Knowledge Sharing, International Joint Conference on Artificial Intelligence (IJCAI '95)*, Montreal, Canada, 19–20 August 1995.
[43] W3C, WS Architecture, http://www.w3.org/TR/2004/NOTE-ws-arch-20040211/#whatis.
[44] G. Wiederhold, Mediators in the architecture of future information systems. *Computer*, **25**(3), 38–49, 1992.
[45] UDDI V2 Specifications, http://www.oasis-open.org/committees/uddi-spec/doc/tcspecs.htm#uddiv2.
[46] DAML-S Services coalition, DAML-S: web service description for the Semantic Web. In I. Horrocks, J.A. Hendler (eds) *Proceedings of the 1st International Semantic Web Conference (ISCW 2002)*, Sardinia, Italy, June 2002, pp. 348–363. Springer-Verlag, Berlin, 2002.
[47] OWL-S, http://www.daml.org/services/owl-s/1.0/owl-s.html.
[48] OWL-S and WSDL, Describing web services, http://www.daml.org/services/owl-s/1.0/owl-s-wsdl.html.
[49] Using WSDL in UDDI Registry, V2.0, http://www.oasis-open.org/committees/uddi-spec/doc/tn/uddi-spec-tc-tn-wsdl-v200-20031104.htm.
[50] UDDI V3 Specifications, http://www.oasis-open.org/committees/uddi-spec/doc/tcspecs.htm#uddiv3.
[51] W3C Semantic Search, http://www.w3.org/2002/05/tap/semsearch/.
[52] W3C, Web Ontology WG's Use Cases and Requirements Document, 2003, http://www.w3.org/TR/2003/WD-webont-req-20030203/#usecase-multimedia.
[53] IST-NoE KnowledgeWeb Project, http://knowledgeweb.semanticweb.org.

11

Multimedia Indexing and Retrieval Using Natural Language, Speech and Image Processing Methods

Harris Papageorgiou, Prokopis Prokopidis, Athanassios Protopapas and George Carayannis

11.1 Introduction

Throughout the chapter, we provide details on implementation issues of practical systems for efficient multimedia retrieval. Moreover, we exemplify algorithms and technologies by referring to practices and results of an EC-funded project called Combined IMage and WOrd Spotting (CIMWOS) [1], developed with the hope that it would be a powerful tool in facilitating common procedures for intelligent indexing and retrieval of audiovisual material. CIMWOS used a multifaceted approach for the location of important segments within multimedia material, employing state-of-the-art algorithms for text, speech and image processing in promoting reuse of audiovisual resources and reducing budgets of new productions.

In the following three sections, we focus on technologies specific to speech, text and image, respectively. These technologies incorporate efficient algorithms for processing and analysing relevant portions from various digital media and thus generating high-level semantic descriptors in the metadata space. After proposing an architecture for the integration of all results of processing, we present indicative evaluation results in the context of CIMWOS. Relevant Content-based Information Retrieval (CBIR) research prototypes and commercial systems are briefly presented in Section 11.7.

Multimedia Content and the Semantic Web Edited by Giorgos Stamou and Stefanos Kollias

11.2 Audio Content Analysis

Audio content analysis for multimedia indexing and retrieval refers to a number of audio features gathered from acoustic signals. These audio features combined with visual descriptors have effectively been exploited in scene content analysis, video segmentation, classification and summarization [2]. Audio features are extracted in the short-term frame level and the long-term clip level (usually called window). A frame is defined as a group of adjacent samples lasting about 10–40 ms, within which the audio signal is assumed to be stationary. Frame-level features are:

- *Time-domain features*: volume, zero crossing rate, pitch.
- *Frequency-domain features* (based on the fast fourier transform (FFT) of the samples in a frame): spectrum of an audio frame, the ratio of the energy in a frequency subband to the total energy (ERSB), the mel-frequency cepstrum coefficients (MFCC) widely used for speech recognition and speaker recognition.

Clip-level features capture the temporal variation of frame-level features on a longer time scale and include volume-based features like the VSTD-standard deviation of the volume over a clip, the ZSTD proposed by Liu et al. [3] and pitch-based features like the SPR, which is the percentage of frames in a clip that have similar pitch as the previous frames.

Audio tracks are further segmented and classified into a number of classes such as speech, silence, noise, music and environmental sound (recognition of shots, cries, explosions, door slams etc.). State-of-the-art continuous speech recognizers perform on the speech segments, generating speech transcriptions. Producing a transcript of what is being said, and determining who is speaking and for how long are all challenging problems, mainly due to the continuous nature of an audio stream: segments of diverse acoustic and linguistic nature exhibit a variety of problematic acoustic conditions, such as spontaneous speech (as opposed to read or planned speech), limited bandwidth (e.g. telephone interviews), and speech in the presence of noise, music and/or background speakers. Such adverse background conditions lead to significant degradation in the performance of speech recognition systems if appropriate countermeasures are not taken. Likewise, segmentation of the continuous speech stream into homogeneous sections (with respect to acoustic/background conditions, speaker and/or topic) poses serious problems. Successful segmentation, however, forms the basis for further adaptation and processing steps. Adaptations to the varied acoustic properties of the signal or to a particular speaker, and enhancements to the segmentation process, are generally acknowledged [4] as key research areas that will result in rendering indexing systems usable for actual deployment. This is reflected in the amount of effort and the number of projects dedicated to the advancement of the current state-of-the-art in these areas [5].

A speech processing subsystem (SPS) usually comprises a set of technology components responsible for a particular task, arranged in a pipeline. These tasks typically include speaker change detection (SCD), automatic speech recognition (ASR), speaker identification (SID) and speaker clustering (SC). To save processing time, SC and SID can be run in parallel to ASR. Input to the overall system is the audio stream, while the final output is a set of audio descriptors containing speech transcriptions and identified speakers. A typical SPS architecture is depicted in Figure 11.1, while each specific task is presented in detail in the following subsections.

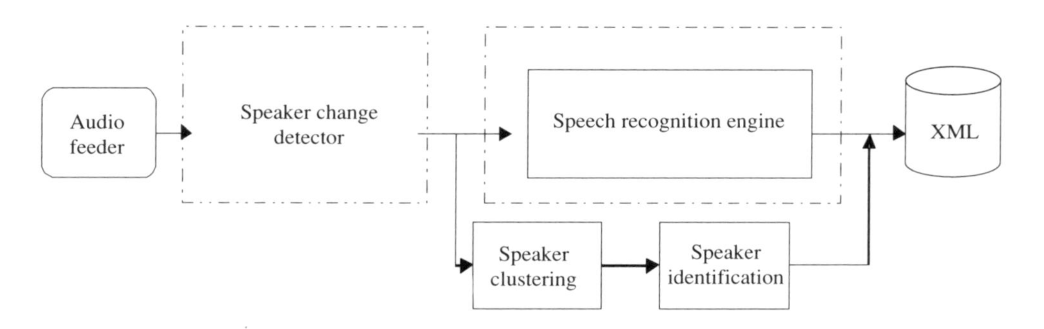

Figure 11.1 Speech processing subsystem

11.2.1 Speaker Change Detection

SCD aims at extracting only the speech portions of the incoming audio stream, and at partitioning those sections into homogeneous sections of speech. The technology used for SCD involves initially running the speech recognizer with a phone-level model (PLD), which is a set of broad phonetic classes of speech sounds like vowels, nasals, obstruents, as well as some non-speech sounds like music, laughter etc.

Using the information produced by the PLD, the SCD module determines, during runtime, likely boundaries of speaker turns based on non-speech portions (e.g. silence, music, background noise) in the signal. Parameters used in this decision are usually empirically determined, as they differ from language to language and even from domain to domain.

11.2.2 Speaker Clustering and Identification

The SC module is responsible for grouping predetermined chunks of audio (the SCD output described above) into a number of clusters. Resulting clusters ideally represent utterances produced by a single speaker. SID uses a predefined set of target models to identify the speaker of a segment or, if identification is not possible, to determine the speaker's gender. The set of target speakers typically comprises anchor speakers of TV stations as well as persons of public interest such as politicians. Current SID technology employs Gaussian mixture models (GMM), trained on corpora of speaker-annotated transcriptions. On the other hand, SC does not use any trained models, but rather a variant of the generalized likelihood ratio criterion at runtime. Evaluation of the CIMWOS SID module on predefined test sets produced 97% precision and 97% recall figures for English, 93% precision and 93% recall for French, and 92% precision and 89% recall for Greek.

11.2.3 Automatic Speech Recognition

In most state-of-the-art speech processing systems [6, 7], ASR is performed via a large vocabulary, speaker-independent, gender-independent, continuous speech recognizer. The recognizers typically use hidden markov models (HMM) with GMM for acoustic modelling (AM) and n-gram models estimated on large corpora for language modelling (LM).

The task for the ASR is to convert the information passed on to it by the SCD stage into time-tagged text. This is often done in a multi-stage approach by finding the most likely sequence of words, given the input audio stream. Each level of processing uses increasingly complex acoustic and language models, thereby keeping a balance between model complexity, search space size and quality of the results. The AMs used by different stages are HMM-based mixture models with various degrees and kinds of tying (sharing of model parameters). They form a representation of the acoustic parameters (acoustics, phonetics, environment, audio source, speaker and channel characteristics, recording equipment etc.). The LM is an n-gram model with different types of back-off and smoothing. It contains a representation of the words included in a vocabulary, and of what words and sequences of words are likely to occur together or follow each other. The recognizer's vocabulary consists of several thousand wordforms and their pronunciation models. Using wordforms as opposed to words means that plural and singular forms (or any kind of declinational or conjugational forms) of a word are considered separate entities in the vocabulary.

ASR in CIMWOS uses a multi-stage approach to provide for flexibility and best use of resources. During *decoding*, in a first step, a *fast-match* is performed using the simplest (and thus fastest) models. Then a *detailed-match* follows on the results of fast-match. Finally a *rescoring pass* is applied to the results of the second stage. This staged approach allows for a dramatical reduction in search space while at the same time recognizer performance and accuracy are kept high. Evaluation of the CIMWOS ASR module on predefined test sets for English and French broadcast news (BNs) yielded word error rates (WERs) of 29% and 27%, respectively. For Greek BNs a 33% WER is reported. Figure 11.2 shows a sample portion of an XML file automatically created by the speech processing subsystem in CIMWOS.

```
<?xml version="1.0" encoding="UTF-8" ?>
<SpeechAnnotation project="CIMWOS">
<Header type="SpeechRecognitionMetadata">
  <Media id="RTBF_20011012_1930_News" xml:lang="fr">
  ....
  <Passage id=" p77 " speaker="male 14" gender ="male"
mediaRelIncrTimePoint="131864" mediaIncrDuration="1420"
xml:lang="fr">
       <Word id="w3541" mediaRelIncrTimePoint="131864"
mediaIncrDuration="13" confidence="0.99">Dans</Word>
       <Word id="w3542" mediaRelIncrTimePoint="131878"
mediaIncrDuration="8" confidence="0.99">le</Word>
       <Word id="w3543" mediaRelIncrTimePoint="131887"
mediaIncrDuration="17" confidence="0.99">hall</Word>
       <Word id="w3544" mediaRelIncrTimePoint="131905"
mediaIncrDuration="10" confidence="0.99">de</Word>
       <Word id="w3545" mediaRelIncrTimePoint="131916"
mediaIncrDuration="48" confidence="0.99">départ</Word>
       <Word id="w3546" mediaRelIncrTimePoint="131965"
mediaIncrDuration="5" confidence="0.99">de</Word>
       <Word id="w3547" mediaRelIncrTimePoint="131971"
mediaIncrDuration="4" confidence="0.99">l'</Word>
       <Word id="w3548" mediaRelIncrTimePoint="131976"
mediaIncrDuration="108" confidence="0.99">aéroport</Word> ...
   </Passage>
  ....
  </Media>
</SpeechAnnotation>
```

Figure 11.2 Segment of speech processing generated metadata

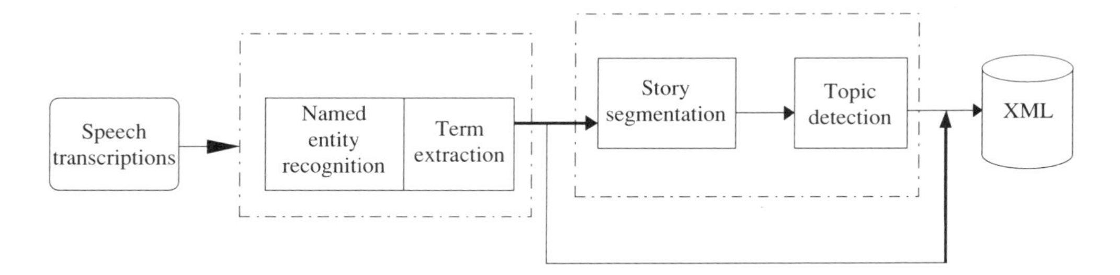

Figure 11.3 Text processing subsystem

11.3 Text Processing Subsystem

Following audio processing, text processing tools are applied on the textual data produced by the speech processing subsystem, in an attempt to enable a retrieval system with the ability to answer questions like what topic a passage is about, or which organizations are mentioned. Named entity detection (NED), term extraction (TE), story segmentation (SD) and topic detection (TD) are tasks that can be included in a text processing pipeline like the one depicted in Figure 11.3.

11.3.1 Named Entity Detection and Term Extraction

The task of a NED module in the context of a multimedia retrieval system is to identify named locations, persons and organizations, as well as dates and numbers in the speech transcriptions automatically produced by the SPS. NED might involve an initial preprocessing phase where sentence boundary identification, and part-of-speech tagging are performed on textual input. State-of-the-art NED implementations involve lookup modules that match lists of NEs and trigger words against the text, hand-crafted and automatically generated pattern grammars, maximum entropy modelling, HMM models, decision-tree techniques, SVM classifiers etc. [8].

As an example, the NED module for the Greek language in CIMWOS [9] is a combination of list-based matching and parsing with a grammar of rules compiled into finite-state transducers. In a collection of TV news broadcasts, this module identified 320 NEs, compared to the 555 identified by human annotators using a suitable annotation tool (Figure 11.4). There were 235 correct guesses, 85 false positives and 320 missed instances. Although promising precision scores were obtained for persons and locations, (Figure 11.5), lower recall in all NE types is due to:

- greater out-of-vocabulary (OOV) rate in the ASR for Greek;
- missing proper names in the vocabulary of the ASR engine;
- the diversity of domains found in broadcasts, leading to a large number of names and NED indicators that were not incorporated in the NED module's resources.

The TE task involves the identification of single or multi-word indicative keywords (index terms) in the output of the speech processing system. Systems for automatic term extraction using both linguistic and statistical modelling are reported in the literature [10]. Linguistic

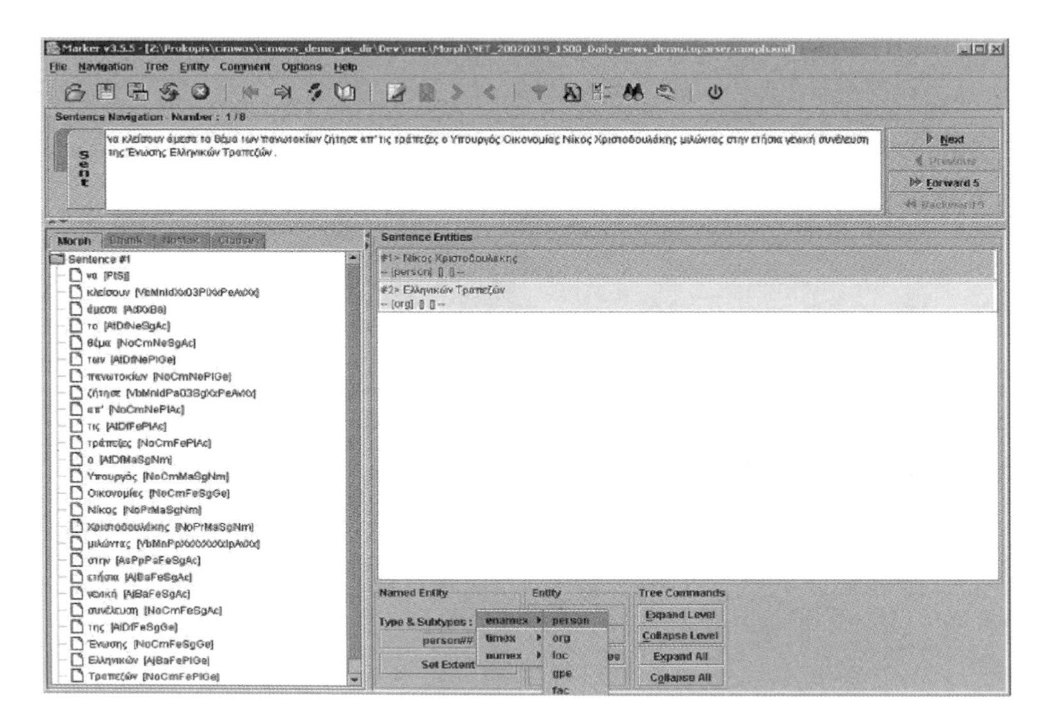

Figure 11.4 Named entity annotation tool

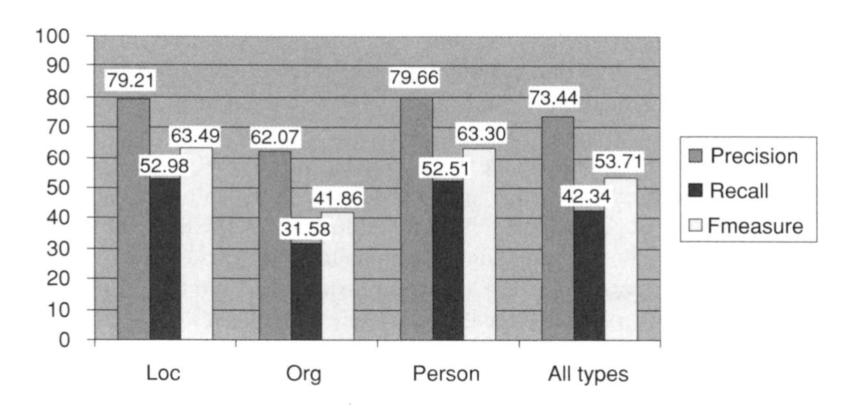

Figure 11.5 Performance of CIMWOS NED module for each NE type

processing is usually performed through an augmented term grammar, the results of which are statistically filtered using frequency-based scores.

The term extractor in CIMWOS was used to automatically identify terms in a testing corpus of manually annotated transcriptions, scoring a 60.28% recall and a 34.80% precision. To obtain higher precision scores, stricter statistical filtering needs to be applied. Nevertheless, a relaxed statistical filtering like the one currently used produces higher recall measures, a factor which plays a more important role in an information retrieval context.

11.3.2 Story Detection and Topic Classification

Story detection and topic classification modules often employ the same set of models, trained on an annotated corpus of stories and their manually associated topics. The basis of these technologies is a generative, mixture-based hidden Markov model that includes one state per topic, as well as one state modelling general language; that is, words not specific to any topic. Each state models a distribution of words given the particular topic. After emitting a single word, the model re-enters the beginning state and the next word is generated. At the end of a story the final state is reached. Detection is performed running the resulting models on a sliding window of fixed size, thereby noting the change in topic-specific words as the window moves on. The result of this phase is a set of 'stable regions' in which topics change only slightly, or not at all. Building on the story boundaries, sections of text are classified according to a set of topic models.

In the Informedia project at Carnegie Mellon University [11] each video document is partitioned into story units based on text, image and audio metadata jointly. Silence periods are identified and subsequently aligned with the nearest shot break indicating the boundary of a story unit. Text markers such as punctuation are taken into consideration in case closed caption is available. In the MAESTRO system [12] developed by SRI, topic boundaries are detected based on prosodic information extracted from the speech waveforms (pause and pitch patterns) combined with word usage statistics. The MITRE Broadcast News Navigator (BNN) incorporates a story segmentation module based on finite state machines [13].

11.4 Image Processing Subsystem

Image processing and understanding is an important field of research where close ties between academic and commercial communities have been established. In the multimedia content analysis framework, a typical image processing subsystem (IPS) like the one presented here consists of modules performing video segmentation and keyframe extraction, face detection and face identification, object identification and video text detection and recognition.

11.4.1 Video Segmentation and Keyframe Extraction

The segmentation of video sequences is a prerequisite for a variety of image processing applications. Video streams consist of many individual images, called frames, generally considered as the smallest unit to be concerned with when segmenting a video. An uninterrupted video stream generated by one camera is called a shot (for example, a camera following an aeroplane, or a fixed camera focusing on an anchorperson), while a shot cut is the point at which shots change within a video sequence. The video segmentation task involves detecting shot cuts and thus partitioning raw material into shots. Under the assumption that frames within a shot have a high degree of similarity for each shot, a few representative frames are selected, referred to as keyframes. Each keyframe represents a part of the shot called subshot. The subdivision of a shot into subshots occurs when, for example, there is an abrupt camera movement or zoom operation, or when the content of the scene is highly dynamic so that a single keyframe no longer suffices to describe the content of the entire shot. Keyframes contain most of the static information present in a shot, so that subsequent modules like face detection and object identification can focus on keyframes only without scanning the whole video sequence.

In order to detect shot cuts and select keyframes, different methods have been developed for measuring the differences between consecutive frames and applying adaptive thresholding on motion and texture cues. The performance of the CIMWOS video segmentation module was evaluated on some of the news broadcasts available in the project's collection. A recall rate of 99% and a precision of 97% were observed in the case of abrupt shot cuts. For smooth shot cuts, performance was lower and highly dependent on the video content, yet still comparable to the state of the art.

The performance of the keyframe selection is much harder to quantify, as there is no objective groundtruth available: what constitutes an appropriate keyframe is a matter of semantic interpretation by humans, and cannot be resolved on the basis of low-level image cues. Nevertheless, judging from comments by test users, the CIMWOS keyframe extraction module was indeed able to reduce the number of redundant keyframes while keeping important information available.

11.4.2 Face Detection and Face Identification

Given an arbitrary image, the goal of face detection is to determine whether or not there are any faces in the image, and, if present, to return the face dimensions in the keyframe. Face detection is a challenging task, since several factors influence the appearance of the face in the image. These include identity, pose (frontal, half-profile, profile), presence or absence of facial features such as beards, moustaches and glasses, facial expression, occlusion and imaging conditions. Face detection has been, and still is, a very active research area within the computer vision community. It is one of the few attempts to recognize from a set of images a class of objects for which there is a great deal of within-class variability. It is also one of the few classes of objects for which this variability has been captured using large training sets of images.

Recent experiments by neuroscientists and psychologists show that face identification is a dedicated process in the human brain. This may have encouraged the view that artificial face identification systems should also be face-specific. Automatic face identification is a challenging task and has recently received significant attention. Rapidly expanding research in the field is based on recent developments in technologies such as neural networks, wavelet analysis and machine vision. Face identification has a large potential for commercialization in applications involving authentication, security system access, and advanced video surveillance. Nevertheless, in spite of expanding research, results are far from perfect, especially in uncontrolled environments, because of lighting conditions and variations, different facial expressions, background changes and occlusion problems. One of the most important challenges in face identification is to distinguish between intra-personal variation (in the appearance of a single individual due to different facial expressions, lighting etc.) and extra-personal variation (between different individuals).

In CIMWOS, the face detection and face identification modules associate detected faces occurring in video streams with names [14] (Figure 11.6). Both modules are based on support

Figure 11.6 Face detection results

vector machine models trained on an extensive database of facial images with a large variation in pose and lighting conditions. Additionally, a semantic base consisting of important persons that should be identified has been constructed. During identification, images extracted from keyframes are compared to each model. At the decision stage, the scores resulting from the comparison are used either to identify a face or to reject it as 'unknown'.

In order to be able to compare the CIMWOS FD performance with other systems' results reported in the literature, the module was evaluated against the MIT/CMU dataset [15], where groundtruth information for frontal face detection is readily available. A correct detection rate of 90% (with 1 false detection in 10. 000 windows) was obtained. A qualitative evaluation on the CIMWOS news broadcasts database shows that similar results are obtained in that context, if the extremely small or out-of-plane rotated faces are ignored.

11.4.3 Video Text Detection and Recognition

Text recognition in images and video aims at integrating advanced optical character recognition (OCR) technologies and text-based search, and is now recognized as a key component in the development of advanced video and image annotation and retrieval systems. Unlike low-level image features (such as colour, texture or shape), text usually conveys direct semantic information on the content of the video, like a player's or speaker's name, location and date of an event etc. However, text characters contained in video are usually of low resolution, of any colour or greyscale value (not always white), embedded in complex background. Experiments show that applying conventional OCR technology directly to video text leads to a poor recognition rate. Therefore, efficient location and segmentation of text characters from the background is necessary to fill the gap between images or video documents and the input of a standard OCR system.

Figure 11.7 Video text detection and recognition

In CIMWOS, the video text detection and recognition module [16, 17] is based on a statistical framework using state-of-the-art machine learning tools and image processing methods (Figure 11.7). It consists of four modules:

- Text detection, aiming at roughly and quickly finding image blocks that may contain a single line of text characters.
- Text verification, using a support vector machine model, to remove false alarms.
- Text segmentation, attempting to extract pixels from text images belonging to characters with the assumption that they have the same colour/greyscale value. This method uses a Markov random field model and an expectation maximization algorithm for optimization.
- Finally, all hypotheses produced by the segmentation algorithm are processed by the OCR engine. A string selection is made based on a confidence value, computed on the basis of character recognition reliability and a bigram language model.

The CIMWOS video text detector was evaluated on 40 minutes of video consisting of 247 text strings, 2899 characters and 548 words. Characters were correctly extracted and recognized in 94% of all cases, while a 92% recognition rate was reached as far as words are concerned. Precision showed that more than 95% of extracted characters correspond to the characters in the groundtruth. In the 2899 characters, 7% of characters were scene text characters and 93% were captions.

11.4.4 Object Identification

The object identification (OI) problem can be defined as the task of identifying a non-deformable man-made object in a 'recognition view', given knowledge accumulated from a set of previously seen 'learning views'. Until recently, visual object recognition was limited to planar (flat) objects, seen from an unknown viewpoint. The methods typically computed numerical geometric invariants from combinations of easily extractable image points and lines. These invariants were used as a model of the planar object for subsequent recognition. Some systems went beyond planar objects, but imposed limits on the possible viewpoints or on the nature of the object. Probably the only approach capable of dealing with general 3D objects and viewpoints is the 'appearance-based' one, which nevertheless requires a very large amount of example views and has fundamental problems in dealing with cluttered recognition views and occlusion [18]. A system capable of dealing with general 3D objects, from general viewing directions, is yet to be proposed.

Since 1998, a small number of systems have emerged that seem to have the potential for reaching the general goal. These are all based on the concept of 'region', defined as a small, closed area on the object's surface. In CIMWOS, the object surface is decomposed into a large number of regions automatically extracted from the images [19]. These regions are extracted from several example views (or frames of a movie) and both their spatial and temporal relationships are observed and incorporated in a model. This model can be gradually learned as new example views (or video streams) are acquired. The power of such a proposal consists fundamentally in two points: first, the regions themselves embed many small, local pieces of the object at the pixel level, and can reliably be put in correspondence along the example views. Even in the case of occlusion or clutter in the recognition view, a subset of the object's regions will still be present. The second strong point is that the model captures the spatio-temporal order inherent in the set of individual regions, and requires it to be present in the recognition view. In this way the model can reliably and quickly accumulate evidence about the identity of the object in the recognition view, even with only a small number of recognized regions.

Thanks to the good degree of viewpoint invariance of the regions, and to the strong model and learning approach developed, the object recognition module in CIMWOS copes with 3D objects of general shape, requiring only a limited number of learning views [20]. Moreover, it can recognize objects from a wide range of previously unseen viewpoints, in possibly cluttered, partially occluded, views. Based on experiments conducted on the project's database (searching for 13 objects in 325 keyframes, on average, for each object), a 83.7% precision and a 96.4% recall were obtained.

11.5 Integration Architecture

Due to the prevalence of XML as a medium for data exchange between applications, all processing output in the three modalities (audio, image and text) can converge to a textual XML metadata annotation document following standard MPEG-7 descriptors. These annotations can be further processed, merged, synchronized and loaded into the multimedia database. A merging component amalgamates various XML annotations to create a self-contained object compliant with a database scheme.

The CIWMOS architecture follows an N-tier scenario by integrating a data services layer (storage and retrieval of metadata), a business services layer incorporating all remote multimedia processors (audio, video and text intelligent engines) and a user services layer which basically includes the user interface (UI) and web access forms. Web services (SOAP) could be used as the integration protocol to access heterogeneous and loosely-coupled distributed technologies. The CIMWOS architecture is depicted in Figure 11.8.

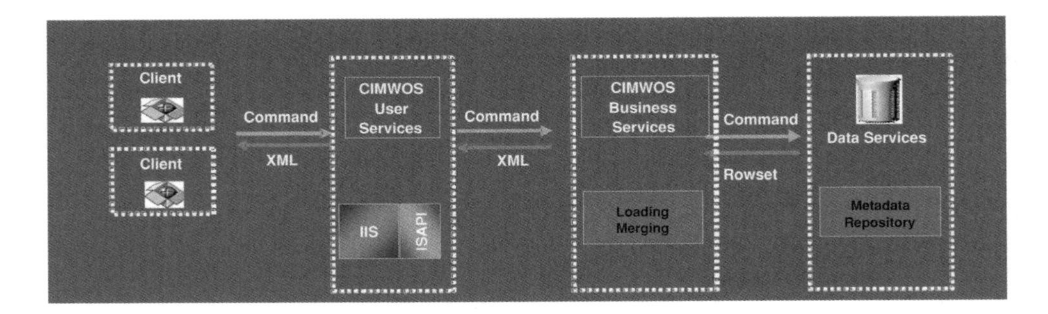

Figure 11.8 The CIMWOS architecture

11.5.1 Indexing, Search and Retrieval

The basic approach to indexing and retrieval is to apply speech, image and text technologies to automatically produce textual metadata and then to use information retrieval techniques on these metadata. The CIMWOS retrieval engine [21] is based on a weighted Boolean model. After query submission, the retrieval engine is invoked with the set of criteria combined flexibly with standard Boolean operators. A matching operation computes the similarity between the query and each passage to determine which ones contain the given set of combined query terms. The calculated similarity measure takes into consideration three different factors: metadata-level weights based on the overall precision of each multimedia processor, value-level statistical confidence measures produced by the engines and tf*idf scores for all textual elements. Finally, passages are ranked based on the result of similarity computation.

The basic retrieval unit is the passage (Figure 11.9), which has the role of a document in a traditional system. Passages are defined on the speech transcriptions—that is, on the

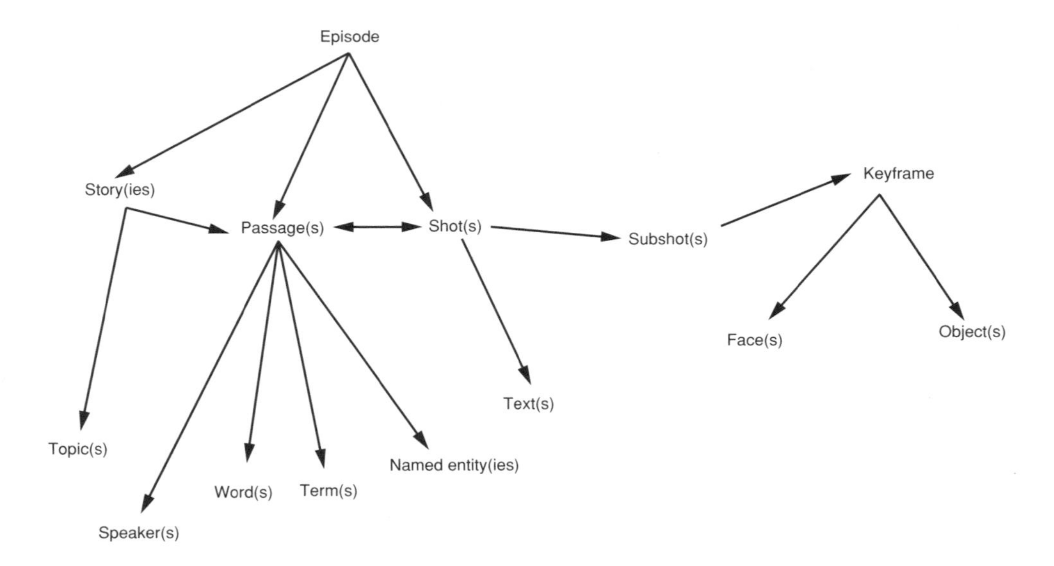

Figure 11.9 CIMWOS indexing schema

post-processed output of the ASR—therefore they correspond to some level of audio segmentation. Other processing streams, that is, text and video, will typically result in segments (stories and shots, respectively) not aligned with passages. This lack of correspondence with the original multimedia object has implications for both indexing and retrieval. The solution given must be based on the eventual use of the material, therefore the needs of the users should be taken into account. Referring to the example of a user searching for Putin co-occurring with Clinton on a particular crisis topic, it is necessary for a search and retrieval system to define the temporal extent over which the existence of both annotations counts as co-occurrence, meaning that the relevant time segment is to be retrieved. For news broadcasts, a news story is the appropriate conceptual unit. However, in the merged annotations one has access to processed outputs from the various streams, which do not map onto the desired unit in a well-defined manner. Thus the selection of passages is made to satisfy both practical constraints (availability, typical duration) and use requirements (to contain all relevant information and nothing irrelevant, as much as possible).

The passage is indexed on a set of textual features: words, terms, named entities, speakers and topics. Each passage is linked to one or multiple shots, and each shot is indexed on another set of textual features: faces, objects and video text. By linking shots to passages, each passage is assigned a broader set of features to be used for retrieval. Passages are represented as sets of features and retrieval is based on computed similarity in the feature space.

A video clip can take a long time to be transferred, e.g. from the digital video library to the user. In addition, it takes a long time for someone to determine whether a clip meets his or her needs. Returning half an hour of video when only one minute is relevant is much worse than returning a complete book when only one chapter is needed. Since the time to scan a video cannot be dramatically shorter than the real time of the video, it is important to give users only the material they need.

Multimedia information retrieval systems typically provide various visualized summaries of video content where the user can preview a specific segment before downloading it. In Informedia, information summaries can be displayed at varying detail, both visually and textually. Text summaries are displayed for each news story through topics and titles. Visual summaries are given through thumbnails, filmstrips and dynamic video skims [11]. In CIMWOS, while skimming a passage, the end user can view its associated metadata, the transcribed speech and results of all processing components, as well as a representative sequence of thumbnails. These thumbnails can act as indicative cues on how relevant the segment content is to the user query. Each thumbnail corresponds to a keyframe that is recognized and extracted by the video segmentation module. Results can also include more pieces of bibliographical information, such as the title of the video to which the passage belongs and the duration of the passage. Finally, the user can play the passage via streaming.

11.6 Evaluation

Apart from the evaluation of each module contributing metadata annotations, assessing the overall system response to user queries is of course of extreme importance for a multimedia information retrieval (MIR) system. Suggestions and guidelines on evaluation procedures have been put under testing in the framework of international contests. For example, the TREC 2002 Video track [22] included three sessions focusing on shot boundary detection, extraction of segments containing semantic features like ‘People’, ‘Indoors Location’ etc., and search for particular topics from a handcrafted list.

In what follows, a brief overview of the CIMWOS retrieval exercise is given. The CIMWOS system and its overall performance have been evaluated in two distinct phases. During the first phase, we tested video search and retrieval of passages relevant to a particular topic. We repeated the retrieval task during the second phase, this time on video material for which boundaries of relevant stories for each topic had been previously identified by human annotators. Our testing material consisted of Greek news broadcasts, produced by state and private TV networks between March 2002 and July 2003. For the first phase we used 15 videos (henceforth, Collection A) of a total duration of approximately 18 hours, captured in BETA SP and transcribed in MPEG-2 format. During the second phase we used 15 news broadcasts (henceforth, Collection B) that amounted to approximately 17 hours of video, captured via standard PC TV cards in MPEG-2 format.

The overall retrieval of the system was tested in both 'interactive' and 'manual' search modes. In interactive search, users are familiar with the test collection, and have full access to multiple interim search results. On the other hand, in the case of manual search, users with knowledge of the query interface but no direct or indirect knowledge of the search test set or search results are given the chance to translate each topic to what they believe to be the most effective query for the system being tested. In each phase, a user familiar with the test collection was responsible for generating a set of topics that s/he judged to be of interest, opting for topics that were represented in more than one news broadcast of the collection. After the list was finalized, the user had to manually locate all video sequences relevant to each topic, thus allowing developers to associate start and end timecodes with each topic.

During the search task, users were given the chance to translate each topic of the list to what they believed to be the most effective queries for the system being tested, using combinations of criteria like Terms, Named Entities and/or tree Text. Users formed five queries, on average, for each topic. Using the HTML interface (Figure 11.10), they were able to browse the results in order to intuitively assess their overall satisfaction with the database results. Nevertheless, results were also saved in an XML format as in Figure 11.11, where the topic 'earthquake prediction' has been translated into a query comprising the word *πρόγνωση* (*prediction*) and a name (the name of a Greek geology professor).

In their initial reactions when searching for a topic in the videos of Collection A, users reported that passages returned by the system were too short and fragmentary. The explanation was that passages were based on automatic segmentation by the ASR engine, based on hints from speaker changes. A different approach was taken while testing on Collection B, in order to produce more intuitive results for end users.

A user was again responsible for manually identifying relevant segments in each of the videos of the collection. These segments corresponded to the stories of each news broadcast. Following that, we aligned the ASR transcription with the manually identified stories. Each video segment was also assigned a description, derived from a list of topics created during the training phase of the Greek ASR module. Thus, while it was possible to have passages consisting of just one shot during the first phase, this counterintuitive phenomenon was absent in the semi-automatically annotated material of Collection B.

The XML files corresponding to each user's queries were assembled and tested against groundtruth data. In Table 11.1, we show results for both video collections. We also tested the system's response when we did not take into account results that scored less than 60% in the ranking system. This filtering had a negative effect on the system's recall in the case of Collection B. Nevertheless, it significantly increased the system's precision with data from both sets.

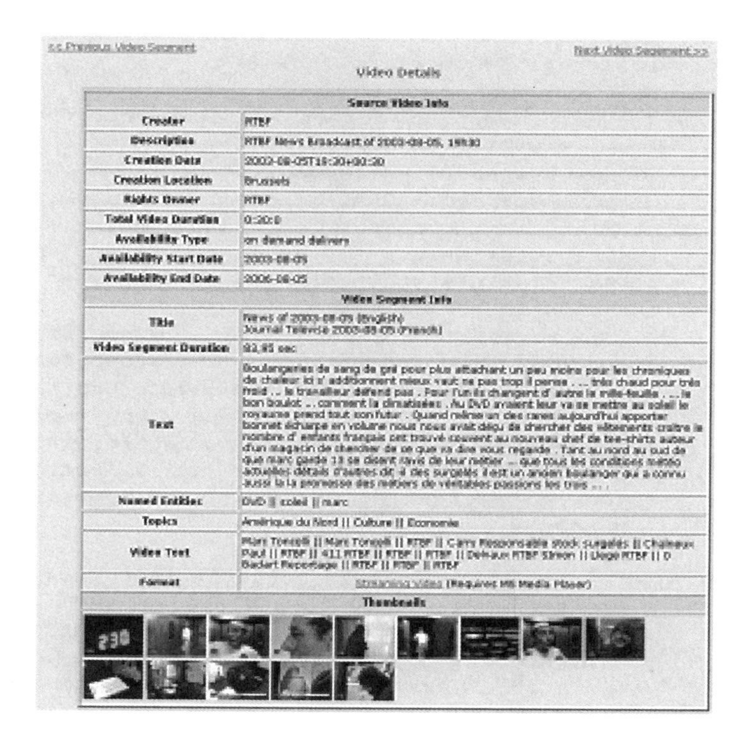

Figure 11.10 CIMWOS HTML interface

11.7 Related Systems

Many multimedia indexing and retrieval systems have been developed over the past decade. Most of these systems aim at dynamically processing, indexing, organizing and archiving digital content [23, 24]. A brief overview of the most relevant systems is given below.

News on demand (NOD) systems have made their appearance in the past decade, integrating image, speech and language technology, and taking advantage of cross-modal analysis in order to monitor news from TV, radio and text sources and provide personalized newscasts [13].

MIT's Photobook [25] was an early example that allowed retrieval from textures, shapes and human faces. Chabot [26], developed by UCBerkeley, is another early system. It provided a combination of text-based and colour-based access to a collection of photographs. The system, which was later renamed Cypress, and is now known as the DWR picture retrieval system [27], has been incorporated into the Berkeley Digital Library project. Berkeley has continued similar research with its Blobworld software [28], which has one of the more sophisticated region segmentation functionalities. Noteworthy are recent developments at Berkeley, which allow finding of specific object classes like horses and naked people. VisualSEEK [29] was the first of a series of systems developed at Columbia University. A strong point was that queries were quite flexible in the spatial layout specifications that they could accept. WebSEEK [30] was a further development, aimed at facilitating queries over the web. Emphasis is on colour, but in another prototype called VideoQ [31], motion was added. Also, relative layouts of regions with specific colours or textures can be spotted. This allows for some primitive

```xml
<?xml version="1.0" encoding="UTF-8"?>
 <results>
   <criteria    Language="Greek"    Text="πρόγνωση"
      Named_Entities="Παπαζάχος"
      Topics="γεωλογία,μετεωρολογία"
      Logical_Operator="OR" />
   <passage          score="100%"          number="1"
      media_title="News        of        2002-05-24"
      passage_id="p44"startpoint_in_sec="529,05"
      duration_in_sec="15,76">
     <content>αναστάτωση  αλλά  και  ανησυχίες
        προκαλεί στην κοινή γνώμη δημοσιοποίηση
        έκθεσης του Βασίλη του καθηγητή Βασίλη
        Παπαζάχου  για  μεσοπρόθεσμη  πρόγνωση
        ισχυρός σεισμός σε τέσσερις περιοχές της
        Ελλάδας ο κύριος Παπαζάχος με δηλώσεις
        του  ο  υπερασπίστηκε  την  επιστημονική
        πέραση.</content>
     <speaker>male 22</speaker>
     <media            id="NET_20020524_1800_News"
        creation_date="2002-05-24T18:00+02:00"
        creator="NET"            rights_owner="NET"
        availability_type="ondemand delivery">
       <mediatitle>News      of      2002-05-24
          </mediatitle>
       <source duration="1:74:52,48" />
     </media>
     <thumbnails number="1" />
   </passage>
   ...
  </results>
```

Figure 11.11 Query results in XML

forms of object detection, e.g. finding the American flag on the basis of the alternation of red and white stripes, with a blue region at the top left corner. The MARS project at the University of Illinois [32] put emphasis on control by the user, and introduced extensive relevance feedback mechanisms. Feature weights were set dynamically. An example of European CBIR

Table 11.1 Retrieval results on Greek video collections

	Precision	Recall	F-measure
Collection A	34.75	57.75	43.39
Collection A + 60% filter	45.78	53.52	49.35
Collection B	44.78	50.24	47.36
Collection B + 60% filter	64.96	37.07	47.20

technology is the Surfimage system from INRIA [33]. It has a similar philosophy to the MARS system, using multiple types of image features which can be combined in different ways, and offering sophisticated relevance feedback. The RSIA system co-developed by ETH Zurich and DLR [34] was dedicated to search in large satellite image databases. It supports remote access through the web and is suitable for texture features, learned as the user indicates positive and negative examples of the kind of regions searched for. The NeTra system [35] uses similar cues to many of the foregoing systems, i.e. colour, texture, shape and spatial location information, but as in Blobworld a more sophisticated segmentation function is included. Synapse [36] is a rather different kind of system, in that it bases its similarity judgements on whole image matching. Query By Image Content [37] is a component of the IBM DB2 database product, performing image retrieval based on example images and selected colour and texture patterns.

Convera [38] provides scene change detection and can send audio tracks to any SAPI-compliant recognition engine. It also provides a closed captioning/teletext extractor and an SMPTE timecode [39] reading module, but these are more 'decoders' than true 'recognition engines', as this information is already available in digital form. On its part, Virage [40] claims to be more comprehensive, including face recognition, on-screen text recognition, speech and speaker, as well as intelligent segmentation.

11.8 Conclusion

Multimedia information retrieval systems address real needs of multimedia producers, archivists and others who need to monitor and/or index audiovisual content. By utilizing vast amounts of information accumulated in audio and video, and by employing state-of-the-art speech, image and text processing technologies, a MIR system can become an invaluable assistant in efficient reuse of multimedia resources, and in reducing new production expenses.

These systems open new possibilities and provide enabling technology for novel application areas and services. Use of markup languages (MPEG-7, MPEG-21) for annotation of audiovisual content can promote standardization and unification of access procedures to large archives and multiple repositories, thus assisting the rise of multimedia digital libraries, and improving the working conditions and productivity of people involved in media and television, video, news broadcasting, show business, advertisement and any organization that produces, markets and/or broadcasts video and audio programmes.

Acknowledgements

The CIMWOS project was supported by shared-cost research and technological development contract IST-1999-12203 with the European Commission. Project work was done at the Institute for Language & Speech Processing, Athens, Greece (coordination, integration, text processing, search and retrieval), Katholieke Universiteit Leuven, Belgium (video segmentation, face detection), Eidgenoessische Technische Hochschule Zurich, Switzerland (scene and object recognition), Sail Labs Technology AG, Vienna, Austria (audio processing, speech recognition, text processing), Canal+ Belgique, Brussels, Belgium (user requirements and validation) and Institut Dalle Molle d'Intelligence Artificielle Perceptive, Martigny, Switzerland (speech recognition, video text detection and recognition, face recognition).

References

[1] H. Papageorgiou, A. Protopapas, CIMWOS: a multimedia, multimodal and multilingual indexing and retrieval system. In E. Izquierdo (ed.) *Digital Media Processing for Multimedia Interactive Services, Proceedings of the 4th European Workshop on Image Analysis for Multimedia Interactive Services*, Queen Mary, University of London, 9–11 April 2003, pp. 563–568. World Scientific, Singapore, 2003.

[2] Y. Wang, Z. Liu, J.-C. Huang, Multimedia content analysis: using both audio and visual clues. *IEEE Signal Processing Magazine*, **17**(6), 12–36, 2000.

[3] Z. Liu, Y. Wang, T. Chen, Audio feature extraction and analysis for scene segmentation and classification. *Journal of VLSI Signal Processing Systems for Signal, Image, and Video Technology*, **20**(1), 61–79, 1998.

[4] J.-L. Gauvain, R. De Mori, L. Lamel, Advances in large vocabulary speech recognition. *Computer Speech and Language*, **16**(1), 1–3, 2002.

[5] J.-L. Gauvain, L. Lamel, G. Adda, Audio partitioning and transcription for broadcast data indexation. *Multimedia Tools and Applications*, **14**, 187–200, 2001.

[6] F. Kubala, J. Davenport, H. Jin, D. Liu, T. Leek, S. Matsoukas, D. Miller, L. Nguyen, F. Richardson, R. Schwartz, J. Makhoul, The 1997 BBN BYBLOS System applied to Broadcast News Transcription. In *Proceedings of the DARPA Broadcast News Transcription and Understanding Workshop*, Lansdowne, VA, 8–11 February 1998. National Institute of Standards and Technology, Gaithersburg, MD, 1998. Available at: http://www.nist.gov/speech/publications/darpa98/pdf/eng110.pdf.

[7] J.L. Gauvain, L. Lamel, G. Adda, Transcribing broadcast news for audio and video indexing. *Communications of the ACM*, **43**(2), 64–70, 2000.

[8] E.F. Tjong Kim Sang, Introduction to the CoNLL-2002 shared task: language-independent named entity recognition. In D. Roth, A. van den Bosch (eds) *CoNLL-2002, Sixth Conference on Natural Language Learning*, Taipei, Taiwan, 31 August –1 September 2002. Available at: http://cnts.uia.ac.be/conll2002/proceedings.html.

[9] I. Demiros, S. Boutsis, V. Giouli, M. Liakata, H. Papageorgiou, S. Piperidis, Named entity recognition in Greek texts. In *Proceedings of the Second International Conference on Language Resources and Evaluation, LREC2000*, Athens, Greece, 31 May–2 June 2000, pp. 1223–1228. ELRA, 2000.

[10] C. Jacquemin, E. Tzoukermann, NLP for term variant extraction: synergy between morphology, lexicon and syntax. In T. Strzalkowski (ed.) *Natural Language Information Retrieval*. Kluwer, Dordrecht, 1999.

[11] A. Hauptmann, M. Smith, Text, speech, and vision for video segmentation: the Informedia Project. In *AAAI Fall 1995 Symposium on Computational Models for Integrating Language and Vision*. American Association for Artificial Intelligence, 1995.

[12] SRI Maestro Team, MAESTRO: conductor of multimedia analysis technologies. *Communications of the ACM*, **43**(2), 57–63, 2000.

[13] A. Merlino, D. Morey, M. Maybury, Broadcast news navigation using story segments. In *Proceedings of the Fifth ACM International Conference on Multimedia*, Seattle, WA, 9–13 November 1997, pp. 381–391. ACM, Seattle, WA, 1997.

[14] F. Cardinaux, C. Sanderson, S. Marcel, Comparison of MLP and GMM classifiers for face verification on XM2VTS. In J. Kittler, M.S. Nixon (eds) *Proceedings of the 4th International Conference on Audio- and Video-Based Biometric Person Authentication (AVBPA)*, Guildford, 9–11 June 2003. Springer-Verlag, Berlin, 2003.

[15] The Combined MIT/CMU Test Set with Ground Truth for Frontal Face Detection, http://vasc.ri.cmu.edu/idb/html/face/frontal_images/index.html.

[16] D. Chen, H. Bourlard, J.-Ph. Thiran, Text identification in complex background using SVM. In *Proceedings of the International Conference on Computer Vision and Pattern Recognition*, Kauai, Hawaii, December 2001, vol. 2, pp. 621–626. IEEE, 2001.

[17] J.M. Odobez, D. Chen, Robust video text segmentation and recognition with multiple hypotheses. In *Proceedings of the International Conference on Image Processing, ICIP 2002*. IEEE, 2002.

[18] V. Ferrari, T. Tuytelaars, L. Van Gool, Wide-baseline muliple-view correspondences. In *Proceedings of the International Conference on Computer Vision and Pattern Recognition ICIP 2003*, Madison, WI, June 2003. IEEE, 2003.

[19] T. Tuytelaars, A. Zaatri, L. Van Gool, H. Van Brussel, Automatic object recognition as part of an integrated supervisory control system. In *Proceedings of IEEE Conference on Robotics and Automation, ICRA00*, San Francisco, CA, April 2000, pp. 3707–3712. IEEE, 2000.

[20] V. Ferrari, T. Tuytelaars, L. Van Gool, Real-time affine region tracking and coplanar grouping. In *Proceedings of the International Conference on Computer Vision and Pattern Recognition*, Kauai, Hawaii, December 2001, vol. 2. IEEE, 2001.

[21] H. Papageorgiou, A. Protopapas, T. Netousek, Retrieving video segments based on combined text, speech and image processing. In *Proceedings of the Broadcast Engineering Conference*, April 2003, pp. 177–182. National Association of Broadcasters, 2003.

[22] A.F. Smeaton, Paul Over, The TREC-2002 Video track report. In E.M. Voorhees, L.P. Buckland (eds) *The Eleventh TExt Retrieval Conference (TREC 2002)*, Gaithersburg, MD, 19–22 November 2002. NIST Special Publication 500-251. National Institute of Standards and Technology, Gaithersburg, MD, 2002. Available at: http://trec.nist.gov/pubs/trec11/t11_proceedings.html.

[23] N. Dimitrova, H.-J. Zhang, B. Shahraray, I. Sezan, T. Huang, A. Zakhor, Applications of video-content analysis and retrieval. *IEEE Multimedia*, **9**(3), 42–55, 2002.

[24] *IEEE Multimedia*, Special issue: Content-Based Multimedia Indexing and Retrieval, **9**(2), 18–60, 2002.

[25] A. Pentland, R.W. Picard, S. Sclaroff, Photobook: content-based manipulation of image databases. *International Journal of Computer Vision*, **18**(3), 233–254, 1996.

[26] V.E. Ogle, M. Stonebraker, Chabot: retrieval from a relational database of images. *IEEE Computer*, **28**(9), 164–190, 1995.

[27] DWR photo database, http://elib.cs.berkeley.edu/photos/dwr/about.html.

[28] Blobworld, http://elib.cs.berkeley.edu/photos/blobworld.

[29] J.R. Smith, S.-F. Chang, Tools and techniques for color image retrieval. In I.K. Sethi, R.C. Jain (eds) *Storage and Retrieval for Still Image and Video Databases IV*, SPIE vol. 2670, pp. 426–437. SPIE, 1996.

[30] J.R. Smith and S.-F. Chang, Image and video search engine for the world wide web. In I.K. Sethi, R.C. Jain (eds) *Storage and Retrieval for Image and Video Databases V*, SPIE vol. 3022, pp. 84–95. SPIE, 1997.

[31] S.-F. Chang, W. Chen, H.J. Meng, H. Sundaram, D. Zhong, VideoQ: an automated content-based video search system using visual cues. In *Proceedings of International Multimedia Conference*, Seattle, WA, November 1997, pp. 313–324. Addison-Wesley, Reading, MA, 1997.

[32] T.S. Huang, S. Mehrotra, K. Ramchandran, Multimedia Analysis and Retrieval System (MARS) project. In P.B. Heidorn, B. Sandore (eds) *Proceedings of the 33rd Annual Clinic on Library Application of Data Processing: Digital Image Access and Retrieval*, Urbana, IL, March 1996, pp. 100–117. University of Illinois, 1997.

[33] C. Nastar, M. Mitschke, C. Meilhac, N. Boujemaa, Surfimage: a flexible content-based image retrieval system. In *Proceedings of International Multimedia Conference*, Bristol, 12–16 September 1998, pp. 339–344. Addison-Wesley, Harlow, 1998.

[34] K. Seidel, M. Datcu, G. Schwarz, L. Van Gool, Advanced remote sensing information system at ETH Zurich and DLR/DFD Oberpfaffenhofen. In IEEE International Geoscience and Remote Sensing Symposium IGARSS '99, Hamburg, Germany, pp. 2363–2365. IEEE, 1999.

[35] W.Y. Ma, B.S. Manjunath, NeTra: a toolbox for navigating large image databases. In *Proceedings of International Conference on Image Processing*, Santa Barbara, CA, October 1997, vol. 1, pp 568–571. IEEE, 1997.

[36] R. Manmatha, S. Ravela, A Syntactic characterization of appearance and its application to image retrieval. In B.E. Rogowitz, T.N. Pappas (eds) *Human Vision and Electronic Imaging II*, SPIE vol. 3016, pp. 484–495. SPIE, 1997.

[37] Query By Image Content, http://wwwqbic.almaden.ibm.com.

[38] Convera, http://www.convera.com.

[39] Society of Motion Picture and Television Engineers, http://www.smpte.org.

[40] Virage, http://www.virage.com.

12

Knowledge-Based Multimedia Content Indexing and Retrieval

Manolis Wallace, Yannis Avrithis, Giorgos Stamou and Stefanos Kollias

12.1 Introduction

By the end of the last century the question was not whether digital archives are technically and economically viable, but rather how digital archives would be *efficient* and *informative*. In this framework, different scientific fields such as, on the one hand, development of database management systems, and, on the other hand, processing and analysis of multimedia data, as well as artificial and computational intelligence methods, have observed a close cooperation with each other during the past few years. The attempt has been to develop intelligent and efficient human–computer interaction systems, enabling the user to access vast amounts of heterogeneous information, stored in different sites and archives.

It became clear among the research community dealing with content-based audiovisual data retrieval and new emerging related standards such as MPEG-21 that the results to be obtained from this process would be ineffective, unless major focus were given to the semantic information level, defining what most users desire to retrieve. It now seems that the extraction of semantic information from audiovisual-related data is tractable, taking into account the nature of useful queries that users may issue and the context determined by user profiles [1].

Additionally, projects and related activities supported under the R&D programmes of the European Commission have made significant contributions to developing:

- new models, methods, technologies and systems for creating, processing, managing, networking, accessing and exploiting digital content, including audiovisual content;
- new technological and business models for representing information, knowledge and know-how;
- applications-oriented research, focusing on publishing, audiovisual, culture, education and training, as well as generic research in language and content technologies for all applications areas.

Multimedia Content and the Semantic Web Edited by Giorgos Stamou and Stefanos Kollias

In this chapter a novel platform is proposed that intends to exploit the aforementioned ideas in order to offer user friendly, highly informative access to distributed audiovisual archives. This platform is an approach towards realizing the full potential of globally distributed systems that achieve information access and use. Of primary importance is the approach's contribution to the Semantic Web [2]. The fundamental prerequisite of the Semantic Web is 'making content machine-understandable'; this happens when content is bound to some formal description of itself, usually referred to as 'metadata'. Adding 'semantics to content' in the framework of this system is achieved through algorithmic, intelligent content analysis and learning processes.

The system closely follows the developments of MPEG-7 [3–5] and MPEG-21 [6] standardization activities, and successfully convolves technologies in the fields of computational intelligence, statistics, database technology, image/video processing, audiovisual descriptions and user interfaces, to build, validate and demonstrate a novel intermediate agent between users and audiovisual archives. The overall objective of the system is to be a stand-alone, distributed information system that offers enhanced search and retrieval capabilities to users interacting with digital audiovisual archives [7]. The outcome contributes towards making access to multimedia information, which is met in all aspects of everyday life, more effective and more efficient by providing a user-friendly environment.

The chapter is organized as follows. In Section12.2 we provide the general architecture of the proposed system. We continue in Section 12.3 by presenting the proprietary and standard data models and structures utilized for the representation and storage of knowledge, multimedia document information and profiles. Section 12.4 presents the multimedia indexing algorithms and tools used in offline mode, while Section 12.5 focuses on the operation of the system during the query. Section 12.6 is devoted to the personalization actions of the system. Finally, Section 12.7 provides experimental results from the actual application of the proposed system and Section 12.8 discusses the directions towards which this system will be extended through its successor R&D projects.

12.2 General Architecture

The general architecture is provided in Figure 12.1, where all modules and subsystems are depicted, but the flow of information between modules is not shown for clarity. More detailed information on the utilized data models and on the operation of the subsystems for the two main modes of system operation, i.e. *update mode* and *query mode*, are provided in the following sections. The system has the following features:

- Adopts the general features and descriptions for access to multimedia information proposed by MPEG-7 and other standards such as emerging MPEG-21.
- Performs dynamic extraction of high-level semantic description of multimedia documents on the basis of the annotation that is contained in the audiovisual archives.
- Enables the issuing of queries at a high semantic level. This feature is essential for unifying user access to multiple heterogeneous audiovisual archives with different structure and description detail.
- Generates, updates and manages users' profile metadata that specify their preferences against the audiovisual content.
- Employs the above users' metadata structures for filtering the information returned in response to their queries so that it better fits user preferences and priorities.

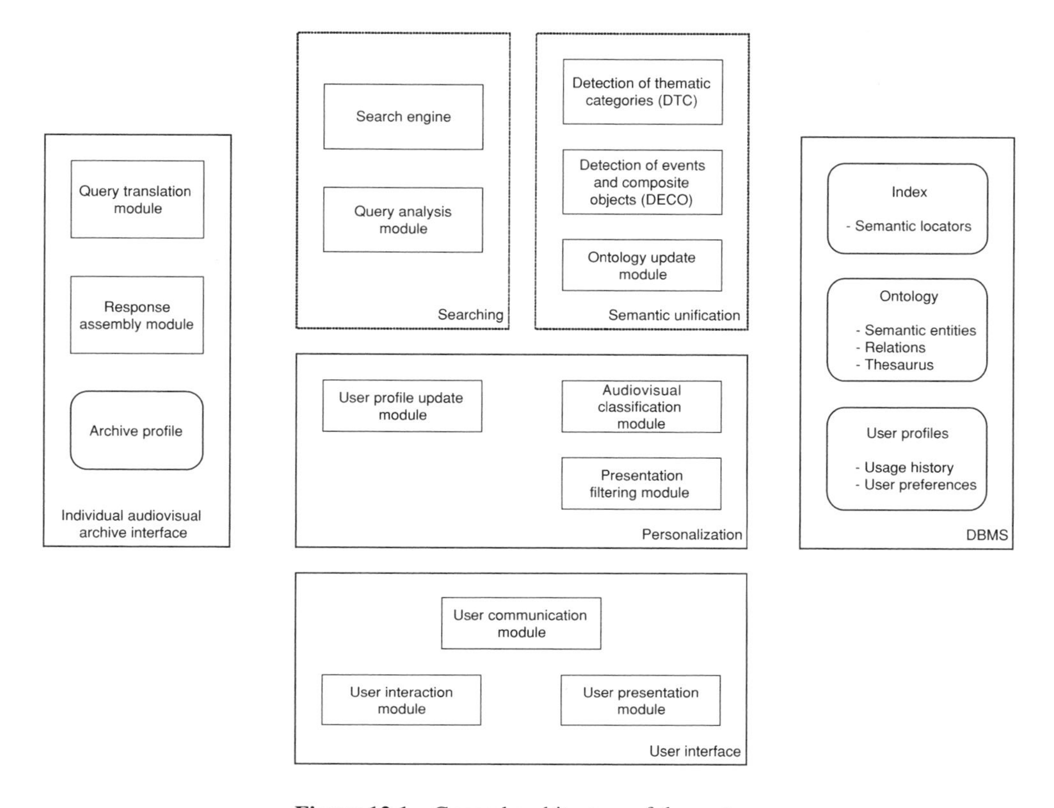

Figure 12.1 General architecture of the system

- Gives users the ability to define and redefine their initial profile.
- Is capable of communicating with existing audiovisual archives with already developed systems with proprietary (user and system) interfaces.
- User interfaces employ platform-independent tools targeting both the Internet and WWW and broadcast type of access routes.

Additionally, it is important that the system has the following features related to user query processing:

- *Response time*: internal intelligent modules may use semantic information available in the DBMS (calculated by *Detection of Thematic Categories* (DTC), *Detection of Events and Composite Objects* (DECO) and the *UserProfile Update Module*) to locate and rank multimedia documents very quickly, without querying individual audiovisual archives. In cases where audiovisual unit descriptions are required, query processing may be slower due to the large volume of information. In all cases it is important that the overall response time of the system is not too long as perceived by the end user.
- *Filtering*: when a user specifies a composite query, it is desirable that a semantic query interpretation is constructed and multimedia documents are filtered as much as possible

according to the semantic interpretation and the user profile, in order to avoid the overwhelming responses of most search engines.

- *Exact matching*: in the special cases where the user query is simple, e.g. a single keyword, the system must return all documents whose description contains the keyword; no information is lost this way.
- *Ranking*: in all cases retrieved documents must be ranked according to the user's preferences and their semantic relevance to the query, so that the most relevant documents are presented first.
- *Up-to-date information*: since the system is designed for handling a large number of individual audiovisual archives whose content may change frequently, the DBMS must be updated (either in batch updates or in updates on demand) to reflect the most recent archive content.

The description of the subsystems' functionality follows the distinction between the two main modes of operation. In *query mode*, the system is used to process user requests, and possibly translate and dispatch them to the archives, and assemble and present the respective responses. The main internal modules participating in this mode are the *query analysis*, *search engine*, *audiovisual classification* and *presentation filtering* modules.

An additional *update mode* of operation is also necessary for updating the content description data. The general scope of the update mode of operation is to adapt and enrich the DBMS used for the unified searching and filtering of audiovisual content. Its operation is based on the *semantic unification* and the *personalization* subsystems. The semantic unification subsystem is responsible for the construction and update of the *index* and the *ontology,* while the personalization subsystem updates the *user profiles*. In particular, a batch update procedure can be employed at regular intervals to perform DTC and DECO on available audiovisual units and update the database. Alternatively, an *update on demand* procedure can be employed whenever new audiovisual units are added to individual archives to keep the system synchronized at all times. Similar choices can be made for the operation of the user profile update module. The decision depends on speed, storage and network traffic performance considerations. The main internal modules participating in the update mode are *DTC*, *DECO*, *ontology update* and *user profile update*.

In the following we start by providing details on the utilized data structures and models, continue by describing the functionality of the objective subsystems operating in offline and online mode, where additional diagrams depict detailed flow of information between modules, and conclude with the presentation of the personalization methodologies.

12.3 The Data Models of the System

The system is aimed to operate as a mediator, providing to the end user unified access to diverse audiovisual archives. Therefore, the mapping of the archive content on a uniform data model is of crucial importance. The specification of the model itself is a challenging issue, as the model needs to be descriptive enough to adequately and meaningfully serve user queries, while at the same time being abstract and general enough to accommodate the mapping of the content of any audiovisual archive. In the following we provide an overview of such a data model, focused on the support for semantic information services.

12.3.1 The Ontology

The ontology of the system comprises a set of description schemes (DSs) for the definition of all semantic entities and their relations. It actually contains all knowledge of semantic information used in the system. The ontology, among other actions, allows:

- storing in a structured manner the description of semantic entities and their relations that experts have defined to be useful for indexing and retrieval purposes;
- forming complex concepts and events by the combination of simple ones through a set of previously specified relations;
- expanding the user query by looking for synonyms or related concepts to those contained in the semantic part of the query.

To make the previous actions possible, three types of information are included in the ontology:

- *Semantic entities*: entities such as thematic categories, objects, events, concepts, agents and semantic places and times are contemplated in the encyclopedia. All normative MPEG-7 semantic DSs are supported for semantic entities whereas the treatment of thematic categories as semantic entities is unique to the system, so additional description schemes are specified.
- *Semantic relations*: the relations linking related concepts as well as the relations between simple entities to allow forming more complex ones are specified. All normative MPEG-7 semantic DSs are supported for semantic relations.
- *A thesaurus*: it contains simple views of the complete ontology. Among other uses, it provides a simple way to associate the words present in the semantic part of a query to other concepts in the encyclopedia. For every pair of semantic entities (SEs) in the ontology, a small number of semantic relations are considered in the generation of the thesaurus views; these relations assess the type and level of relationship between these entities. This notion of a thesaurus is unique to this system and, therefore, additional DSs are specified.

An initial ontology is manually constructed possibly for a limited application domain or specific multimedia document categories. That is, an initial set of *semantic entities* is created and structured using the experts' assessment and the supported *semantic relations*. The *thesaurus* is then automatically created.

A similar process is followed in the ontology update mode, in which the knowledge experts specify new semantic entities and semantic entity relations to be included in the encyclopedia. This is especially relevant when the content of the audiovisual archives is dramatically altered or extended.

Semantic entities

The semantic entities in the ontology are mostly media abstract notions in the MPEG-7 sense. Media abstraction refers to having a single semantic description of an entity (e.g. a soccer player) and generalizing it to multiple instances of multimedia content (e.g. a soccer player from any picture or video). As previously mentioned, entities such as thematic categories, objects, events, concepts, agents and semantic places and times are contemplated in the ontology, and

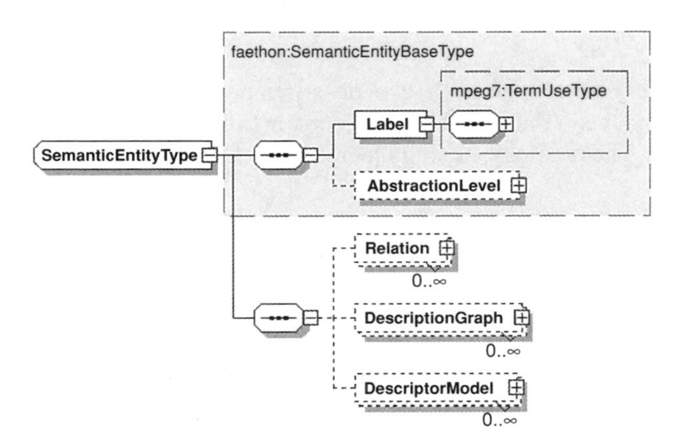

Figure 12.2 The SemanticEntityType. Textual descriptions are supported through Labels and composite objects are described through the DescriptionGraphs

all normative MPEG-7 semantic DSs are supported for SEs. Semantic entities are structured in the SemanticEntities DS (Figure 12.2).

An SE is composed of:

- a textual annotation including synonyms and different language representations;
- zero to several Description Graphs (DGs) relating the various SEs that are associated to the SE and linked by their valued SRs. DGs provide a means for 'semantic definition' of the entity.

Very simple SEs do not require a DG but are only described by their corresponding terms (e.g. ball).

Semantic relations

As previously mentioned, all normative MPEG-7 semantic DSs are supported for semantic relations. Additionally, the definition of custom, system proprietary semantic relations is supported via the utilization of the generic SemanticRelationType (Figure 12.3). In order to make the storage of the relations more compact and to allow for some elementary ontological consistency checks, the relations' mathematical properties, such as symmetry, transitivity and type of transitivity, reflexivity etc., are also stored in the ontology. Using them the ontology update tools can automatically expand the contained knowledge by adding implied and inferred semantic relations between SEs, and validate new information proposed by the knowledge experts against that already existing in the ontology.

An important novelty of the ontology utilized in this system, when compared to the current trend in the field of ontological representations, is the inherent support of degrees in all semantic relations (Figure 12.4). Fuzziness in the association between concepts provides greater descriptive power, which in turn allows for more 'semantically meaningful' analysis of documents, user queries and user profiles. As a simple example of the contribution of this fuzziness in the descriptive of the resulting ontology, consider the concepts of car, wheel and

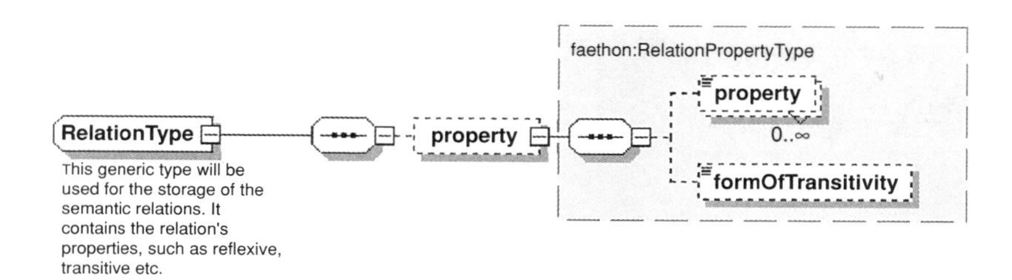

Figure 12.3 The RelationType. In the generic type only the relation properties are required

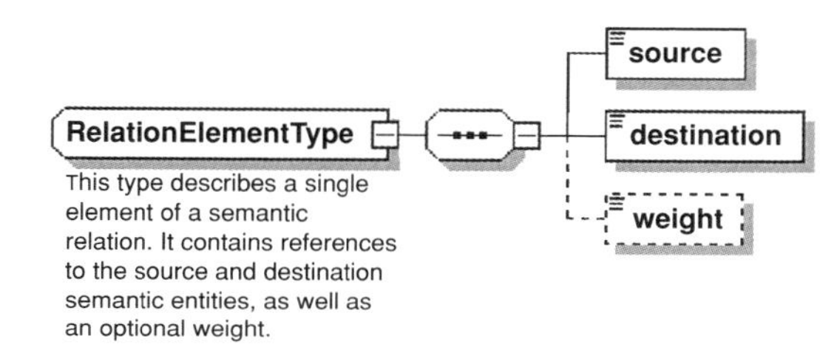

Figure 12.4 The RelationElementType. All semantic relations between pairs of semantic entities are described using this type. The weight, although optional, is of major importance for this system

rubber. Although the inclusion between them is obvious, it is clear that the inclusion of wheel in car semantically holds more than that of rubber in car. Using degrees of membership other than one, and applying a sub-idempotent transitive closure, i.e. an operation that will allow the relation of car and rubber to be smaller than both the relation between car and wheel and the relation between wheel and rubber, we acquire a much more meaningful representation.

Thesaurus

The description of the relationships among the various SEs in the ontology using a single semantic relation forms a graph structure. The graph nodes correspond to all SEs in the encyclopedia, whereas graph links represent the type and degree of relationship between the connected nodes. Combining all the relations in one graph, in order to acquire a complete view of the available knowledge, results in a very complex graph that cannot really provide an easy to use view of an application domain.

Simplified views of this complex graph structure are represented in the ontology by means of the thesaurus. Since the concept of thesaurus is unique to this system, additional DSs are specified; in order to make the representation more flexible, the same structure as the one used for the distinct semantic relations of the ontology is also utilized for the representation of the ontological views in the thesaurus (Figure 12.5).

All the information in the thesaurus can be obtained by tracking the links among different SEs through the SemanticEntities and SemanticRelations DSs contained in the ontology, based on the thesaurus generation rules, specifying which relations to utilize for each view, and in which way to combine them, as well as the relation properties. Actually, this is the way in

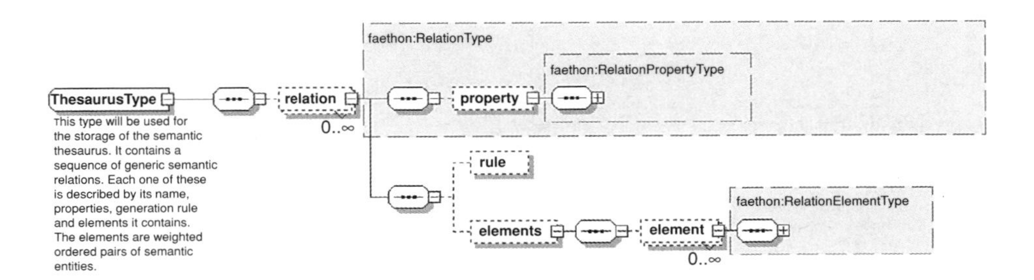

Figure 12.5 The ThesaurusType. Initially only the rule and property fields are filled. In ontology update mode, the ontology update module uses them as input, together with the distinct semantic relations, in order to automatically generate the semantic views stored in the relation element fields

which the thesaurus is initially created and periodically updated in the encyclopedia update module.

The usefulness of the thesaurus is that it codes the information in a simpler, task-oriented manner, allowing faster access.

12.3.2 Index

The index is the heart of the unified access to various archives, as it collects the results of the document analysis taking place in the framework of the semantic unification process. Specifically, the index contains sets of document locators (links) for each SE (thematic category, object, event, concept, agent, semantic place or semantic time) in the ontology (Figure 12.6). Links from thematic categories to multimedia documents are obtained by the DTC procedure (mapping the abstract notions to which each multimedia document is estimated to be related to the thematic categories in the ontology) while links to the remaining SEs are provided by the DECO procedure (mapping the simple and composite objects and events detected in each multimedia document to their corresponding semantic entities in the ontology).

The index is used by the search engine for fast and uniform retrieval of documents related to the semantic entities specified in, or implied by, the query and the user profile. Document locators associated to index entities may link to complete audiovisual documents, objects, still images or other video decomposition units that may be contained in the audiovisual databases (Figure 12.7).

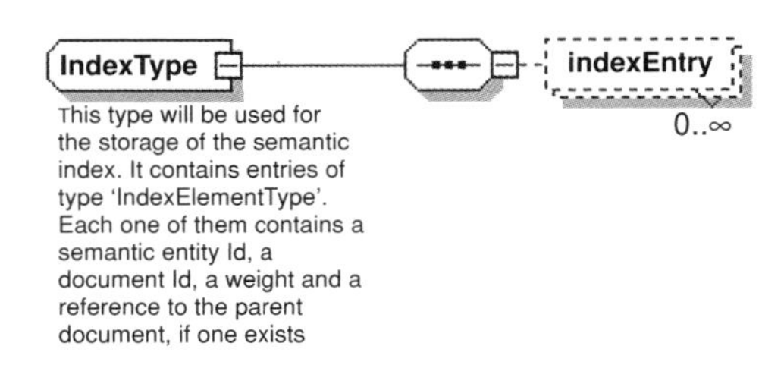

Figure 12.6 The IndexType comprises a sequence of entries, each one referring to a distinct semantic entity–document pair

```
</complexType><complexType name="IndexEntryType">
    <attribute name="semanticEntity" type="IDREF" use="required"/>
    <attribute name="document" type="string" use="required"/>
    <attribute name="weight" type="mpeg7:zeroToOneType" use="optional"/>
    <attribute name="parentDocument" type="string" use="optional"/>
</complexType>
```

Figure 12.7 The Index EntryType; it cannot be displayed graphically, as all of its components are included as attributes rather than child elements. Entities are represented using their unique id in the ontology and documents using a URL, the detailed format of which may be custom to the specific archive. Attribute weight provides for degrees of association, while attribute parentDocument provides for decomposition of multimedia documents into their semantic spatio-temporal components

12.3.3 User Profiles

User profiles contain all user information required for personalization. The contents of the user profiles are decomposed into the *usage history* and the *user preferences*. Profiles are stored using UserProfile Ds, which contain a UserPreferences DS and possibly a UsageHistory DS (Figures 12.8 and 12.9). The UsageHistory DS is only used in dynamic (i.e. not static) profiles.

Usage history

All of the actions users perform while interacting with the system are important for their profile and are therefore included in their usage history (Figure 12.10). When the user logs on to the

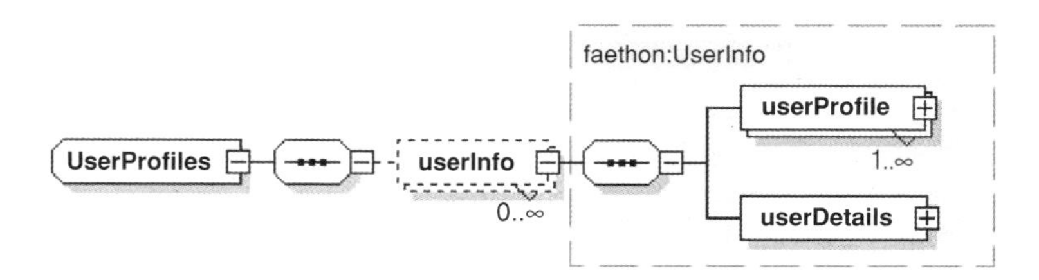

Figure 12.8 As already mentioned, a user may have more than one profile. Distinct profiles of the same user are grouped together via the UserInfo DS

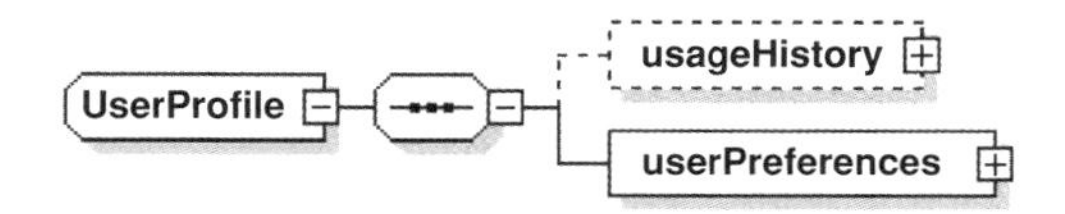

Figure 12.9 The UserProfile DS. The usageHistory part is only utilized for dynamic profiles, i.e. when the user has allowed the system to monitor user actions and based on them to automatically update user preferences

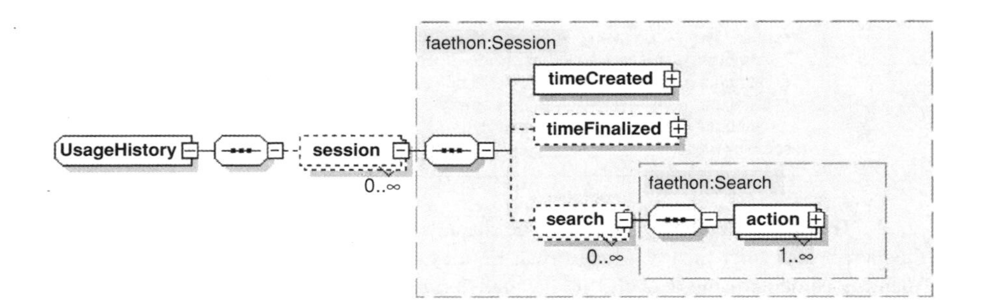

Figure 12.10 The UsageHistory DS. Each action is formed as a selection among new query, request for structural information about the document, request for metadata of the selected document or document segment, or request for the actual media

system a new *session* starts. The session ends when the user logs out, terminates the client program or changes his/her *active profile* (i.e. the profile he/she is currently using).

Within a session a user may try to satisfy a single or more of his/her needs/requests. Each one of those attempts is called a *search*. The search is a complex multi-step procedure; each one of the possible steps is an *action*. Different types of actions are supported by the system; these include formulation of a *query*, request for *structural* or *meta* information and request for the *media* itself.

Usage history contains records of sessions that belong to the same profile, stored using the Session DS. This DS may contain information concerning the time it was created (i.e. the time the session started) as well as the time it was finalized (i.e. the time the session was terminated). It also contains an ordered set of Search DSs. Their order is equivalent to the order in which the corresponding searches were performed by the user. Search DS, as implied by its name, is the structure used to describe a single search. It contains an ordered ser of Action DSs. Since different searches are not separated by a predefined event (as logging on) it is up to the system to separate the user's actions into different searches. This is accomplished by using query actions as separators but could also be tackled using a more complex algorithm, which might for example estimate the relevance between consequent queries. Action DSs may be accompanied by records of the set of documents presented to the user at each time. Such records need not contain anything more than document identifiers for the documents that were available to the user at the time of his/her action, as well as their accompanying ranks (if they were also presented to the user). Their purpose is to indicate what the user was reacting to.

User preferences

User preferences are partitioned into two major categories. The first one includes *metadata*-related and *structural* preferences while the second contains *semantic* preferences (Figures 12.11 and 12.12). The first category of preferences contains records indicating user preference for creation, media, classification, usage, access and navigation (e.g. favourite actors/directors

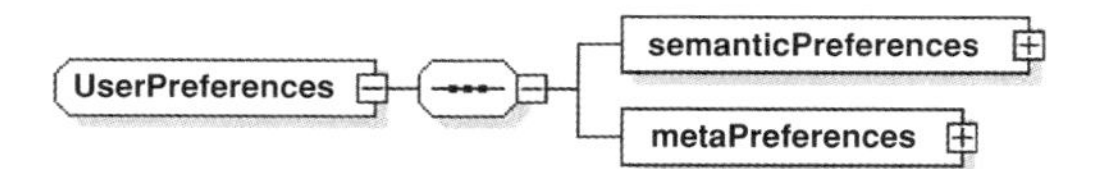

Figure 12.11 The UserPreferences DS. Although the metadata part is supported, the main emphasis of the system is on the semantic portion of the user preferences

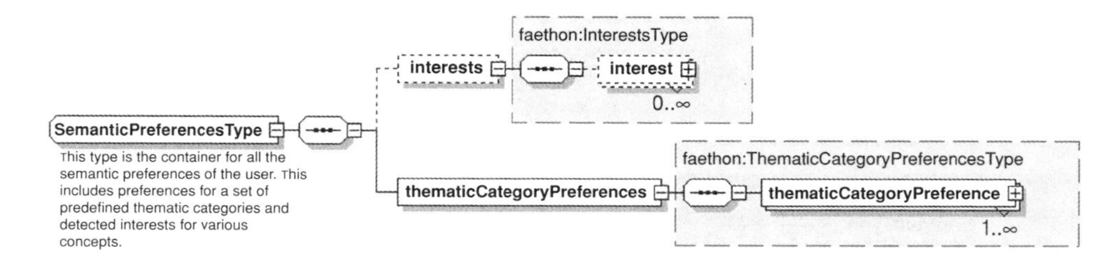

Figure 12.12 Preferences are grouped in preference degrees for the predefined categories and custom, automatically mined, interests

or preference for short summaries). Semantic preferences may again be divided into two (possibly overlapping) categories. The first contains records of *thematic categories*, thus indicating user preference for documents related to them. The second, which we may refer to as *interests*, contains records of *simple* or *composite semantic entities* or *weighted sets of semantic entities*, thus indicating user preference for documents related to them. Both metadata-related and semantic preferences are mined through the analysis of usage history records and will be accompanied by weights indicating the *intensity* of the preference. The range of valid values for these weights may be such as to allow the description of 'negative' intensity. This may be used to describe the user's dislike(s).

Metadata-related and structural preferences are stored using the UserPreferences DS, which has been defined by MPEG-7 for this purpose. Still, it is the semantic preferences that require the greater attention, since it is at the semantic level that the system primarily targets. Semantic preferences are stored using the system proprietary SemanticPreferences DS.

This contains the semantic interests, i.e. degrees of preference for semantic entities and degrees of preference for the various predefined thematic categories. Out of those, the thematic categories, being more general in nature, (i) are related to more documents than most semantic entities and (ii) are correctly identified in documents by the module of DTC, which takes the context into consideration. Thus, degrees of preference for thematic categories are mined with a greater degree of certainty than the corresponding degrees for simple semantic entities and shall be treated with greater confidence in the process of personalization of retrieval than simple interests. For this reason it is imperative that thematic categories are stored separately from interests. The SemanticPreferences DS contains a ThematicCategoryPreferences DS (Figure 12.13), which corresponds to the user's preferences concerning each of the predefined thematic categories, as well as an Interests DS (Figure 12.14), which contains mined interests for more specific entities in the ontology. Static profiles, either predefined

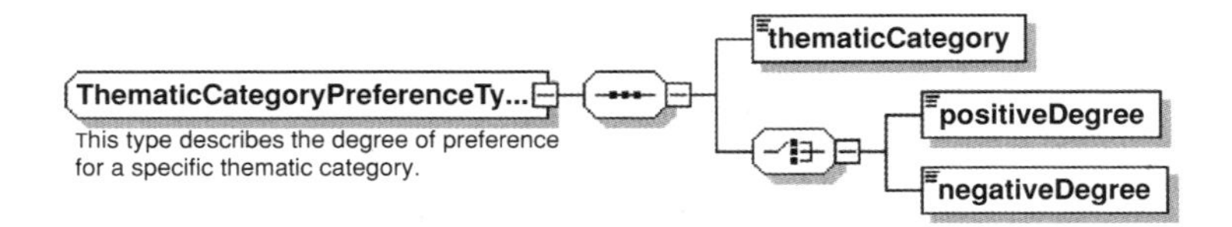

Figure 12.13 The ThematicCategoryPreferenceType allows both for preference and dislike degrees for a given topic/thematic category

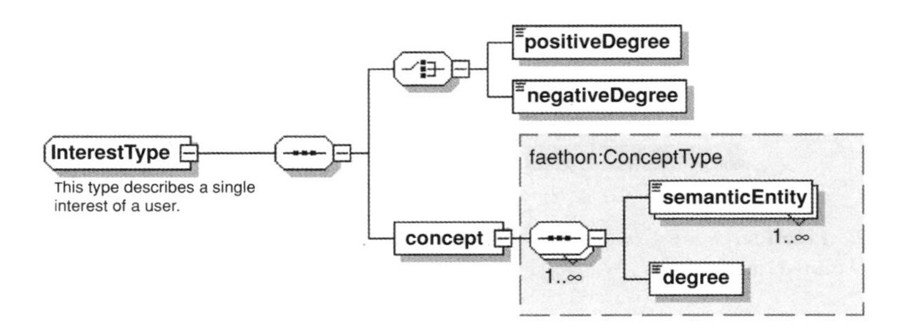

Figure 12.14 The InterestType provides for the representation of complex notions and composite objects in the form of fuzzy sets of semantic entities

by experts or defined by the end users themselves, only contain preferences for thematic categories.

The ThematicCategoryPreferences DS contains a record for each thematic category in the ontology. This entry contains a thematic category identifier and a weight indicating the intensity of the user's preference for the specific thematic category. When, on the other hand, it comes to the representation of more specific, automatically estimated user interests, such a simple representation model is not sufficient [8].

For example, let us examine how an error in estimation of interests affects the profiling system and the process of retrieval, in the cases of positive and negative interests. Let us suppose that a user profile is altered by the insertion of a positive interest that does not actually correspond to a real user interest. This will result in consistent selection of irrelevant documents; the user reaction to these documents will gradually alter the user profile by removing this preference, thus returning the system to equilibrium. In other words, miscalculated positive interests are gradually removed, having upset the retrieval process only temporarily.

Let us now suppose that a user profile is altered by the insertion of a negative interest that does not correspond to a real user dislike. Obviously, documents that correspond to it will be down-ranked, which will result in their consistent absence from the set of selected documents; therefore the user will not be able to express an interest in them, and the profile will not be re-adjusted.

This implies that the personalization process is more sensitive to errors that are related to negative interests, and therefore such interests need to be handled and used with greater caution. Therefore, negative interests need to be stored separately than positive ones, so that they may be handled with more caution in the process of personalized retrieval.

Let us also consider the not rare case in which a user has various distinct interests. When the user poses a query that is related to one of them, then that interest may be used to facilitate the ranking of the selected documents. Usage of interests that are unrelated to the query may only be viewed as addition of noise, as any proximity between selected documents and these interests is clearly coincidental, in the given context. In order to limit this inter-preference noise, we need to be able to identify which interests are related to the user query, and to what extent. Thus, distinct positive interests need to be stored separately from each other as well.

Following the above principles, the Interests DS contains records of the interests that were mined from this profile's usage history; each of these records is composed of an interest intensity value as well as a description of the interest (i.e. the semantic entities that compose it and the degree to which they participate to the interest). Simple and composite semantic entities can be described using a single semantic entity identifier. Weighted sets can easily be described as a sequence of semantic entity identifiers accompanied by a value indicating the degree of membership.

12.3.4 Archive Profiles

The main purpose of an audiovisual archive profile is to provide a mapping of an archive's custom multimedia document DS to the system's unified DSs. Each archive profile contains all necessary information for the construction of individual queries related to metadata, and particularly mapping of creation, media, usage, syntactic, access and navigation description schemes. Therefore, the structure of archive profiles is based on the multimedia document description schemes. Semantic description schemes are included as they are handled separately by the semantic unification subsystem.

In contrast to the ontology, the index and the user profiles, the archive profiles are stored at the distinct *audiovisual archive interfaces* and not in the central DBMS (Figure 12.15).

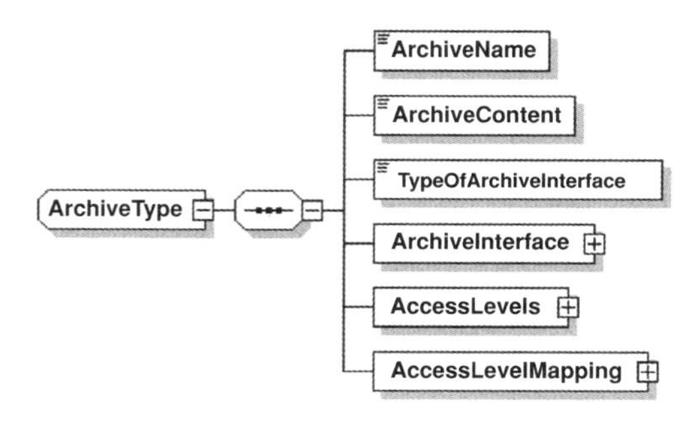

Figure 12.15 The information stored locally at the archive profile allows for the automatic translation of system queries to a format that the custom content management application of the archive can parse, as well as for the translation of the response in the standardized data structures of the system

12.4 Indexing of Multimedia Documents

As we have already mentioned, the main goal of the system is to provide to the end users uniform access to different audiovisual archives. This is accomplished by mapping all audiovisual content to a semantically unified index, which is then used to serve user queries. The update mode of operation, in addition to the analysis of usage history for the update of user preferences, is charged with the effort to constantly adapt to archive content changes and enrich the index used for the unified searching and filtering of audiovisual content.

The index is stored in the DBMS as an XML file containing pairs of semantic entities and documents or document segments, and possibly degrees of association. This structure, although sufficient as far as its descriptive power is concerned, does not allow for system operation in a timely manner. Therefore, a more flexible format is used to represent the index information in main memory; the chosen format employs binary trees to represent the index as a fuzzy binary relation between semantic entities and documents (in this approach each document segment is treated as a distinct document). This model allows for $O(logn)$ access time for the documents that are related to a given semantic entity, compared to a complexity of $O(n)$ for the sequential access to the stored XML index [9]. It is worth mentioning that although thematic categories have a separate and important role in the searching process, they are a special case of other concepts, and thus they are stored in the index together with other semantic entities.

The modules that update the semantic entities in the index and their links to the audiovisual units are DECO and DTC (Figure 12.16). The former takes the multimedia document descriptions as provided by the individual archive interfaces and maps them to semantic entities' definitions in the ontology, together with a weight representing the certainty with which the system has detected the semantic entities in question. Furthermore, it scans the audiovisual units and searches for composite semantic structures; these are also linked together in the index. The latter accepts as input the semantic indexing of each document, as provided in the DBMS, and analyses it in order to estimate the degree to which the given document is related to each one of the predefined thematic categories.

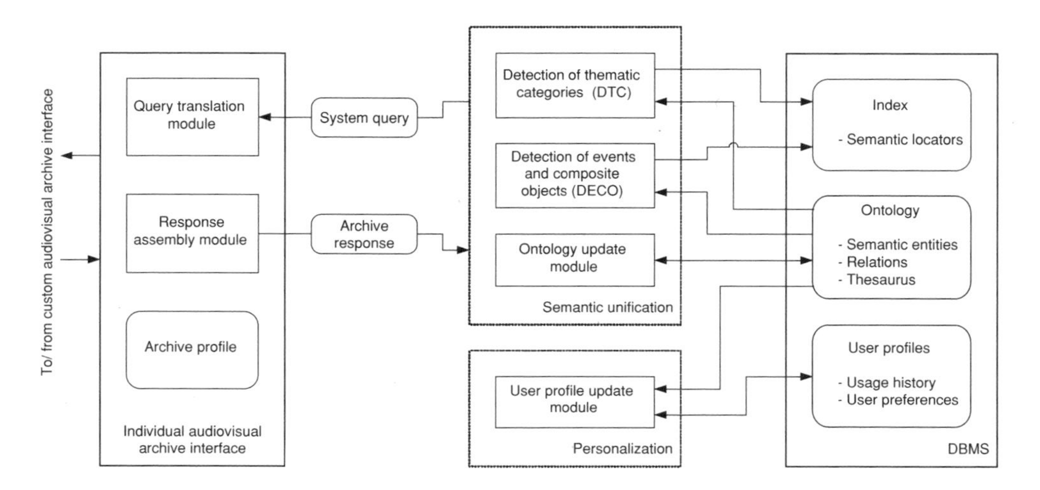

Figure 12.16 The system at update mode of operation

As already mentioned, all update procedures may be performed globally for the entire content of the audiovisual archives at regular intervals or whenever the audiovisual content of an archive is updated. In the latter case, which is preferable due to low bandwidth and computational cost, the update process is *incremental*, i.e. only the newly inserted audiovisual unit descriptions need to be retrieved and processed.

Prior to any indexing process, the content of the ontology may be updated with the aid of the *ontology update* module. The main goal of this module is to update the thesaurus according to any changes in the detailed semantic relations of the ontology, as DECO and DTC rely on correct input from the thesaurus views of knowledge in order to operate. Moreover, the definitions of semantic entities of the encyclopedia need to be updated, especially when the content of the audiovisual archives is dramatically changed.

12.4.1 Detection of Events and Composite Objects

The DECO is executed in a two-pass process for each multimedia document that has to be indexed. In the first pass the audiovisual archive is queried and the full description of a document that has not yet been indexed, or whose description has been altered, is retrieved. The individual archive interface assures that the structure that arrives at the central system is compliant with the MPEG-7 multimedia content description standard, thus allowing a unified design and operation of the indexing process to follow. The DECO module scans the MPEG-7 description and identifies mentioned semantic entities by their definitions in the ontology. Links between these semantic elements and the document in question are added to the index; weights are added depending on the location of the entity in the description scheme and the degree of matching between the description and the actual entity definition in the ontology.

In the second pass, the DECO module works directly on the semantic indexing of documents, attempting to detect events and composite objects that were not directly encountered in the document descriptions, but the presence of which can be inferred from the available indexing information. The second pass of the DECO process further enriches the semantic indexing of the documents.

Although the importance of the DECO as a stand-alone module is crucial for the operation of the overall system, one may also view it as a pre-processing tool for the following DTC procedure, since the latter uses the detected composite objects and events for thematic categorization purposes.

12.4.2 Detection of Thematic Categories

The DTC performs a matching between the archived material and the predefined thematic categories. It takes as input the indexing of each multimedia document, or document segment, as provided in the index by the DECO module, and analyses it in order to estimate the degree to which the document in question is related to each one of the thematic categories. Although the output of DTC is also stored in the index, as is the output of DECO, an important difference exists between the two: the weights in the output of DECO correspond to degrees of confidence, while the degrees in the output of DTC correspond to estimated degrees of association. Another important difference between the DECO and DTC modules is that whereas DECO searches for any semantic link between multimedia documents and semantic entities, DTC limits its operation to the case of thematic categories.

What makes the predefined categories, and accordingly the DTC process, so important, is the fact that through them a unified representation of multimedia documents originating from different audiovisual archives is possible. Thus, they have a major contribution to the semantic unification and unified access of diverse audiovisual sources, which is the main goal of the system.

12.4.3 Indexing Algorithms

The DTC and DECO run in offline time. They first run when the encyclopedia and audiovisual archive documents are constructed to create the index. Every time the audiovisual archives are enriched with new documents, or the annotation of existing documents is altered, the DTC and DECO run in order to update the index accordingly, processing only the updated segments of the audiovisual archives. Every time the ontology is updated the DTC and DECO run for all the audiovisual archives, and all the documents in each archive, in order to create a new index; an incremental update is not appropriate, as the new entities and new thesaurus knowledge views will result in different analysis of the document descriptions. In the following we provide more details on the methodologies utilized by these modules in the process of document analysis, after the first pass of DECO has completed, having provided an elementary semantic indexing of multimedia content.

The utilized view of the knowledge

The semantic encyclopedia contains *110 000* semantic entities and definitions of numerous MPEG-7 semantic relations. As one might expect, the existence of many relations leads to the dividing of the available knowledge among them, which in turn results in the need for the utilization of more relations than one for the meaningful analysis of multimedia descriptions. On the other hand, the simultaneous consideration of multiple semantic relations would pose an important computational drawback for any processing algorithm, which is not acceptable for a system that hopes to be able to accommodate large numbers of audiovisual archives and multimedia documents. Thus, the generation of a suitable view T in the thesaurus is required. For the purpose of analysing multimedia document descriptions we use a view that has been generated with the use of the following semantic relations:

- Part P, inverted.
- Specialization Sp.
- Example Ex. $Ex(a,b) > 0$ indicates that b is an example of a. For example, a may be 'player' and b may be 'Jordan'.
- Instrument Ins. $Ins(a,b) > 0$ indicates that b is an instrument of a. For example, a may be 'music' and b may be 'drums'.
- Location Loc, inverted. $L(a,b) > 0$ indicates that b is the location of a. For example, a may be 'concert' and b may be 'stage'.
- Patient Pat. $Pat(a,b) > 0$ indicates that b is a patient of a. For example, a may be 'course' and b may be 'student'.
- Property Pr, inverted. $Pr(a,b) > 0$ indicates that b is a property of a. For example, a may be 'Jordan' and b may be 'star'.

Thus, the view T is calculated as:

$$T = (Sp \cup P^{-1} \cup Ins \cup Pr^{-1} \cup Pat \cup Loc^{-1} \cup Ex)^{(n-1)}$$

The $(n-1)$ exponent indicates $n-1$ compositions, which are guaranteed to establish the property of transitivity for the view [10]; it is necessary to have the view in a closed transitive form, in order to be able to answer questions such as 'which entities are related to entity x?' in $O(logn)$ instead of $O(n^2)$ times, where $n = 110\,000$ is the count of known semantic entities. Alternatively, a more efficient methodology, targeted especially to sparse relations, can be utilized to ensure transitivity [9]. Based on the semantics of the participating relations, it is easy to see that T is ideal for the determination of the topics that an entity may be related to, and consequently for the analysis of multimedia content based on its mapping to semantic entities through the index.

The notion of context

When using an ontological description, it is the context of a term that provides its truly intended meaning. In other words, the true source of information is the co-occurrence of certain entities and not each one independently. Thus, in the process of content analysis we will have to use the common meaning of semantic entities in order to best determine the topics related to each examined multimedia document. We will refer to this as their *context*; in general, the term *context* refers to whatever is common among a set of elements. Relation T will be used for the detection of the context of a set of semantic entities, as explained in the remaining of this subsection.

As far as the second phase of the DECO and the DTC are concerned, a document d is represented only by its mapping to semantic entities via the semantic index. Therefore, the context of a document is again defined via the semantic entities that are related to it. The fact that relation T is (almost) an ordering relation allows us to use it in order to define, extract and use the context of a document, or a set of semantic entities in general.

Relying on the semantics of relation T, we define the context $K(s)$ of a single semantic entity $s \in S$ as the set of its antecedents in relation T, where S is the set of all semantic entities contained in the ontology. More formally, $K(s) = T(s)$, following the standard superset–subset notation from fuzzy relational algebra [9]. Assuming that a set of entities $A \subseteq S$ is crisp, i.e. all considered entities belong to the set with degree one, the context of the group, which is again a set of semantic entities, can be defined simply as the set of their common antecedents:

$$K(A) = \bigcap K(s_i),\ s_i \in A$$

Obviously, as more entities are considered, the context becomes narrower, i.e. it contains fewer entities and to smaller degrees:

$$A \supset B \rightarrow K(A) \subseteq K(B)$$

When the definition of context is extended to the case of fuzzy sets of semantic entities, this property must still hold. Taking this into consideration, we demand that, when A is a normal fuzzy set, the 'considered' context $\mathcal{K}(s)$ of s, i.e. the entity's context when taking its degree of participation in the set into account, is low when the degree of participation $A(s)$ is high, or when the context of the crisp entity $K(s)$ is low. Therefore:

$$cp(\mathcal{K}(s)){=}cp(K(s)){\cap}\ (S \cdot A(s))$$

where cp is an involutive fuzzy complement. By applying de Morgan's law, we obtain:

$$\mathcal{K}(s){=}K(s) \cup cp(S \cdot A(s))$$

Then the overall context of the set is again easily calculated as:

$$K(A) = \bigcap \mathcal{K}(s_i),\ s_i \in A$$

Considering the semantics of the T relation and the process of context determination, it is easy to realize that when the entities in a set are highly related to a common meaning, the context will have high degrees of membership for the entities that represent this common meaning. Therefore, the height of the context $h(K(A))$, i.e. the greatest membership degree that appears in it, may be used as a measure of the semantic correlation of entities in set A. We will refer to this measure as *intensity* of the context.

Fuzzy hierarchical clustering and topic extraction

Before detecting the topics that are related to a document d, the set of semantic entities that are related to it needs to be clustered, according to their common meaning. More specifically, the set to be clustered is the support of the document:

$$^{0+}d = \{s \in S : I(s, d) > 0\}$$

where $I{:}S \rightarrow D$ is the index and D is the set of indexed documents.

Most clustering methods belong to either of two general categories, partitioning and hierarchical. Partitioning methods create a crisp or fuzzy clustering of a given data set, but require the number of clusters as input. Since the number of topics that exist in a document is not known beforehand, partitioning methods are inapplicable for the task at hand; a hierarchical clustering algorithm needs to be applied. Hierarchical methods are divided into agglomerative and divisive. Of those, the first are more widely studied and applied, as well as more robust. Their general structure, adjusted for the needs of the problem at hand, is as follows:

1. When considering document d, turn each semantic entity $s \in {}^{0+}d$ into a singleton, i.e. into a cluster c of its own.
2. For each pair of clusters c_1, c_2 calculate a degree of association $CI(c_1,c_2)$. The CI is also referred to as cluster similarity measure.
3. Merge the pair of clusters that have the best CI. The best CI can be selected using the *max* operator.
4. Continue at step 2 until the termination criterion is satisfied. The termination criterion most commonly used is the definition of a threshold for the value of the best degree of association.

The two key points in hierarchical clustering are the identification of the clusters to merge at each step, i.e. the definition of a meaningful measure for *CI*, and the identification of the optimal terminating step, i.e. the definition of a meaningful termination criterion.

When clustering semantic entities, the ideal association measure for two clusters c_1, c_2 is one that quantifies their semantic correlation. In the previous we have defined such a measure: the intensity of their common context $h(K(c_1 \cup c_2))$. The process of merging should terminate when the entities are clustered into sets that correspond to distinct topics. We may identify this case by the fact that no pair of clusters will exist with a common context of high intensity. Therefore, the termination criterion shall be a threshold on the *CI*.

This clustering method, being a hierarchical one, will successfully determine the count of distinct clusters that exist in ${}^{0+}d$. Still, it is inferior to partitioning approaches in the following senses:

1. It only creates crisp clusters, i.e. it does not allow for degrees of membership in the output.
2. It only creates partitions, i.e. it does not allow for overlapping among the detected clusters.

Both of the above are great disadvantages for the problem at hand, as they are not compatible with the task's semantics: in real life, a semantic entity may be related to a topic to a degree other than 1 or 0, and may also be related to more than one distinct topics. In order to overcome such problems, we apply a method for fuzzification of the partitioning. Thus, the clusters' scalar cardinalities will be corrected, so that they may be used later on for the filtering of misleading entities.

Each cluster is described by the crisp set of semantic entities $c \subseteq {}^{0+}d$ that belong to it. Using those, we may construct a fuzzy classifier, i.e. a function C_c that measures the degree of correlation of a semantic entity s with cluster c. Obviously a semantic entity s should be considered correlated with c, if it is related to the common meaning of the semantic entities in it. Therefore, the quantity

$$Cor_1\ (c,s) = h(K(c \cup \{s\}))$$

is a meaningful measure of correlation. Of course, not all clusters are equally compact; we may measure cluster compactness using the similarity among the entities they contain, i.e. using the intensity of the clusters' contexts. Therefore, the aforementioned correlation measure needs to be adjusted, to the characteristics of the cluster in question:

$$C_c(s) = \frac{Cor_1(c, s)}{h(K(c))} = \frac{h(K(c \cup \{s\}))}{h(K(c))}$$

Using such classifiers, we may expand the detected crisp partitions, to include more semantic entities and to different degrees. Partition c is replaced by cluster c^{fuzzy}:

$$c^{fuzzy} = \sum_{s \in {}^{0+}d} s/C_c(s)$$

Obviously $c^{fuzzy} \supseteq c$.

The process of fuzzy hierarchical clustering has been based on the crisp set ${}^{0+}d$, thus ignoring fuzziness in the semantic index. In order to incorporate this information when calculating the 'final' clusters that describe a document's content, we adjust the degrees of membership for them as follows:

$$c^{final}(s) = t(c^{fuzzy}(s), I(s, d)), \forall s \in {}^{0+}d$$

where t is a t-norm. The semantic nature of this operation demands that t is an Archimedean norm [11]. Each one of the resulting clusters corresponds to one of the distinct topics of the document. Finally, once the fuzzy clustering of entities in a multimedia document's indexing has been performed, DTC and DECO can use the results in order to produce their own semantic output.

In order for DTC to determine the topics that are related to a cluster c^{final}, two things need to be considered: the scalar cardinality of the cluster $|c^{final}|$ and its context. Since context has been defined only for normal fuzzy sets, we need to first normalize the cluster as follows:

$$c^{normal}(s) = \frac{c^{final}(s)}{h(c^{final}(s))}, \forall s \in {}^{0+}d$$

Obviously, semantic entities that are not contained in the context of c^{normal} cannot be considered as being related to the topic of the cluster. Therefore:

$$R_T(c^{final}) \subseteq R_T{}^*(c^{normal}) = w(K(c^{normal}))$$

where w is a weak modifier. Modifiers, which are also met in the literature as *linguistic hedges*, are used to adjust mathematically computed values so as to match their semantically anticipated counterparts.

In the case where the semantic entities that index document d are all clustered in a unique cluster c^{final}, then $R_T(d) = R_T{}^*(c^{normal})$ is a meaningful approach. On the other hand, when multiple clusters are detected, then it is imperative that cluster cardinalities are considered as well.

Clusters of extremely low cardinality probably only contain misleading entities, and therefore need to be ignored in the estimation of $R_T(d)$. On the contrary, clusters of high cardinality almost certainly correspond to the distinct topics that d is related to, and need to be considered in the estimation of $R_T(d)$. The notion of 'high cardinality' is modelled with the use of a 'large' fuzzy number $L(\cdot)$. $L(a)$ is the truth value of the proposition 'a is high', and, consequently, $L(|b|)$ is the truth value of the preposition 'the cardinality of cluster b is high'.

The set of topics that correspond to a document is the set of topics that correspond to each one of the detected clusters of semantic entities that index the given document.

$$R_T(d) = c^{final} \in G(R_T(c^{final}))$$

where $\cup$ is a fuzzy co-norm and G is the set of fuzzy clusters that have been detected in d. The topics that are related to each cluster are computed, after adjusting membership degrees according to scalar cardinalities, as follows:

$$R_T(c^{final}) = R_T{}^*(c^{normal}) \cdot L(|c^{final}|)$$

It is easy to see that $R_T(s,d)$ will be high if a cluster c^{final}, whose context contains s, is detected in d, and additionally, the cardinality of c^{final} is high and the degree of membership of s in the context of the cluster is also high (i.e. if the topic is related to the cluster and the cluster is not comprised of misleading entities).

The DECO module, on the other hand, relies on a different view of the ontology that is constructed using only the specialization and example relations in order to take advantage of the findings of the fuzzy clustering of index terms. In short, DECO relates to each document the entities that are in the context of the detected clusters. In this framework the context is estimated using the same methodology as above, but instead of the T view of the knowledge we utilize one that contains only information extracted from the example, part and specialization relations.

12.5 Query Analysis and Processing

At the online mode of operation the system receives user queries from the end-user interfaces and serves them in a semantic and timely manner, based primarily on the information stored in the index and the ontology (Figure 12.17). Specifically, the semantic part of the query is analysed by the query analysis module in order to be mapped to a suitable set of semantic entities from the ontology; the entities of this set can then be mapped by the search engine to the corresponding multimedia documents, as the latter are indicated by the index. In the cases

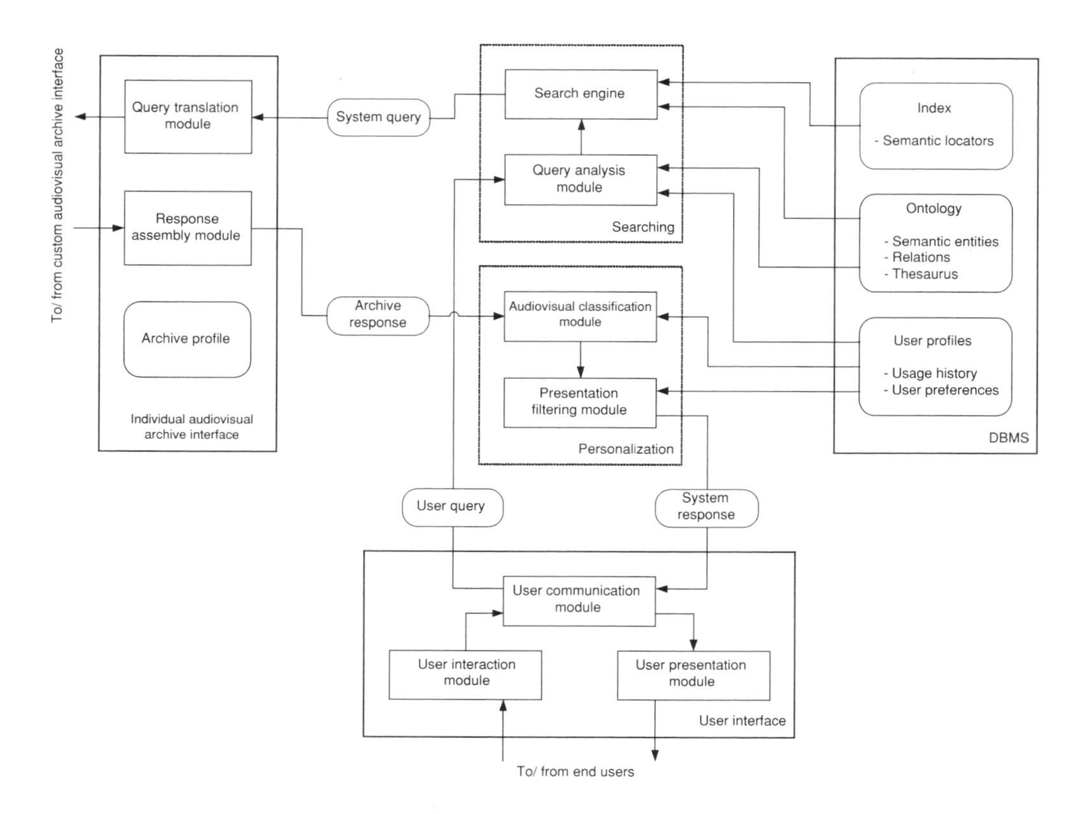

Figure 12.17 The system at query mode of operation

that the user query contains a structural part as well, or when the metadata part of the user profile is to be used during the personalization of the response, the audiovisual archives may have to be queried as well for the MPEG-7 annotations of the selected multimedia documents (it is the archive interface that takes care of the translation of the archive's custom DS to the MPEG standard, based on the information stored in the archive profile).

The weighted set of documents selected through this process is then adapted to the user that issued the query by the personalization subsystem, using the preferences defined in the active user profile. Of course, as the system aims to be the mediator for searches in audiovisual archives, it also supports the consideration of metadata in all the steps of searching and personalizing the results; still, the emphasis and novel contribution is found in the ability for semantic treatment of the user query, the multimedia documents and the user profiles, as it is exactly this characteristic that allows for the unified access to multiple and diverse audiovisual archives.

Focusing more on the searching procedure itself, we start by clarifying that both the user query and the index are fuzzy, meaning that the user can supply the degree of importance for each term of the query, and that the set of associated semantic entities for each document also contains degrees of association, as provided by the DECO and DTC modules. Consequently, the results of the searching procedure will also have to be fuzzy [12]; the selected documents are sorted by estimated degree of relevance to the user query, and in a later step according to relevance to the user preferences, and the best matches are presented (first) to the user.

It is possible that a query does not match a given index entry, although the document that corresponds to it is relevant to the query. For example, a generalization of a term found in a document may be used in the query. This problem is typically solved with the use of a fuzzy thesaurus containing, for each term, the set of its related ones. The process of enlarging the user's query with the associated terms is called query expansion; it is based on the associative relation A of the thesaurus, which relates terms based on their probability of coexisting in a document [13, 14].

To make query expansion more intelligent, it is necessary to take into account the meaning of the terms [15]. In order to be able to use the notion of context, as defined in the previous subsection, to estimate and exploit the common meaning of terms in the query, we need to map the query to the set of semantic entities in the ontology; this task is referred to as query interpretation, as it extracts the semantics of the terms of the user query. Finally, the utilization of a statistically generated associative thesaurus for query expansion, although a common and generally accepted practice in textual information retrieval, is avoided in this work, as this approach is known to overpopulate the query with irrelevant terms, thus lowering the precision of the response [16]; instead, we define and use a view of the ontology that is based strictly on partially ordering fuzzy relations, such as the specialization, the part and the example relation; the ordering properties of the considered relations make the resulting view more suitable for the definition and estimation of the context of a set of semantic entities.

12.5.1 Context-Sensitive Query Interpretation

As we have already mentioned, the definitions of semantic entities in the ontology contain sequences of labels, each one providing a different textual form of the semantic entity, possibly in more than one language. Matching those to the terms in the user query, we can acquire the semantic representation of the query. Of course, in most cases this is far from trivial:

the mapping between terms and semantic entities is a many-to-many relation, which means that multiple possible semantic interpretations exist for a single textual query. As a simple example, let us consider the case of the term 'element'. At least two distinct semantic entities correspond to it: 'element1', which is related to chemistry, and 'element2', which is related to XML. Supposing that a user query is issued containing the term 'element', the system needs to be able to automatically determine to which semantic entity in the ontology the term should be mapped, in order to retrieve the corresponding multimedia documents from the index.

In the same example, if the remaining terms of the query are related to chemistry, then it is quite safe to suppose that the user is referring to semantic entity 'element1' rather than to semantic entity 'element2'. This implies that the context of the query can be used to facilitate the process of semantic entity determination in the case of ambiguities. However, the estimation of the query context, as described in the previous section, needs as input the representation of the query as a fuzzy set of entities, and thus cannot be performed before the query interpretation is completed.

Consequently, query interpretation needs to take place simultaneously with context estimation. We follow the following method: let the textual query contain the terms $\{t_i\}$ with $i = 1, \ldots, T$. Let also t_i be the textual description of semantic entities $\{s_{ij}\}$ with $j = 1, \ldots, T_i$. Then, there exist $N_Q = \prod_i T_i$ distinct combinations of semantic entities that may be used for the representation of the user's query; for each one of those we calculate the corresponding context.

As already explained, the intensity of the context is a semantic measure of the association of the entities in a set. Thus, out of the candidate queries $\{q_k\}$, where $k = 1, 2, \ldots, N_Q$, the one that produces the most intense context is the one that contains the semantic entities that are most related to each other; this is the combination that is chosen as output of the process of query interpretation:

$$q = q_i \in \{q_1, \ldots, q_{N_Q}\} : h(q_i) \geq h(q_j) \forall q_j \in \{q_1, \ldots, q_{N_Q}\}$$

This semantic query interpretation is exhaustive, in the sense that it needs to consider all possible interpretations of a given query. Still, this is not a problem in the framework where it is applied as:

- queries do not contain large numbers of terms;
- the number of distinct semantic entities that may have a common textual description is not large;
- the gain in the quality of the semantic content of the interpreted query, as indicated by the difference in the precision of the system response, is largely more important than the added computational burden.

12.5.2 Context-Sensitive Query Expansion

The process of query expansion enriches the semantic query, in order to increase the probability of a match between the query and the document index. The presence of several semantic entities in the query defines a context, which we use in order to meaningfully direct the expansion process, so that it generates expanded queries that provide enhanced recall in the result, without suffering the side effect of poor precision.

As will become obvious from the presentation of the process of matching the query to the index, optimal results can only be acquired if the origin of the new entities in the expanded query is known; in other words, we will need to know to which entity in the initial query each new entity corresponds. Thus, in query expansion, we replace each semantic entity s with a set of semantic entities $X(s)$; we will refer to this set as the expanded semantic entity.

In a more formal manner, we define the expanded entity as $X(S_i) = \sum_i s_{ij}/x_{ij}$, using the sum notation for fuzzy sets; the weight x_{ij} denotes the degree of significance of the entity s_j in $X(s_i)$. We compute it using the semantic query q, the context $K(q)$ of the query, and the *In* relation of the thesaurus; the *In* relation has resulted from the combination of the *Sp*, P and the *Ex* relations as $In = (Sp \cup P^{-1} \cup Ex)^{(n-1)}$.

In a query expansion that does not consider the context, the value of x_{ij} should be proportional to the weight w_i and the degree of inclusion $I(s_i, s_j)$. Therefore, in that case we would have $x_{ij} = w_{ij} = w_i In(s_i, s_j)$. In a context-sensitive query expansion, on the other hand, x_{ij} increases monotonically with respect to the degree to which the context of s_j is relative to the context of the query. We use the value:

$$h_j = \max\left(\frac{h(In(s_j) \cap K(q))}{h_q}, h_q\right)$$

to quantify this relevance. We additionally demand that the following conditions be satisfied by our query expansion method:

- x_{ij} increases monotonically with respect to w_{ij}
- $h_j = 1 \rightarrow x_{ij} = w_{ij}$
- $h_q = 0 \rightarrow x_{ij} = w_{ij}$
- $h_q = 1 \rightarrow x_{ij} = w_{ij} h_j$
- x_{ij} increases monotonically with respect to h_j.

Thus, we have:

$$x_{ij} = \max(h_j, c(h_q)) w_{ij} = w_i In(s_i, s_j) \max(h_j, c(h_q))$$

The fuzzy complement c in this relation is Yager's complement with a parameter of 0.5.

12.5.3 Index Matching

The search engine, supposing that the query is a crisp set (i.e. entities in the query do not have weights) and that no expansion of the query has preceded, uses the semantic query q, which is a fuzzy set of semantic entities, and the document index I, which is a fuzzy relation between the set of semantic entities S and the set of documents D, to produce the result r; r is again a fuzzy set on D. When the query is comprised of a single semantic entity s, then the result is simply the respective row of I, i.e. $r(q) = I(s)$. When, on the other hand, the query contains more than one semantic entity, then the result is the set of documents that contain all the semantic entities, or, more formally:

$$r(q) = \bigcap_{s_i \in q} r(s_i)$$

Generalizing this formula to the case when query expansion has preceded, we should include in the result set only documents that match all the expanded entities of the query. Therefore, it is imperative that independent search results are first computed for each expanded entity separately, and then combined to provide the overall result of the search process.

Considering the way expanded entities are calculated, it is rather obvious that a document should be considered to match the expanded entity when it matches any of the terms in it. Moreover, the percentage of semantic entities that a document matches should not make a difference (it is the same if the document matches only one or all of the semantic entities in the same expanded entity, as this simply indicates that the document is related to just one of the entities in the original query). Consequently, we utilize the *max* operator in order to join results for a single expanded entity:

$$r\left(X(s_i)\right) = \bigcup_{s_j \in X(s_i)} r(s_j)$$

or, using a simpler notation:

$$r\left(X(s_i)\right) = X(s_i) \circ I$$

On the other hand, results from distinct entities are treated using an intersection operator, i.e. only documents that match all of the entities of the query are selected.

$$r(q) = \bigcap_{s_i \in q} r\left(X(s_i)\right)$$

Unfortunately, this simple approach is limiting; it is more intuitive to select the documents that match all of the terms in the query first, followed by those documents that match fewer of the query terms. The effect of this limitation becomes even more apparent when the initial query is not crisp, i.e. when the absence of an entity that was assessed as unimportant by the user prevents an otherwise relevant document from being included in the results.

Thus, we follow a more flexible approach for the combination of the results of the matching of distinct entities with the semantic index. Specifically, we merge results using an ordered weighted average operator [17, 18], instead of the *min* operator. The selection of weights for the OWA operator is a monotonically increasing one. The required flexibility is achieved by forcing the degree of the last element to be smaller than one. Thus, the chosen family of OWA operators behaves as a 'soft' intersection on the intermediate results.

12.6 Personalization

Due to the massive amount of information that is nowadays available, the process of information retrieval tends to select numerous documents, many of which are barely related to the user's wish [19]; this is known as *information overload*. The reason is that an automated system cannot acquire from the query adequate information concerning the user's wish. Traditionally, information retrieval systems allow the users to provide a small set of keywords describing their wishes, and attempt to select the documents that best match these keywords. Although the information contained in these keywords rarely suffices for the exact determination of user wishes, this is a simple way of interfacing that users are accustomed to; therefore, we need to investigate ways to enhance retrieval, without altering the way they specify their request. Consequently, information about the user wishes needs to be found in other sources.

Personalization of retrieval is the approach that uses the information stored in user profiles, additionally to the query, in order to estimate the user's wishes and thus select the set of relevant documents [20]. In this process, the query describes the user's current search, which is the *local interest* [21], while the user profile describes the user's preferences over a long period of time; we refer to the latter as *global interest.*

12.6.1 Personalization Architecture

During the query mode (Figure 12.17), the audiovisual classification module performs ranking (but not filtering) of the retrieved documents of the archive response based on semantic preferences contained within the user profiles. The semantic preferences consist of user interests and thematic categories preferences. At the presentation filtering module further ranking and filtering is performed according to the metadata preferences such as creation, media, classification, usage, access and navigation preferences (e.g. favourite actors/directors or preference for short summaries).

The entire record of user actions during the search procedure (user query, retrieved documents, documents selected as relevant) is stored in the usage history of the specific user; this information is then used for tracking and updating the user preferences. The above actions characterize the user and express his/her personal view of the audiovisual content. The user profile update module takes these transactions as input during update mode (Figure 12.16) of operation and, with the aid of the ontology and the semantic indexing of the multimedia documents referred to in the usage history, extracts the user preferences and stores them in the corresponding user profile.

12.6.2 The Role of User Profiles

When two distinct users present identical queries, they are often satisfied by different subsets of the retrieved documents, and to different degrees. In the past, researchers have interpreted this as a difference in the perception of the meaning of query terms by the users [20]. Although there is definitely some truth in this statement, other more important factors need to be investigated.

Uncertainty is inherent in the process of information retrieval, as terms cannot carry unlimited information [21], and, therefore, a limited set of terms cannot fully describe the user's wish; moreover, relevance of documents to terms is an ill-defined concept [22]. The role of personalization of information retrieval is to reduce this uncertainty, by using more information about the user's wishes than just the local interest.

On the other hand, the user profile is not free of uncertainty either, as it is generated through the constant monitoring of the user's interaction; this interaction contains inherent uncertainty which cannot be removed during the generation of the user profile. Nevertheless, user profiles tend to contain less uncertainty than user queries, as long as the monitoring period is sufficient and representative of the user's preferences.

Therefore, a user profile may be used whenever the query provides incomplete or insufficient information about the user and their local interest. However, it is the query that describes the user's local preference, i.e. the scope of their current interaction. The profile is not sufficient on its own for the determination of the scope of the current interaction, although it contains valuable information concerning the user's global interest. Therefore, the user profile cannot totally dominate over the user query in the process of information retrieval.

The above does not imply that the degree to which the query dominates the retrieval process may be predefined and constant. Quite the contrary, it should vary in a manner that optimizes the retrieval result, i.e. in a manner that minimizes its uncertainty. We may state this more formally by providing the following conditions:

1. When there is no uncertainty concerning the user query, the user profile must not interfere with the process of information retrieval.
2. When the uncertainty concerning a user query is high, the user profile must be used to a great extent, in order to reduce this uncertainty.
3. The degree to which the user profile is used must increase monotonically with respect to the amount of uncertainty that exists in the user query, and decrease monotonically with respect to the amount of uncertainty that exists in the user profile.

The above may be considered as minimal guidelines or acceptance criteria for the way a system exploits user profiles in the process of information retrieval.

12.6.3 *Audiovisual Content Classification*

The audiovisual classification module receives as input a weighted set of documents. It is the set of documents retrieved based on the user's actions. Ranking has already taken place based on the query itself, producing an objective set of selected documents. The goal of the audiovisual classification module is to produce a subjective set of selected documents, i.e. a set that is customized to the preferences of the user who posed the query.

As has already been mentioned, preferences for thematic categories and semantic interests are utilized in different manners by the module of audiovisual classification. We elaborate on both in the following.

Exploitation of Preferences for Thematic Categories

Through the index, each document that the system handles has been related to the system's predefined thematic categories to various degrees by the DTC module. Moreover, for each user profile, degrees of preference are mentioned for all of the predefined thematic categories. These may be manually predefined by the user or by an expert when referring to static user profiles, or created based solely on the monitoring of user actions in the case of dynamic user profiles.

The audiovisual classification module examines each document in the set of result of the search independently. The user's degree of preference for each one of the thematic categories that are related to a document is checked. If at least one of them is positive, negative preferences are ignored for this document. The document is promoted (its rank is increased), to the degree that it is related to some thematic category and that thematic category is of interest to the user. The re-ranking is performed using a parameterized weak modifier, where the intensity of the preference sets the parameter. A typical choice is:

$$r'(q)_d = \sqrt[1+x]{r(d)}$$

where x is given by:

$$x = h\left(I^{-1}(d) \cap P_T^+\right)$$

and P_T^+ are the positive thematic category preferences of the user. In other words, when the document is related to a thematic category to a high degree, and the preference of the user for that category is intense, then the document is promoted in the result. If the preference is less intense or if the document is related to the preference to a smaller extent, then the adjusting of the document's rank is not as drastic.

Quite similarly, when the document is only related to negative preferences of the user, then a strong modifier is used to adjust the document's ranking in the results:

$$r'(q)_d = (r(d))^{1+x}$$

where x is given by:

$$x = h\left(I^{-1}(d) \cap P_T^-\right)$$

and P_T^- are the negative thematic category preferences of the user.

Exploitation of Semantic Interests

Semantic interests offer a much more detailed description of user preferences. Their drawback is that they are mined with a lesser degree of certainty and they are more sensitive to context changes. This is why they are utilized more moderately.

The simple and composite entities contained in each document need to be compared with the ones contained in the user's profile. This comparison, though, needs to be performed in a context-sensitive manner. Specifically, in the case that no context can be detected in the query, the whole set of user interests is considered. If, on the other hand, the query context is intense, interests that do not intersect with the context should not be considered, thus eliminating inter-preference noise. Thus, semantic interests can only refine the contents of the result set moderately, always remaining in the same general topic. Ranks are updated based on similarity measures and relativity to context, as well as the preferences' intensities.

The relevance of an interest to the context of the query is quantified using the intensity of their common context, while the adjusting of ranks is performed similarly to the case of thematic categories, with x defined as:

$$x = \max(x_i)$$
$$x_i = h\left(I^{-1}(d) \cap P_i^+\right) \cdot h_{K(q) \cap P_i^+}$$

where P_i^+ is one of the positive interests of the user.

As far as negative interests are concerned, there is no need to consider the context before utilizing them. What is needed, on the other hand, is to make sure that they do not overlap with any of the in-context positive interests, as this would be inconsistent. Within a specific query context one may demand, as a minimum consistency criterion, that the set of considered interests does not contain both positive and negative preferences for the same semantic entities.

When 'correcting' the view of the user profile that is acquired by removing out-of-context interests, the following need to be obeyed:

- Positive interests are generally extracted with greater confidence. Therefore, positive interests are treated more favourably than negative ones, in the process of creating a consistent view of the profile.
- Obviously, if only positive interests correspond to a specific semantic entity, then their intensities must not be altered. Likewise, if only a negative interests corresponds to a specific semantic entity, then its intensity must not be altered.
- In general, the intensities of positive preferences should increase monotonically with respect to their original intensity, and decrease monotonically with respect to the original intensity of the corresponding negative preference, and vice versa.

These guidelines lead to the generation of a valid, i.e. consistent, context-sensitive user profile [8].

12.6.4 Extraction of User Preferences

Based on the operation of the DECO and DTC modules, the system can acquire in an automated manner and store in the index the fuzzy set of semantic entities and thematic categories (and consequently topics) that are related to each document. Still, this does not render trivial the problem of semantic user preference extraction. What remains is the determination of the following:

1. Which of the topics that are related to documents in the usage history are indeed of interest to the user and which are found there due to coincidental reasons?
2. To which degree is each one of these topics of interest to the user?

As far as the main guidelines followed in the process of preference extraction are concerned, the extraction of semantic preferences from a set of documents, given their topics, is quite similar to the extraction of topics from a document, given its semantic indexing. Specifically, the main points to consider may be summarized in the following:

1. A user may be interested in multiple, unrelated topics.
2. Not all topics that are related to a document in the usage history are necessarily of interest to the user.
3. Documents may have been recorded in the usage history that are not of interest to the user in some way (these documents were related to the local interest of the user at the time of the query, but are not related to the user's global interests.)

Clustering of documents

These issues are tackled using similar tools and principles to the ones used to tackle the corresponding problems in multimedia document analysis and indexing. Thus, once more, the basis on which the extraction of preferences is built is the context. The common topics of documents are used in order to determine which of them are of interest to the user and which exist in the usage history coincidentally.

Moreover, since a user may have multiple preferences, we should not expect all documents of the usage history to be related to the same topics. Quite the contrary, similarly to semantic entities that index a document, we should expect most documents to be related to just one of the user's preferences. Therefore, a clustering of documents, based on their common topics, needs to be applied. In this process, documents that are misleading (e.g. documents that the user chose to view once but are not related to the user's global interests) will probably not be found similar with other documents in the usage history. Therefore, the cardinality of the clusters may again be used to filter out misleading clusters.

For reasons similar to those in the case of thematic categorization, a hierarchical clustering algorithm needs to be applied. Thus, the clustering problem is reduced to the selection of merging and termination criteria. As far as the former is concerned, two clusters of documents should be merged if they are referring to the same topics. As far as the latter is concerned, merging should stop when no clusters remain with similar topics.

What is common among two documents $a,b \in D$, i.e. their common topics, can be referred to as their common context. This can be defined as:

$$K(a, b) = I^{-1}(a) \cap I^{-1}(b)$$

A metric that can indicate the degree to which two documents are related is, of course, the intensity (height) of their common context. This can be extended to the case of more than two documents, in order to provide a metric that measures the similarity between clusters of documents:

$$Sim(c_1, c_2) = h(K(c_1, c_2))$$

where $c_1,c_2 \subseteq H^+ \subseteq D$ and:

$$K(c_1, c_2) = \bigcap_{d \in c_1 \cup c_2} I^{-1}(d)$$

$$H = \{H^+, H^-\}$$

where H is a view of the usage history, comprising H^+, the set of documents that the user has indicated interest for, and H^-, the set of documents for which the user has indicated some kind of dislike. *Sim* is the degree of association for the clustering of documents in H^+. The termination criterion is again a threshold on the value of the best degree of association.

Extraction of interests and of preferences for thematic categories

The topics that interest the user, and should be classified as positive interests, or as positive preferences for thematic categories, are the ones that characterize the detected clusters. Degrees of preference can be determined based on the following parameters:

1. The cardinality of the clusters. Clusters of low cardinality should be ignored as misleading.
2. The weights of topics in the context of the clusters. High weights indicate intense interest. This criterion is only applicable in the case of user interests.

Therefore, each one of the detected clusters c_i is mapped to a positive interest as follows:

$$U_i^+ = L(c_i) \cdot K(c_i)$$
$$K(c_i) = \bigcap_{d \in c_i} I^{-1}(d)$$

where U_i^+ is the interest and $L(c_i)$ is a 'large' fuzzy number. When it comes to the case of thematic categories, they are generally extracted with higher degrees of confidence, but a larger number of documents need to correspond to them before the preference can be extracted. More formally, in the case of thematic categories the above formula becomes:

$$P_{T_i} = w(L'(c_i) \cdot K(c_i))$$

where w is a weak modifier and L' is a 'very large' fuzzy number.

The information extracted so far can be used to enrich user requests with references to topics that are of interest to the user, thus giving priority to related documents. What it fails to support, on the other hand, is the specification of topics that are known to be uninteresting for the user, as to filter out, or down-rank, related documents. In order to extract such information, a different approach is required.

First of all, a document's presence in H^- has a different meaning than its presence in H^+. Although the latter indicates that at least one of the document's topics is of interest to the user, the former indicates that, most probably, all topics that are related to the document are uninteresting to the user.

Still, topics may be found in H^- for coincidental reasons as well. Therefore, negative interests should be verified by the repeated appearance of topics in documents of H^-:

$$U^- = \sum s_i / u_i^-$$
$$u_i^- = L\Big(\sum_{d \in H^-} I(s_i, d) \Big)$$

Finally, both for positive and negative thematic category preferences and interests, due to the nature of the document analysis and document clustering processes, multiple semantic entities with closely related meanings are included in each preference and to similar degrees. In order to avoid this redundancy, a minimal number of semantic entities have to be selected and stored for each preference. This is achieved by forming a maximum independent set of entities in each preference, with semantic correlation (as shown by the height of the common context) indicating the proximity between two semantic entities. As initially connected we consider the pairs of entities whose common context has a height that exceeds a threshold.

12.7 Experimental Results

This section describes the quantitative performance analysis and evaluation of the audiovisual document retrieval capabilities of the proposed system, essentially verifying the responses to specific user queries against 'ground truth' to evaluate retrieval performance. In the sequel, the methodology followed for constructing the ground truth, carrying out the experiments and analysing the results is outlined. The overall results are presented and conclusions are drawn.

12.7.1 Methodology

Methodology for information retrieval performance evaluation

Performance characterization of audiovisual content retrieval often borrows from performance figures developed over the past 30 years for probabilistic text retrieval. Landmarks in the text retrieval field are the books [23], [24] and [25]. Essentially all information retrieval (IR) is about cluster retrieval: the user having specified a query would like the system to return some or all of the items, either documents, images or sounds, that are in some sense part of the same semantic cluster, i.e. the relevant fraction of the database with respect to this query for this user. The ideal IR system would quickly present the user some or all of the relevant material and nothing more. The user would value this ideal system as being either 100% effective or being without (0%) error.

In practice, IR systems are often far from ideal: the query results shown to the user, i.e. the finite list of retrieved items, generally are incomplete (containing some retrieved relevant items but without some missed relevant items) and polluted (with retrieved but irrelevant items). The performance is characterized in terms of precision and recall. *Precision* is defined as the number of retrieved relevant items over the number of total retrieved items. *Recall* is defined as the number of retrieved relevant items over the total number of relevant items:

$$p = precision = \frac{relevant\ retrieved\ items}{retrieved\ items}$$

$$r = recall = \frac{relevant\ retrieved\ items}{relevant\ items}$$

The performance for an 'ideal' system is to have both high precision and high recall. Unfortunately, they are conflicting entities and cannot practically assume high values at the same time. Because of this, instead of using a single value of precision and recall, a *precision–recall* (PR) graph is typically used to characterize the performance of an IR system. This approach has the disadvantage that the length, or *scope*, of the retrieved list, or visible size of the query results, is not displayed in the performance graph, whereas this scope is very important to the user because it determines the amount of items to be inspected and therefore the amount of time (and money) spent in searching. The scope is the main parameter of economic effectiveness for the user of a retrieval system. Moreover, even though well suited for purely text-based IR, a PR graph is less meaningful in audiovisual content retrieval systems where recall is consistently low or even unknown, in cases where the ground truth is incomplete and the cluster size is unknown. In these cases the *precision–scope* (PS) graph is typically employed to evaluate retrieval performance.

In [26], another performance measure is proposed: the *rank* measure, leading to *rank–scope* (RS) graphs. The rank measure is defined as the average rank of the retrieved relevant items. It is clear that the smaller the rank, the better the performance. While PS measurements only care if a relevant item is retrieved or not, RS measurements also care about the rank of that item. Caution must be taken when using RS measurements, though. If system A has higher precision and lower rank measurements than system B, then A is definitely better than B, because A not only retrieves more relevant images than B, but also all those retrieved images are closer to the top in A than in B. But if both precision and rank measurements of A are higher than those of B, no conclusion can be made.

Equally important is the degradation due to a growing database size, i.e. lowering the fraction of relevant items resulting in overall lower precision–recall values. A comparison between two information retrieval systems can only be done well when both systems are compared in terms of equal *generality*:

$$g = generality = \frac{relevant\ items}{all\ items}$$

Although there is a simple method of minimizing the number of irrelevant items (by minimizing the number of retrieved items to zero) and a simple one to minimize the number of missed relevant items (by maximizing the number of retrieved items up to the complete database), the optimal length of the result list depends upon whether one is satisfied with finding one, some or all relevant items.

The parameterized *error* measure of [23]:

$$E = Error = 1 - \frac{1}{a(1/p) + (1-a)(1/r)}$$

is a normalized error measure where a low value of a favours recall and a high value of a favours precision. E will be 0 for an ideal system with both precision and recall values at 1 (and in that case irrespective of a). The setting of $a = 0.5$ is typically chosen, a choice giving equal weight to precision and recall and giving rise to the normalized symmetric difference as a good single number indicator of system performance. Moreover, an intuitive best value of 1 (or 100%) is to be preferred; this is easily remedied by inverting the [1,0] range. Thus, *effectiveness* is defined as:

$$e = effectiveness = 1 - E(a = 0.5) = \frac{1}{(1/2p) + (1/2r)}$$

Evaluation procedure

Based on the above methodology and guidelines for retrieval performance evaluation, a series of experiments was carried out to evaluate the system's retrieval performance. Evaluation was based on ground truth in a well-defined experimental setting allowing the recovery of all essential parameters. The evaluation test bed was the prototype of the experimental system developed in the framework of the FAETHON IST project, which served as a mediator for unified semantic access to five archives with documents annotated in different languages and using diverse data structures. The five archives were the Hellenic Broadcast Corporation (ERT) and Film Archive Greece (FAG) from Greece, Film Archiv Austria (FAA) and Austrian Broadcasting Corporation (ORF) and Alinary from Italy.

The first step was to develop the ground truth against which all retrieval results had to be compared in order to measure retrieval performance. The ground truth in general included a set of semantic test queries and the corresponding sets of 'ideal' system responses for each query. There are three actions involved in this process:

1. Since the content of the five participating archives belongs in general to varying thematic categories, the set of queries had to relate to concepts that were common in all, or most, archives, so that corresponding responses were sufficiently populated from all archives.

2. Once the set of test queries was specified, the 'ideal' set of responses (list of audiovisual documents to be returned) had to be specific with corresponding degrees of confidence, or equivalently ranked. This was repeated for each test query and for each participating archive.
3. Finally, in order to include personalization in retrieval performance evaluation, separate response sets were prepared for a limited number of pre-specified user profiles, differing in their semantic preferences.

Due to the large size of the archives, the existing knowledge base and user profiles, caution was taken to limit the number of test queries and user preferences, and even use a subset only of existing archive content. All ground-truth information was manually generated so the above selections were crucial in making the test feasible in terms of required effort.

Subsequently, the test queries were fed into the system and corresponding responses were recorded from all archives and for each specified user profile; these were automatically tested against ground-truth data in order to make comparisons. The latter were performed according to the performance evaluation criteria and measures specified above. In particular, for each test query:

1. the retrieved documents had been directly recorded;
2. the relevant documents, with associated degrees of relevance, were available from ground-truth data;
3. the relevant retrieved documents were calculated as the intersection of the two above sets of documents;
4. the total number of all documents in the system index was a known constant.

Thus, all quantities required for the calculation of precision, recall, generality error and effectiveness were available. Additionally to the above described methodology, wherever relevance or confidence values were available, such as in the list of retrieved documents, all cardinality numbers, or total number of documents, were replaced by the respective sums of degrees of relevance.

Finally, all precision and recall measurements were recorded for each experiment, i.e. for each test query and user profile. Average precision–recall values were calculated per query and user profile, and corresponding PR graphs were drawn and studied. The overall results are presented and conclusions on the system's retrieval performance are drawn.

12.7.2 Experimental Results and Conclusions

Experimental settings

Following the procedure described above, which is in turn based on the methodology of the previous subsection, we are going to calculate the quadruple $\{p, r, g, e\}$. The number of all audiovisual documents is a known constant, i.e. $d = 1005$. Because of the reasons mentioned in the evaluation procedure above, the parameterization could not be as extensive as theory demands. So the following compromises were made, in order to achieve reliable results within a feasible evaluation period of time:

1. Only three different user profiles were taken into account; one without any semantic preferences, a second with interest in politics and a third with interest in sports.

2. The ground truth was built manually for the following semantic queries: 'Navy', 'Football match', 'Elections', 'Aircraft' and 'Olympic games'.
3. Since the system returns the same audiovisual documents for a user query irrelative to the user's semantic preferences and only changes their degree of relevance (audiovisual classification re-ranks documents, it does not filter them), we consider that the retrieved documents are only these which have a degree above 30%. So, depending on the user's preferences, we retrieve a different number of documents.

Based on the above constraints, we executed the five semantic queries for each one of the three user profiles; thus we acquired fifteen different result lists. In Tables 12.1–12.5 we demonstrate the results, grouped by each query.

Retrieval results

Query 1: 'Navy'
For this specific query, we see that the results do not vary a lot among the three different user profile (Table 12.1). This is expected, since the word 'Navy' semantically is not related more to one of the two pre-selected semantic user preferences. We also notice that the system tends to respond with less accuracy in favour of better recall numbers.

Query 2: 'Football Match'
This time the query is related to the thematic category 'sports and athletics', which makes distinguishable better results for the user whose semantic preferences are set to 'sports'. This can be seen from the higher effectiveness number (Table 12.2).

Query 3: 'Elections'
The user who made the semantic query 'Elections' expects to retrieve some audiovisual content related to elections in the first place and to politics in extension. Consequently the user with preference in the topic 'politics' gets both higher precision and recall indices than the user

Table 12.1 The estimated parameters are demonstrated for each user profile for the semantic query 'Navy' against the ground truth

Profile	Relevant	Retrieved	Relative and retrieved	p	r	e	g
None	18	26	16	0.615	0.889	0.727	
Politics	18	50	15	0.750	0.833	0.789	0.018
Sports	18	22	15	0.682	0.833	0.750	

Table 12.2 The estimated parameters are demonstrated for each user profile for the semantic query 'Football match' against the ground truth

Profile	Relevant	Retrieved	Relative and retrieved	p	r	e	g
None	59	58	42	0.724	0.712	0.718	
Politics	59	48	41	0.854	0.695	0.765	0.059
Sports	59	86	52	0.605	0.881	0.717	

Table 12.3 The estimated parameters are demonstrated for each user profile for the semantic query 'Elections' against the ground truth

Profile	Relevant	Retrieved	Relative and retrieved	*p*	*r*	*e*	*g*
None	49	56	35	0.625	0.714	0.667	
Politics	49	51	40	0.784	0.816	0.800	0.049
Sports	49	39	31	0.795	0.633	0.705	

with no special interests, and the effectiveness index is higher than that of all the other users (Table 12.3).

Query 4: 'Aircraft'

In this query, we observed lower figures for the precision (Table 12.4). In other words, the system returned among the relevant documents many other irrelevant (according to the ground truth). This is depicted in Figure 12.18, with the points corresponding to this query being in the lower right part of the diagram.

Query 5: 'Olympic games'

In comparison to the previous queries this one, 'Olympic games', is related to more audiovisual documents in all five archives, a fact apparent from the generality index as well. This time we had results with higher precision compared to the recall for the queries performed by the user with no special interest and the user with interest in politics (Table 12.5), something which is also shown in Figure 12.18, where the corresponding points in the graph are in the left upper part of the diagram.

Figure 12.18 summarizes all of the above queries in the form of a PR graph. In the same figure we demonstrate the three different user profiles used. Five points were drawn, since we performed five queries. The lines were drawn after polynomial interpolation and with the use of statistical techniques, in order to show the exponential decrease of the PR graph. Although few sound conclusions can be drawn from a diagram that has resulted from so few queries, one can easily make at least two observations:

- System responses are generally better when user queries are also considered, as in this case more information is available to the system for the selection of relevant documents.
- The PR diagrams are generally located in the upper right corner of the 0–100 space, a fact that is not common in PR diagrams, thus indicating that the utilization of ontological semantic

Table 12.4 The estimated parameters are demonstrated for each user profile for the semantic query 'Aircraft' against the ground truth

Profile	Relevant	Retrieved	Relative and retrieved	*p*	*r*	*e*	*g*
None	17	29	16	0.552	0.941	0.696	
Politics	17	22	15	0.682	0.882	0.780	0.017
Sports	17	20	13	0.650	0.765	0.696	

Table 12.5 The estimated parameters are demonstrated for each user profile for the semantic query 'Olympic games' against the ground truth

Profile	Relevant	Retrieved	Relative and retrieved	*p*	*r*	*e*	*g*
None	95	66	55	0.833	0.579	0.683	
Politics	95	62	56	0.903	0.589	0.712	0.095
Sports	95	105	79	0.752	0.832	0.790	

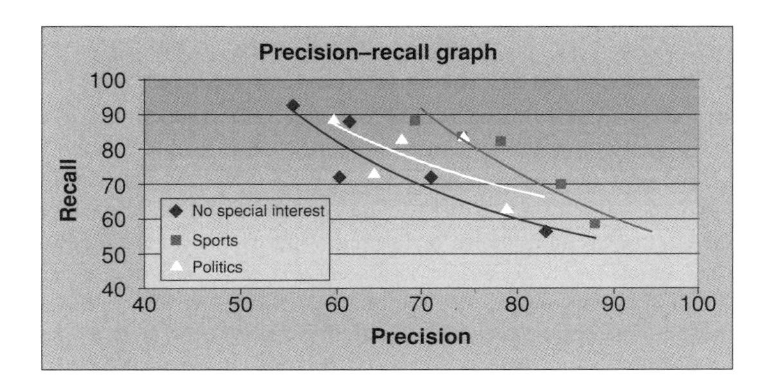

Figure 12.18 The estimated parameters from the five queries are demonstrated for each user profile against the ground truth

knowledge in the processing of user queries, documents and profiles can greatly contribute to the enhancement of the acquired results.

12.8 Extensions and Future Work

The key aspect of the FAETHON developments has been the generation and use of metadata in order to provide advanced content management and retrieval services. The web will change drastically in the following years and become more and more multimedia enabled, making already complex content management tasks even more complex and requiring solutions based on Semantic Web technologies. Unlike today, content itself will be a commodity in a future web, making the use of metadata essential. Content providers, for instance, will have to understand the benefits obtained from the systematic generation of metadata; service providers will have to accept metadata as the basis on which to build new services; and the producers of software tools for end users will redirect their imagination towards more appropriate integration of application software with web content, taking advantage of metadata. These developments clearly present some challenging prospects, in technological, economic, standardization and business terms.

Another interesting perspective of FAETHON's developments is the personalization, based on usage history, of the results of content retrieval. Personalization software is still in its infancy, which means there are no turnkey solutions. Solutions using agent technologies still have a lot of hurdles to overcome. To improve this scenario, additional technology approaches need to be evaluated and areas of improvement identified. In both perspectives, clearly FAETHON made some interesting steps on the correct route, and its developments are currently influencing the next research activities in the area of semantic-based knowledge systems.

The long-term market viability of multimedia services requires significant improvements to the tools, functionality and systems to support target users. *aceMedia* seeks to overcome the barriers to market success, which include user difficulties in finding desired content, limitations in the tools available to manage personal and purchased content, and high costs to commercial content owners for multimedia content processing and distribution, by creation of means to generate semantic-based, context- and user-aware content, able to adapt itself to users' preferences and environments. aceMedia will build a system to extract and exploit meaning inherent to the content in order to automate annotation and to add functionality that makes it easier for all users to create, communicate, find, consume and reuse content.

aceMedia targets knowledge discovery and embedded self-adaptability to enable content to be self-organising, self-annotating, self-associating, more readily searched (faster, more relevant results), and adaptable to user requirements (self-reformatting). aceMedia introduces the novel concept of the Autonomous Content Entity (ACE), which has three layers: content, its associated metadata, and an intelligence layer consisting of distributed functions that enable the content to instantiate itself according to its context (e.g. network, user terminal, user preferences). The ACE may be created by a commercial content provider, to enable personalized self-announcement and automatic content collections, or may be created in a personal content system in order to make summaries of personal content, or automatically create personal albums of linked content.

Current multimedia systems and services do not support their target users well enough to imagine their long-term market expansion, without significant improvements to the tools, functionality and systems available to the user. The aceMedia project sets out to offer solutions to the barriers to market success, which include:

- users being unwilling to sign up for commercial multimedia services when they are unable to readily find desired content, and are limited in the tools available to manage that content once purchased;
- commercial content owners unwilling to invest resources (usually staff) in content provision due to the high costs associated with multimedia content processing and distribution;
- individual users of multimedia acquisition and storage systems being unable to manage their ever-growing personal content collections, but the only tools available to assist them meet only a part of their needs, and the complexity of such tools usually sites them in the realm of the professional user

To address these problems, aceMedia focuses on generating value and benefits to end users, content providers, network operators and multimedia equipment manufacturers, by introducing, developing and implementing a system based on an innovative concept of knowledge-assisted, adaptive multimedia content management, addressing user needs. The main technological objectives are to discover and exploit knowledge inherent to the content in order to make content more relevant to the user, to automate annotation at all levels, and to add functionality to ease content creation, transmission, search, access, consumption and reuse. In addition, available user and terminal profiles, the extracted semantic content descriptions and advanced mining methods will be used to provide user and network adaptive transmission and terminal optimized rendering.

The current World Wide Web is, by its function, the syntactic web where structure of the content has been presented while the content itself is inaccessible to computers. The next generation of the web (the Semantic Web) aims to alleviate such problems and provide specific solutions targeting the concrete problems. Web resources will be much easier and more readily accessible by both humans and computers with the added semantic information in a machine-understandable and machine-processable fashion. It will have much higher impact on e-work and e-commerce than the current version of the web. There is, however, still a long way to go to transfer the Semantic Web from an academic adventure into a technology provided by the software industry.

Supporting this transition process of ontology technology from academia to industry is the main and major goal of *Knowledge Web*. This main goal naturally translates into three main objectives given the nature of such a transformation:

1. Industry requires immediate support in taking up this complex and new technology. Languages and interfaces need to be standardized to reduce the effort and provide scalability to solutions. Methods and use cases need to be provided to convince and to provide guidelines for how to work with this technology.
2. Important support to industry is provided by developing high-class education in the area of the Semantic Web, web services and Ontologies.
3. Research on ontologies and the Semantic Web has not yet reached its goals. New areas such as the combination of the Semantic Web with web services realizing intelligent web services require serious new research efforts.

In a nutshell, it is the mission of Knowledge Web to strengthen the European software industry in one of the most important areas of current computer technology: Semantic Web enabled e-work and e-commerce. Naturally, this includes education and research efforts to ensure the durability of impact and support of industry.

References

[1] A. Delopoulos, S. Kollias, Y. Avrithis, W. Haas, K. Majcen, Unified intelligent access to heterogeneous audio-visual content. In *Proceedings of International Workshop on Content-Based Multimedia Indexing (CBMI '01)*, Brescia, Italy, 19–21 September 2001.

[2] J. Hunter, Adding multimedia to the Semantic Web: building an MPEG-7 ontology. In *Proceedings of First Semantic Web Working Symposium, SWWS '01*, Stanford University, CA, July 2001.

[3] F. Nack, A. Lindsay, Everything you wanted to know about MPEG-7: part 1. *IEEE Multimedia*, **6**(3), 65–77, 1999.

[4] F. Nack, A. Lindsay, Everything you wanted to know about MPEG-7: Part 2. *IEEE Multimedia*, **6**(4), 64–73, 1999.

[5] T. Sikora, The MPEG-7 visual standard for content description—an overview. *IEEE Transactions on Circuits and Systems for Video Technology*, special issue on MPEG-7, **11**(6), 696–702, 2001.

[6] ISO/IEC JTC1/SC29/WG11 N5231, MPEG-21 Overview, Shanghai, October 2002.

[7] G. Akrivas, G. Stamou, Fuzzy semantic association of audiovisual document descriptions. In *Proceedings of Int errational Workshop on Very Low Bitrate Video Coding (VLBV)*, 2001.

[8] M. Wallace, G. Akrivas, G. Stamou, S. Kollias, Representation of user preferences and adaptation to context in multimedia content-based retrieval. In *Workshop on Multimedia Semantics, Accompanying 29th Annual Conference on Current Trends in Theory and Practice of Informatics (SOFSEM)*, Milovy, Czech Republic, November 2002.

[9] M. Wallace, S. Kollias, Computationally efficient incremental transitive closure of sparse fuzzy binary relations. In *Proceedings of the IEEE International Conference on Fuzzy Systems (FUZZ-IEEE)*, Budapest, Hungary, July 2004. IEEE, 2004.

[10] G. Klir, B. Yuan, *Fuzzy Sets and Fuzzy Logic, Theory and Applications*. Prentice Hall, Eaglewood Cliffs, NJ, 1995.

[11] M. Wallace, G. Akrivas, G. Stamou, Automatic thematic categorization of documents using a fuzzy taxonomy and fuzzy hierarchical clustering. In *Proceedings of the IEEE International Conference on Fuzzy Systems (FUZZ-IEEE)*, St Louis, MO, May 2003. IEEE, 2003.

[12] D.H. Kraft, G. Bordogna, G. Passi, Fuzzy set techniques in information retrieval. In J.C. Berdek, D. Dubois, H. Prade (eds) *Fuzzy Sets in Approximate Reasoning and Information Systems.* Kluwer Academic, Boston, MA, 2000.

[13] S. Miyamoto, *Fuzzy Sets in Information Retrieval and Cluster Analysis*. Kluwer Academic, Dordrecht, 1990.

[14] W.-S. Li, D. Agrawal, Supporting web query expansion efficiently using multi-granularity indexing and query processing. *Data and Knowledge Engineering*, **35**(3), 239–257, 2000.

[15] D.H. Kraft, F.E. Petry, Fuzzy information systems: managing uncertainty in databases and information retrieval systems. *Fuzzy Sets and Systems*, **90**, 183–191, 1997.

[16] G. Akrivas, M. Wallace, G. Andreou, G. Stamou, S. Kollias, Context-sensitive semantic query expansion. In *IEEE International Conference on Artificial Intelligence Systems (ICAIS)*, Divnomorskoe, Russia, September 2002. IEEE, 2002.

[17] R.R. Yager, J. Kacprzyk, *The Ordered Weighted Averaging Operators: Theory and Applications*, pp. 139–154. Kluwer Academic, Norwell, MA, 1997.

[18] R.R. Yager, Families of OWA operators. *Fuzzy Sets and Systems*, **59**, 125–148, 1993.

[19] P.M. Chen, F.C. and Kuo, An information retrieval system based on a user profile. *Journal of Systems and Software,* **54**, 3–8, 2000.

[20] C.L. Barry, User-defined relevance criteria: an exploratory study. *Journal of the American Society for Information Science*, **45**, 149–159, 1994.

[21] C.H. Chang, C.C. Hsu, Integrating query expansion and conceptual relevance feedback for personalized Web information retrieval. *Computer Networks and ISDN Systems*, **30**, 621–623, 1998.

[22] D.H. Kraft, G. Bordogna, G. Passi, Information retrieval systems: where is the fuzz? In *Proceedings of IEEE International Conference on Fuzzy Systems*, Anchorage, Alaska, May 1998. IEEE, 1998.

[23] C.J. van Rijsbergen, *Information Retrieval*, 2nd edn. Butterworths, London, 1979.

[24] G. Salton, M.J. McGill, *Introduction to Modern Information Retrieval*, McGraw-Hill, New York, 1982.

[25] R.A. Baeza-Yates, B.A. Ribeiro-Neto, *Modern Information Retrieval.* ACM Press/Addison-Wesley, Reading, MA, 1999.

[26] J. Huang, S. Kumar, M. Mitra,W.-J. Zhu, R. Zabih, Image indexing using color correlogram. In *Proceedings of IEEE Conference on Computer Vision and Pattern Recognition*. IEEE, 1997.

13

Multimedia Content Indexing and Retrieval Using an Object Ontology

Ioannis Kompatsiaris, Vasileios Mezaris and Michael G. Strintzis

13.1 Introduction

In recent years, the accelerated growth of digital media collections, both proprietary and on the web, has underlined the need for the development of human-centred tools for the efficient access and retrieval of visual information. As the amount of information available in the form of multimedia content continuously increases, the necessity for efficient methods for the retrieval of visual information becomes increasingly evident. Additionally, the continuously increasing number of people with access to such collections further dictates that more emphasis be put on attributes such as the user-friendliness and flexibility of any multimedia content retrieval scheme. These facts, along with the diversity of available content collections, varying from *restricted*, e.g. medical databases and satellite photo collections, to general purpose collections, which contain heterogeneous images and videos, and the diversity of requirements regarding the amount of knowledge about the content that should be used for indexing, have led to the development of a wide range of solutions. They have also triggered the development of relevant standards allowing object-based functionalities (MPEG-4), efficient access to content (MPEG-7) and supporting the whole multimedia chain (MPEG-21). Current indexing schemes employ descriptors ranging from low-level features to higher-level semantic concepts [1]. Low-level features are machine-oriented and can be automatically extracted (e.g MPEG-7 compliant descriptors [2]), whereas high-level concepts require manual annotation of the medium or are restricted to specific domains.

The very first attempts at multimedia content retrieval were focused mainly on still images and were based on exploiting existing image captions to classify images according to predetermined classes or to create a restricted vocabulary [3]. Although relatively simple and

Multimedia Content and the Semantic Web Edited by Giorgos Stamou and Stefanos Kollias

computationally efficient, this approach has several restrictions mainly deriving from the use of a restricted vocabulary that neither allows for unanticipated queries nor can be extended without re-evaluating the possible connection between each item in the database and each new addition to the vocabulary. Additionally, such keyword-based approaches assume either the pre-existence of textual annotations (e.g. captions) or that annotation using the predetermined vocabulary is performed manually. In the latter case, inconsistency of the keyword assignments among different indexers can also hamper performance. Recently, a methodology for computer-assisted annotation of image collections was presented [4].

To overcome the limitations of the keyword-based approach, the use of the visual content has been proposed. This category of approaches utilizes the visual contents by extracting low-level indexing features. In this case, pre-processing of multimedia data is necessary as the basis on which indices are extracted. The pre-processing is of *coarse granularity* if it involves processing of images or video frames as a whole, whereas it is of *fine granularity* if it involves detection of objects within an image or a video frame [5]. Then, relevant images or video segments are retrieved by comparing the low-level features of each item in the database with those of a user-supplied sketch or, more often, a key image/key frame that is either selected from a restricted image set or is supplied by the user (*query-by-example*). One of the first attempts to realize this scheme is the *Query by Image Content* system [6, 7]. Newer contributions to query-by-example include UCSB's *NeTra* [8, 9], UIUC's *Mars* [10], MIT's *Photobook* [11], Columbia's *VisualSEEK* [12] and ITI's *Istorama* [13]. They all employ the general framework of query-by-example, demonstrating the use of various indexing feature sets either in the *image* or in the *region* domain. Interesting work is presented in a recent addition to this group, Berkeley's *Blobworld* [14, 15], which clearly demonstrates the improvement in query results attained by querying using region-based indexing features rather than global image properties, under the query-by-example scheme. The main drawback of all the above approaches is that, in order to start a query, the availability of an appropriate key image is assumed; occasionally, this is not feasible, particularly for classes of images or videos that are underrepresented in the database.

Hybrid methods exploiting both keywords and the visual content have also been proposed [16–18]. In [16], the use of *probabilistic multimedia objects* (*multijects*) is proposed; these are built using hidden Markov models and necessary training data. Significant work was recently presented on unifying keywords and visual contents in image retrieval. The method of [18] performs semantic grouping of keywords based on user relevance feedback to effectively address issues such as word similarity and allow for more efficient queries; nevertheless, it still relies on pre-existing or manually added textual annotations. In well-structured specific domain applications (e.g. sports and news broadcasting) domain-specific features that facilitate the modelling of higher level semantics can be extracted [1, 5]. A priori knowledge representation models are used as a knowledge base that assists semantic-based classification and clustering. In [19], *semantic entities*, in the context of the MPEG-7 standard, are used for knowledge-assisted video analysis and object detection, thus allowing for semantic-level indexing. However, the need for accurate definition of semantic entities using low-level features restricts this class of approaches to domain-specific applications and prohibits non-experts from defining new semantic entities. In [20] the problem of bridging the gap between low-level representation and high-level semantics is formulated as a probabilistic pattern recognition problem. In [21, 22] hybrid methods extending the query-by-example strategy are developed.

This work attempts to address the problem of retrieval in generic multimedia content collections, where no possibility of structuring a domain-specific knowledge base exists, without

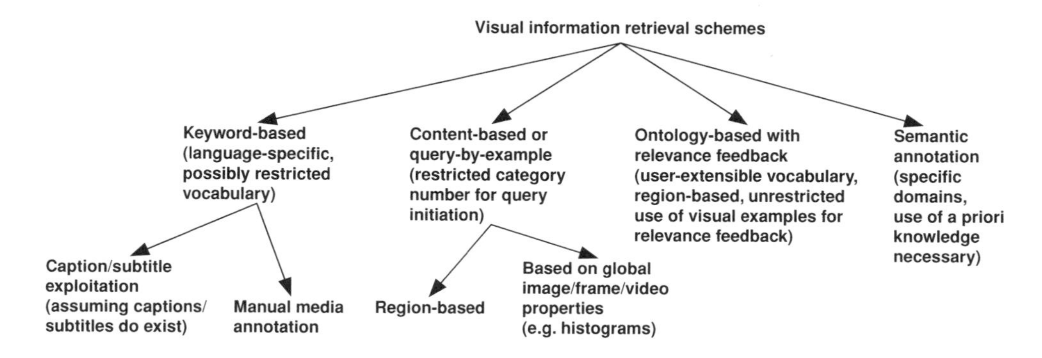

Figure 13.1 Overview of visual information retrieval techniques

imposing restrictions such as the availability of key images or textual annotations. The adopted object-based approach employs still-image and video segmentation tools that enable the time-efficient and unsupervised analysis of visual information to spatial or spatio-temporal objects, thus enabling 'content-based' access and manipulation via the extraction of MPEG-7 compliant low-level indexing features for each object. To take further advantage of the user-friendly aspects of the object-based approach, the low-level indexing features for the objects can be associated with higher-level concepts that humans are more familiar with. This is achieved by introducing the use of *ontologies* [23–25], which define a formal language for the structuring and storage of the high-level features, facilitate the mapping of low-level to high-level features and allow the definition of relationships between pieces of multimedia information. In the proposed approach, this ontology paradigm is coupled with a relevance feedback mechanism, based on support vector machines, to allow for precision in retrieving the desired content. The resulting retrieval scheme provides flexibility in defining the desired semantic object/keyword and bridges the gap between automatic semantic indexing for specific domains and query-by-example approaches (Figure 13.1). Therefore, it is endowed with the capability of querying large collections of heterogeneous content in a human-centred fashion.

The chapter is organized as follows: The general architecture of the proposed multimedia content retrieval scheme is developed in Section 13.2. A brief overview of the employed image segmentation algorithm is presented in Section 13.3. The algorithm for the real-time, unsupervised spatio-temporal segmentation of video sequences in the compressed domain is described in Section 13.4. Section 13.5 presents the MPEG-7 compliant low-level indexing features that are calculated for every object. In Section 13.6, the use of ontologies and their importance in associating low-level features and high-level concepts in a flexible manner are discussed. The employed relevance feedback technique is presented in Section 13.7. Section 13.8 contains an experimental evaluation of the developed methods, and finally, conclusions are drawn in Section 13.9.

13.2 System Architecture

As already mentioned, an object-based approach to image and video retrieval has been adopted; thus, the process of inserting a piece of visual information into the database starts by applying a segmentation algorithm to it, so as to break it down into a number of spatial or spatio-temporal objects. After the segmentation process, a set of MPEG-7 compliant low-level indexing features is calculated for each formed object. These arithmetic features compactly describe attributes such as the color, shape and position/motion of the object.

The calculated low-level indexing features are machine-centred rather than human-centred. For this reason, they are subsequently translated to intermediate-level descriptors qualitatively describing the object attributes, such as 'slightly oblong', that humans are more familiar with. The intermediate-level descriptors that can be used for this qualitative description form a simple vocabulary, the object ontology. Since these qualitative descriptors roughly describe the object, as opposed to the low-level features, they will be used only for ruling out objects that are irrelevant to the ones desired by the user in a given query, while accurate object ranking will still be based on the low-level features. Nevertheless, the whole system is designed to hide the existence of low-level features from the user; thus the user has to manipulate only intermediate-level descriptors, in contrast to most other systems.

For the proposed system to be able to rule out, during the query process, spatial or spatio-temporal objects irrelevant to the desired semantic object, one has to additionally supply a rough definition of the latter. This definition is easily formulated using the intermediate-level descriptors of the simple object ontology. In this way, objects irrelevant to a given query can be discarded by comparing their intermediate-level qualitative descriptors with those of the desired semantic object. This procedure is described schematically in Figure 13.2.

The comparison of intermediate-level descriptors is the first step in executing a query, made of one or more keywords (semantic objects) already defined by the system supervisor using the object ontology or defined by the user during the query procedure. For this purpose, the user can assign the proper intermediate-level descriptions to the keyword through a user interface. Additional constraints regarding the desired temporal and spatial relations of the semantic objects (e.g. object A 'before' and 'left' of object B) can be defined using the simple *query ontology* employed. The output of this query is a collection of multimedia items which contain potentially relevant spatial/spatio-temporal objects, whose relevance cannot be quantitatively

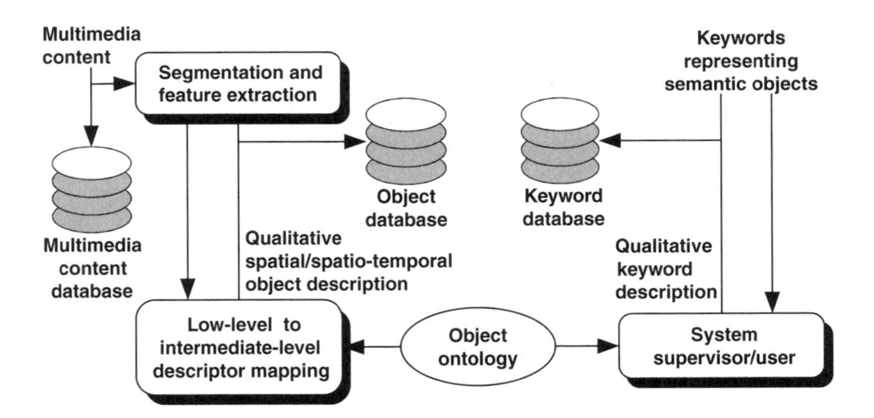

Figure 13.2 Indexing system overview. Low-level and intermediate-level descriptor values for the objects are stored in the object database; intermediate-level descriptor values for the user-defined keywords (semantic objects) are stored in the keyword database

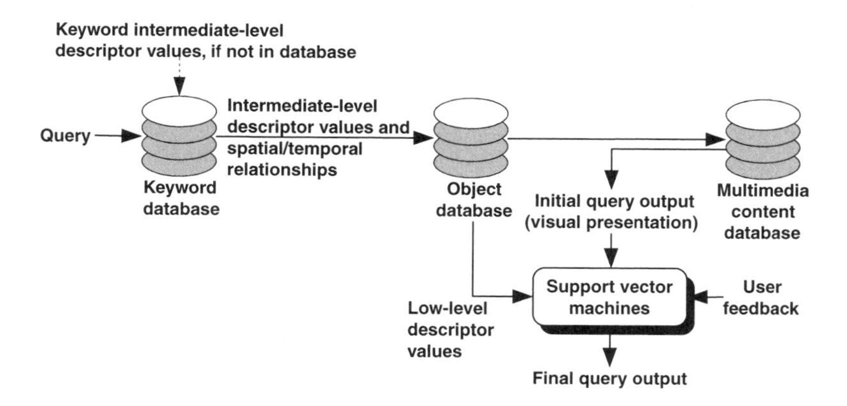

Figure 13.3 Query process overview

expressed at this point. Therefore, they are presented to the user in random order. The user then evaluates the members of a small subset of the retrieved collection, marking relevant objects simply by checking the appropriate 'relevant' box. By submitting this relevance feedback, one or two support vector machines are trained and subsequently rank according to relevance all objects returned by the initial query, using their low-level features. This relevance feedback process can then be repeated, to further enhance the output of the query. The query procedure is graphically described in Figure 13.3.

13.3 Still-Image Segmentation

13.3.1 Still-Image Segmentation System Overview

The segmentation system described in this section is based on a novel variant of the K-Means-with-connectivity-constraint algorithm (KMCC) [26, 27], a member of the popular K-Means family. The KMCC variant presented is an algorithm that classifies the pixels into regions taking into account the intensity, texture and position information associated with each pixel. The KMCC algorithm is coupled with an initial clustering procedure, based on breaking down the image into square blocks and assigning an intensity feature vector and a texture feature vector to each block, which makes unnecessary any user intervention.

The overall segmentation algorithm consists of the following stages:

- Stage 1: Extraction of the intensity and texture feature vectors corresponding to each pixel.
- Stage 2: Initial estimation of the number of regions and their spatial, intensity and texture centres.
- Stage 3: Conditional filtering using a moving average filter.
- Stage 4: Final classification of the pixels, using the KMCC algorithm.

13.3.2 Color and Texture Features

For every pixel $\mathbf{p} = [p_x \; p_y]$, $p_x = 1, \ldots, x_{max}$, $p_y = 1, \ldots, y_{max}$, where x_{max}, y_{max} are the image dimensions in pixels, a color feature vector and a texture feature vector are calculated.

The color features used are the three intensity coordinates of the CIE L*a*b* color space. What makes CIE L*a*b* more suitable for the proposed algorithm than the widely used RGB color space is perceptual uniformity: the CIE L*a*b* is approximately perceptually uniform, i.e. the numerical distance in this color space is approximately proportional to the perceived color difference [28]. The color feature vector of pixel $\mathbf{p}$, $\mathbf{I}(\mathbf{p})$ is defined as:

$$\mathbf{I}(\mathbf{p}) = [I_L(\mathbf{p})\ I_a(\mathbf{p})\ I_b(\mathbf{p})] \tag{13.1}$$

In order to detect and characterize texture properties in the neighbourhood of each pixel, the discrete wavelet frames (DWF) decomposition, proposed by Unser [29], is used. This is a method similar to the discrete wavelet transform (DWT), which uses a filter bank to decompose each intensity component of the image to a set of subbands. The main difference between the two methods is that in the DWF decomposition the output of the filter bank is not subsampled. The DWF approach has been proven to decrease the variability of the estimated texture features, thus improving classification performance [29]. In this method, the filter bank used is based on the lowpass Haar filter $H(z) = \frac{1}{2}(1 + z^{-1})$, which, despite its simplicity, has been demonstrated to perform surprisingly well for texture segmentation. The resulting texture feature vector of pixel $\mathbf{p}$ is denoted by $\mathbf{T}(\mathbf{p})$.

13.3.3 Initial Clustering

Similarly to any other variant of the K-Means algorithm, the KMCC algorithm requires initial values: an initial estimation of the number of regions in the image and their spatial, intensity and texture centres (all of these initial values can and are expected to be altered during the execution of the algorithm). In order to compute them, the image is broken down into square, non-overlapping blocks of dimension $f \times f$. In this way, a reduced image composed of a total of L blocks, $b_l, l = 1, \ldots, L$, is created. A color feature vector $\mathbf{I}^b(b_l) = [I_L^b(b_l)\ I_a^b(b_l)\ I_b^b(b_l)]$ and a texture feature vector $\mathbf{T}^b(b_l)$ are then assigned to each block; their values are estimated as the averages of the corresponding features for all pixels belonging to the block. The distance between two blocks is defined as follows:

$$D^b(b_l, b_n) = \|\mathbf{I}^b(b_l) - \mathbf{I}^b(b_n)\| + \lambda_1 \|\mathbf{T}^b(b_l) - \mathbf{T}^b(b_n)\| \tag{13.2}$$

where $\|\mathbf{I}^b(b_l) - \mathbf{I}^b(b_n)\|$, $\|\mathbf{T}^b(b_l) - \mathbf{T}^b(b_n)\|$ are the Euclidean distances between the block feature vectors. In our experiments, $\lambda_1 = 1$, since experimentation showed that using a different weight λ_1 for the texture difference would result in erroneous segmentation of textured images if $\lambda_1 \ll 1$ and non-textured images if $\lambda_1 \gg 1$. The value $\lambda_1 = 1$ was experimentally shown to be appropriate for a variety of textured and non-textured images; small deviations from this value have little effect on the segmentation results.

The number of regions of the image is initially estimated by applying a variant of the maximin algorithm [30] to this set of blocks. The distance C between the first two centres identified by the maximin algorithm is indicative of the intensity and texture contrast of the particular image. Subsequently, a simple K-Means algorithm is applied to the set of blocks, using the information produced by the maximin algorithm for its initialization. Upon convergence, a recursive four-connectivity component labelling algorithm [31] is applied, so that a total of K' connected regions s_k, $k = 1, \ldots, K'$ are identified. Their intensity, texture and spatial centres, $\mathbf{I}^s(s_k)$, $\mathbf{T}^s(s_k)$ and $\mathbf{S}(s_k) = [S_x(s_k)\ S_y(s_k)]$, $k = 1, \ldots, K'$, are calculated as follows:

$$\mathbf{I}^s(s_k) = \frac{1}{M_k} \sum_{\mathbf{p} \in s_k} \mathbf{I}(\mathbf{p}), \tag{13.3}$$

$$\mathbf{T}^s(s_k) = \frac{1}{M_k} \sum_{\mathbf{p} \in s_k} \mathbf{T}(\mathbf{p}), \tag{13.4}$$

$$\mathbf{S}(s_k) = \frac{1}{M_k} \sum_{\mathbf{p} \in s_k} \mathbf{p}, \tag{13.5}$$

where M_k is the number of pixels belonging to region s_k: $s_k = \{\mathbf{p}_1, \mathbf{p}_2, \ldots, \mathbf{p}_{M_k}\}$.

13.3.4 Conditional Filtering

An image may contain parts in which intensity fluctuations are particularly pronounced, even when all pixels in these parts of the image belong to a single object. In order to facilitate the grouping of all these pixels in a single region based on their texture similarity, a moving average filter is used to alter the intensity information of the corresponding pixels. The decision of whether the filter should be applied to a particular pixel **p** or not is made by evaluating the norm of the texture feature vector **T(p)** (Section 13.3.2); the filter is not applied if that norm is below a threshold τ. The output of the conditional filtering module can thus be expressed as:

$$\mathbf{J}(\mathbf{p}) = \begin{cases} \mathbf{I}(\mathbf{p}) & \text{if } \|\mathbf{T}(\mathbf{p})\| < \tau \\ \frac{1}{f^2} \sum \mathbf{I}(\mathbf{p}) & \text{if } \|\mathbf{T}(\mathbf{p})\| \geq \tau \end{cases} \tag{13.6}$$

Correspondingly, region intensity centres calculated similarly to equation (13.3) using the filtered intensities **J(p)** instead of **I(p)** are symbolized $\mathbf{J}^s(s_k)$.

An appropriate value of threshold τ was experimentally found to be:

$$\tau = \max\{0.65 \cdot T_{max}, 14\} \tag{13.7}$$

where T_{max} is the maximum value of the norm $\|\mathbf{T}(\mathbf{p})\|$ in the image. The term $0.65 \cdot T_{max}$ in the threshold definition serves to prevent the filter from being applied outside the borders of textured objects, so that their boundaries are not corrupted. The constant bound 14, on the other hand, is used to prevent the filtering of images composed of chromatically uniform objects; in such images, the value of T_{max} is expected to be relatively small and would correspond to pixels on edges between objects, where filtering is obviously undesirable.

The output of the conditional filtering stage is used as input by the KMCC algorithm.

13.3.5 The K-Means with Connectivity Constraint Algorithm

Clustering based on the K-Means algorithm, originally proposed by McQueen [32], is a widely used region segmentation method [33] which, however, tends to produce unconnected regions. This is due to the propensity of the classical K-Means algorithm to ignore spatial information about the intensity values in an image, since it only takes into account the global intensity or color information. Furthermore, previous pixel classification algorithms of the K-Means family do not take into account texture information. In order to alleviate these problems, we propose the use of a novel variant of the KMCC algorithm. In this algorithm the *spatial proximity* of each region is also taken into account by defining a new centre for the K-Means algorithm and by integrating the K-Means with a component labelling procedure. In addition to that, texture features are combined with the intensity and position information to permit efficient handling of textured objects.

The KMCC algorithm applied to the pixels of the image consists of the following steps:

- Step 1: The region number and the region centres are initialized, using the output of the initial clustering procedure described in Section 13.3.3.
- Step 2: For every pixel **p**, the distance between **p** and all region centres is calculated. The pixel is then assigned to the region for which the distance is minimized. A generalized distance of a pixel **p** from a region s_k is defined as follows:

$$D(\mathbf{p}, s_k) = \|\mathbf{J}(\mathbf{p}) - \mathbf{J}^s(s_k)\| + \lambda_1 \|\mathbf{T}(\mathbf{p}) - \mathbf{T}^s(s_k)\| + \lambda_2 \frac{\bar{M}}{M_k} \|\mathbf{p} - \mathbf{S}(s_k)\| \qquad (13.8)$$

  where $\|\mathbf{J}(\mathbf{p}) - \mathbf{J}^s(s_k)\|$, $\|\mathbf{T}(\mathbf{p}) - \mathbf{T}^s(s_k)\|$ and $\|\mathbf{p} - \mathbf{S}(s_k)\|$ are the Euclidean distances between the pixel feature vectors and the corresponding region centres; pixel number M_k of region s_k is a measure of the area of region s_k, and $\bar{M}$ is the average area of all regions, $\bar{M} = \frac{1}{K}\sum_{k=1}^{K} M_k$. The regularization parameter λ_2 is defined as $\lambda_2 = 0.4 \cdot \frac{C}{\sqrt{x_{max}^2 + y_{max}^2}}$, while the choice of the parameter λ_1 has been discussed in Section 13.3.3.
- Step 3: The connectivity of the formed regions is evaluated; those which are not connected are broken down to the minimum number of connected regions using a recursive four-connectivity component labelling algorithm [31].
- Step 4: Region centres are recalculated (equations (13.3)–(13.5)). Regions with areas below a size threshold ξ are dropped. In our experiments, the threshold ξ was equal to 0.5% of the total image area. This is lower than the minimum accepted region size ψ, which in our experiments was equal to 0.75% of the total image area. The latter is used to ensure that no particularly small, meaningless regions are formed. Here, the slightly lower threshold ξ is used to avoid dropping, in one iteration of the KMCC algorithm, regions that are close to threshold ψ and are likely to exceed it in future iterations. The number of regions K is also recalculated, taking into account only the remaining regions.
- Step 5: Two regions are merged if they are neighbours and if their intensity and texture distance is not greater than an appropriate merging threshold:

$$D^s(s_{k_1}, s_{k_2}) = \|\mathbf{J}^s(s_{k_1}) - \mathbf{J}^s(s_{k_2})\| + \lambda_1 \|\mathbf{T}^s(s_{k_1}) - \mathbf{T}^s(s_{k_2})\| \leq \mu \qquad (13.9)$$

  Threshold μ is image-specific, defined in our experiments by

$$\mu = \begin{cases} 7.5 & \text{if } C < 25 \\ 15 & \text{if } C > 75 \\ 10 & \text{otherwise} \end{cases} \qquad (13.10)$$

  where C is an approximation of the intensity and texture contrast of the particular image, as defined in Section 13.3.3.

- Step 6: Region number K and region centres are re-evaluated.
- Step 7: If the region number K is equal to the one calculated in Step 6 of the previous iteration and the difference between the new centres and those in Step 6 of the previous iteration is below the corresponding threshold for all centres, then stop, else goto Step 2. If index 'old' characterizes the region number and region centres calculated in Step 6 of the previous iteration, the convergence condition can be expressed as $K = K^{old}$ and:

$$\|\mathbf{J}^s(s_k) - \mathbf{J}^s\left(s_k^{old}\right)\| \leq c_I, \quad \|\mathbf{T}^s(s_k) - \mathbf{T}^s\left(s_k^{old}\right)\| \leq c_T, \quad \|\mathbf{S}(s_k) - \mathbf{S}\left(s_k^{old}\right)\| \leq c_S$$

  for $k = 1, \ldots, K$. Since there is no certainty that the KMCC algorithm will converge for any given image, the maximum allowed number of iterations was chosen to be 20; if this is exceeded, the method proceeds as though the KMCC algorithm had converged.

The output of the still-image segmentation algorithm is a set of regions which, as can be seen from Section 13.8, mostly correspond to spatial objects depicted in the image. Even in those cases where correspondence is not completely accurate, most indexing features (e.g. color, shape) describing the object can be reliably extracted and used in the query process.

13.4 Spatio-temporal Segmentation of Video Sequences

13.4.1 Video Sequences Segmentation System Overview

Several approaches have been proposed in the literature for video segmentation [34]. Most of these operate in the uncompressed pixel domain [27, 35, 36], which provides them with the potential to estimate object boundaries with pixel accuracy but requires that the processed sequence be fully decoded before segmentation can be performed. As a result, the usefulness of such approaches is usually restricted to non-real-time applications; this is due to the high computational complexity resulting from the large number of pixels that have to be processed. Often the need also arises for motion feature extraction [37–39] using block matching algorithms. Real-time pixel-domain methods [40] are usually applicable only on head-and-shoulder sequences [27] (e.g. videoconference applications) or are based on the assumption that the background is uniformly colored, an assumption not always valid in practice.

To counter these drawbacks of pixel-domain approaches, compressed domain methods have been proposed for spatio-temporal segmentation. However, some of them, although significantly faster than most pixel-domain algorithms, cannot operate in real time [41,42]. In [43] translational motion vectors are accumulated over a number of frames and the magnitude of the displacement is calculated for each macroblock; macroblocks are subsequently assigned to regions by uniformly quantizing the magnitude of the displacement. In [44,45], translational motion vectors and DC coefficients are clustered. In [46], segmentation is performed using AC/DC DCT coefficients only; foreground/background classification is based on thresholding the average temporal change of each region, while the macroblock motion vectors are not used. In [47], a method is developed for tracking manually identified moving objects in the compressed stream based on macroblock motion vectors.

To allow efficient indexing of large video databases, an algorithm for the real-time, unsupervised spatio-temporal segmentation of video sequences in the compressed domain is proposed in this section (Figure 13.4). Only I- and P-frames are examined, since they contain all

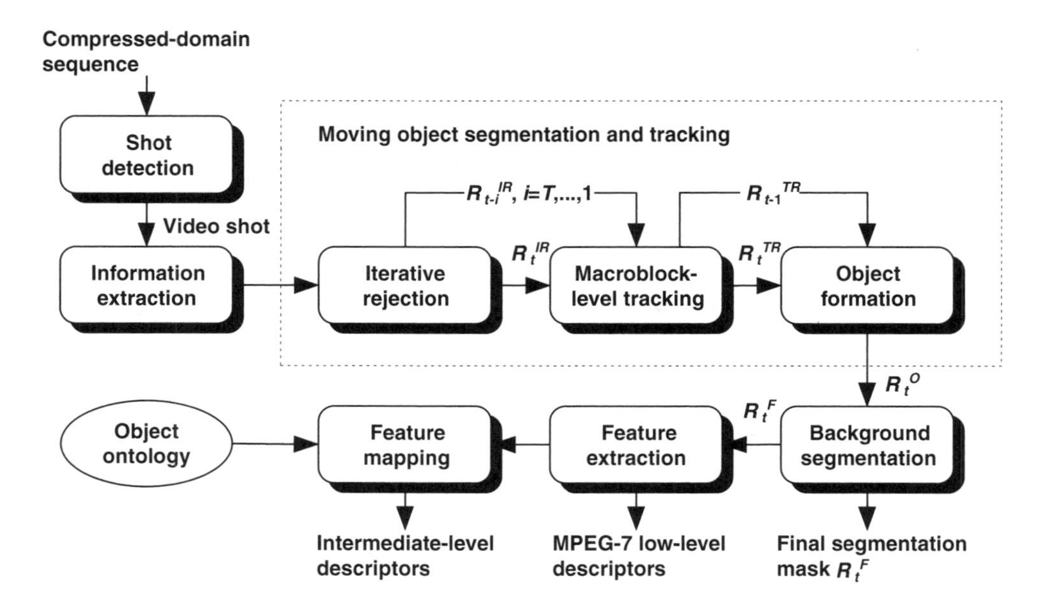

Figure 13.4 Overview of the compressed-domain spatio-temporal segmentation algorithm and the feature extraction procedure

information that is necessary for the proposed algorithm; this is also the case for most other compressed-domain algorithms. The bilinear motion model [48] is used to model the motion of the camera (equivalently, the perceived motion of static background) and, wherever necessary, the motion of the identified moving objects. This is seen to produce much better results than the simple clustering techniques using pure translational motion, which have been used by previous methods. An iterative rejection scheme [49] and temporal consistency constraints are employed to deal with the fact that motion vectors extracted from the compressed stream may not represent accurately the actual object motion. Both foreground and background spatio-temporal objects are identified. This is useful, for example, in retrieval applications, where instead of querying for a compound background, one is allowed to query for its constituent objects, such as sky, sea, mountain etc. The proposed spatio-temporal segmentation algorithm is applied to shots; shot detection is performed using the method of [50], chosen because of its computational simplicity.

13.4.2 Compressed-domain Information Extraction

The information used by the proposed segmentation algorithm is extracted from MPEG-2 [51] sequences during the decoding process. Specifically, motion vectors are extracted for the P-frames and are used for foreground/background segmentation and for the subsequent identification of different foreground objects. Since P-frames are coded using motion information from I-frame to P-frame or from P-frame to P-frame, their motion information provides a clearer indication of the motion of an object compared to motion information derived from temporally adjacent frames. In order to derive motion information for the I-frames, averaging of the motion vectors of the P-frames that are temporally adjacent to the given I-frame is performed, rather than block matching employed in previous work on compressed-domain tracking [47].

In order to further segment the background to its constituent objects (e.g. sky, grass etc.), the use of color information is essential; this is due to the fact that the background has already been identified, using the motion information, as a non-connected region of uniform motion. The color information extracted for the purpose of background segmentation is restricted to the DC coefficients of the macroblocks, corresponding to the Y, Cb and Cr components of the MPEG color space. A single DC coefficient is used to describe luminance (Y) information for every macroblock. Since DC coefficients are present only in I-frames, they can be extrapolated for P-frames using the method in [52]; alternatively, motion information can be used for temporal tracking in P-frames of the background regions formed in I-frames using color information. The latter technique is employed in the proposed work.

13.4.3 Moving Object Segmentation and Tracking

Overview

The extraction of spatio-temporal moving objects is the key challenge in any video segmentation algorithm. Before proceeding with the detailed discussion of each step of the moving object segmentation and tracking algorithm, the notion of *spatio-temporal objects* is defined.

A spatio-temporal object o_q is a set of temporally adjacent spatial regions o_q^t, $t = t_1, t_1 + 1, \ldots, t_2$, $t_1 < t_2$, all of which are non-empty ($o_q^t \neq \emptyset$, $\forall t \in [t_1, t_2]$) and which for $t \in [t_1 + 1, t_2]$ have been associated with o_q by temporal tracking of spatial region $o_q^{t_1}$, using the framework described below, under 'Spatio-temporal object formation', and in Section 13.4.4 for foreground and background spatio-temporal objects, respectively.

$$o_q = \left\{ o_q^{t_1}, \ldots, o_q^{t_2} \right\}.$$

The proposed algorithm for moving object extraction is based on exploiting the motion information (motion vectors) of the macroblocks and consists of three main steps:

- Step 1: Iterative macroblock rejection is performed in a frame-wise basis to detect macroblocks with motion vectors deviating from the single rigid plane assumption. As a result, certain macroblocks of the current frame are activated (marked as possibly belonging to the foreground).
- Step 2: The temporal consistency of the output of iterative rejection over a number of frames is examined, to detect activated macroblocks of the current frame that cannot be tracked back to activated macroblocks for a number of previous frames. These are excluded from further processing (deactivated). This process is based on temporal tracking of activated macroblocks using their motion vectors.
- Step 3: Macroblocks still activated after step 2 are clustered to connected regions that are in turn assigned to either pre-existing or newly appearing spatio-temporal objects, based on the motion vectors of their constituent macroblocks. Spatial and temporal constraints are also applied to prevent the creation of spatio-temporal objects inconsistent with human expectation (e.g single-macroblock objects or objects with undesirably small temporal duration).

Table 13.1 Masks used during foreground object segmentation

Symbol	Description
R_t^{IR}	Foreground/background mask, output of the iterative rejection process of Section 13.4.3, 'Iterative macroblock rejection'
R_t^{TR}	Foreground/background mask, output of the macroblock-level tracking of Section 13.4.3 'Macroblock-level tracking'
$R_{t-i}^{temp}, i = 0, \ldots, T$	Temporary foreground/background masks, used during macroblock-level tracking
R_t^I	Mask created from R_t^{IR} by clustering foreground macroblocks to connected regions s_k^t, $k = 1, \ldots, \kappa^t$ (Section 13.4.3, 'Spatio-temporal object formation')
R_t^O	Foreground spatio-temporal object mask, output of the spatio-temporal object formation process of Section 13.4.3, 'Spatio-temporal object formation'

The above steps are explained in more detail in the sequel. An overview of the different segmentation masks used in this process is shown in Table 13.1.

Iterative macroblock rejection

Iterative rejection is a method originally proposed in [53] for global motion estimation using the output of a block matching algorithm (BMA) and a four-parameter motion model. In [49], the method was extended to the estimation of the eight parameters of the bilinear motion model, used in turn for the retrieval of video clips based on their global motion characteristics. This method is based on iteratively estimating the parameters of the global-motion model using least-square estimation and rejecting those blocks whose motion vectors result in larger than average estimation errors. The iterative procedure is terminated when one iteration leaves the set of rejected blocks unaltered. The underlying assumption is that the background is significantly larger than the area covered by the moving objects; thus, the application of the iterative rejection scheme to the entire frame results in motion vectors affected by local (object) motion being rejected, and global motion (camera motion or, equivalently, the perceived motion of still background) being estimated.

In this work, iterative rejection based on the bilinear motion model is used to generate the mask R_t^{IR}, indicating which macroblocks have been rejected at time t (or activated, from the segmentation objective's point of view). This is the first step of foreground/background segmentation. Rejected (activated) macroblocks are treated as potentially belonging to foreground objects. Compared to classical methods based on examining the temporal change of color features [27] for purposes of fast raw-domain foreground/background segmentation, the employed method of iterative rejection enjoys high efficiency, especially when dealing with sequences captured by a moving camera.

Although this is a fast and relatively simple method for detecting macroblocks that belong to the foreground, several macroblocks may be falsely activated. This may be due to inaccurate estimation of motion vectors from the compressed stream or to inability of the motion-model to accurately capture the undergoing global motion. In order to identify and discard falsely activated macroblocks, the temporal consistency of the output of iterative rejection over a number of previous frames is examined, as discussed in the next subsection.

Macroblock-level tracking

In order to examine the temporal consistency of the output of iterative rejection, activated macroblocks are temporally tracked using the compressed-domain motion vectors. The temporal tracking is based upon the work presented in [47], where objects are manually marked by selecting their constituent macroblocks and these objects are subsequently tracked in the compressed domain using the macroblock motion vectors. However, in contrast to the method in [47], the proposed method requires no human intervention for the selection of the macroblocks to be tracked. A shortcoming of the method in [47] is the need for block matching in order to extract motion features for the I-frames. This is avoided in the present work by averaging the motion vectors of the P-frames that are temporally adjacent to the given I-frame, as already discussed in Section 13.4.2.

More specifically, let $\tau(.)$ be the tracking operator realizing the tracking process of [47], whose input is a macroblock at time t and whose output is the corresponding macroblock or macroblocks at time $t + 1$. This correspondence is established by estimating the overlapping of the examined macroblock with its spatially adjacent ones, determined using the displacement indicated by its motion vector. Then, the operator $\mathcal{T}(.)$ is defined as having a mask (such as R_t^{IR}) as input, applying the $\tau(.)$ operator to the set of all foreground macroblocks of that mask, and outputting the corresponding mask at time $t + 1$.

Let R_t^{TR} denote the output foreground/background mask derived via macroblock-level tracking, using masks R_{t-i}^{IR}, $i = T, \ldots, 0$. The derivation of mask R_t^{TR}, using the operator $\mathcal{T}(.)$ to evaluate and enforce the temporal consistency of the output of iterative rejection over T frames, can be expressed as:

$$R_{t-T}^{temp} = R_{t-T}^{IR}$$

$$\text{for } i = T, \ldots, 1, \quad R_{t-i+1}^{temp} = \mathcal{T}\left(R_{t-i}^{temp}\right) \cap R_{t-i+1}^{IR}$$

$$R_t^{TR} = R_t^{temp}$$

where $\cap$ denotes the intersection of foreground macroblocks and R_{t-i}^{temp}, $i = T, \ldots, 0$ is a set of temporary foreground/background segmentation masks.

It is important to observe that the above process does not lead to infinite error propagation: if a macroblock is falsely assigned to the background, this will affect at most the T subsequent frames. Further, it may affect only the current frame, since the tracking process, as explained in [47], results in the inflation of tracked regions (in this case, the foreground part of the foreground/background mask). The efficiency of macroblock-level tracking in rejecting falsely activated macroblocks is demonstrated in Figure 13.5 for frame 220 of the 'penguin' sequence.

Spatio-temporal object formation

After the rejection of falsely activated macroblocks, as described in the previous subsection, the remaining macroblocks are clustered to connected foreground regions and subsequently assigned to foreground spatio-temporal objects. Clustering to connected regions s_k^t, $k = 1, \ldots, \kappa^t$ is performed using a four-connectivity component labelling algorithm [31]; this results in the creation of mask R_t^I. As will be discussed in the sequel, this does not necessarily imply that each of these connected spatial regions in R_t^I belongs to a single spatio-temporal object o_q. Only for the first frame of the shot, in the absence of a previous object mask R_t^O (i.e. the output

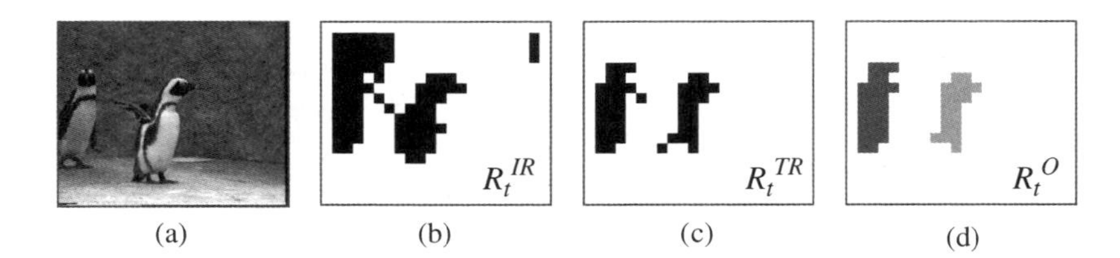

Figure 13.5 Frame 220 of the 'penguin' sequence: (a) original image, (b) output of iterative rejection R_t^{IR}, (c) activated macroblocks after macroblock-level tracking for $T = 4$ (mask R_t^{TR}) and (d) final results showing the two spatio-temporal objects present in this frame (mask R_t^{O}). The usefulness of macroblock-level tracking in rejecting falsely activated macroblocks is evident

of the region formation and tracking step, expressing the spatio-temporal object membership of each macroblock), each connected region is assumed to correspond to a single object:

$$\text{for } t = 0, \; R_t^O = R_t^I$$

To determine whether a given spatial region belongs to one or more pre-existing spatio-temporal objects or to a newly appearing one, and to eventually create the object mask R_t^O, motion projection is performed by applying the tracking operator $\tau(.)$ to the macroblocks of each spatio-temporal object of mask R_{t-1}^O. Thus, every connected region s_k^t of mask R_t^I can be assigned to one of the following three categories:

1. A number of macroblocks $M_{k,q}$, $M_{k,q} \geq S_{k,q}$, of s_k^t have been assigned to spatio-temporal object o_q in mask $\mathcal{T}(R_{t-1}^O)$, and no macroblock of s_k^t has been assigned to a spatio-temporal object o_m, $m \neq q$.
2. A number of macroblocks $M_{k,q}$, $M_{k,q} \geq S_{k,q}$, of s_k^t have been assigned to spatio-temporal object o_q in mask $\mathcal{T}(R_{t-1}^O)$, and one or more macroblocks of s_k^t have been assigned to different spatio-temporal objects, namely o_m, $m = 1, \ldots, M$.
3. There is no spatio-temporal object o_q in mask $\mathcal{T}(R_{t-1}^O)$ having $M_{k,q}$ macroblocks of s_k^t, $M_{k,q} \geq S_{k,q}$, assigned to it.

The parameter $S_{k,q}$ in the definition of the above categories is estimated for every pair of a spatial region s_k^t of mask R_t^I and the projection of a spatio-temporal object o_q in mask $\mathcal{T}(R_{t-1}^O)$. Let M_k^s, M_q^o denote the size in macroblocks of the examined pair (s_k^t and motion projection of object o_q in mask $\mathcal{T}(R_{t-1}^O)$ respectively). Then, the parameter $S_{k,q}$ is calculated as follows:

$$S_{k,q} = a \cdot \frac{M_k^s + M_q^o}{2} \tag{13.11}$$

The value of the parameter a was set to 0.5 on the basis of experimentation.

Obviously, the spatial regions s_k^t classified in the third category can not be associated with an existing spatio-temporal object; therefore, each region of this category forms a new spatio-temporal object.

Similarly, the spatial regions s_k^t classified in the first category can only be associated with a single spatio-temporal object o_q. However, more than one spatial region may be associated

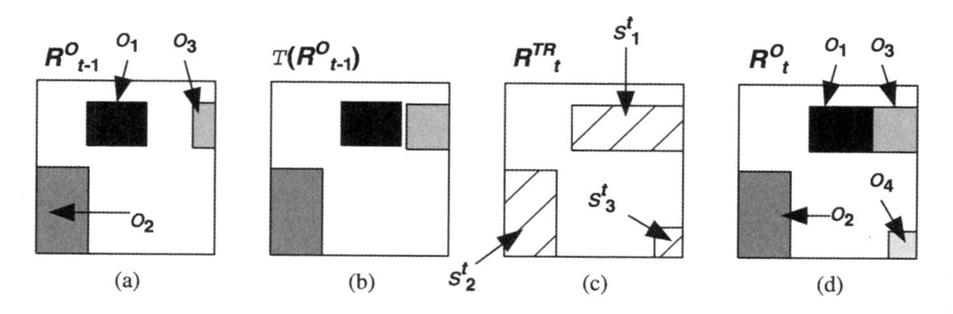

Figure 13.6 A synthetic example of the spatio-temporal object formation process of Section 13.4.3. Using the object mask for the previous frame, R^O_{t-1} (a), mask $\mathcal{T}(R^O_{t-1})$ (b) is created. Comparison of the latter with mask R^{TR}_t (c) reveals the presence of one spatial region (s^t_1) that belongs to the second category (can be associated with spatio-temporal objects o_1 and o_3), one that belongs to the first category (s^t_2, can be associated only with o_2) and one that belongs to the third category (s^t_3). Treating the three different cases as described in Section 13.4.3, the object mask R^O_t (d) is formed

with the same spatio-temporal object. In this case, the larger spatial region becomes part of o_q, while the rest are discarded (their macroblocks are assigned to the background). This procedure is intended to deal with objects breaking up: the fragments that are discarded at time t, if they actually correspond to moving objects, will clearly be assigned to category 3 at time $t + 1$, and in turn be identified as new spatio-temporal objects.

As for the regions s^t_k classified in the second category, the initial correspondence of specific macroblocks belonging to s^t_k with spatio-temporal objects is employed so as to estimate the parameters of the bilinear motion model for each of the competing objects. Subsequently, each macroblock is assigned to the object for which the motion estimation error is minimized. This process elegantly handles moving objects that become spatially adjacent. The possibility of merging two adjacent objects, leaving unaltered any masks created at time $t - i$, $i > 0$, is also examined by estimating the parameters of their common motion model (at time t) and comparing the corresponding mean-square error with those of the motion models estimated for each object separately.

The process of moving object segmentation and tracking is terminated by imposing application-oriented restrictions regarding the size and temporal duration of valid moving objects, if any such restrictions exist. Generally, the removal of very small objects and objects of very small temporal duration is beneficial, since they are most likely to be false objects and are of little use in the proposed indexing and retrieval application. The above procedure is illustrated using real and synthetic masks in Figure 13.5 and 13.6, respectively.

Pixel-domain boundary refinement

In specific applications, the information that can be extracted from a segmentation mask of macroblock-level accuracy may be insufficient (e.g. for the extraction of shape descriptors of high accuracy). In this case, pixel-domain processing of a partially decompressed sequence may be required, to extract object masks of pixel accuracy. This can be achieved using the color features of pixels in the area of each moving object and a Bayes classifier for two-class separation (moving object/background) to reclassify all pixels in that area or a portion of them, in a fashion similar to that of [54]. Pixel-accuracy masks created using this refinement method are presented in the experimental results section.

13.4.4 Background Segmentation

After foreground spatio-temporal objects have been extracted, background segmentation is performed based on classifying the remaining macroblocks (assigned to the background) to one of a number of background spatio-temporal objects. This task is performed using two distinct steps, each dealing with one of the different types of the examined frames. These steps are preceded at the beginning of the shot, by a procedure for the estimation of the number of background objects that should be created. The result of the background segmentation is a final segmentation mask R_t^F.

Background segmentation begins by applying the *maximin* algorithm [30] to the color DC coefficients of the first frame, which is an I-frame. The maximin algorithm employs the Euclidean distance in the YCrCb color space to identify radically different colors; these indicate the presence of different background spatio-temporal objects. Its output is the number of estimated objects and their corresponding colors, which are used to initiate the clustering process.

In I-frames, background macroblocks are clustered to background objects using the K-Means algorithm [30, 32], where K is the number of objects estimated by the maximin algorithm. For the first frame of the shot, the color centres are initialized using the output of the maximin algorithm, while for the subsequent I-frames the color centres of the resulting objects in the previous I-frame are used for initialization. The connectivity of the K objects is enforced using a recursive component labelling algorithm [31] to identify small non-connected parts, which are subsequently assigned to a spatially adjacent object on the basis of color similarity. The connectivity constraint is useful in accurately estimating the position of each object, which could otherwise comprise non-connected parts scattered in the entire frame.

In P-frames, the absence of color information can be dealt with by using the macroblock motion vectors and a previous final mask R_{t-1}^F. Temporal tracking is then performed as discussed in previous sections. Macroblocks that are associated via the tracking process with more than one background object are assigned to the one for which the motion information indicates a stronger association (i.e a higher degree of overlapping with it, as in Section 13.4.3), while those not associated with any object are assigned to one on the basis of spatial proximity.

13.5 MPEG-7 Low-level Indexing Features

As soon as segmentation masks are produced for each frame of each video shot or for each image, a set of descriptor values useful in querying the database are calculated for each spatial or spatio-temporal object. Standardized MPEG-7 descriptors are used, to allow for flexibility in exchanging indexing information with other MPEG-7 compliant applications. The different MPEG-7 descriptors used in this work are summarized in Table 13.2.

As can be seen from Table 13.2, each object property need not be associated with a single descriptor. For example, in the case of object motion, two different motion trajectories are calculated for each foreground spatio-temporal object, as the result of the use of two different coordinate systems (values 'Local' and 'Integrated' of *Spatial 2D Coordinates* descriptor). In the latter case, the use of a fixed, with respect to the camera, reference point for the coordinate system allows the categorization (e.g. fast or slow, direction) of foreground object motion

Table 13.2 Set of used MPEG-7 descriptors

Descriptor	Image/Video entity described
Motion Activity	Shot (video only)
Dominant Color	Spatial/spatio-temporal object
GoF/GoP Color	Spatial/spatio-temporal object
Contour Shape	Spatial/spatio-temporal object
Motion Trajectory using 'Local' coordinates	Spatial/spatio-temporal object
Motion Trajectory using 'Integrated' coordinates	Foreground spatio-temporal object

even in the presence of a moving camera. Regarding spatial and background spatio-temporal objects, using 'Local' coordinates is more appropriate, since the goal is not to extract speed characterization or motion direction but rather their qualitative space-localization in the frame. To facilitate the policing of spatial relations between foreground and background objects, a trajectory using 'Local' coordinates is calculated for the former, as well.

Two MPEG-7 'color' descriptors are also used; unlike motion descriptors, they both apply to all objects. This duality serves the purpose of satisfying the diverse requirements set by the general architecture of the retrieval scheme: low-level descriptors should be easy to map to intermediate-level qualitative descriptors (e.g. names of basic colors) and still permit accurate retrieval. A few most-dominant colors of the *Dominant Color* descriptor are most appropriate for associating with color names, whereas when using the low-level descriptors directly, color histograms (*GoF/GoP Color*) demonstrate better retrieval performance [55]; they also have the advantage of being compatible with the L2 norm used as part of the employed relevance feedback mechanism.

13.6 Object-based Indexing and Retrieval using Ontologies

13.6.1 Overview

With the exception of a few MPEG-7 descriptors, such as Motion Activity, which are fairly high level, most standardized descriptors are low-level arithmetic ones, chosen so as to ensure their usefulness in a wide range of possible applications. These descriptors, however, are not suitable for being directly manipulated by the user of an indexing and retrieval scheme, e.g. for defining the color of a desired object. When examining the specific application of object-based image and video indexing, it is possible to alleviate this problem by translating certain low-level arithmetic values to intermediate-level descriptors qualitatively describing the object attributes; the latter are preferable, since humans are more familiar with manipulating qualitative descriptors than arithmetic values.

Extending the approach in [56, 57], the values of the intermediate-level descriptors used for this qualitative description form a simple vocabulary, the *object ontology*. Ontologies are tools for structuring knowledge [23–25]. An ontology may be defined as the specification of a representational vocabulary for a shared domain of discourse, which may include definitions of classes, relations, functions and other objects [58]. Ontologies are primarily used in text retrieval [59]. Recently, their usefulness in image retrieval was demonstrated in [60], where the use of background knowledge contained in ontologies in indexing and searching collections of photographs was explored and appropriate tools for facilitating manual annotation of images and for querying in collections of such annotated images were developed. Facial expression recognition is another possible application of ontologies [61]. In the proposed scheme, ontologies are used to facilitate the mapping of low-level descriptor values to higher-level semantics. An object ontology and a query ontology are employed to enable the user to form, respectively, a simple qualitative description of the desired objects and their spatial and temporal relationships; in parallel, a qualitative description of each spatial/spatio-temporal object in the database is automatically estimated using the object ontology, as discussed in the next section.

13.6.2 Object Ontology

In this work, ontologies [59, 60] are employed to allow the user to query an image and video collection using semantically meaningful concepts (semantic objects), without the need for performing manual annotation of visual information. A simple *object ontology* is used to enable the user to describe semantic objects, like 'rose', using a vocabulary of intermediate-level descriptor values. These are automatically mapped to the low-level descriptor values calculated for each spatial/spatio-temporal object in the database, thus allowing the association of keywords representing semantic objects (e.g. the 'rose' keyword) and potentially relevant spatial/spatio-temporal objects. The simplicity of the employed object ontology make it applicable to generic video collections without requiring the correspondence between spatial/spatio-temporal objects and relevant descriptors to be defined manually. This object ontology can be expanded so as to include additional descriptors corresponding either to low-level properties (e.g. texture) or to higher-level semantics, which, in domain-specific applications, could be inferred either from the visual information itself or from associated information (e.g. subtitles).

The object ontology is presented in Figure 13.7, where the possible intermediate-level descriptors and descriptor values are shown. Each value of these intermediate-level descriptors is mapped to an appropriate range of values of the corresponding low-level, arithmetic descriptor. With the exception of color (e.g. 'black') and direction (e.g. 'low→high') descriptor values, the value ranges for every low-level descriptor are chosen so that the resulting intervals are equally populated. This is pursued to prevent an intermediate-level descriptor value from being associated with a plurality of spatial/spatio-temporal objects in the database, since this would

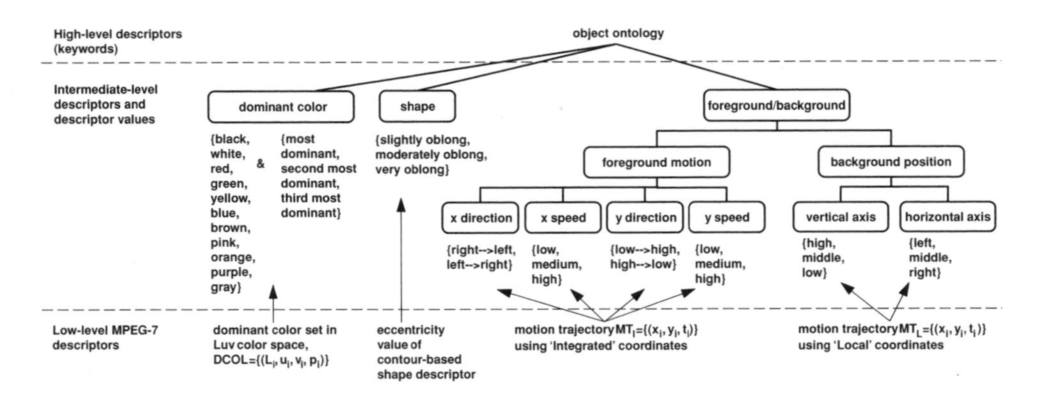

Figure 13.7 Object ontology. The correspondence between low-level MPEG-7 descriptors and intermediate-level descriptors is shown

render it useless in restricting a query to the potentially most relevant ones. Overlapping, up to a point, of adjacent value ranges, is used to introduce a degree of fuzziness to the descriptor values; for example, both 'slightly oblong' and 'moderately oblong' values may be used to describe a single spatial or spatio-temporal object.

Let $D_{q,i}$ be the q-th descriptor value (e.g. 'slightly oblong') of intermediate-level descriptor D_i (e.g. 'shape') and $R_{q,i} = [L_{q,i}, H_{q,i}]$ be the range of values of the corresponding arithmetic descriptor z_i. Given the probability density function $pdf(z_i)$ and the factor V expressing the degree of overlapping of adjacent value ranges, the requirement that value ranges should be equally populated defines lower and upper bounds $L_{q,i}$, $H_{q,i}$ which are easily calculated by equations (13.12)–(13.14).

$$L_{1,i} = L_i, \qquad \int_{L_{q-1,i}}^{L_{q,i}} pdf(z_i)dz_i = \frac{1-V}{Q_i - V \cdot (Q_i - 1)}, \quad q = 2, \ldots, Q_i, \qquad (13.12)$$

$$\int_{L_{1,i}}^{H_{1,i}} pdf(z_i)dz_i = \frac{1}{Q_i - V \cdot (Q_i - 1)}, \qquad (13.13)$$

$$\int_{H_{q-1,i}}^{H_{q,i}} pdf(z_i)dz_i = \frac{1-V}{Q_i - V \cdot (Q_i - 1)}, \quad q = 2, \ldots, Q_i, \qquad (13.14)$$

where Q_i is the number of descriptor values defined for the examined descriptor D_i (for example, for 'shape', $Q_i = 3$), and L_i is the lower bound of the values of variable z_i. The overlapping factor V was selected equal to $V = 0.25$ in our experiments.

Regarding color, a correspondence between the 11 basic colors [62], used as color descriptor values, and the values of the HSV color space is heuristically defined. More accurate correspondences based on the psychovisual findings of e.g. [62] and others are possible, as in [63, 64]; however this investigation is beyond the scope of the present work. Regarding the direction of motion, the mapping between values for the descriptors 'x direction', 'y direction' and the MPEG-7 *Motion Trajectory* descriptor is based on the sign of the cumulative displacement of the foreground spatio-temporal objects.

13.6.3 Query Ontology

In order to enable the formulation of descriptive queries, a simple *query ontology* is defined. As described in Figure 13.8, the definition of the query ontology allows the submission of either single-keyword or dual-keyword queries; however, one can easily extend the query ontology to allow for multiple-keyword queries. Desired temporal and spatial relationships between the queried objects can be expressed, along with the specification of the desired motion activity of the shot, according to the MPEG-7 *Motion Activity* descriptor. The temporal relations defined in [65] are employed in this work, along with simple spatial relationship descriptors defining the desired position of the second object with respect to the first.

As soon as a query is formulated using the query ontology, the intermediate-level descriptor values associated with each desired semantic object/keyword are compared to those of each spatial/spatio-temporal object contained in the database. Descriptors for which no values have been associated with the desired semantic object are ignored; for each remaining descriptor, spatial/spatio-temporal objects not sharing at least one descriptor value with those assigned

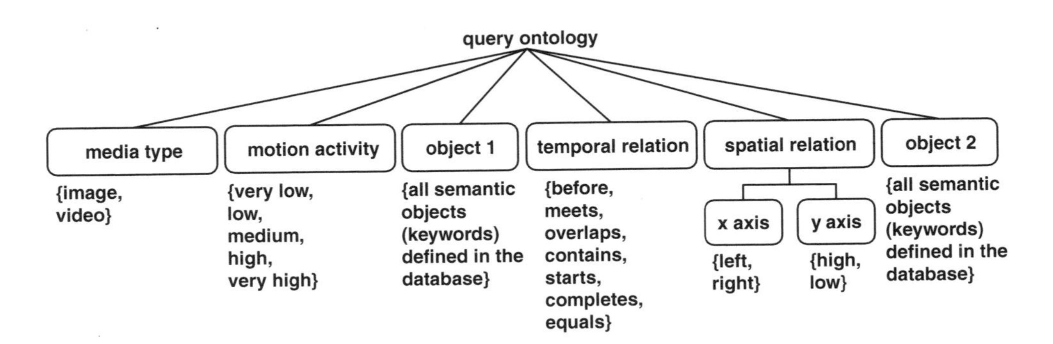

Figure 13.8 Query ontology

to the desired semantic object are deemed irrelevant. In the case of dual-keyword queries, the above process is performed for each desired semantic object separately and only multimedia items containing at least two distinct potentially relevant spatial/spatio-temporal objects, one for each keyword, are returned; if desired spatial or temporal relationships between the semantic objects have been defined, compliance with these constraints is checked using the corresponding low-level descriptors, in order to further reduce the number of potentially relevant media returned to the user.

13.7 Relevance Feedback

Relevance feedback is a powerful tool for interactive visual content retrieval [66,67]; however, most state-of-the-art systems adopting the query-by-example paradigm do not make use of such a technique. [6–9, 11–15] In [68], a newer edition of [11], the use of relevance feedback is introduced; the employed relevance feedback scheme relies on extensive human–computer interaction, in the form of manual outlining of regions of interest, rather that unsupervised segmentation, and manual assignment of predefined labels. In [69, 70], the introduction of relevance feedback based on term reweighting and a *query expansion model* to the system of [10] is discussed. Other recent work on relevance feedback for visual information retrieval is presented in [66,71–76]; in all cases, however, the query initiation process is not examined or it is assumed that the query commences using an appropriate key image [71], possibly uploaded by the user [66,72,74]. In [74], attempts to address the problem of reduced flexibility deriving from the key-image restriction are based on the automatic generation of a set of modified images, based on the user-supplied one; these can subsequently be evaluated by the user. This technique is named *intra-query modification*.

After narrowing down the search to a set of potentially relevant spatial/spatio-temporal objects, as discussed in the previous sections, relevance feedback is employed to produce a qualitative evaluation of the degree of relevance of each spatial/spatio-temporal object. The employed mechanism is based on a method proposed in [76], where it is used for image retrieval using global image properties under the query-by-example scheme. This method combines support vector machines (SVM) [77] with a constrained similarity measure (CSM) [76]. SVMs employ the user-supplied feedback (training samples) to learn the boundary separating the two classes (positive and negative samples, respectively). Each sample (in our case, spatial/spatio-temporal object) is represented by its low-level descriptor vector $\mathbf{F}$, as discussed in the sequel.

Following the boundary estimation, the CSM is employed to provide a ranking; in [76], the CSM employs the Euclidean distance from the key image used for initiating the query for images inside the boundary (images classified as relevant) and the distance from the boundary for those classified as irrelevant. Under the proposed scheme, no key image is used for query initiation; the CSM is therefore modified so as to assign to each spatial/spatio-temporal object classified as relevant the minimum of the Euclidean distances between it and all positive training samples (i.e spatial/spatio-temporal objects marked as relevant by the user during relevance feedback).

The above relevance feedback technique was realized using the SVM software libraries of [78]. The Gaussian radial basis function is used as a kernel function by the SVM, as in [76], to allow for nonlinear discrimination of the samples. The low-level descriptor vector **F** is composed of the 256 values of the histogram (*GoF/GoP Color* descriptor) along with the *eccentricity* value of the *Contour Shape* descriptor and either the speed or the position in the x- and y-axies. The speed is used only for foreground spatio-temporal objects and providing that the queried semantic object has been defined by the user as being one such. In any other case (when the queried object is defined as a background one or when foreground/background is not specified, thus spatial objects are also potential matches), the position of the spatial/spatio-temporal object is included in **F**. In the case of dual-keyword queries, two different SVMs are independently trained and the overall rank is calculated as the sum of the two ranks. This relevance feedback process can be repeated as many times as necessary, each time using all previously supplied training samples.

Furthermore, it is possible to store the parameters of the trained SVM and the corresponding training set for every keyword that has already been used in a query at least once. This endows the system with the capability to respond to anticipated queries without initially requiring any feedback; in a multi-user (e.g. web-based) environment, it additionally enables different users to share knowledge, either in the form of semantic object descriptions or in the form of results retrieved from the database. In either case, further refinement of retrieval results is possible by additional rounds of relevance feedback.

13.8 Experimental Results

The proposed algorithms were tested on a media collection comprising 5000 images from the Corel library [79] and 812 video shots created by digitizing parts of movies and collecting video clips available on the Internet. Application of the segmentation algorithm of Section 13.3 to the 5000 still images resulted in the creation of a database containing 34 433 spatial objects. The average segmentation time for the 192×128 pixel images of the collection was 65.5 s on an 800 Mhz Pentium III. Improved time-efficiency can be achieved utilizing the method of [54].

Objective evaluation of still-image segmentation quality was performed using synthetic images, created using the reference textures of the VisTex database [80], and natural images of the Corel gallery. The employed evaluation criterion is based on the measure of *spatial accuracy* proposed in [81] for foreground/background masks. For the purpose of evaluating still-image segmentation results, each reference object r_q of the reference mask is associated with a different spatial object s_k on the basis of region overlapping (i.e. s_k is chosen so that $r_q \cap s_k$ is maximized). Then, the spatial accuracy of the segmentation is evaluated by separately considering each reference object as a foreground reference object and applying the criterion of [81] for the pair $\{r_q, s_k\}$; during this process, all other reference objects are treated as background. A weighted sum of misclassified pixels for each reference object is the output of this process.

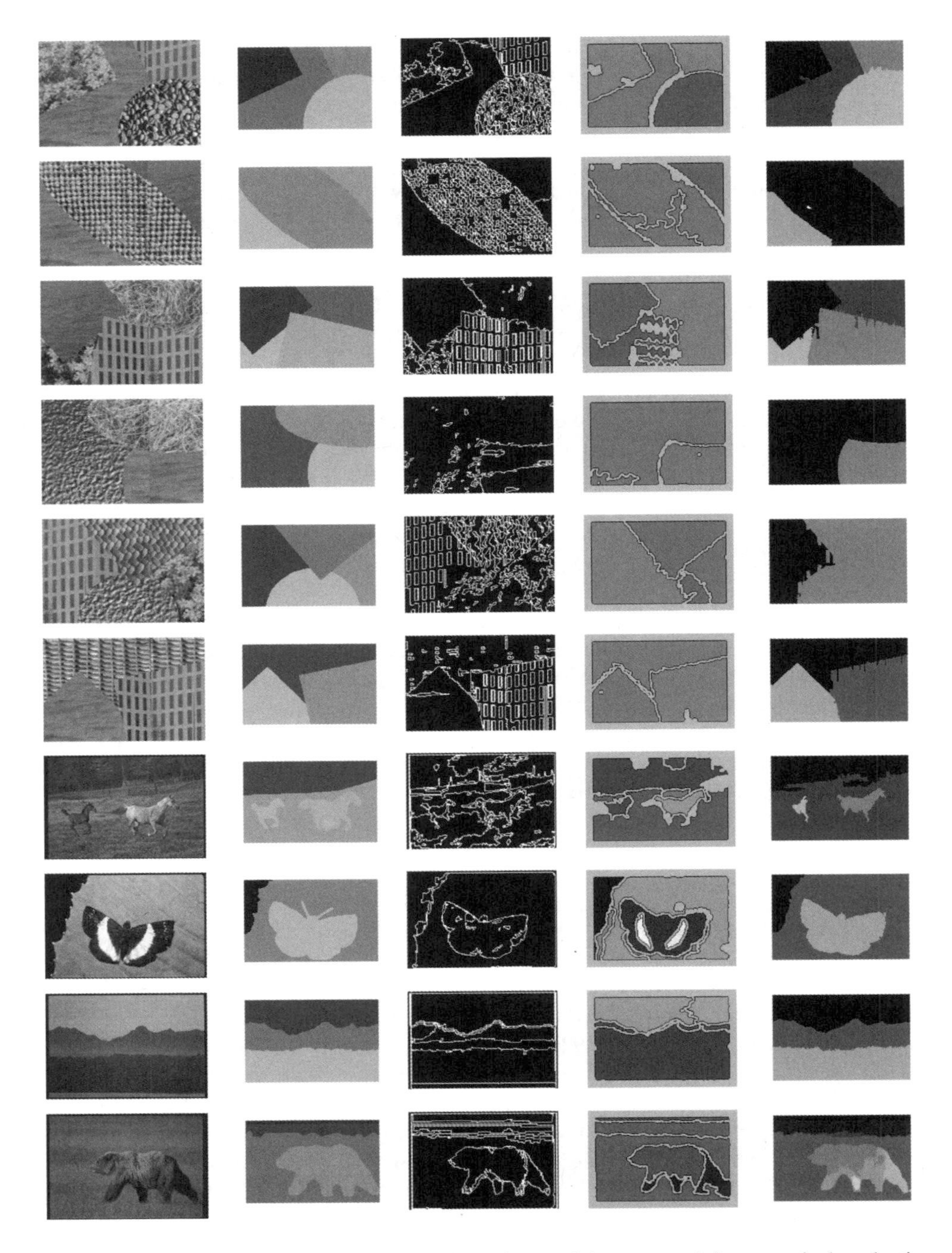

Figure 13.9 Segmentation results for synthetic and natural images used for numerical evaluation. Synthetic images were created using the reference textures of MIT's VisTex database. Reference masks are shown in the second column; results for a modified RSST, the Blobworld algorithm and the proposed algorithm are shown in columns 3 to 5, respectively

Table 13.3 Numerical evaluation of the segmentations of Figure 13.9

Images	RSST	Blobworld	KMCC
Synth1	105.339744	12.188144	**1.260071**
Synth2	187.123016	40.027074	**1.787774**
Synth3	105.995881	45.812201	**2.167452**
Synth4	78.353790	56.613260	**42.442787**
Synth5	136.206447	**34.720163**	50.283481
Synth6	73.851239	10.601577	**1.197819**
Horse	222.936757	55.687840	**18.226707**
Butterfly	22.572128	48.468529	**7.800168**
Sunset	68.794582	89.307062	**5.722744**
Bear	86.269010	**55.090216**	60.948571

The sum of these error measures for all reference objects is used for the objective evaluation of segmentation accuracy; values of the sum closer to zero indicate better segmentation. The test images used for objective evaluation are presented in Figure 13.9, along with their reference masks and results of the algorithm proposed here (KMCC), the Blobworld algorithm [15] and a modified RSST algorithm [82]. The values of the evaluation metric for the images of Figure 13.9 are shown in Table 13.3. The superiority of the proposed algorithm in producing spatial objects with accurate boundaries is demonstrated in Figure 13.9 and is numerically verified.

The video segmentation algorithm of Section 13.4 was subsequently applied to the 812 video shots of the collection, resulting in the creation of 3058 spatio-temporal objects. Here, results of the real-time compressed-domain segmentation algorithm are presented for two known test sequences, namely the 'table-tennis' (Figure 13.10) and 'Penguin' (Figure 13.11) sequences in CIF format. Segmentation masks both before (R_t^O, second column of Figures 13.10 and 13.11) and after the background segmentation (R_t^F, third column of Figures 13.10 and 13.11) are presented, to clearly demonstrate the foreground and background objects identified in the compressed stream by the proposed algorithm. Results after the application of pixel-domain boundary refinement (Section 13.4.3) to the moving objects are also presented in the aforementioned figures. It is seen that the proposed algorithm succeeds in extracting the actual foreground objects depicted in the sequences. No over-segmentation is caused by the proposed approach and thus the formation of meaningful spatio-temporal objects is facilitated. Additionally, very few false objects are created. Moving objects that have halted, as the rightmost penguin in Figure 13.11, are assigned new labels when they resume moving.

The proposed video segmentation approach imposes little additional computational burden to the computational complexity of a standard MPEG decoder. Excluding any processes of the MPEG decoder, the proposed compressed-domain segmentation algorithm requires on average 5.02 ms per processed CIF format I- or P-frame on an 800 Mhz Pentium III. This translates to almost 600 frames per second considering the presence of two consecutive B-frames between every two I- or P-frames, which is typical for MPEG-2 sequences and is the case for the employed test media. The pixel-domain boundary refinement of Section 13.4.3 requires on average 0.48 s per processed I- or P-frame. The latter procedure is, however, not necessary for applications like the indexing and retrieval scheme proposed in this work.

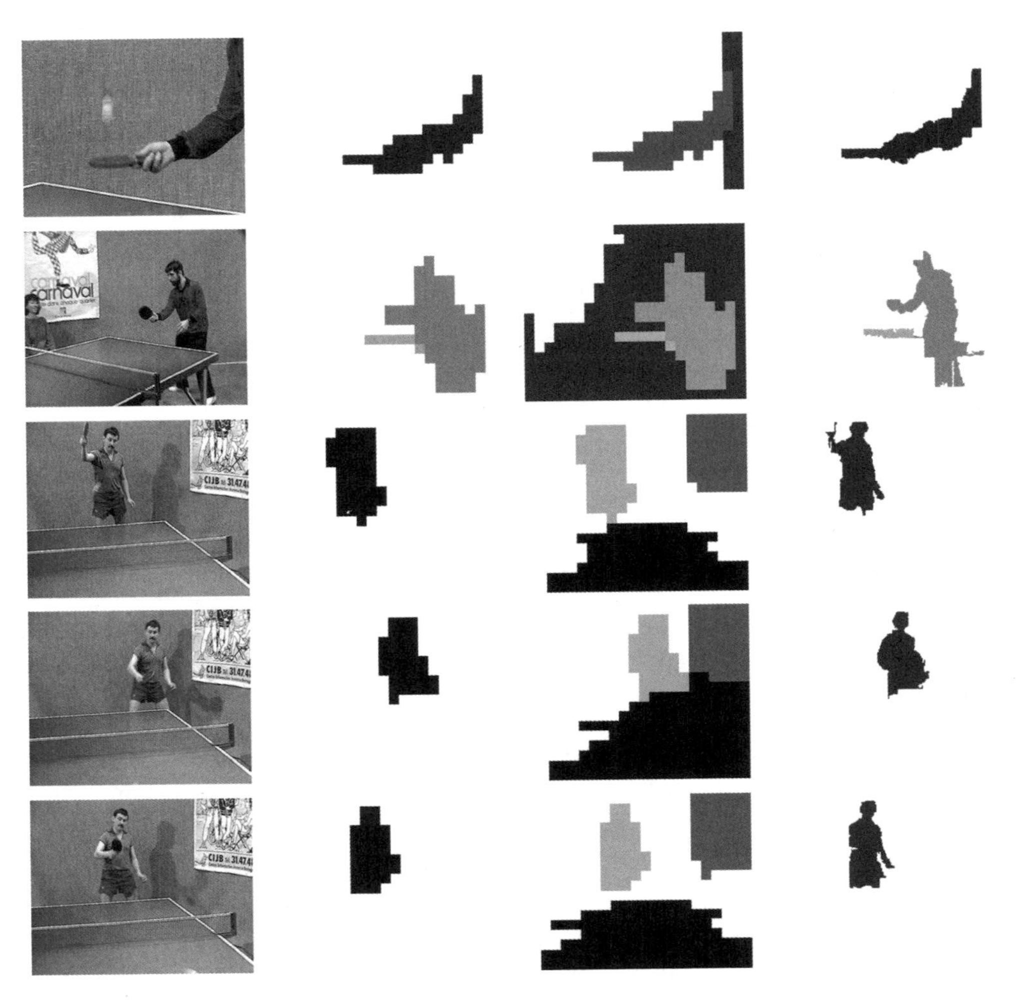

Figure 13.10 Results of moving-object detection, final mask after background segmentation, and moving objects after pixel-domain boundary refinement for the 'Table-tennis' sequence, frames 10, 88, 154, 214 and 265

Following segmentation, MPEG-7 low-level descriptors were calculated for each of the created spatial/spatio-temporal objects, as described in Section 13.5. Subsequently, the mapping between these low-level descriptors and the intermediate-level descriptors defined by the object ontology was performed; this was done by estimating the low-level descriptor lower and upper boundaries corresponding to each intermediate-level descriptor value, as discussed in Section 13.6.2. Since a large number of heterogeneous spatial/spatio-temporal objects was used for the initial boundary calculation, future insertion of heterogeneous images and video clips to the database is not expected to significantly alter the proportion of spatial/spatio-temporal objects associated with each descriptor value; thus, the mapping between low-level and intermediate-level descriptors is not to be repeated, regardless of future insertions.

Figure 13.11 Results of moving-object detection, final mask after background segmentation, and moving objects after pixel-domain boundary refinement for the 'Penguin' sequence, frames 1, 7, 175, 220 and 223

The next step in the experimentation with the proposed system was to use the object ontology to define, using the available intermediate-level descriptors, high-level concepts, i.e. semantic objects (keywords). Since the purpose of the first phase of each query is to employ these definitions to reduce the data set by excluding obviously irrelevant spatial/spatio-temporal objects, the definitions of semantic objects need not be particularly restrictive; this is convenient from the user's point of view, since the user can not be expected to have perfect knowledge of the color, shape and motion characteristics of the object sought in the database [83]. Two such definitions, namely for the 'blue car' and 'brown horse' keywords, are illustrated in Figure 13.12.

As soon as a descriptive definition of the desired object or objects is supplied, such as the ones presented in Figure 13.12, one can employ the query ontology to form and submit a query. Several experiments were conducted using single-keyword or dual-keyword queries, to

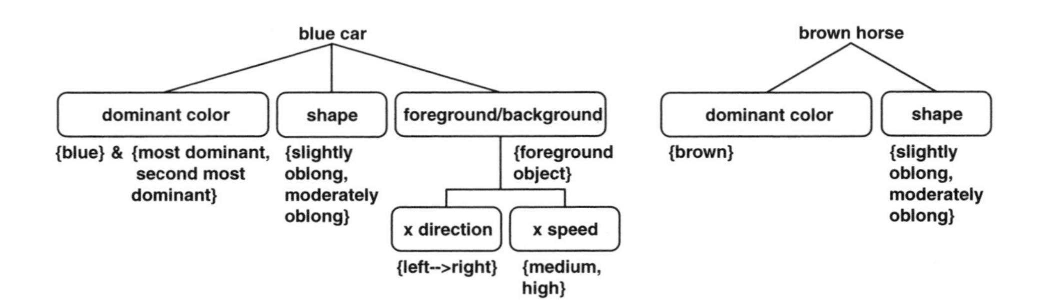

Figure 13.12 Exemplary definitions of semantic objects using the object ontology, employed in retrieval experiments

retrieve media belonging to particular classes, e.g. containing eagles, roses, cars etc. Performing ontology-based querying resulted in initial query results being produced by excluding the majority of objects in the database, that were found to be clearly irrelevant, as shown in Figure 13.13(a) for a 'blue car' query.

As a result, one or more pages of 15 randomly-selected, potentially relevant spatial/spatio-temporal objects were presented to the user to be manually evaluated; this resulted in the 'relevant' check-box being checked for those that were actually relevant. Subsequently, one or two SVMs were trained and were used for ranking the potentially relevant objects, as discussed in Section 13.7. In our experiments, this first round of relevance feedback was followed by a second one and in this case, the objects that were manually evaluated at the previous round were also used for the SVM training. Usually, evaluating two pages of potentially relevant objects was found to be sufficient. Results after the first or second application of relevance feedback are presented in Figure 13.13 for a 'blue car' query and in Figure 13.14 for a 'brown horse—grass' query; precision–recall diagrams are shown in Figure 13.15, along with those for the results after the first relevance feedback stage. Note that in all experiments, each query was submitted five times, to allow for varying performance due to different randomly chosen media sets being presented to the user. The term *precision* is defined as the fraction of retrieved images which are relevant, and the term *recall* as the fraction of relevant images which are retrieved [15].

For comparison purposes, experiments were also conducted using the query-by-example paradigm and global histograms, which were introduced in [84] and have been used widely ever since. For videos, one key frame from each shot was used for histogram calculation. The histograms were based on bins of width 20 in each dimension of the L*a*b* color space. Again, each query was submitted five times, each time using a different, randomly selected key image belonging to the desired class. Precision–recall diagrams for these experiments are also presented in Figure 13.15. On comparing them to the results of the proposed scheme, it can be seen that even a single stage of relevance feedback generally yields significantly better results. Further, the application of a second stage of relevance feedback leads to significant improvement. Additionally, the diagrams calculated for the global histogram method rely on the assumption that it is possible to provide the user with an appropriate key image/key frame, so as to enable initiation of the query; the validity of this assumption is, however, questionable, particularly when it comes to under-represented classes of multimedia items.

Query results for "blue_car": 1 to 15

(a)

Query results for "blue_car": 1 to 15

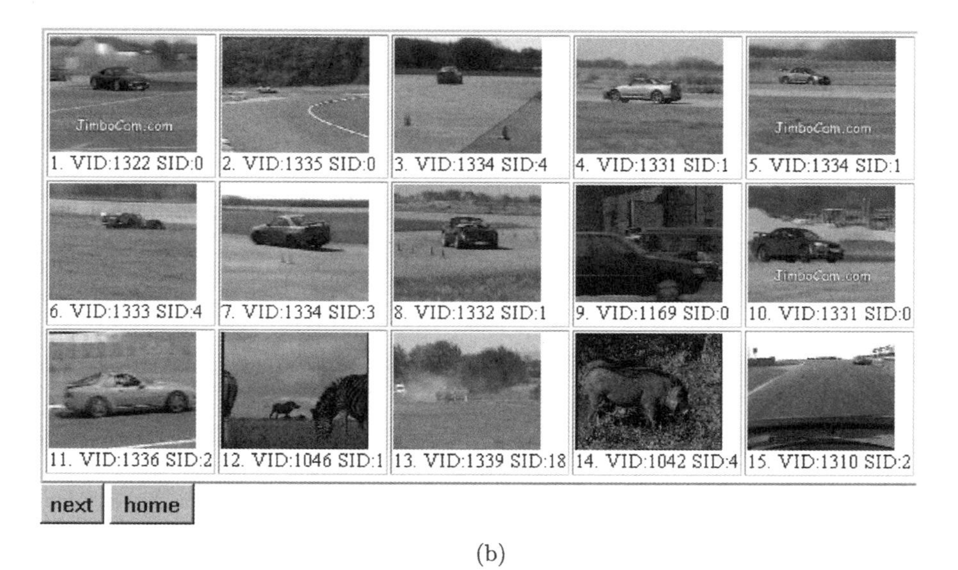

(b)

Figure 13.13 Results for a 'blue car' query, where 'blue car' has been defined as a foreground object: (a) potentially relevant visual content, identified using the intermediate-level descriptors, and (b) results after one round of relevance feedback

Query results for "brown_horse": 1 to 15

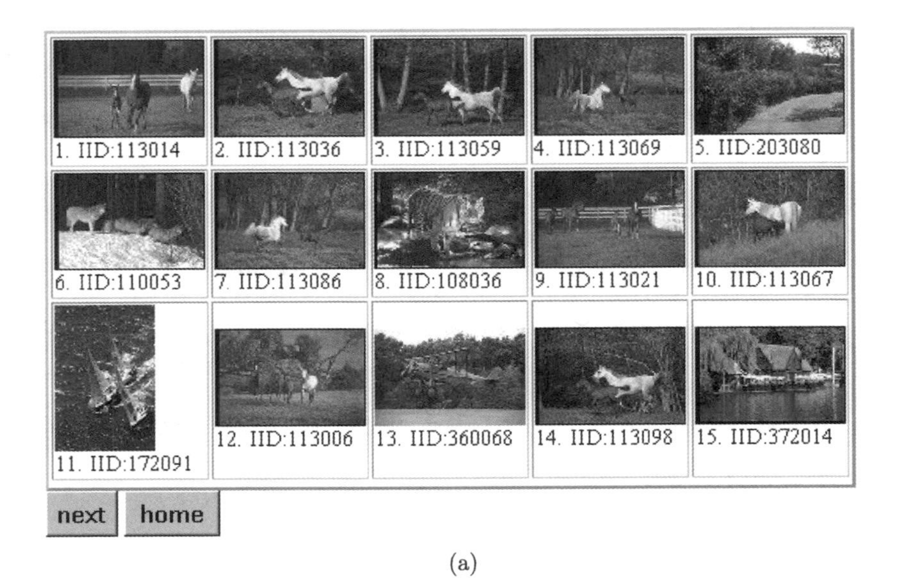

(a)

Query results for "brown_horse": 1 to 15

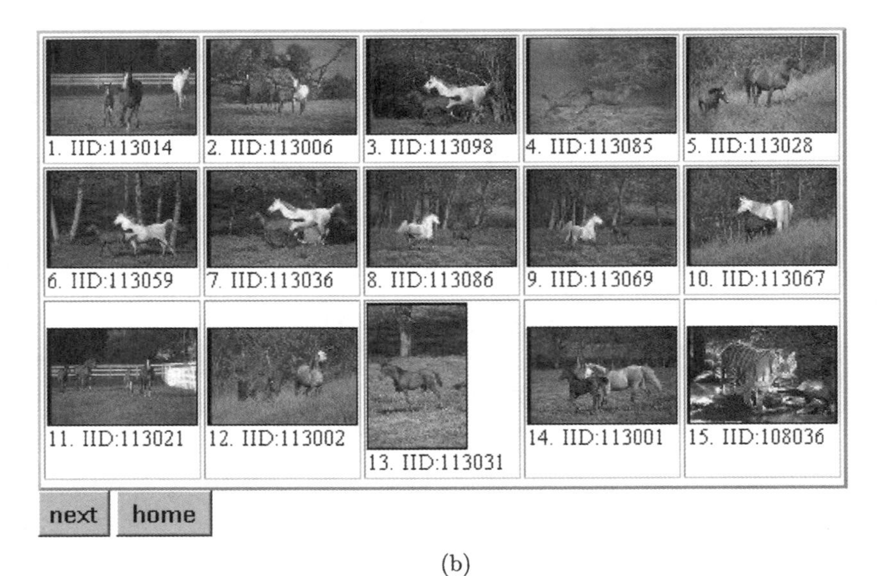

(b)

Figure 13.14 Results for a 'brown horse—grass' query, where desired media type has been specified as 'image' using the query ontology: results after one (a) and two (b) rounds of relevance feedback

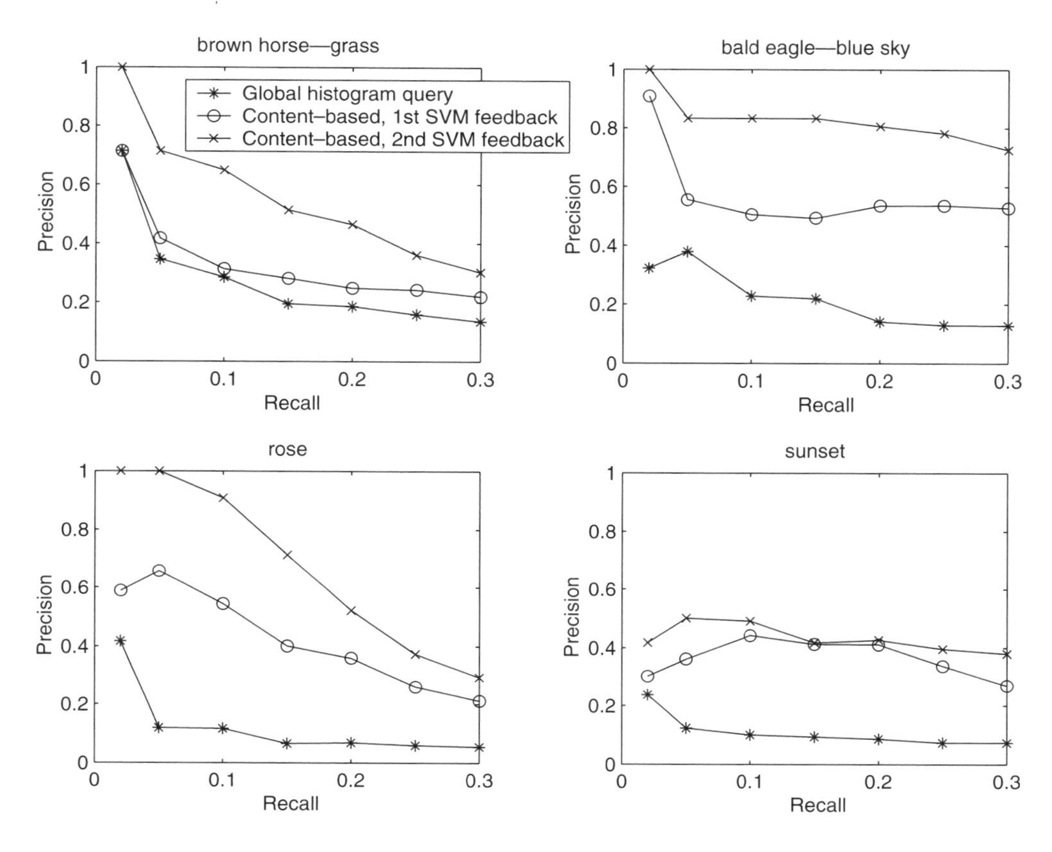

Figure 13.15 Precision—recall diagrams for two dual-keyword and two single-keyword queries, and comparison with global histogram method

13.9 Conclusions

A methodology was presented in this chapter for the flexible and user-friendly retrieval of visual information, combining a number of image processing and machine learning tools. These include a time-efficient still-image segmentation algorithm and a real-time compressed-domain video segmentation algorithm, both capable of unsupervised operation; simple ontologies defining intermediate-level descriptors; and a relevance feedback mechanism based on support vector machines. The resulting methodology is applicable to generic image and video collections, where no possibility of structuring a domain-specific knowledge base exists. The proposed scheme overcomes the restrictions of conventional methods, such as the need for the availability of key images/key frames or manually generated textual annotations, and requires no manual tuning of weights, thus offering flexibility and user friendliness. This, along with the time-efficiency and the unsupervised operation of the employed segmentation algorithms, renders it suitable for the manipulation of large volumes of visual data. Experiments conducted on a large collection of images and video clips demonstrate the effectiveness of this approach.

Acknowledgement

This work was supported by the EU project SCHEMA 'Network of Excellence in Content-Based Semantic Scene Analysis and Information Retrieval' (IST-2001-32795).

References

[1] A. Yoshitaka, T. Ichikawa, A survey on content-based retrieval for multimedia databases. *IEEE Transactions on Knowledge and Data Engineering*, **11**(1), 81–93, 1999.

[2] S.-F. Chang, T. Sikora, A. Puri, Overview of the MPEG-7 standard. *IEEE Transactions on Circuits and Systems for Video Technology*, special issue on MPEG-7, **11**(6), 688–695, 2001.

[3] S. Christodoulakis, M. Theodoridou, F. Ho, M. Papa, A. Pathria, Multimedia document presentation, information extraction, and document formation in MINOS: a model and a system. *ACM Transactions on Office Information Systems*, **4**(4), 345–383, 1986.

[4] C. Zhang, T. Chen, An active learning framework for content-based information retrieval. *IEEE Transactions on Multimedia*, **4**(2), 260–268, 2002.

[5] W. Al-Khatib, Y.F. Day, A. Ghafoor, P.B. Berra, Semantic modeling and knowledge representation in multimedia databases. *IEEE Transactions on Knowledge and Data Engineering*, **11**(1), 64–80, 1999.

[6] C. Faloutsos, R. Barber, M. Flickner, J. Hafner, W. Niblack, D. Petkovic, W. Equitz, Efficient and effective querying by image content. *Journal of Intelligent Information Systems*, **3**(3/4), 231–262, 1994.

[7] M. Flickner, H. Sawhney, W. Niblack, J. Ashley, Q. Huang, B. Dom, M. Gorkani, J. Hafner, D. Lee, D. Petkovic, D. Steele, P. Yanker, Query by image and video content: the QBIC system. *Computer*, **28**(9), 23–32, 1995.

[8] B.S. Manjunath, W.Y. Ma, Texture features for browsing and retrieval of image data. *IEEE Transactions on Pattern Analysis and Machine Intelligence*, **18**(8), 837–842, 1996.

[9] W.Y. Ma, B.S. Manjunath, NeTra: A toolbox for navigating large image databases. *Multimedia Systems*, **7**(3), 184–198, 1999.

[10] T.S. Huang, S. Mehrotra, K. Ramchandran, Multimedia Analysis and Retrieval System (MARS) project. In *Proceedings of the 33rd Annual Clinic on Library Application of Data Processing—Digital Image Access and Retrieval*, University of illinors, Urbana-Champaign, 1L, 24–26 March 1996.

[11] A. Pentland, R. Picard, S. Sclaroff, Photobook: content-based manipulation of image databases. *International Journal of Computer Vision*, **18**(3), 233–254, 1996.

[12] J.R. Smith, S.-F. Chang, VisualSEEK: a fully automated content-based image query system. In *Proceedings of the Fourth ACM International Conference on Multimedia*, Boston, MA, pp. 87–98. ACM Press, New York, 1997.

[13] I. Kompatsiaris, E. Triantafillou, M.G. Strintzis, Region-based color image indexing and retrieval. In *Proceedings of IEEE International Conference on Image Processing*, Thessaloniki, Greece, October 2001. IEEE, 2001.

[14] S. Belongie, C. Carson, H. Greenspan, J. Malik, Color- and texture-based image segmentation using EM and its application to content-based image retrieval. In *Proceedings of the Sixth International Conference on Computer Vision*, Mumbai, India, January 1998. IEEE, 1998.

[15] C. Carson, S. Belongie, H. Greenspan, J. Malik, Blobworld: image segmentation using expectation-maximization and its application to image querying. *IEEE Transactions on Pattern Analysis and Machine Intelligence*, **24**(8), 1026–1038, 2002.

[16] M.R. Naphade, T. Kristjansson, B. Frey, T.S. Huang, Probabilistic multimedia objects (multijects): a novel approach to video indexing and retrieval in multimedia systems. In *Proceedings IEEE International Conference on Image Processing*, 1998, vol. 3, pp. 536–540. IEEE, 1998.

[17] Y. Lu, C. Hu, X. Zhu, H. Zhang, Q. Yang, A unified framework for semantics and feature based relevance feedback in image retrieval systems. In *Proceedings of ACM Multimedia Conference*, pp. 31–37. ACM, 2000.

[18] X.S. Zhou, T.S. Huang, Unifying keywords and visual contents in image retrieval. *IEEE Multimedia*, **9**(2), 23–33, 2002.

[19] G. Tsechpenakis, G. Akrivas, G. Andreou, G. Stamou, S.D. Kollias, Knowledge-assisted video analysis and object detection. In *Proceedings of the European Symposium on Intelligent Technologies, Hybrid Systems and their implementation on Smart Adaptive Systems (Eunite02)*, Algarve, Portugal, September 2002.

[20] M.R. Naphade, I.V. Kozintsev, T.S. Huang, A factor graph framework for semantic video indexing. *IEEE Transactions on Circuits and Systems for Video Technology*, **12**(1), 40–52, 2002.

[21] W. Chen, S.-F. Chang, VISMap: an interactive image/video retrieval system using visualization and concept maps. In *Proceedings of IEEE International Conference on Image Processing*, Thessaloniki, Greece, October 2001, vol. 3, pp. 588–591. IEEE, 2001.

[22] S.S.M. Chan, L. Qing, Y. Wu, Y. Zhuang, Accommodating hybrid retrieval in a comprehensive video database management system. *IEEE Transactions on Multimedia*, **4**(2), 146–159, 2002.

[23] V. Kashyap, K. Shah, A.P. Sheth, Metadata for building the multimedia patch quilt. In V.S. Subrahmanian, S. Jajodia (eds) *Multimedia Database Systems: Issues and Research Direction*, pp. 297–319. Springer-Verlag, Berlin, 1996.

[24] B. Chandrasekaran, J.R. Josephson, V.R. Benjamins, What are ontologies, and why do we need them? *IEEE Intelligent Systems*, **14**(1), 20–26, 1999.

[25] S. Staab, R. Studer, H.-P. Schnurr, Y. Sure, Knowledge processes and ontologies. *IEEE Intelligent Systems*, **16** (1), 26–34, 2001.

[26] I. Kompatsiaris, M.G. Strintzis, Content-based representation of colour image sequences. In *IEEE International Conference on Acoustics, Speech, and Signal Processing 2001 (ICASSP 2001)*, Salt Lake City, UT, 7–11 May 2000. IEEE, 2000.

[27] I. Kompatsiaris, M.G. Strintzis, Spatiotemporal segmentation and tracking of objects for visualization of videoconference image sequences. *IEEE Transactions on Circuits and Systems for Video Technology*, **10**(8), 1388–1402, 2000.

[28] S. Liapis, E. Sifakis, G. Tziritas, Color and/or texture segmentation using deterministic relaxation and fast marching algorithms. In *Proceedings of the International Conference on Pattern Recognition*, September 2000, vol. 3, pp. 621–624. IEEE, 2000.

[29] M. Unser, Texture classification and segmentation using wavelet frames. *IEEE Transactions on Image Processing*, **4**(11), 1549–1560, 1995.

[30] N.V. Boulgouris, I. Kompatsiaris, V. Mezaris, D. Simitopoulos, M.G. Strintzis, Segmentation and content-based watermarking for color image and image region indexing and retrieval. *EURASIP Journal on Applied Signal Processing*, **2002**(4), 418–431, 2002.

[31] R. Jain, R. Kasturi, B.G. Schunck, *Machine Vision*. McGraw-Hill, New York, 1995.

[32] J. McQueen, Some methods for classification and analysis of multivariate observations. In *Proceedings of the 5th Berkeley Symposium on Mathematical Statistics and Probability*, 1967, vol. 1, pp. 281–296. University of California Press, Berkeley, CA.

[33] I. Kompatsiaris, M.G. Strintzis, 3D representation of videoconference image sequences using VRML 2.0. In *European Conference for Multimedia Applications Services and Techniques (ECMAST'98)*, Berlin, Germany, May 1998, pp. 3–12.

[34] P. Salembier, F. Marques, Region-based representations of image and video: segmentation tools for multimedia services. *IEEE Transactions on Circuits and Systems for Video Technology*, **9**(8), 1147–1169, 1999.

[35] E. Sifakis, G. Tziritas, Moving object localisation using a multi-label fast marching algorithm. *Signal Processing: Image Communication*, **16**, 963–976, 2001.

[36] N. O'Connor, S. Sav, T. Adamek, V. Mezaris, I. Kompatsiaris, T.Y. Lui, E. Izquierdo, C.F. Bennstrom, J.R. Casas, Region and object segmentation algorithms in the Qimera segmentation platform. In *Proceedings of the Third International Workshop on Content-Based Multimedia Indexing (CBMI03)*, 2003.

[37] A.A. Alatan, L. Onural, M. Wollborn, R. Mech, E. Tuncel, and T. Sikora, Image sequence analysis for emerging interactive multimedia services—the European COST 211 framework. *IEEE Transactions on Circuits Systems for Video Technology*, **8**(7), 19–31, 1998.

[38] E. Izquierdo, J. Xia, R. Mech, A generic video analysis and segmentation system. In *Proceedings of IEEE International Conference on Acoustics, Speech, and Signal Processing*, 2002, vol. 4, pp. 3592–3595. IEEE, 2002.

[39] V. Mezaris, I. Kompatsiaris, M.G. Strintzis, Video object segmentation using Bayes-based temporal tracking and trajectory-based region merging. *IEEE Transactions on Circuits and Systems for Video Technology*, **14**(6), 782–795, 2004.

[40] E. Izquierdo, M. Ghanbari, Key components for an advanced segmentation system. *IEEE Transactions on Multimedia*, **4**(1), 97–113, 2002.

[41] R. Wang, H.-J. Zhang, Y.-Q. Zhang, A confidence measure based moving object extraction system built for compressed domain. In *Proceedings of IEEE International Symposium on Circuits and Systems*, 2000, vol. 5, pp. 21–24. IEEE, 2000.

[42] R.V. Babu, K.R. Ramakrishnan, Compressed domain motion segmentation for video object extraction. In *Proceedings of IEEE International Conference on Acoustics, Speech, and Signal Processing*, 2002, vol. 4, pp. 3788–3791. IEEE, 2002.

[43] M.L. Jamrozik, M.H. Hayes, A compressed domain video object segmentation system. In *Proceedings of IEEE International Conference on Image Processing*, 2002, vol. 1, pp. 113–116. IEEE, 2002.

[44] H.-L. Eng, K.-K. Ma, Spatiotemporal segmentation of moving video objects over MPEG compressed domain. In *Proceedings of IEEE International Conference on Multimedia and Expo*, 2000, vol. 3, pp. 1531–1534. IEEE, 2000.

[45] N.V. Boulgouris, E. Kokkinou, M.G. Strintzis, Fast compressed-domain segmentation for video indexing and retrieval. In *Proceedings of the Tyrrhenian International Workshop on Digital Communications (IWDC 2002)*, September 2002, pp. 295–300.

[46] O. Sukmarg, K.R. Rao, Fast object detection and segmentation in MPEG compressed domain. In *Proceedings of IEEE TENCON 2000*, 2000, vol. 3, pp. 364–368.

[47] L. Favalli, A. Mecocci, F. Moschetti, Object tracking for retrieval applications in MPEG-2. *IEEE Transactions on Circuits and Systems for Video Technology*, **10**(3), 427–432, 2000.

[48] S. Mann, R.W. Picard, Video orbits of the projective group: a simple approach to featureless estimation of parameters. *IEEE Transactions on Image Processing*, **6**(9), 1281–1295, 1997.

[49] T. Yu, Y. Zhang, Retrieval of video clips using global motion information. *Electronics Letters*, **37**(14), 893–895, 2001.

[50] V. Kobla, D.S. Doermann, K.I. Lin, Archiving, indexing, and retrieval of video in the compressed domain. In *Proceedings of SPIE Conference on Multimedia Storage and Archiving Systems*, vol. 2916, 1996, pp. 78–89. SPIE, 1996. C.-C. Jay Kuo (ed.)

[51] MPEG-2, Generic Coding of Moving Pictures and Associated Audio Information, technical report, ISO/IEC 13818, 1996.

[52] B.-L. Yeo, Efficient processing of compressed images and video, PhD Thesis, Princeton University, NJ, 1996.

[53] G.B. Rath, A. Makur, Iterative least squares and compression based estimations for a four-parameter linear global motion model and global motion compensation. *IEEE Transactions on Circuits and Systems for Video Technology*, **9**(7), 1075–1099, 1999.

[54] V. Mezaris, I. Kompatsiaris, M.G. Strintzis, A framework for the efficient segmentation of large-format color images. In *Proceedings IEEE International Conference on Image Processing*, 2002, vol. 1, pp. 761–764. IEEE, 2002.

[55] B.S. Manjunath, J.-R. Ohm, V.V. Vasudevan, A. Yamada, Color and texture descriptors. *IEEE Transactions on Circuits and Systems for Video Technology, special issue on MPEG-7*, **11**(6), 703–715, 2001.

[56] V. Mezaris, I. Kompatsiaris, M.G. Strintzis, Ontologies for object-based image retrieval. In *Proceedings of the Workshop on Image Analysis for Multimedia Interactive Services*, London, April 2003.

[57] V. Mezaris, I. Kompatsiaris, M.G. Strintzis, An ontology approach to object-based image retrieval. In *Proceedings of IEEE International Conference on Image Processing (ICIP03)*, Barcelona, Spain, September 2003. IEEE, 2003.

[58] T. Gruber, A translation approach to portable ontology specifications. *Knowledge Acquisition*, **5**(2), 199–220, 1993.

[59] P. Martin, P.W. Eklund, Knowledge retrieval and the World Wide Web. *IEEE Intelligent Systems*, **15**(3), 18–25, 2000.

[60] A.T. Schreiber, B. Dubbeldam, J. Wielemaker, B. Wielinga, Ontology-based photo annotation. *IEEE Intelligent Systems*, **16**(3), 66–74, 2001.

[61] A. Raouzaiou, N. Tsapatsoulis, V. Tzouvaras, G. Stamou, S.D. Kollias, A hybrid intelligence system for facial expression recognition. In *Proceedings of European Symposium on Intelligent Technologies, Hybrid Systems and their Implementation on Smart Adaptive Systems (Eunite02)*, Algarve, Portugal, September 2002.

[62] B. Berlin, P. Kay, Basic color terms: their universality and evolution, University of California, Berkeley, CA, 1969.

[63] J.M. Lammens, A computational model of color perception and color naming, PhD Thesis, University of Buffalo, NY, 1994.

[64] A. Mojsilovic, A method for color naming and description of color composition in images. In *Proceedings of IEEE International Conference on Image Processing (ICIP02)*, Rochester, New York, September 2002. IEEE, 2002.

[65] Y.F. Day, S. Dagtas, M. Iino, A. Khokhar, A. Ghafoor, Spatio-temporal modeling of video data for on-line object-oriented query processing. In *Proceedings of the International Conference on Multimedia Computing and Systems*, May 1995, pp. 98–105.
[66] Y. Rui, T.S. Huang, M. Ortega, S. Mehrotra, Relevance feedback: a power tool for interactive content-based image retrieval. *IEEE Transactions on Circuits and Systems for Video Technology*, **8**(5), 644–655, 1998.
[67] X.S. Zhou, T.S. Huang, Exploring the nature and variants of relevance feedback. In *Proceedings of IEEE Workshop on Content-Based Access of Image and Video Libraries*, 2001, pp. 94–101. IEEE, 2001.
[68] T.P. Minka, R.W. Picard, Interactive learning with a 'Society of Models'. *Pattern Recognition*, **30**(4), 565–581, 1997.
[69] Y. Rui, T.S. Huang, S. Mehrotra, Content-based image retrieval with relevance feedback in MARS. In *Proceedings of IEEE International Conference on Image Processing*, 1997, vol. 2, pp. 815–818. IEEE, 1997.
[70] K. Porkaew, S. Mehrotra, M. Ortega, K. Chakrabarti, Similarity search using multiple examples in MARS. In *Proceedings of the International Conference on Visual Information Systems*, 1999, pp. 68–75.
[71] M.E.J. Wood, N.W. Campbell, B.T. Thomas, Iterative refinement by relevance feedback in content-based digital image retrieval. In *Proceedings of ACM Multimedia*, 1998, pp. 13–20.
[72] N.D. Doulamis, A.D. Doulamis, S.D. Kollias, Nonlinear relevance feedback: improving the performance of content-based retrieval systems. In *Proceedings of IEEE International Conference on Multimedia and Expo (ICME 2000)*, 2000, vol. 1, pp. 331–334.
[73] F. Jing, M. Li, H.-J. Zhang, B. Zhang, Region-based relevance feedback in image retrieval. In *Proceedings of IEEE International Symposium on Circuits and Systems*, 2002, vol. 4, pp. 145–148. IEEE, 2002.
[74] G. Aggarwal, A.T.V. Ghosal, S. Ghosal, An image retrieval system with automatic query modification. *IEEE Transactions on Multimedia*, **4**(2), 201–214, 2002.
[75] S. Tong, E. Chang, Support vector machine active learning for image retrieval. In *Proceedings of ACM International Conference on Multimedia*, 2001, pp. 107–118.
[76] G.-D. Guo, A.K. Jain, W.-Y. Ma, H.-J. Zhang, Learning similarity measure for natural image retrieval with relevance feedback. *IEEE Transactions on Neural Networks*, **13**(4), 811–820, 2002.
[77] V.N. Vapnik, *Statistical Learning Theory*, John Wiley & Sons, New York, 1998.
[78] C.-C. Chang, C.-J. Lin, LIBSVM: a library for support vector machines, 2001, Software available at `http://www.csie.ntu.edu.tw/~cjlin/libsvm`
[79] Corel stock photo library, Corel Corp., Ontario, Canada.
[80] MIT Vision Texture (VisTex) database, `http://www-white.media.mit.edu/vismod/imagery/VisionTexture/vistex.html`.
[81] P. Villegas, X. Marichal, A. Salcedo, Objective evaluation of segmentation masks in video sequences. In *Proceedings of the Workshop on Image Analysis for Multimedia Interactive Services*, Berlin, May 1999.
[82] N. O'Connor, T. Adamek, S. Sav, N. Murphy, S. Marlow, QIMERA: a software platform for video object segmentation and tracking. In *Proceedings of the Workshop on Image Analysis For Multimedia Interactive Services*, London, April 2003.
[83] M.R. Naphade, T.S. Huang, Extracting semantics from audio-visual content: the final frontier in multimedia retrieval. *IEEE Transactions on Neural Networks*, **13**(4), 793–810, 2002.
[84] M. Swain, D. Ballard, Color indexing. *International Journal of Computer Vision*, **7**(1), 11–32, 1991.

14

Context-Based Video Retrieval for Life-Log Applications

Kiyoharu Aizawa and Tetsuro Hori

14.1 Introduction

The custom of writing a diary is common all over the world. This fact shows that many people like to log their everyday lives. However, to write a complete diary, a person must recollect and note what was experienced without missing anything. For an ordinary person, this is impossible.

It would be nice to have a secretary who observed your everyday life and wrote your diary for you. In the future, a wearable computer may become such a secretary-agent. In this chapter, we aim at the development of a 'life-log agent' (that operates on a wearable computer). The life-log agent logs our everyday life on storage devices instead of paper, using multimedia such as a small camera instead of a pencil [1,2] (Figure 14.1).

There have been a few attempts to log a person's life in the area of wearable computing [2], video retrieval [3,4], and databases [5]. A person's experiences or activities can be captured from different points of view. In [5], a person's activities such as web browsing are completely recorded. We focus on capturing our experiences by wearable sensors including a camera. In our previous work [3], we used brain waves of a person that were simultaneously recorded with video images to retrieve scenes of interest to him or her. Our capture system was described in [4]. In this chapter, we describe our latest work, which uses more sensors and is able to retrieve using more contexts. The interface of the retrieval system will be described in detail.

14.2 Life-Log Video

A life-log video can be captured using a small wearable camera with a field of view equivalent to the user's field of view. By continuously capturing the life-log video, personal experiences of everyday life can be recorded by video, which is a most popular medium. Instead of writing a diary every day, a person can simply order the life-log agent to start capturing a life-log

Multimedia Content and the Semantic Web Edited by Giorgos Stamou and Stefanos Kollias

Figure 14.1 A conventional diary and the life-log agent

video at the beginning of every day. Listed below are the potential advantages of life-log videos:

- we can catch the best moments that we want to keep forever;
- we can vividly reproduce and recollect our experiences via video;
- we can remember things we would otherwise forget;
- we can see what we did not see;
- we can prove what we did and what we did not do.

However, in realizing the life-log agent, we face problems with the hardware and also with browsing videos.

14.2.1 Hardware Problems

First, regarding the camera for capturing the life-log video, small CCD/CMOS cameras are already available, and even cellular phones with such a camera are on the market. Such small cameras could lead to an environment where photographs or videos could be captured always and anywhere.

Next, regarding the computer for processing the captured life-log video, we can easily obtain a very small and highly efficient computer. In the near future, we can expect a pocket-sized PC that can digitize and process video signals in real time while capturing the video.

Next, regarding the storage capacity of the storage device in which we store our life-log videos, how huge would be the amount of video data that we could capture for our entire life? Assuming we capture video for 16 hours per day for 70 years, the amounts of video data are listed in Table 14.1.

Suppose its quality is almost equivalent to the quality of a TV phone, we would require 11 Tbytes to record 70 years. Even today, we can buy a lunch box-sized 100 Gbyte hard disk drive (HDD) for less than $US150. Thus, if we have 100 of them, their capacity is almost enough for 70 years. The progress of capacity improvement of HDD is very fast. It should be feasible to contain videos for 70 years. in a single lunch box-sized HDD in the near future. (The captured video would first be stored in a wearable device and then occasionally moved to a huge storage.)

Table 14.1 Amounts of life-log video data

Quality of video	Bit rate	Data size for 70 years (Tbytes)
TV phone quality	64 kbps	11
Video CD quality	1 Mbps	183
Broadcasting quality	4 Mbps	736

14.2.2 Problems in Browsing Life-Log Videos

The hardware problems are likely to be solvable. On the other hand, the problems that may arise when a person browses life-log videos are serious.

For a conventional written diary, a person can look back on a year at its end by reading the diary, and will soon finish reading the diary and will easily review events in the year.

However, watching life-log videos is a critical problem. It would take another year to watch the entire life-log video for one year.

Then, although it is surely necessary to digest or edit life-log videos, this takes even more time. Therefore, it is most important to be able to process a vast quantity of video data automatically. The life-log agent must be able to help so that we can easily retrieve and browse the desired scenes in the videos.

Conventional video editing systems

Recently, because their prices have fallen, the use of digital camcorders has spread widely. The number of people who sometimes record the lives of their families using digital camcorders is increasing. These are (partial) life-log videos. Unlike videos in television broadcasting or movies, the videos that a user records personally contain many redundant portions. Accordingly, the videos must be digested or edited. However, video editing is not easy for a general user, although a variety of systems for editing videos are on the market.

Such systems can index videos semi-automatically. Almost all of them digest or edit videos by processing the various features grasped from the image or audio signals. For example, they may utilize colour histograms extracted from the image signals. However, even if they utilize such information, computers cannot understand the contents of the videos, and they can seldom help their users to retrieve and browse the desired scenes in life-log videos easily. Therefore, the video retrieval system cannot respond to queries of various forms from the user. In addition, such image signal processing has a very high calculation cost, so the processing takes a long time unless a very expensive computer is used.

Although there have been various studies of life-log video retrieval systems, almost all of them use the visual feature-based techniques described above, for example [6,7].

Our proposed solution to this problem

Regarding videos such as television broadcasting and movies, a user only receives and sees them. But life-log videos are captured by a user. Therefore, as the life-log video is captured, various data other than video and audio can be recorded simultaneously. Using this information,

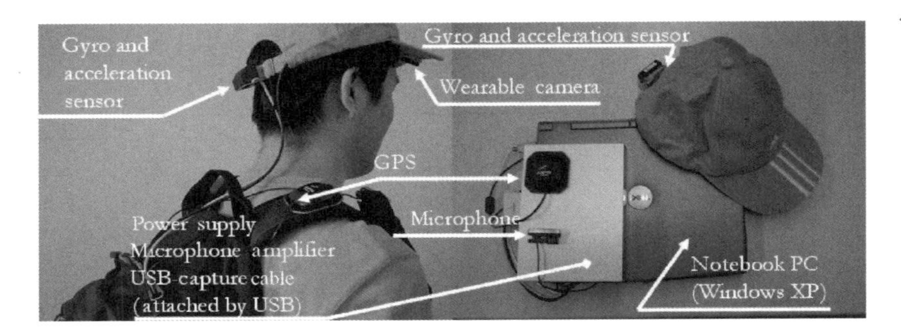

Figure 14.2 The life-log agent system

computers may be able to understand more fully the contents of videos, in an approach completely different from conventional video retrieval technologies. A similar approach has been taken in wearable computing research; for example, skin conductivity is captured in [1].

We attempt to use contexts captured by sensors. We aim at the development of a life-log agent with a retrieval system (for life-log videos) that can respond to queries in more natural form from its user, for example, 'I want to watch the video that was captured when I was running at Ueno Zoological Gardens' or 'I want to watch the video that was captured when I was watching that interesting movie two weeks ago'.

14.3 Capturing System

Our life-log agent is a system that can capture data from a wearable camera, a microphone, a brainwave analyser, a GPS receiver, an acceleration sensor and a gyro sensor synchronously. The main component of the system is a notebook PC. All sensors are attached to the PC through USB, serial ports and PCMCIA slots, as shown in Figure 14.2.

All data from the sensors are recorded directly by the PC. A software block diagram is shown in Figure 14.3. In particular, video signals are encoded into MPEG-1, and audio signals into MP3 using Direct Show. To simplify recording, we modified the sensors to receive their power from the PC, and customized their device drivers for the life-log agent system. The user can start capturing the life-log video simply by clicking a button.

14.4 Retrieval of Life-Log Video

A user wants to find desired scenes efficiently from a vast quantity of life-log videos. The agent provides a convenient interface for browsing and retrieving the life-log videos efficiently. Of course, the agent has the general functions of standard media player software, for example, play, stop, pause, fast-forward, and fast-rewind (the appearance of the user interface is shown in Figure 14.12). Recording of the life-log video introduced in the previous section is also performed using this interface.

In the following, we consider some ideas for making the life-log video retrieval efficient. Simultaneously, we introduce our implementation.

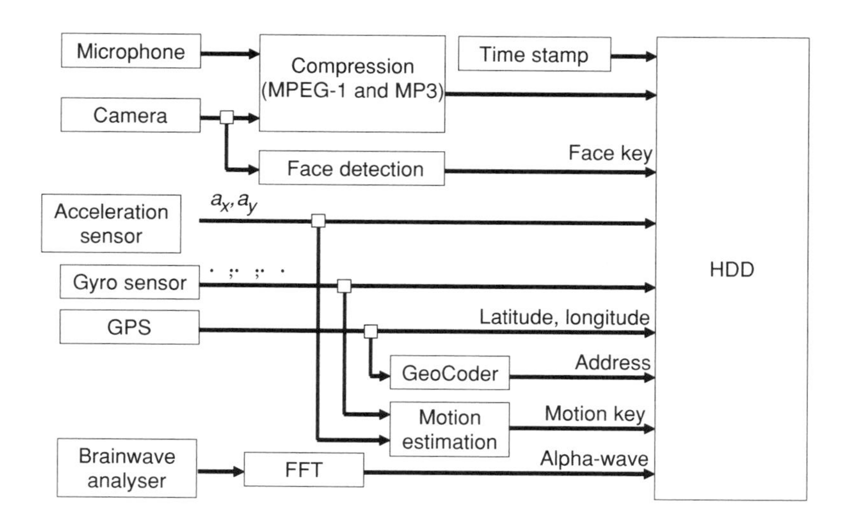

Figure 14.3 Block diagram (when recording)

14.4.1 Human Memory Recollection

We save many experiences as a vast quantity of memories over many years of life while arranging and selecting them, and we can quickly retrieve and utilize the requisite information from our memory. Considering these processes, when we want to recollect past experiences, information about contexts plays an important role. Some psychology research says that we use contexts as keys for recollection, and it can be assumed that we manage our memories based on contexts at the time. When we want to remember something, we can often use such contexts as keys, and recall the memories by associating them with these keys (Figure 14.4). For example, to recollect the scenes of a conversation, the typical keys used in the memory recollection process are such context information as 'where', 'with whom', 'when'. A user may put the following query to the life-log agent: 'I talked with Kenji while walking at a shopping centre in Shinjuku on a cloudy day in mid-May. The conversation was very interesting! I want to see the video of our outing to remember the contents of the conversation'. (In this chapter, we will refer to this query as 'Query A' later.)

Contexts are classified into two categories, subjective and objective. A subjective context expresses the user's feeling, for example, 'I'm interested' or 'I'm excited'. On the other hand, an objective context expresses information, for example, about where the user is, what the user is doing, the time or the weather.

As we said above, in conventional video retrieval the low-level features of image and audio signals of the videos are used as keys for retrieval. Probably, they will not be suitable for queries compatible with the way we query memories as in Query A. In addition, to process audiovisual signals, extensive calculations are required.

Data from the brainwave analyser, the GPS receiver, the acceleration sensor and the gyro sensor correlate highly with the user's contexts. Our life-log agent estimates its user's contexts from various sensor data, and uses them as keys for video retrieval. Thus, the agent retrieves life-log videos by imitating the way a person recollects experiences. It is conceivable that by

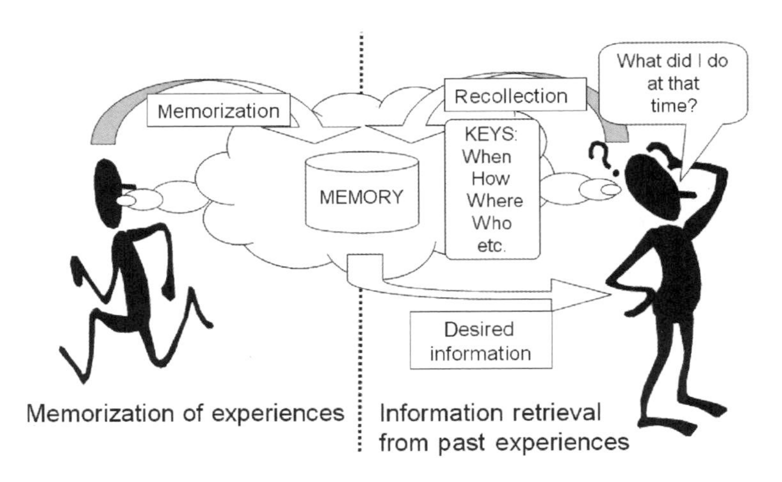

Figure 14.4 Human ability for memory recollection

using such context information in addition to audiovisual data, the agent can produce more accurate retrieval results than by using only audiovisual data. Moreover, each input from these sensors is a one-dimensional signal, and the computational cost for processing them is low.

14.4.2 Keys Obtained from Motion Data

Our life-log agent acquires its user's x-directional acceleration a_x and y-directional acceleration a_y from the acceleration sensor, and α, β and γ, respectively the angles around the z, y and x axes, from the gyro sensor. The agent calculates angular velocities by differentiating three outputs from the gyro sensor, and creates five-dimensional feature vectors in conjunction with two outputs from the acceleration sensor at the rate of 30 samples per second.

The 60-sample feature vectors (equivalent to the number of samples for two seconds) are quantized by the K-Means method and are changed into a symbol sequence. The agent gives the observed symbol sequence to a hidden markov model (HMM), which beforehand has learned various motions of its user, and estimates the motion, for example, walking, running or stopping, by identifying the model that outputs the observed symbol sequence with the highest probability. Please read our previous paper [4] for details.

Such information about motion conditions of the user can be a useful key for video retrieval. In Query A, the conversation that the user wants to remember was held while walking. This kind of retrieval key is helpful in finding the conversation scene from life-log videos.

The agent enumerates the time when the motion of its user changed, as shown in Figure 14.5. However, it is very hard for the user to understand only by enumerating motion and time. The agent shows scaled-down frame images so that the user can recollect the contents of the video at the time. Such information is of course related to information about the position in the video stream where the user's motion changed. If the user double-clicks a frame image, the agent will start playing the video from that scene.

In addition, HMM-based estimation has been studied by many researchers [7]. According to the results of their experiments, the HMM-based method shows a high correlation between

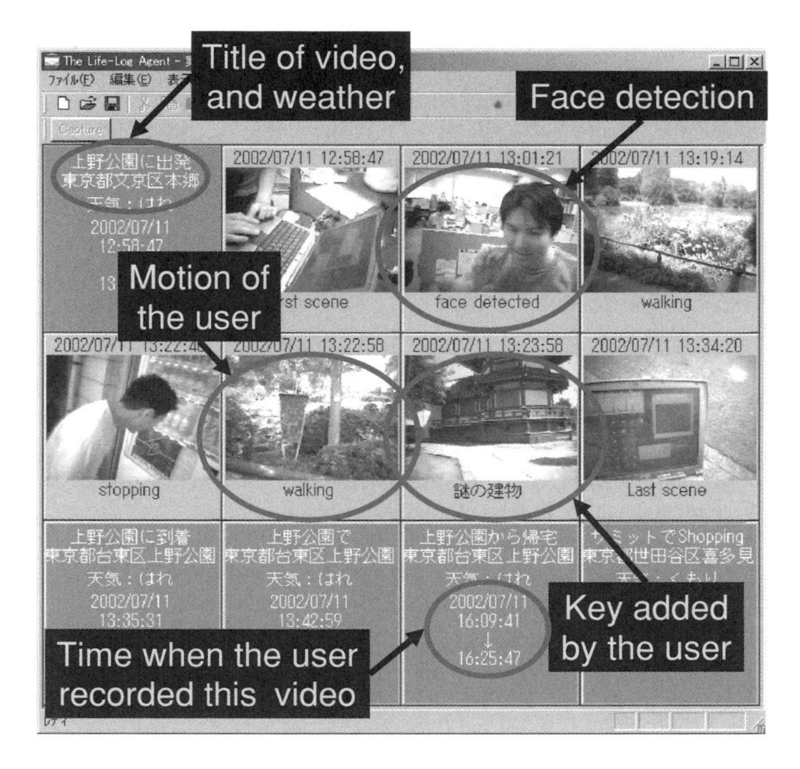

Figure 14.5 Interface for managing videos

actual conditions and estimated conditions. However, because these studies used only image and audio features of videos as observation sequences for HMMs, they are not very robust against environmental changes. For example, for videos captured in darkness, the estimation accuracy will fall extremely. In contrast to their work, we use motion sensors. Hence, we expect to achieve high robustness against environmental changes.

14.4.3 Keys Obtained from Face Detection

Our life-log agent detects a person's face in life-log videos by processing the colour histogram of the video image. Although we do not introduce the details in this paper, we show an example of the results of such processing in Figure 14.6. To reduce calculation cost, the method only uses very easy processing of the colour histogram. Accordingly, even if there is no person in the image, when skin colour domains are included predominantly, the agent detects wrongly.

The agent shows its user the frame images and the time of the scene in which the face was detected, as shown in Figure 14.5. If it is a wrong detection, the user can ignore it and can also delete it. If the image is detected correctly, the user can look at it and judge who it is. Therefore, identification of a face is unnecessary and simple detection is enough here. The images displayed are of course related to information about the position in the video stream where the face was detected. If a frame image is double-clicked, the agent will start playing

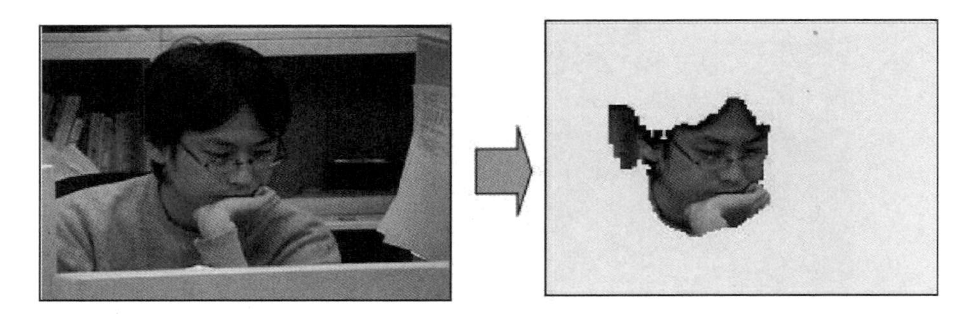

Figure 14.6 A result of face detection

the scene from the video. Thus, the user can easily access the video which was captured when he was with someone he wants. Although face detection is video signal processing, because it is simplified, it does not require much calculation.

Such information about face detection can be used as a key for video retrieval. In Query A, the conversation that the user wants to remember was held with Kenji. This kind of retrieval key is helpful in finding the conversation scene from life-log videos.

14.4.4 Keys Obtained from Time Data

Our life-log agent records the time when capturing its user's life-log video by asking the operating system of the wearable computer on which the agent runs for the present time. The contents of videos and the time at which they were captured are automatically associated. The user can know the time when each video was recorded and the time of each key of each video, as shown in Figure 14.5. Moreover, as shown in Figure 14.7, the user can know the present time in the video under playback, and by moving the slider in the figure he can traverse the time rapidly or rewind the time. Thus, the user can access the video, for example, that was captured at 3:30 p.m. on 2 May easily and immediately.

Such information about time can be used as a key for video retrieval. In Query A, the conversation that the user wants to remember was held in mid-May. This kind of retrieval key is helpful in finding the conversation scene from life-log videos.

14.4.5 Keys Obtained from Weather Information

By referring to data on the Internet, our life-log agent records the present weather in its user's location automatically when capturing a life-log video. The agent can connect to the Internet using the PHS network of NTT-DoCoMo almost anywhere in Japan. During retrieval, the agent informs the user of the weather at the time of recording each video, as shown in Figure 14.5. Thus, the user can choose a video that was captured on a fine, cloudy, rainy or tempestuous day easily and immediately.

Such information about weather can be used as a key for video retrieval. In Query A, the conversation that the user wants to remember was held on a cloudy day. This kind of retrieval key is helpful in finding the conversation scene from life-log videos.

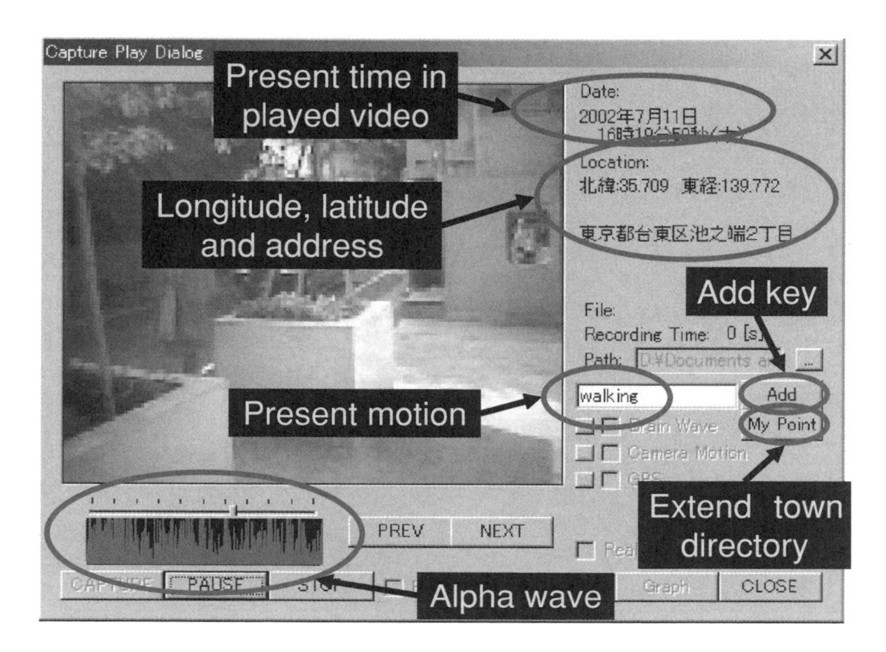

Figure 14.7 Interface for playing the video

14.4.6 Keys Obtained from GPS Data

From the GPS signal, our life-log agent acquires information about the position of its user as longitude and latitude when capturing the life-log video. The contents of videos and the location information are automatically associated.

Longitude and latitude information are one-dimensional numerical data that identify positions on the Earth's surface relative to a datum position. Therefore, they are not intuitively readable for users. However, the agent can convert longitude and latitude into addresses with hierarchical structure using a special database, for example, '7-3-1, Hongo, Bunkyo-ku, Tokyo, Japan'. The result is information familiar to us, and we can use it as a key for video retrieval as shown in Figure 14.7.

They also become information that we can intuitively understand by plotting latitude and longitude information on a map as the footprints of the user, and thus become keys for video retrieval. 'What did I do when capturing the life-log video?' A user may be able to recollect it by seeing his or her footprints.

The agent draws the user's footprint in the video under playback using a thick light-blue line, and draws other footprints using thin blue lines on the map, as shown in Figure 14.8. By simply dragging the mouse on the map, the user can change the area displayed on the map. The user can also order the map to display the other area by clicking arbitrary addresses of all the places where a footprint was recorded, as shown in the tree in Figure 14.8. The user can watch the desired scenes by choosing an arbitrary point of footprints. Thus, it becomes easy to immediately access a scene that was captured in an arbitrary place. For example, the user can access the video that was captured in Shinjuku-ku, Tokyo.

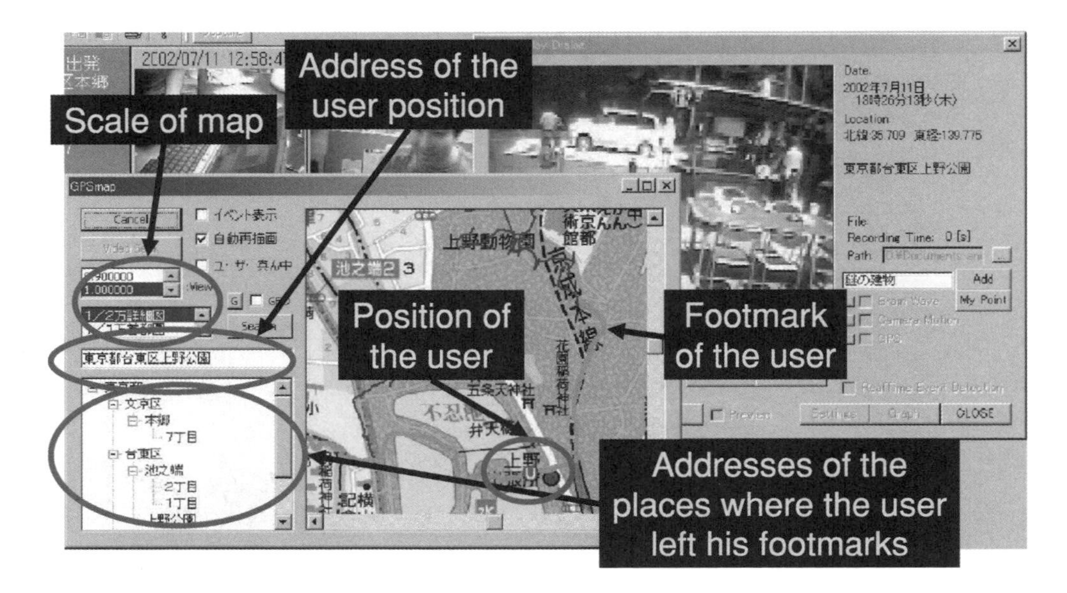

Figure 14.8 Interface for retrieval using a map

Moreover, the agent has a town directory database. The database has a vast amount of information about one million or more public institutions, stores, companies, restaurants and so on in Japan. Except for individual dwellings, the database covers almost all places in Japan including small shops or small companies that individuals manage (we can use it only in Tokyo, Yokohama, Nagoya and Osaka at the present stage). In the database, each site has information about its name, its address, its telephone number and its category. Examples of the contents of the database are listed in Table 14.2.

Table 14.2 The contents of the database

Institution	Details
Institution A	Name: Ueno Zoological Gardens Address: X1-Y1-Z1, Taito-ku, Tokyo Telephone number: A1-BBB1-CCC1 Category: zoo
Store B	Name: Summit-store (Shibuya store) Address: X2-Y2-Z2, Shibuya-ku, Tokyo Telephone number: A2-BBB2-CCC2 Category: supermarket
Company C	Name: Central Japan Railway Company Address: X3-Y3-Z3, Nagoya-shi, Aichi Telephone number: AA3-BB3-CCC3 Category: railroad company
Restaurant D	Name: McDonald's (Shinjuku store) Address: X4-Y4-Z4, Shinjuku-ku, Tokyo Telephone number: A4-BBB4-CCC4 Category: hamburger restaurant

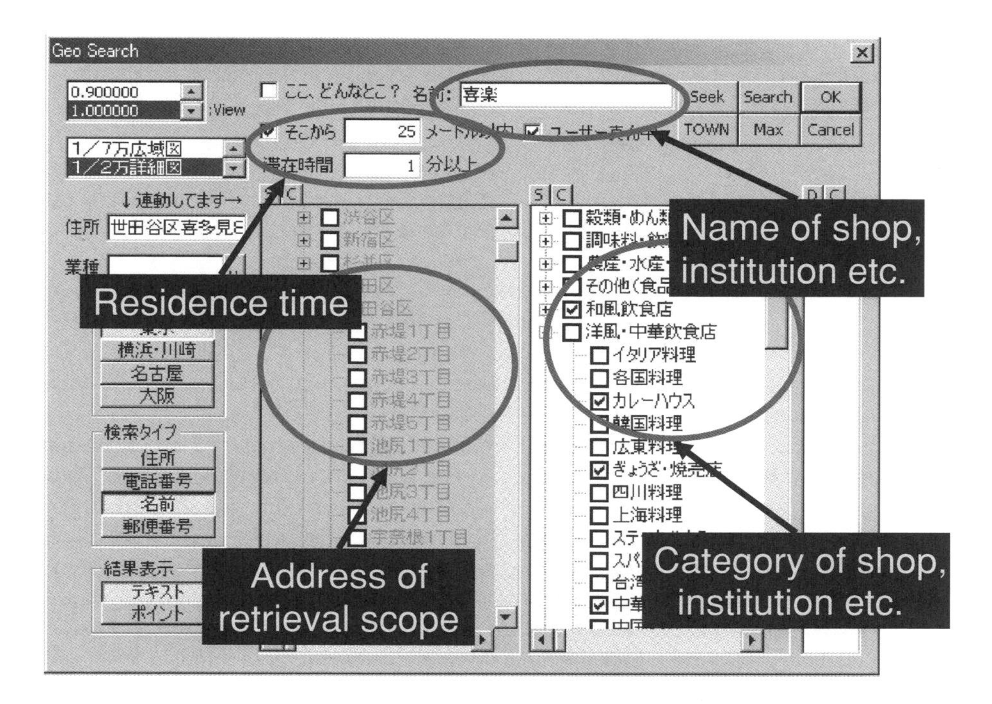

Figure 14.9 Retrieval using the town directory

As explained previously, the contents of videos and information about the user's position are associated. Furthermore, the agent can associate mutually the information about its user's position and this database. The user can enter the name of a store or an institution, or can input the category as shown in Figure 14.9. The user can also enter both. For example, we assume that the user wants to review the scene in which he or she visited the supermarket called 'Shop A', and enters the category-keyword 'supermarket'. To filter retrieval results, the user can also enter the rough location of Shop A, for example, 'Shinjuku-ku, Tokyo'. Because the locations of all the supermarkets visited must be indicated in the town directory database, the agent accesses the town directory, and finds one or more supermarkets near footprints including Shop A. The agent then shows the user the formal names of all the supermarkets visited and the times of visits as retrieval results. Probably he chooses Shop A from the results. Finally, the agent knows the time of the visit to Shop A, and displays the desired scene.

Thus, the agent can respond to the following queries correctly: 'I want to see the video that was captured when I had an ache at a dentist's rooms', 'I want to see the video that was captured when I was seeing a movie in Shinjuku' and 'I want to see the video that was captured when I was eating hamburgers at a McDonald's one week ago'. The scenes that correctly correspond to the queries shown above are displayed.

However, the agent may make mistakes, for example, to the third query shown above. Even if the user has not actually gone into a McDonald's but has passed in front of it, the agent will enumerate that event as one of the retrieval results. To cope with this problem, the agent investigates whether the GPS signal was received for a time following the event. If the GPS became unreceivable, it is likely that the user went into McDonald's. The agent investigates the length of the period when the GPS was unreceivable, and equates that to the time spent in McDonald's. If the GPS did not become unreceivable at all, the user most likely did not go into McDonald's.

Such information about the user's position is very convenient as a key for video retrieval. In Query A, the conversation that the user wants to remember was held at a shopping centre in Shinjuku. This kind of retrieval key is helpful in finding the conversation scene from life-log videos.

However, a place that the user often visits may not be registered in the town directory database, for example, a company that did not exist when the database was created (the database we use was created in 2000). To cope with this problem, the agent enables its user to extend this database by a simple operation. A place visited can be manually registered in the database when watching the video by clicking the button shown in Figure 14.7.

We examined the validity of this retrieval technique. The appearance of the experiment is shown in Figure 14.10. First, we went to Ueno Zoological Gardens, the supermarket 'Summit' and the pharmacy 'Matsumoto-Kiyoshi'. We found that this technique was very effective. For example, when we referred to a name-keyword 'Summit', we found the scene that was captured when the user was just about to enter 'Summit' as the result. When we referred to the category-keyword 'pharmacy', we found the scene that was captured when the user was just about to enter 'Matsumoto-Kiyoshi' as the result, and similarly for Ueno Zoological Gardens. These retrievals were completed very quickly; retrieval from a three-hour video took less than one second.

When recording a video, the agent can also navigate for its user by using the town directory and the map. For example, the user can ask the agent whether there is a convenience store nearby

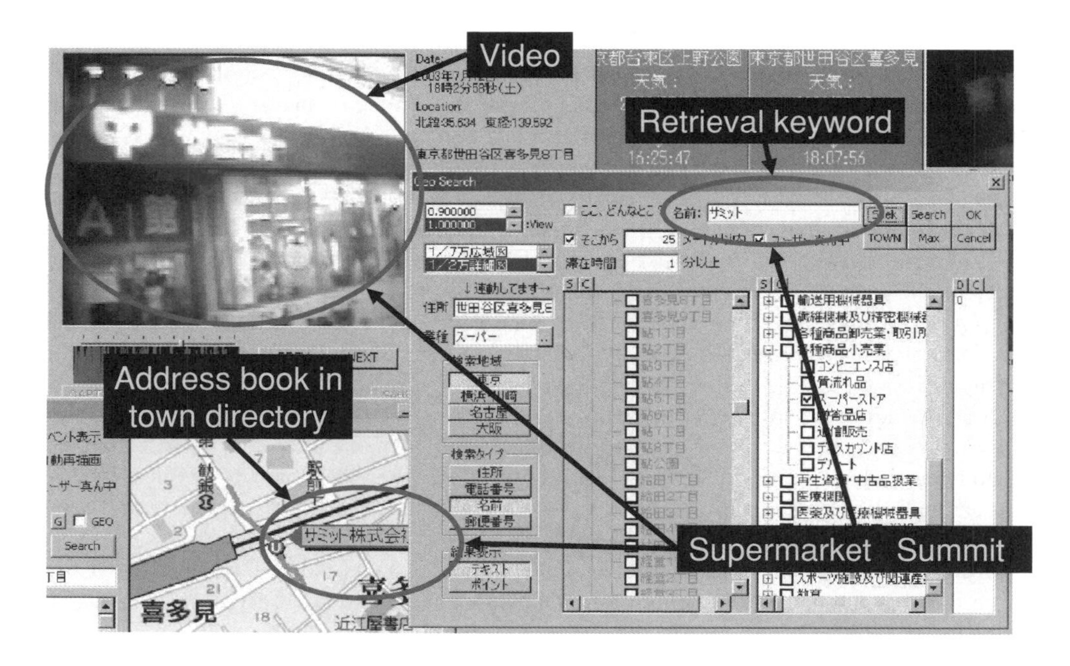

Figure 14.10 The retrieval experiment

and can ask the agent where it is. The agent then draws the store's location on the map. Of course, the user can ask about all the shops, institutions and companies that appear in the town directory.

14.4.7 Keys Added by the User

To label a special event, the user can order the life-log agent to add a retrieval key with a name while the agent is capturing the life-log video, thus identifying a scene that the user wants to remember throughout his or her life by a simple operation. This allows easy access to the video that was captured during a precious experience.

Furthermore, similarly to looking back on a day and writing a diary, the user can also order the agent to add a retrieval key with a label while watching the video; by clicking the button in Figure 14.7, and deleting any existing key. The user can also enter a title for each video, by an easy operation on the interface shown in Figure 14.5. Thus, the agent also supports the work of indexing life-log videos.

These additional keys can be displayed on the map and it becomes quite clear where they happened, as shown in Figure 14.11. The agent associates the key and its position automatically. By double-clicking the key displayed on the map, the scene is displayed.

14.4.8 Keys Obtained from BrainWave Data

In our previous work [3], we used brainwaves to retrieve scenes of personal interest. A subband (8–12 Hz) of brainwaves is named the alpha wave and it clearly shows the person's arousal status. When the alpha wave is small (alpha-blocking), the person is in arousal, or in other

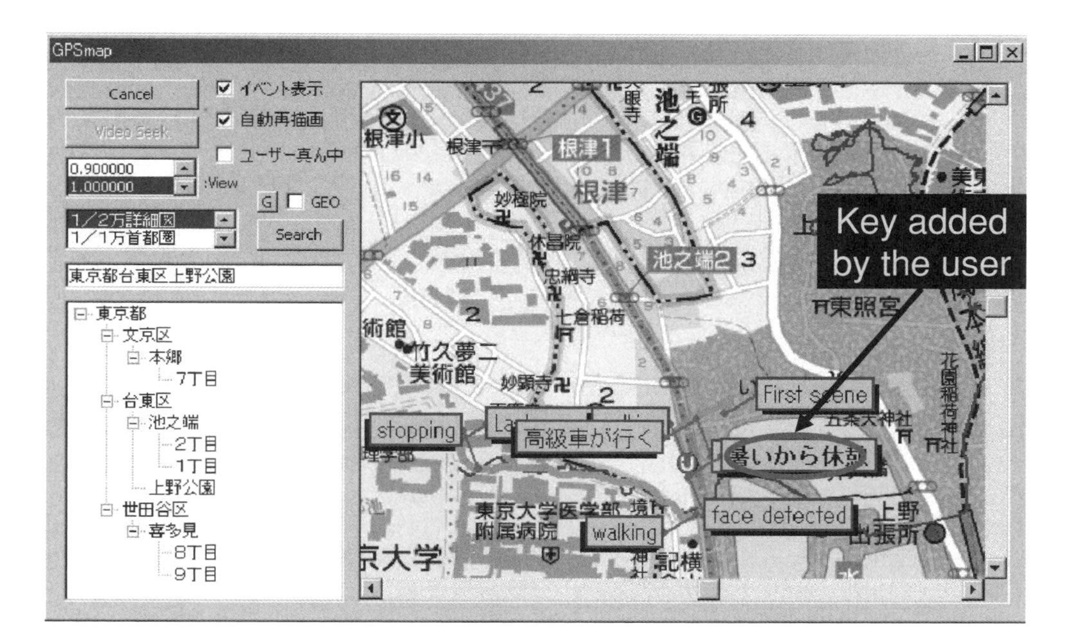

Figure 14.11 Displaying keys on a map

Figure 14.12 Interface of the life-log agent for browsing and retrieving life-log videos

words, is interested in something or pays attention to something. We clearly demonstrated that we can very effectively retrieve a scene of interest to a person using brainwaves in [3].

In Query A, the conversation that the user wants to remember was very interesting. This kind of retrieval key is helpful in finding the conversation scene from life-log videos.

In the current system, we can use brainwave-based retrieval, although it was not always used in our recent experiments. The agent displays the alpha wave extracted from the brain waves of the user, as shown in Figure 14.7.

14.4.9 Retrieval Using a Combination of Keys

The agent creates a huge number of MPEG files containing the user's life-log videos over a long period of time, and manages all of them collectively. The user can also manage them by giving the agent various commands. For example, in Figure 14.5, five videos are managed. The first video is chosen, and the agent shows its keys. By double-clicking a video identified by a grey frame in Figure 14.5, the user can choose another video. The agent shows the keys of the chosen video immediately.

Naturally, as more life-log videos are recorded, more candidates of video retrieval results are likely to be found.

Consider Query A again. The user may have met Kenji many times during some period of time. The user may have gone to a shopping centre many times during the period. The user may have walked many times during the period. The user may have had many interesting experiences during the period. It may have been May many times during the period. It may have been cloudy many times during the period.

Accordingly, if the life-log agent uses only one key among the various keys that we have described when retrieving life-log videos, too many wrong results will appear. However, by using as many different keys as possible, only the desired result may be obtained, or at least most of the wrong results can be eliminated, so it is easier for the user to identify the desired result.

14.5 Conclusions

While developing the life-log agent, we considered various functions that the agent could use, implemented them and added them to the agent. Finally, by using the data acquired from various sensors while capturing videos and combining these data with data from some databases, the agent can estimate its user's various contexts with a high accuracy that does not seem achievable with conventional methods. Moreover, the estimation is quite fast. These are the reasons the agent can respond to video retrieval queries using more natural forms correctly and flexibly.

Sensors, such as a GPS receiver, could be implemented in next-generation digital camcorders. When such a time comes, a context-based video retrieval system similar to what we have described will become popular.

References

[1] J. Healey, R.W. Picard, StartleCam: a cybernetic wearable camera. In *Proceedings of 2nd International Symposium on Wearable Computers (ISWC '98)*, Pittsburgh, PA, 19–20 October 1998, pp. 42–49. IEEE, 1998.

[2] S. Mann, 'WearCam' (the wearable camera): personal imaging systems for long-term use in wearable tetherless computer-mediated reality and personal photo/videographic memory prosthesis. In *Proceedings of 2nd International Symposium on Wearable Computers (ISWC '98)*, Pittsburgh, PA, 19–20 October 1998, pp. 124–131. IEEE, 1998.

[3] K. Aizawa, K. Ishijima, M. Shiina, Summarizing wearable video. In *Proceedings of International Conference on Image Processing (ICIP 2002)*, Thessaloniki, Greece, 7–10 October 2001, vol. 3, pp. 398–401. IEEE, 2001.

[4] Y. Sawahata, K. Aizawa, Wearable imaging system for summarizing personal experiences. In *Proceedings of IEEE International Conference on Multimedia and Expo*, Baltimor, MD, 6–9 July, pp. 45–48. IEEE, 2003.

[5] J. Gemmell, G. Bell, R. Lueder, S. Drucker, C. Wong, MyLifeBits: fulfilling the memex vision. In *Proceedings of the 10th ACM Multimedia Conference*, Juan-les-Pins, France, December 2002, pp. 235–238. ACM, 2002.

[6] H. Aoki, B. Schiele, A. Pentland, realtime personal positioning system for wearable computers. In *Proceedings of 3rd International Symposium on Wearable Computers (ISWC '99)*, San Francisco, CA, 18–19 October 1999, pp. 37–44. IEEE, 1999.

[7] B. Clarkson, A. Pentland, Unsupervised clustering of ambulatory audio and video. In *Proceedings of International Conference on Acoustics, Speech, and Signal Processing (ICASSP '99)*, 15–19 March, vol. 6, pp. 3037–3040. IEEE, 1999.

Index

Multimedia Content and the Semantic Web Edited by Giorgos Stamou and Stefanos Kollias